Ple
renewed

Due

C.

CHILDREN OF PROMETHEUS

Christopher Wills

CHILDREN OF PROMETHEUS

The Accelerating Pace of Human Evolution

ALLEN LANE
THE PENGUIN PRESS

ALLEN LANE
THE PENGUIN PRESS

Published by the Penguin Group
Penguin Books Ltd, 27 Wrights Lane, London W8 5TZ, England
Penguin Putnam Inc., 375 Hudson Street, New York, New York 10014, USA
Penguin Books Australia Ltd, Ringwood, Victoria, Australia
Penguin Books Canada Ltd, 10 Alcorn Avenue, Toronto, Ontario, Canada M4V 3B2
Penguin Books (NZ) Ltd, Private Bag 102902, NSMC, Auckland, New Zealand

Penguin Books Ltd, Registered Offices: Harmondsworth, Middlesex, England

First published in the USA by Perseus Books 1998
First published in Great Britain by Allen Lane The Penguin Press 1999
1 3 5 7 9 10 8 6 4 2

Printed and bound in Great Britain by The Bath Press, Bath

A CIP catalogue record for this book is available from the British Library

ISBN 0–713–99348–0

To my daughter Anne Marie, who will see more
of the future than I will, and who will
help her patients to survive it.

. . . [W]hat other hand than mine
Gave these young Gods in fulness all their gifts?
                    . . . [L]ike forms
Of phantom-dreams, throughout their life's whole length
They muddled all at random, did not know
Houses of brick that catch the sunlight's warmth,
Nor yet the works of carpentry. They dwelt
In hollowed holes, like swarms of tiny ants,
In sunless depths of caverns; and they had
No certain signs of winter, nor of spring
Flower-laden, nor of summer with her fruits . . .
Until I showed the risings of the stars,
And settings hard to recognize. And I
Found Number for them, chief devise of all,
Groupings of letters, Memory's handmaid that,
And mother of the Muses. And I first
Bound in the yoke wild steeds, submissive made
To the collar or men's limbs, that so
They might in man's place bear his greatest toils. . . .

AESCHYLUS, *Prometheus Bound* (460 B.C.)
TR. E. H. PLUMPTRE, 1868

CONTENTS

INTRODUCTION

❖

So dominant is the forest that it is said to be possible for an orang-outang to travel from the south to the north of Borneo without descending from the tree-tops. His only barrier would be the big rivers and, since the majority of these run north and south, they would merely prevent his spread longitudinally east and west.

PATRICK M. SYNGE, *Beauty in Borneo* (1932)

Myth ascribes to Prometheus the gift of fire to humans. His presumptuous transfer of this power from the gods to mere mortals caused him to suffer a terrible fate. The bards first sang that tale at least four thousand years ago: they were trying to provide an explanation, in terms their listeners could understand, for why human beings have taken a path that is so different from those taken by all other animals.

It subtracts nothing from the power of the Promethean legend to realize that our own path has been so different, not because of fire stolen from heaven, but because of the forces of evolution. After all, it was those same evolutionary forces that gave us the ability to devise the Promethean legend itself, along with all the other tales, poems, theories, and inventions that have transformed our existence in the subsequent four thousand years.

Evolution has provided us with the power to accomplish all that change. But are we still evolving, and if so how? What might we be evolving into? These questions have fascinated biologists ever since Charles Darwin. In this book I will explore some surprising recent answers and show not only that we are still evolving, but that our

1

evolution—particularly the evolution of our minds—is actually proceeding at an accelerating pace.

In this respect humans are dramatically different from our close primate relatives, such as chimpanzees, gorillas, and orangutans. They are, without exception, evolving at a much slower rate. The ancient Greeks would have said that they never received the Promethean spark. We now say that, whatever combination of chance and unusual environmental circumstances caused us to start us down our path, that combination never arose in the environment of our close relatives. The scientific revolution has allowed us to explain that process in terms that are a little more specific, and a great deal less poetic, than the Promethean legend.

Whatever that evolutionary spark might have been, why did our primate relatives not receive it? To explore this question, let me begin by examining a close animal relative of ours, one that has undergone very little change over the past two million years—but that is changing now, primarily because of the impact that our own species has had on its remote and isolated world. This relative of ours is evolving slowly, in the way that Darwin first envisioned. Indeed, such slow evolution is the norm among the animals and plants of our planet—humans and a few others are dramatic exceptions. So to make the contrast vivid, let us begin with an example of the norm.

Borneo is the largest and one of the most striking of the seventeen thousand islands that make up the vast Indonesian archipelago. The north and center of the island are dominated by high mountains, but as I flew toward it recently across the Java Sea from the southwest, the mountains were quite invisible, lying far below the misty horizon. My first glimpse of the island was a flat green plain, cut here and there by the threads of rivers and streams.

Travel through this part of Borneo is still primarily by water. A few settlements cling to the edges of the rivers, and large hand-built Dayak boats move majestically between them. In the riverside town of Pangkalanbun, we hired a little boat with a cheerful Dayak crew to take us further inland to nearby Tanjung Puting National Park.

The park occupies a blunt peninsula jutting out from the coast. Here the ground is low and swampy, and the rivers overflow their banks during the rainy season, flooding the forest for hundreds of feet on either side. The area is ideal for tree-dwelling primates who rarely need to come down to ground level. Bands of long-tailed macaques swing through the trees, and even from a boat it is easy to catch glimpses of the tangled branches of old orangutan nests.

At one point a troupe of proboscis monkeys leaped from a tree that hung out over the river and swam furiously across. Their webbed fingers and toes make them some of the most impressive swimmers of the primate world; the danger from crocodiles lends an extra urgency to their natatorial escapades. As they threshed through the water, it appeared to me that they were using a stroke remarkably like the modern crawl—if so, an excellent example of evolutionary convergence with a human behavior!

We headed away from the muddied main river, which has been badly polluted by mercury and tailings from gold mining farther upstream, and quickly found ourselves in untouched wilderness. Now the stream flowed clear, its waters stained a deep orange color from dissolved vegetable matter. It was here, in 1971, that Canadian anthropologist Birute Galdikas established her first camp. The site has since grown into a substantial scientific settlement, built on fairly high ground in the heart of the forest, several hundred meters from the river. For years Galdikas and her coworkers were forced to wade through the leech-infested, waist-high mud and water to get to their settlement, but the camp is now connected to the river by a long wooden causeway.

The camp is named after Louis Leakey, the brilliant paleontologist who founded the study of fossil humans in East Africa. Leakey realized that primate behavior might give important clues to the behavior of our ancestors. He encouraged Galdikas, along with Jane Goodall and the late Dian Fossey, to study the behavior of our closest living relatives in the wild. Goodall chose chimpanzees, and Fossey chose gorillas. Galdikas, the third of Leakey's recruits, turned to the least known of this group of species, the orangutans.

We followed her into the forest in order to observe the orangs as they swung through the high treetops, as much as sixty meters above the ground, and showered us with fragments of unripe durian fruit. Her meticulous field observations have shown that these primates can feed on four hundred different kinds of fruit, so that they are always able to find food somewhere in a healthy and diverse forest. Their survival, however, depends absolutely on the maintenance of that diversity, for in the forests of Southeast Asia many tree species flower and fruit suddenly and unpredictably. The massive outpouring of fruit and seeds overwhelms the birds and animals that feed on them, so that some seeds escape. Orangutans range widely through the forest, searching for these sudden and unpredictable bonanzas.

At least in Tanjung Puting the orangutans can still pursue this ancient way of life. Primarily through Galdikas's efforts, the entire peninsula has now been saved from massive logging—although poachers still nibble at the edges of the park. She has been repeatedly threatened as a result of her efforts and was once briefly kidnapped. It is becoming grimly obvious that hers is ultimately a losing battle, but it is one that is nonetheless worth fighting.

AN EVOLUTIONARY FEEDBACK LOOP

Although Borneo is still one of the most remote and exotic places in the world, it is starting to feel the impact of the modern world. By visiting the island before it changed utterly, I hoped to fit some further pieces into the great mosaic of impressions and ideas about human origins that evolutionary biologists have been putting together ever since Darwin. My ultimate goal was an immodest one—I wanted nothing less than satisfactory answers to some of the great remaining puzzles that confront students of human evolution.

The first of these puzzles has to do with the differences between human beings and our primate relatives, particularly the great apes. Why are they so huge? Why did orangutans, for example, stay unchanged in their forest during much of the time when the human lineage was evolving into the most remarkable set of creatures the planet has ever seen? After all, we too started our evolutionary history in a forest, although it was in Africa rather than Asia. Why were our own ancestors catapulted in new evolutionary directions when the ancestors of the orangutans were not?

A few years earlier I had proposed a partial answer to this question in a book called *The Runaway Brain*. I had suggested that during the last several million years our ancestors were caught up in a runaway evolutionary process. When genetic changes led to changes in their brains and bodies, some of them were selected for, as our ancestors' own activities created an ever more complex and ever more human-oriented environment.

All these genetic and environmental changes have tended to reinforce each other in a feedback loop. Any cultural or environmental change would have selected for individuals who happened to carry genes that made them best able to take advantage of it. As a result, some of them (or their progeny) would have been clever enough to produce still more cultural and environmental changes. This would lead to even more selection, not only for new genetic variants that had arisen by mutation but also for new combinations of old genes that are produced by the genetic shuffling known as recombination that takes place each generation.

Once humans were caught up by this feedback loop, it would have been difficult for them to stop or reverse it. But the nature of the trigger that started our ancestors on this runaway course remains elusive. At the moment, among students of human origins the favored candidate for such a trigger is a sudden alteration in the climate. The problem with this, as was recently pointed out by Mark Ridley in a book review in *The New York Times*, is that "there is (as the detective Poirot would say) too much evidence: too much climatic change . . . some climatic change or other will almost always be associated with any evolutionary event." Sudden climatic alterations are happening all the time, which means that other primates should have been just as affected by them as our own ancestors were.

If it was indeed a climatic change that served as a trigger, then our own ancestors reacted to it in a different way from their great ape relatives. When the ancestors of apes were confronted by a change—perhaps something that caused the forests in which they lived to shrink in size—they did not adapt to that new condition. Instead, they were simply forced to reduce their own range. Rather than conquering their environment, they merely reacted to it.

The initial trigger that started our human ancestors on their unique course must have had about it at least a hint of new worlds to conquer. As many people have suggested, our forebears may have had to invade some new environment because the old one had essentially disappeared or because they had been driven from it. If the conquest of that new environment happened to give some clear advantage to upright posture, tool use, communication skills, or all three, then this shift would have set them on their way. From that point on they never looked back. The choice might initially have been a matter of life or-death, but it eventually turned into a wonderful evolutionary opportunity.

Soon after I arrived in Borneo, I realized why this process had not happened to the orangs: They cannot escape the forest. No other ecosystem in the area can provide them with the abundant plant food that they need. The Borneo orangs are still where they always were simply because they have never needed to cope with an ecological upheaval that forced them to live elsewhere. Had their forest vanished, they too might have joined us on that evolutionary escalator.

The second evolutionary question about which I hoped to find clues in Borneo is a more immediate one, though it is related in many ways to the first: Are we, unlike our primate relatives, still caught up in that feedback loop? Are we still evolving, and if we are, into what?

To answer this question, I have drawn from much research into both human evolution and the evolution of the pathogenic organisms that prey on us, along with many discussions with scientists who are examining behavioral and social change, how our genes work, and the interactions between us and the many other species with which we share the planet. The picture that emerges is one in which our physical evolution is certainly continuing, perhaps at roughly the same rate it always has, although certainly too slowly for us to perceive any change in our lifetimes. But at the same time, exploding cultural change is laying the groundwork for a far more rapid evolution of our mental capabilities in the future. This evolution, too, will not be visible in a single human lifetime, although it has proceeded at an accelerating pace during our evolutionary history and is likely to take place even more quickly in the centuries ahead.

Rapid environmental change causes the evolution of the organisms that are caught up in it to accelerate. By this principle, there can be no doubt that we are still evolving: we are altering our environment, and invading new ones, at an accelerating speed. Sometimes the alterations have all the earmarks of disaster. In the year since I left Borneo, for example, dramatic changes have taken place on the island. During the summer and fall of 1997, dozens of massive forest fires raced through logging-damaged areas of the jungle. The destruction they wrought is still being assessed, but it looks as if the Bornean rain forest is even more vulnerable than had been previously supposed.

The only beneficiaries of this disaster, which choked huge areas of Southeast Asia in a life-threatening smog, are oil-palm companies that set many of the fires and then smugly watched them go out of control, in order to gain for themselves new acres on which to plant their valuable crops.

In a hundred years human activities will have altered Borneo beyond recognition. The current changes have already severely affected the orangutan populations. Reports emerging from eastern Borneo tell of more than a hundred new orphan orang babies that have ended up in the villages, their parents killed by the fires or by poachers.

The inevitable loss of the Bornean ecosystem is a small part of a worldwide tidal wave of changes, not all of which will be entirely destructive. The next century will see far more changes than the last one, even though the last century has certainly been no slouch in the change department. A hundred years from now our culture and technology will be changed almost beyond recognition. We can only hope that it will give us the capability of undoing some of the thoughtless damage that we are now inflicting on our planet.

The evolution of orangutans has been slow, and they seem unable to adapt to environmental change to the same extent that we can—although they are being forced now to undergo at least some genetic changes. But because they were never caught up in the runaway-brain feedback loop, they will need our help in order to survive the destruction that we have caused.

If all this change is affecting the evolution of the hapless orangutans in their forest, it has a far greater impact on our own species. Humans have accelerated the pace of evolutionary change everywhere, and, at the forefront of that change, we are altering ourselves more rapidly than any other species.

This bald statement, of course, goes against conventional wisdom. Evolution, as we all know, is slow, and the rate of human-induced environmental change must surely be too swift for the glacial pace of evolution to keep up. But the conventional wisdom is wrong. Our gene pool, the huge collection of genes that together makes up all the genetic information in our species, is indeed responding to all this environmental change. In this book I intend to explore with you how this is happening.

PREVENTING SPECIATION

Orangutans, our third closest living relatives after chimpanzees and gorillas, are strikingly intelligent. They do not exhibit the same extensive social interactions and use of tools as chimpanzees, but on the nearby island of Sumatra they are more social than they are on Borneo and they have been observed to use and modify a number of types of tools in the wild. Indeed, a large project is now under way at Washington's National Zoo to determine whether orangs are as capable as chimpanzees of learning and comprehending language. So far, they seem to be scoring very high.

Orangs' intelligence extends to other areas as well. Known as the Houdinis of the primate world, they are notorious among zookeepers for their ability to escape from apparently secure cages by using a great variety of tools. Many airlines refuse to transport them, regardless of the precautions that have been taken to keep them confined.

Orangs are also becoming dangerously rare. It is estimated that about thirty thousand still thrive on Borneo, but only about a tenth as many have managed to survive on much more heavily impacted island of Sumatra. Human activity has already driven the orangs of Java and peninsular Malaysia to extinction.

Birute Galdikas made plain to me that, in spite of their confinement to the shrinking forests, we are forcing orangs into a new evolutionary direction. We are doing this by dramatically short-circuiting the evolutionary process that leads to the appearance of new species. Before humans appeared on the scene, the orangutans of Borneo and Sumatra, separated for millions of years, had been on their way to becoming different species. Now we are reversing that process, with consequences that cannot be predicted.

In spite of their separation, orangs from the two islands look very similar. In the zoological equivalent of a police lineup, even a trained primatologist would have a hard time distinguishing them. There are nonetheless invisible differences between the two populations, including small variations in their chromosomes that can be seen only under careful microscopic examination. DNA has been used to determine the approximate time at which the single ancestral group of orangs was split into two. Although there is a large range of error on this time estimate, the split appears to have taken place at least two million years ago.

These molecular investigations show that the two groups of orangs really are distinct, and as a result they have been categorized as separate subspecies. Biologists of the "splitter" persuasion have attempted to raise the two groups to full species status, though so far they have been unsuccessful.

Such quibbles might seem of interest only to taxonomists, that subgroup of biologists who spend their lives naming species. But a more important issue has to do with conservation: If two species of orangutan really exist, then this provides a powerful reason for preserving the habitats of both of them.

It is surprising that the orangs of the two islands have been separated for so long because the islands themselves have not always been separated geographically. When the Pleistocene series of ice ages began, about 1.64 million years ago, the sea level dropped and rose again repeatedly. Whenever the glaciers began to spread far to the north, dropping sea levels caused Borneo, Sumatra, and most of the other islands of the Indonesian archipelago to become joined together into a huge land mass. Orangs must have had ample opportunities to migrate among the islands during these episodes.

On the other hand, the land bridges that temporarily joined the islands may have been environmentally unfriendly places for the highly specialized orangutans. Or, even before the ice ages began, subtle behavioral differences may have accumulated between orangs in various parts of the great archipelago. These differences might have been enough to prevent any hybridization when the islands grew together and the animals were able to meet.

Any such behavioral differences that they have managed to evolve, however, can easily be overcome in captivity. Bornean and Sumatran orangs have mated readily in zoos and have produced a number of offspring that appear to be perfectly normal (although these offspring have not been tested for survival under fully wild conditions).

The zoo matings took place before the clear genetic differences between the Bornean and Sumatran subspecies were found. The discovery of these differences has thrown the zoos of the developed world into a state of confusion. In many but not all zoos, the two groups are now kept carefully separate. Some programs have been started to sterilize the hybrid offspring that have already been born, but in other cases the hybrids have simply been sold to less scrupulous zoos. Many other zoos, particularly the numerous private zoos in Asia, have made no effort to keep the two groups separate.

All this might not matter to the genetic integrity of the wild populations, were it not for the growing illegal trade in baby orangs. This trade is particularly rampant in Sumatra but is also growing in Borneo. Poachers kill the mothers that they track down in the forest, then take their babies to Java, where they sell them to unscrupulous dealers for about five hundred dollars each. In Semarang, in northern Java, I encountered a number of people who told me in detail about this ugly trade. Many of the infant orangs are sold to rich Taiwanese who keep them as house pets or put them in private zoos.

Cute baby orangs, like tiny kittens, inevitably grow up. Few households can withstand the rampages of a clever two-to-four-hundred-pound animal, particularly one that can bite through a human limb as if it were a wet noodle. Most house-pet orangs are killed when they become inconveniently large, but a few are returned to Java or Sumatra. Sometimes the babies are intercepted by the police during the kidnapping process and taken back to what the authorities suppose must be their native haunts. Galdikas, like other scientists at research stations on both Borneo and Sumatra, finds more and more of her time is consumed by the need to care for the growing numbers of orphaned orangs appearing on her doorstep.

Most of the orphans, accustomed to the plentiful food provided by humans, have tended to remain near Camp Leakey even after they have been released. But Galdikas has succeeded in returning a few to the forest. One, named Priscilla, was among the orangs that we observed swinging through the treetops about two kilometers from Camp Leakey. Priscilla's return was a brilliant success, for through binoculars we could see her tiny baby, Popeye, clinging tightly to her body. He had been born after his mother was returned to the wild.

None of the orangutan rehabilitation projects have had the resources to sort out the returned orangs genetically, to see which are from Borneo and which are from Sumatra. Some inadvertent gene flow has probably already taken place between the populations on the two islands. More is certain to occur.

One evening during my visit, as the rain poured down outside, Galdikas and I talked at some length about the probable consequences of this recent gene flow. The story is complicated. Both Borneo and Sumatra are huge islands, among the largest in the world, and the genetics of most of the orang populations that inhabit them has been little explored.

Galdikas suspects that gene flow might already have taken place among the islands, perhaps during the glacial maxima, long before the arrival of humans. The very orangs that she has been working with living as they do relatively close to Java, peninsular Malaysia, and Sumatra—might have been in the middle of this flow. A recent visitor to her camp, who had observed many Sumatran and Bornean orangs, concluded that her animals seemed to fall somewhere between the northeastern Bornean orangs and those of Sumatra. But this anecdotal observation was based on appearance and behavioral differences that are difficult to quantify.

Questions abound, and finding some of the answers may take years. How much gene flow has already taken place between the islands? How much will take place as orang populations are rescued by conservationists from soon-to-be-destroyed forests and transferred to other parts of the islands? The accidental gene flow that has already occurred, through the recent introduction of Sumatran orangs into Borneo and vice versa, is likely to be dwarfed by the gene flow caused by the ecological upheavals to come.

Any slight deleterious effects of the new gene flow, however, are probably the least of the orangs' current problems. The recent fires are an extreme manifestation of the invasion of Borneo by foreign lumber and plantation companies. These companies are accelerating their wholesale destruction of Borneo's coastal forests and are pushing inexorably farther inland. In another century any remaining orangs will be living in fragments of rain forest, perhaps covered by air-conditioned domes for the convenience of tourists. Any slight differences between the rainforest environments of Borneo and Sumatra will be dwarfed by the immense changes that are going to take place in the orangs' entire universe.

As the evening wore on, Galdikas and I speculated that, in view of the coming ecological disturbances, introducing some new genes into the Bornean and Sumatran populations might even be a good thing. In the days when the orang populations had adapted to slightly different environments on the two islands, such gene flow would almost certainly have been harmful. But during the present period of blindingly rapid change, it might now be advantageous. Gene flow, and the resulting new mixes of genes, might actually help orangs adapt to the few small altered remnants of their world that we will allot to them. In order to survive, orangs may need all the genes they can get. The more genetic variants their populations have, the more likely that adaptive gene combinations will arise and aid in their subsequent adaptation.

During my slogs through the humid leech-infested swamps, as I craned my neck to glimpse orangs in the sunlight and breezes of the high forest canopy far above, I reminded myself that these primates have been subjected to some of the same evolutionary forces that have shaped our own species, but that the outcome has been very different. We humans have, quite without meaning to, literally taken over and begun to redirect orang evolution, as we have with so many other species—including our own.

Like that of orangs, our ancestral lineage has been subdivided again and again. Sometimes this subdivision has resulted in the appearance of two or more separate species. Sometimes, as is now happening with orangs, the process has been aborted as two partially separated gene pools came together again. What happened when partially separated groups of our own ancestors were reunited?

Evolutionary theory suggests that such blending or fusion events, because they produce new combinations of genes, should help to accelerate evolution. Indeed, blending is perhaps the most powerful evolutionary process affecting us at present. Although no human groups are as genetically different as the Sumatran and Bornean orangs, some have been separated for 100,000 years and perhaps more. Throughout the world, however, different human races and groups are coming together. The consequences of this mixing are likely to be profound—and generally positive.

By the time I left Borneo, I had a much clearer idea why, in spite of their remarkable intelligence, orangutans did not follow human beings (or precede us) on an evolutionary feedback loop of the type that led to our own runaway brains. They had never been faced with the kind of sudden environmental changes that must have challenged our ancestors. The sadly battered state of the island itself vividly brought home to me how we are forcing the orangs' evolution in new and unpredictable directions.

What remained was the central question of how we humans are forcing our own evolution. The story of how we are affecting the Bornean orangutans forms just one small part of the answer to this question. Besides the orang story, other pieces that I have assembled in order to tell this tale include some recent fossil finds in a deep ancient cave in northern Spain, the way that people survive on the windswept reaches of the Tibetan plateau, a well-concealed evolutionary drama that is taking place in the hushed corridors of Whitehall, a superb adaptation of some of our primitive relatives to the rainforest of Africa's Côte d'Ivoire, and a glimpse of our frenetic future in the pell-mell world of extreme sports.

I hope you enjoy reading about the answer to this question as much as I have enjoyed putting it together, and that this book will leave you with a new appreciation of the power of evolution to change our species.

Many people have contributed to the book, including Juan Luis Arsuaga, Kurt Benirschke, Jack Bradbury, José M. Bermudez de Castro, Dan Bricker, Eudald Carbonell, Rusty Gage, Birute Galdikas, Tony Goldberg, Jean-Jacques Hublin, Michael Marmot, Richard Moxon, David Metzgar, Jon Singer, Ajit Varki, John West, Anne Marie Wills, Elizabeth Wills, David Woodruff, the members of the LOH study group, Helen deBolt, and the students of OSLEP. I would particularly like to thank Harvey Itano and Lorna Moore for detailed help. Ted Case and Pascal Gagneux read the whole book and made many valuable comments. As always, however, I am ready to take the fall for any errors.

Notes are sequestered at the end of the book, where they belong. Although I have tried to define technical terms when they first appear, a glossary is also provided to help the reader thread his or her way through the complicated bits.

PART I

The Many Faces of Natural Selection

ONE

❖

Authorities Disagree

Ever since Darwin, pundits—expert and otherwise—have made pronouncements on whether human beings are still evolving and what might become of us in the future. But these authority figures have historically disagreed enormously. The biggest battle has been fought about whether natural selection will continue to drive human biological evolution, or whether our species will instead degenerate. Surely, if the pressure of natural selection slackens, an accumulation of harmful mutations will sully our gene pool. These mutations, which are appearing now at much the same rate as they always have in the past, might soon overwhelm the weakened ability of natural selection to remove them.

Bald statements of this fear are less common now than they were before political correctness became a concern. In the 1880s Charles Darwin's cousin Francis Galton founded the Eugenics Society to encourage the selective breeding of highly intelligent people. In the 1920s and 1930s distinguished scientists such as geneticist H. J. Muller and evolutionist Julian Huxley lent their reputations to predicting the genetic destruction of our species. Huxley, in his 1948 book *Man in the Modern World*, made this grim prognosis:

> The net result is that many deleterious mutations can and
> do survive, and the tendency to degradation of the germ-
> plasm can manifest itself . . . Humanity will gradually

destroy itself from within, will decay in its very core and essence, if this slow and relentless process is not checked.

In 1950 Muller coined the term *genetic load* to express the burden of harmful mutations carried by our species, and he predicted dire consequences if nothing was done to remove them.

Some of this degenerative change is physical and has already taken place, anthropologist C. Loring Brace suggested in 1963:

> [S]ome of the major and formerly unexplained changes which have occurred in human evolution are the results of probable mutation effect. Reduction in the size of the teeth and face and of the skeletal and muscular systems may have been brought about by such a mechanism, as a result of changes in the principal human adaptation mechanism, culture.

Many authorities have claimed that mental degeneration has taken place and will get worse with time because of the accumulation of harmful mutations in the gene pool. They have cited a variety of studies showing how in developed societies people at the top of the socioeconomic ladder have fewer children than those near the bottom. The claim that this tendency will lead to degeneration has given rise to innumerable polemics about the inevitable destruction of our gene pool.

So concerned were these scientists about this perceived trend that they tended to ignore the political consequences of their statements. By the end of World War II, the question of whether our gene pool was accumulating harmful mutations, which perhaps had started out as a legitimate evolutionary inquiry, had become hopelessly tainted by ugly racist rhetoric and by the terrible deeds the Nazis carried out in the name of eugenics. The presumption of degeneracy itself was questioned, and many scientists who had a far less pessimistic view of our genetic future began to speak out.

Still, the theme of species degeneration sometimes bobs up in the mainstream. In 1997 writer Jerry Adler used the Millennium Notebook page of *Newsweek* to proclaim, "For Humans, Evolution Ain't What It Used to Be." In his article he suggested that while telephones might "improve" in the future, we will not!

Adler was at least careful to couch his worries in general terms. But is there any biological evidence for degeneration? Not really. Consider the apparent physical degeneration that concerned anthropologist Brace. Peoples around the world do seem to have become slighter in build, perhaps beginning with the introduction of fire to cook food and accelerating after the agricultural revolution. Our long bones have become lighter, and even our teeth have become smaller. Brace measured the reduction in tooth size in many different populations over time and calculated that it has averaged about one percent per thousand years—although he found much variation from one population to another.

Such changes mean that we are less capable than our ancestors of surviving in a world in which brute strength and the crushing power of mighty teeth are all-important. But these changes are not necessarily degenerative. We have, on the contrary, every reason to suppose that present-day humans are more capable than those ancestors of surviving in a world in which food is plentiful and easy to gather and chew, and in which smaller bodies and teeth are advantageous. For example, brute strength may be incompatible with fine motor skills—perhaps it is possible to select for one but not for both. An environment that places a premium on fine motor skills might then select for the kinds of changes that Brace has measured.

Brace and others have also proposed that apparently "degenerative" changes are the result of a relaxation of natural selection and the consequent accumulation of harmful mutations. In order to fit the "degeneration" model, these mutations would have to make us lighter boned and generally smaller. Brace argued that if there had originally been selection for mutations that made teeth and bones large, then simply by chance subsequent mutations would be likely to make them smaller.

We know from studies of many animals and plants, however, that most mutations are harmful because they introduce noise into the system. Thus, while random mutations might well lead us to have funny-looking jaws and teeth, there is no obvious reason why they would make our jaws and teeth smaller.

If we have simply been accumulating random mutations, moreover, it seems surprising that our jaws and teeth would have become smaller at such a regular rate while at the same time remaining well formed.

We can learn something about such changes and how they happened by examining an extreme example, pituitary dwarfism. When this condition appears in otherwise normal families it can be traced to an underproduction of pituitary growth hormone. A number of drastic mutations can produce this kind of dwarfism, ranging from a defect in a gene that releases growth hormone to defects in genes that result in damage to the pituitary gland as a whole. The harmful consequences of these mutations often extend far beyond dwarfism, so that the mutations do not persist for long in the population.

Pygmies of the African Efe tribe, on the other hand, have been tiny for millennia, and have thrived in their rainforest home. They have the shortest stature of any African pygmies, yet their pituitary function is completely normal. Their dwarfism stems from the fact that the cells in their bodies show a low responsiveness to the growth hormone, which their pituitaries manufacture in normal quantities. Even though the exact mutation has not yet been found, it is already apparent that it is far less damaging than the rare familial dwarfism genes that appear in small numbers throughout our species.

Selection for slight build has also played a role in the domestication of animals. The size of teeth and jaws in domesticated cats has been reduced compared with their wild relatives (as has, incidentally, their brain size). We would not, however, dream of attributing this reduction to the accumulation of mutational noise—cats are too elegant for that. Far more likely, we have selected for cats that do less damage to us and our belongings. In so doing we have presumably drawn on variation that was already present in the wild population from which the cats originally came. Any harmful mutations that appeared during this process would have been selected against—the genes that have produced these changes in cats will probably turn out to be more like the genes that produced short stature in the Efe.

The selective forces that have driven us in the direction of a mild kind of dwarfism are probably the result of a trade-off that gives an advantage to increased or altered brain power and fine motor skills despite decreased bodily strength. These changes, like the changes that took place in the cat population, have drawn on genetic variation that was already present in our ancestors. Not all of our distant ancestors, after all, were huge and burly with large teeth.

Another possible driving force behind such an evolutionary trend, as Darwin himself was the first to realize, might be sexual selection. If standards of desirability in a mate have shifted from sheer size and raw power to something a little bit less intimidating, the resulting sexual selection might explain the decrease in robustness of our species over the last few dozen millennia.

Oddly enough, these possibilities have seldom been suggested as explanations. Later in the book we will look at some suggestive evidence that such trade-offs have taken place, reinforcing the argument that selective pressures, acting on preexisting variation, are far more likely to have produced such a shift than the degeneration of our gene pool through the accumulation of harmful mutations.

If degenerative change is a shaky explanation for our decreased robustness, it is even less persuasive as an explanation for a decrease in our intellectual powers. The ugly, popular, and potentially racist view is that, because the most intellectually fit among us are not reproducing in sufficient numbers, our entire species is becoming stupider and more degenerate. Although this idea can be traced back to the ancient Greeks, it was first articulated compellingly by Francis Galton.

Galton, like most of his contemporaries, was what one might term a genteel racist. He took it for granted that some races were superior and some were inferior. Yet if he is able, from his vantage in heaven, to observe what has happened to his ideas in the ninety years since his death, I feel sure that he must be truly horrified. Even today, racists and supremacists of every stripe (the stripes are usually but not always pale ones) continue to cite Galton to support their belief in the intellectual degeneration of people unlike themselves. They persist even in the face of strong evidence that our species seems, if anything, to be getting smarter over time. In Chapter 12 we will examine the evidence that runs counter to the intellectual degeneracy argument, and the probable reasons for this remarkable and encouraging trend.

Whether or not we are actually getting stupider or more physically degenerate over time, it still seems obvious to most people that natural selection has pretty well stopped, bringing our physical evolution to a halt. This is because our environment is now less severe and dangerous than it once was. At the same time, as many observers have pointed out, this relaxed physical environment has freed us to do other things, so that our cultural evolution has sped up. To be sure, cultural evolution is not the same as physical evolution: a human population may acquire an elaborate culture without any changes in its genes. So it is quite possible that physical evolution has come to a halt at the same time as cultural evolution has roared ahead.

In 1963 the prominent evolutionary biologist Ernst Mayr of Harvard University wrote in his magisterial book *Animal Species and Evolution:*

> To be sure, there may have been an improvement of the brain [over the last 100,000 years] without an enlargement of cranial capacity but there is no real evidence of this. Something must have happened to weaken the selective pressure drastically. We cannot escape the conclusion that man's evolution towards manness suddenly came to a halt. . . . *The social structure of contemporary society no longer awards superiority with reproductive success* [my italics].

In 1991 English geneticist Steve Jones took to the op-ed pages of *The New York Times* to ask, "Is Evolution Over?"

> [M]ost babies born now survive until they themselves have babies, so that existence is less of a struggle than it was. Natural selection involves inherited differences in the chance of surviving that struggle, and as most of us do survive nowadays until we have passed on our genes, the strength of selection has decreased. . . . It may even be that we are near the end of our evolutionary road, that we have got as close to utopia as we ever will.

Mayr and Jones both fell into the trap of assuming that just because natural selection is not as obvious as it once was, it must therefore be coming to a halt. But as we will see, selection is still acting on us, even though the emphasis is now shifting away from obvious physical selective factors to ones that are more psychological.

A slightly less dogmatic view was espoused by physiologist and evolutionist Jared Diamond, in a 1989 article for *Discover* magazine:

> After the Great Leap Forward [approximately 35,000 years ago], cultural development no longer depended on genetic change. Despite negligible changes in our anatomy, there has been far more cultural evolution in the past 35,000 years than in the millions of years before.

Anthropologist Richard Klein, in a 1992 article in *Evolutionary Anthropology*, made the same point.

> [A] major transformation in human behavior occurred about 40,000 years ago. Prior to this time, human form and human behavior evolved hand-in-hand, *very slowly over long time intervals* [my italics]. Afterward, evolution of the human form all but ceased, while the evolution of human behavior or, perhaps more precisely, the evolution of culture, accelerated dramatically.

Quite correctly, both authors make a clear distinction between changes in culture and evolutionary changes in the genes that influence how our brains and our bodies work. They essentially rule out the latter.

Klein is right about the very long period over which we have evolved. As he also correctly points out, many important changes took place in our ancestors during the whole span of four million or so years for which we have a fossil record, while relatively few have taken place recently. But both he and Jones are, I think, wrong about the pace of evolution. In fact, our evolution has been blindingly fast compared with that of our closest relatives, all during that four-million-year period.

Nor is there hard evidence that our physical evolution has slowed down. Even though relatively little change seems to have taken place during the last forty thousand years, this span of time represents only one percent of our fossil record. The argument of Diamond and Klein would be more convincing if we had a really plentiful fossil record of our ancestors throughout our history, so that we could measure just how much change actually took place during various forty-thousand-year periods in the more distant past. But we do not—the gaps in the record are often hundreds of thousands or even millions of years long. And as they themselves admit, some physical change has even taken place during the last forty-thousand years. Was it more or less than the one percent we would expect if the change were spread evenly over our four-million-year fossil history? We simply do not know, and so we really cannot say whether our physical evolution has recently slowed down. Indeed, we cannot rule out the possibility that it has sped up!

By contrast, another group of writers are confident that we are still evolving—although they tend to waffle a bit about how much and in what direction. They do, however, emphasize far more than the others we have quoted the extent to which cultural pressures can, like any other sort of natural selection, act to modify us.

Roger Lewin, in his 1993 book *Human Evolution*, raises the possibility of an invisible kind of selection, one that might affect something other than our bodies:

> Why . . . would social complexity have taken 90,000 years to manifest itself after the origin of anatomically modern humans? One possibility, of course, is that a subtle intellectual evolutionary change may have occurred rather recently in human history, one that does not manifest itself physically.

At the end of his controversial 1975 book *Sociobiology*, which examined how behaviors evolved, Edward O. Wilson expresses this viewpoint even more clearly:

> Starting about 10,000 years ago agriculture was invented and spread, populations increased enormously in density, and the primitive hunter-gatherer bands gave way locally to the relentless growth of tribes, chiefdoms and states. . . . There is no reason to believe that during this final spurt there has been a cessation in the evolution of either mental capacity or the predilection toward special social behaviors. The theory of population genetics and experiments on other organisms show that substantial changes can occur in the span of less than 100 generations, which for man reaches back only to the time of the Roman Empire.

A similar view comes from David Hamburg, again from 1975; but he goes a step further, pointing out that cultural change is in fact continually moving the goal posts. The selective pressures that result from cultural change are themselves changing:

> The poignant dilemma is that ways of fostering survival, self-esteem, close human relationships, and meaningful group membership for hundreds, thousands or even millions of years now often turn out to be ineffective or even dangerous in the new world which man has suddenly created. Some of the old ways are still useful, others are not. They will have to be sorted out, and sorted out soon.

This small sampling of opinion shows that, over the last few decades, our view of the evolutionary future of humans has shifted. The discovery of mutations, and the demonstration in the 1920s that they can be induced by environmental factors, led to many dire predictions that our gene pool would become damaged. The apparent slackening of the pressures of natural selection would seem to have put the brakes on our future adaptive evolution, leaving these harmful mutations to accumulate unchecked. But as Hamburg suggests, selection pressures have actually shifted from the visible to the invisible, from forces that test the resources of our bodies to forces that test the resources of our minds.

WHITHER *HOMO SAPIENS?*

Lord Ronald . . . flung himself upon his horse and rode madly off in all directions.

STEPHEN LEACOCK, *Gertrude the Governess*

Assuming we are evolving, what might we be evolving *into?* Suppose, as I have suggested, that human beings are becoming better adapted to the changing environment that we are creating for ourselves. This is, after all, what evolution would be expected to do. Assuming that this trend continues, what might eventually happen to us?

Science fiction writers commonly conjure up speculative futures in which they suppose that we will evolve over the next few centuries into hairless, dome-headed, spindle-shanked creatures of great intelligence.

Scientists, too, have made such predictions. In 1967 anthropologist C. Loring Brace drew a tongue-in-cheek cartoon of just such a creature (see Figure 1–1). How this remote descendant of ours might possibly hold his head up is unclear. But Brace went the science fiction writers one better, for he hung around his neck something that looks remarkably like a Sony Walkman. Not bad for 1967!

Science fiction writers, of course, make a living by extrapolating trends. The fossil record and other sources of evidence tell us that the bodies of our ancestors have indeed become weaker and less hairy, while their brains have become larger. So these science fiction scenarios certainly seem like logical extensions of trends that we have observed. But will these trends really continue? Does anything about the process of evolution suggest that they will? Or are other things more important—at least in the short term?

To understand what might happen to us, we have to understand something about evolution: Selection-driven evolution does not necessarily always go in a particular direction. Even if, over the next few centuries or millennia, some directional changes do take place, they are likely to be dwarfed by other changes that will take place in response not to overall trends in our environment but rather to its rapid diversification.

As people today find more effective ways to shield themselves from the extremes of the natural world, it might be supposed that our environment is becoming less rather than more diverse. It is fashionable to decry the global Americanization of world culture, for example, which enables us to stay in a Holiday Inn in Marrakesh or to buy a Big Mac in Uzbekistan. Yet most inhabitants of these places have no interest in either activity, and for those who do, their environment has actually become more diverse, not less.

Around the world the number of types of occupations has exploded, as have the opportunities that have opened to people of both sexes and all races. Children of Uzbek sheepherders are now traders on the Tashkent stock exchange, perhaps dealing in mutton futures. Children of Tuareg nomads thrive on city life in Marrakesh (and one of them may even be the manager of that Holiday Inn).

A second thing to remember is that all this change is being paralleled by changes in the human gene pool. These genetic changes are taking place more rapidly than we might think.

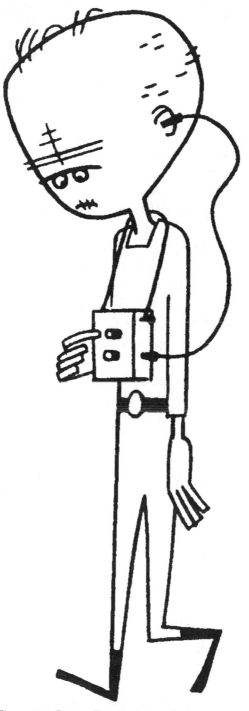

Figure 1–1. Loring Brace's vision of a future man.

Most of us suppose that evolution is always a very slow process. A favorable mutation arises somewhere in a population and then spreads, followed by the eventual appearance of another favorable mutation, its subsequent spread, and so on. Since most newly arisen mutations either have little effect or are unfavorable, for every favorable mutation that does appear and manage to spread, many others must arise and then disappear. The whole process would seem to be achingly slow—most of the substantial evolutionary change that we can follow from the fossil record seems to take millions or tens of millions of years.

If our gene pool responds so slowly even to our rapidly changing environment, the argument runs, then our evolution could not be getting faster. Even if it is going on at the same rate as it did in the past, the argument would be that we are likely to see little result from it in the space of a few centuries or millennia.

Yet surprisingly rapid short-term evolution can take place. In Africa's Lake Victoria, which has appeared and disappeared repeatedly in the past, many new and distinct species of fish have arisen over just the last few thousand years. The current collection of fish species, diverse in their behaviors and appearance, has evolved during the ten thousand years since the lake was completely dry. So closely related are they still that when the lake becomes murky because of human pollution, they lose the visual cues that allow them to distinguish mates of the same species. Hybrid fish are now turning up in increasing numbers in the cloudy waters of the once-pristine lake.

Even in the absence of any new mutations, evolution can happen quickly. Dramatic changes in a population can occur simply through a shift in the frequencies of the alternative forms of genes known as *alleles*.

A simple example will make this plain. Suppose some characteristic of our species is controlled primarily by the activities of four different genes. The characteristic might be something obvious, like skin or hair color, or it might be something more subtle, like the ability to perceive spatial relationships or to imitate a particular pitch of sound. Often there may be a trade-off: sheer muscle strength, as I suggested earlier, may be incompatible with the ability to perform small precise movements.

Suppose further that two alternative alleles of each gene are present in our species. Depending on our parentage, for each gene we could have inherited two copies of one allele, two copies of the other, or one of each. Since there are three possible combinations of the two alleles of each of these genes, there are $(3 \times 3 \times 3 \times 3)$ or $3^4 = 81$ possible combinations of these genes, or *genotypes*, in our population.

We will assume further that the different alleles of each of these genes interact in an additive way. Let us call the two allelic forms of each of these genes 0 and 1. Suppose that allele 0 of each gene results in more of one characteristic, while allele 1 results in more of another. Then, if the character governed a trade-off such as the one between strength and fine motor skills, a person with genotype 00,00,00,00 might be very strong but clumsy, while a person with genotype 11,11,11,11 might be relatively weak but have a high level of the skills needed to chip an elegant stone arrowhead—or repair a watch.

Whether a given population will carry all eighty-one of these combinations, and in what frequency, will depend on the population's history. Suppose that, through chance or selection, one human group ends up with a predominance of 0 alleles. There would be many strong people in that group, and few if any expert watchmakers. If another group ends up with a predominance of 1 alleles, the situation would be reversed. These two groups would be, for this characteristic, very different from each other, and there might be relatively few people in either population with an "average" mixture of the two characters.

Two such populations, each made up of ten thousand people, are shown in Figure 1–2. In the first population the frequency of the 0 alleles for each of the four genes is eighty percent, while in the second the frequency of each 0 allele is only twenty percent. To make things fair, a great deal of environmental "noise" has been added to this genetic picture: the genetic contribution to this characteristic is assumed to be only fifty percent, while the other fifty percent has been supplied entirely at random.

Even with all this random environmental noise, you can see that the two populations are clearly different from each other. Almost nobody in the first population scores an 8, and almost nobody in the second population scores a 0.

Now suppose that these two formerly isolated populations come together, intermarry, and after a few generations become thoroughly mixed. The result, seen in Figure 1–3, is very different. After mixing, the majority of the population is made up of people with average

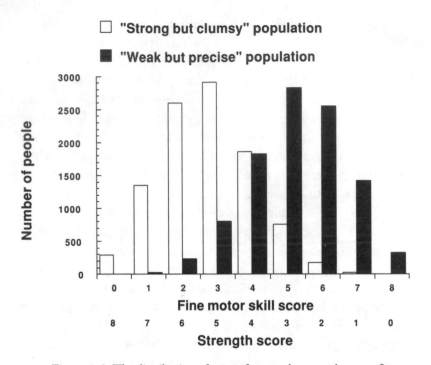

Figure 1–2. The distribution of scores for muscle strength versus fine motor skills in two different populations. In the first population, the frequencies of alleles of four genes that contribute to these abilities are all at 0.2, and in the second they are all at 0.8. Note that, because environmental fluctuations have been added to the picture, it is not possible to determine an individual's genotype from his or her score or *phenotype*.

attainment, and there are some people at each extreme. Unlike the first two populations, all eighty-one gene combinations are found in this mixed population.

The ability of this mixed population to move in different evolutionary directions has been enhanced, for it has a wider range of *phenotypes* than either of the original populations. (The phenotype is simply the appearance and abilities of an organism—what you see, in other words—and is the result of a complex interaction between the genes and the environment.) The people who make up the population will on average be slightly more variable, since the proportion of genes that are heterozygous in an individual will have risen slightly.

It is important to remember that none of this evolutionary change has been brought about through the appearance of new mutations.

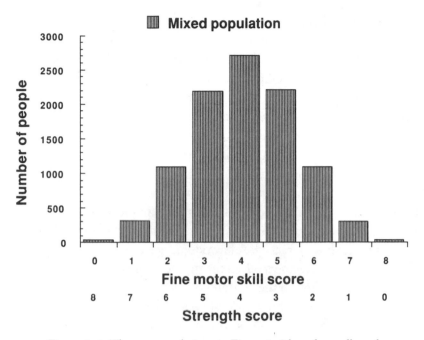

Figure 1–3. The two populations in Figure 1–2 have been allowed to interbreed. This population has a wider range of phenotypes, and therefore a greater evolutionary potential, than either of the earlier populations.

All that has happened is a shift in the frequencies of existing alleles. Of course, the alleles of the four different genes did originally arise by mutation in the past, but none had yet managed to take over entirely in either of the original populations.

Moreover, only four genes are required to produce this clear separation between the original populations. In the human population there are thousands of genes that have different alleles and that influence many different characters. No wonder that all of us, with the exception of identical twins, are so different from each other—and that our species has so much potential for future evolution.

This little exercise was designed to show you that any population can have an enormous evolutionary potential, so long as it happens to be plentifully endowed with genetic variation in the form of a wide assortment of alleles. It is a myth that species need new mutations in order to evolve. At least in the short term, old mutations that have already accumulated will do quite nicely, though if evolution is to continue for long periods new mutations will eventually become necessary.

Nor do we need to wait for one allele to replace another in order to increase this evolutionary potential further. Suppose that, as the result of a mutation, an allele arises that increases fine motor skills at the cost of a further reduction in brute strength. We will call this allele 2. Through chance or selection it rises in frequency until alleles 0, 1, and 2 are all equally frequent. Then a new allele arises at one of the other genes, but this allele enhances strength at the expense of fine motor skills. This allele, too, increases in frequency. What would be the effect on the population?

As you can see in Figure 1–4, the average of the population has changed very little as a result of these new mutations—most of the people still have average characteristics. But the range of abilities in the population has increased. A few people are now very strong, and a few have extreme dexterity. Increasing the range of variation in the population, even though the new alleles have not replaced the others, has increased its evolutionary potential still further.

If both such alleles appear, the population has not changed in any obvious direction, but its genetic variation has increased. Such

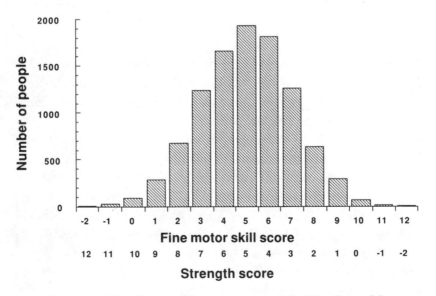

Figure 1–4. The effect of adding a new motor skill allele of one of the four genes and a new strength allele of another. The range of characteristics in the population has increased.

changes are nonetheless evolutionary changes, for they alter the composition of the population's gene pool and increase its range. I contend that such changes make up at least as important a part of the evolutionary process as any obvious directional selection for bigger heads or less hairy bodies.

Now, at the end of the twentieth century, different human groups, carrying different mixes of alleles, are blending together with unparalleled speed in almost every part of the planet. The rate of this blending can only increase in the future. As a result, the evolutionary potential of our species is growing enormously.

SORTING OUT THE POTENTIAL

What will happen to all this evolutionary potential, and whether it will really matter to our survival, is a different question. Even if the genetic variation in our species is increasing, selection is not necessarily acting on it. Without selection, regardless of whether our gene pool is changing, most of us would conclude that we are not truly evolving. It is my task to convince you that much of this variation really does matter.

To begin, we must remember that the history of our species over the last few million years, *and continuing down to the present time*, is one of pell-mell evolutionary change.

Much of this selection has been for sheer diversity. No other species on the planet is as behaviorally diverse as our own. Composers and concert musicians, brilliant mathematicians and engineers, linguists, superb athletes who have the ability to perform physical tasks that are at least the equal of what the most athletic of our primate relatives can accomplish—all these and more show that we have a growing ability both to create and to take advantage of a cultural and environmental diversity that would have been unimaginable a few millennia or even centuries ago.

The evolutionary changes that we are undergoing today are not taking us in one single direction, as those science fiction writers have tended to suppose. Instead, like Stephen Leacock's hero, we are galloping in many directions at once. It is now possible to construct testable models for how all this genetic diversity has been maintained and increased over time. We will see that selection for diversity, particularly behavioral diversity, has been going on for a long time, and that it has been greatly enhanced by the complexity of our current societies.

WHERE IS THE SELECTION?

In 1962 the eminent geneticist Theodosius Dobzhansky published a book called *Mankind Evolving*, which addressed the question of whether we are still evolving and concluded that we are indeed. As a beginning graduate student, I was most impressed by the book. I was delighted soon afterward to meet the author and have the opportunity to work with him.

In retrospect, Dobzhansky timed his book rather badly. Shortly after he wrote it, a flurry of amazing discoveries about our fossil record began, a flurry that has continued and accelerated down to the present and that has totally changed how we think about human prehistory. And decades later new genetic research about our species exploded onto the scene. Nonetheless, many of the matters that concerned him turn out to be timeless themes that we will also address in this book. Perhaps we will push a little closer to some answers.

Much of the last part of Dobzhansky's book directly addressed the question that we are wrestling with in this chapter: Just how much is natural selection likely to be affecting our species at the present time? Here, oddly, he found himself on rather uncertain ground. He could not point to a great many clear-cut examples of such selection.

To be sure, he emphasized that a great variety of diseases still exert selective pressures on human beings, particularly in underdeveloped countries, and that in those regions malnutrition continues to play an important selective role. But he had trouble finding examples of selective pressures acting on people in the industrialized world. He did his best, telling a few tales of selective pressures that technology has introduced. For example, he noted that people who suffer from a genetic disease called porphyria are highly sensitive to certain barbiturates, and that a number of them died after treatment with these otherwise relatively benign drugs. But such examples of selection, as he would have been the first to admit, are unlikely to have had much of an impact on our gene pool. Of course, he wrote at a time when our species' impact on the planet was only just beginning to be perceived—the environmental disaster stories that assault us daily were then unknown.

Dobzhansky did make the important but usually unappreciated point that the gene pool of our species is not simply the result of selection *against* things, such as susceptibility to diseases. Nor is it entirely the result of selection *in favor of* things, such as upright stance, bigger brains, and cleverer hands. Rather, our gene pool is a complex and dynamic collection of alleles that rise and fall in frequency and reshuffle with each generation in ever-shifting alliances. Because there is a diversity of environments, there will be selection for diversity in the gene pool.

Dobzhansky also pointed out that human beings clearly differ from all other species. Because of our intelligence, we have achieved a unique status in the world of living organisms. For millennia we have had the power to make sweeping changes in entire ecosystems—and even to destroy them utterly, as has happened in many parts of the Middle East since the invention of agriculture.

In spite of these insights, it was hard for him to find a convincing answer to the central question: Does our current success at altering the world around us, and our parallel ability to insulate ourselves from the natural world, provide us with immunity to the process of natural selection? Are harmful genes still being removed from our gene pool at the same rate as before? Is our evolutionary potential still increasing?

Today, despite a few contrarian views that echo Dobzhansky's, the received wisdom is that these things are not happening. Our recent conquest of many diseases is the example that is most often trotted out to explain the presumptive disappearance of natural selection.

Infectious diseases have certainly been powerful sources of selection on our species in the past. Yet over the last few centuries, and particularly in the last century, mortality from these diseases has plummeted in the developed world. Americans now live nearly twice as long on average as their great-grandparents did. In much of the world, any healthy child born now has an excellent chance of growing up and having children in turn. The obvious conclusion is that our culture, especially our new technologies, is shielding us from the kinds of traumatic adventures that our millions of ancestors experienced.

Although the argument has some justice, it is less forceful than it first appears. There is no doubt that the obvious impact of natural selection has diminished during our very recent history, particularly for those of us who are lucky enough to live in the developed world—although new pressures have also been introduced, such as allergies to newly invented substances. But the apparent diminution of selection raises two questions.

First, diminished from what? If selection operated on us in the past with particular vigor, then even after its effects have lessened, a great deal of it may still be going on.

The second and more complex question is, what does natural selection really mean in our species? In the industrialized world, people happen not to be dying in large numbers in obvious and premature ways, but this does not mean that selection has necessarily ceased to operate. Natural selection has many more subtle routes by which it can act on the variation-filled gene pool of a complicated species like ourselves.

TWO

◈

Natural Selection Can Be Subtle

Though Nature, red in tooth and claw,
With ravine shrieked against his creed . . .
ALFRED LORD TENNYSON, *In Memoriam* (1850)

Natural selection sometimes moves in mysterious ways, its wonders to perform. To understand how it works, we must carefully examine the concept itself.

INVISIBLE SELECTION

It is a fallacy to assume that the effects of natural selection are always visible. Sometimes they are quite invisible. To take just one example: A couple chooses not to have children. This conscious choice, whatever its reason, ensures that their genes will be lost to the gene pool. The effect is the same for a couple that is unable to bring a baby to full term. The absence of a baby who was never conceived, or who was aborted very early in development, can have a greater effect on the gene pool of our species than the untimely death of an adult who has already reproduced.

There are two general reasons why a baby does not reach term. The first is that at some point, usually early in development, something goes severely wrong with the developing embryo. Estimates vary, but it would appear that between twenty and thirty percent of fertilized eggs never complete development. Some of these failures happen late enough in pregnancy to be detected, but most occur so early that the mother may not even have suspected that she was pregnant.

But in the overwhelming majority of cases the reason an embryo dies is a gross genetic abnormality—missing or extra chromosomes, or even the presence of an entire extra chromosome set. This last severe defect accounts for ten percent of the spontaneous abortions that occur in the first trimester.

The influence of the environment can also be profound. During the 1984 Bhopal disaster in India, huge quantities of the deadly gas methyl isocyanate were released into the atmosphere. In the succeeding weeks and months the numbers of reported spontaneous abortions among the survivors who had been exposed went up almost fourfold, to a quarter of the total number of pregnancies. (The actual rate is certain to have been much higher than the reported rate.) Many of these extra abortions could be traced to chromosome abnormalities, apparently induced by the gas exposure.

Bhopal was an extreme case, but it was not unusual. Most of us are lucky enough not to live in India's ironically named "Golden Corridor," where industrial disasters continue to happen even after Bhopal. But we are all affected to some degree by the rise of chemical waste in our environment.

These chemicals may be applying their own selective pressure on our population, preventing the births of babies who would otherwise have developed normally. We simply do not know the extent of this pressure, or whether it is increasing or decreasing, but we do know that the numbers of accidental large-scale releases of dangerous chemicals are going up dramatically in many parts of the world. What are the evolutionary effects of these mini-Bhopals, and of all the everyday, less spectacular releases of contaminants?

Declining male sperm counts have been much in the news lately. It has been suggested that chemicals produced by the plastics industry that resemble female hormones might be to blame, and such chemicals can indeed cause sperm counts to decline in rats. But we may be worrying about a nonexistent phenomenon. The studies that claim to show declining counts in humans have been strongly critized in evey aspect, and it is now unclear whether the declines are real or merely statistical artifacts.

However, men who already have low sperm counts may be particularly susceptible to the effects of environmental contamination. A recent study of a group of such men in Athens, Greece, showed a significant decline over a seventeen-year period. If sperm count really is falling among such susceptible groups, then this is certainly a brand-new selective pressure that is being brought to bear on at least part of the human population.

The second reason that a baby might not be born is a relatively new one in human societies: Couples can now choose, more than ever before, whether to have a baby at all. This factor adds an entirely new and extremely complex psychological dimension to the influences shaping our gene pool. The genes that contribute to psychological makeup of a population are fair game for selection, and as we will see in Chapter 5, the effects and impact of such genes may well be increasing.

Not all of the invisible selection that alters our gene pool is selection against physically or psychologically harmful genes, or against sperm, eggs, or embryos that have been damaged by the environment. Selection can act to preserve and even to enhance the genetic variation within a population, or to increase its genetic capabilities in other ways.

Such alterations are often missed. When scientists follow the evolution of various plant and animal species over time, they generally measure changes in the average value of some character that is of interest to them. The increase in the human brain size is an obvious example. But far fewer studies follow changes in the variation of a character around its mean. Yet such changes are an integral part of biological evolution. It would be most important to know, for example, how much selection there has been for new alleles of genes that influence our psychological makeup, as the psychological landscape of our societies has become more complex. Although we do not yet know the answer to this question, we are on the point of being able to find out.

IMITATIONS OF NATURAL SELECTION

Invisible natural selection is likely to be changing our gene pool quite rapidly. But its effects will soon be supplemented dramatically by our growing expertise in genetic technology. We will soon have the power to make changes in our genes that resemble the changes caused by natural selection.

The first such changes will be straightforward "patches" that cure defective genes. Early attempts at such gene therapy have met with a notable lack of success, but a new generation of modified viruses that can carry undamaged genes into damaged cells is starting to become available. At the moment gene therapists are concentrating their efforts on putting functioning genes into affected tissues—for example, inserting a normal CFTR gene into lung cells of cystic fibrosis patients or a normal hemoglobin gene into the blood-forming tissue in people with sickle cell anemia. A strict moratorium has been imposed on making such changes in the cells that produce eggs and sperm and that can therefore pass the altered genes on to the next generation.

But once gene-delivery systems become more sophisticated, and more precise gene surgery becomes possible, the moratorium will undoubtedly be abandoned. W. French Anderson, one of the pioneers of gene therapy, is already arguing that research on germ-line therapy should be permitted. Eventually, as the technology gets better, his argument is likely to prevail.

If it can be done neatly and without damaging any other genes in the process, patching a sickle cell anemia gene so that the cured rather than the damaged gene will be passed on to the next generation seems like a very sensible thing to do. But it will be a short step from patching genes that clearly cause specific diseases, to patching genes that cause less obviously harmful conditions, and from there to modifying genes in such a way as to enhance our abilities. Genes that contribute to obesity, small stature, and other stigmatizing conditions are already fair game. As long ago as 1984, transgenic mice carrying a rat growth hormone gene were produced; they reached twice the size of their littermates. And, recently, mice with a defective gene that results in obesity and diabetes have been given successful gene therapy in the form of a virus carrying the gene.

These manipulations in rats and mice point the way to such manipulations in humans. Genes discovered in these animals are almost certain to have human equivalents. Human genes that increase muscle strength and even mental abilities will soon be discovered, and totally new modifications of these genes will quickly be invented. Success stories involving such genes will soon sweep aside any moral scruples people might have about altering our gene pool.

It is well to remember two things, however. First, such manipulations are very different from the natural process of evolution. Many of the changes that genetic surgeons carry out never have been tested by natural selection. There is nothing to prevent some of those changes from reducing a patient's long-term chances of survival or reproduction, just as the current craze for tongue studs and for nose, navel, and nipple rings increases the chance of infection.

What we consciously do to ourselves will always be, to one degree or another, in conflict with the often-invisible and hard-to-comprehend processes of natural selection. But we are sure to do it anyway. Many athletes, for example, currently use dangerous steroids to increase body mass. More than half of competitive athletes interviewed have said that even if the steroids were absolutely certain to cause them to die young, they would still use them, because winning is so important to them.

Second, the human gene pool is a very big place. With 5.5 billion people and 200,000 genes per person, the genes that are changed by deliberate intervention will simply be a drop in the bucket. Genetic changes driven by natural selection will continue to dwarf the efforts of even the most indefatigable gene therapists, at least for the foreseeable future. But chemical means of enhancing our bodies and our minds will also become commonplace in the future, and as we will see, these too can have an evolutionary effect.

Human beings have already been carrying out such gene manipulations on domesticated plants and animals, in some cases for millennia. Most of these changes were accomplished simply by artificial selection. The history of such efforts is surprisingly long: recent evidence shows that our ancestors domesticated dogs on two separate occasions, tens of thousands and perhaps more than a hundred thousand years ago.

The results of artificial selection have often been grotesque. When ears of corn and wheat are left unharvested in the field, they simply fall to the ground near the parent plant and rot as they germinate. If milk cows are left unmilked for a few days, they will die of burst udders. Cattle are now being bred for looks—but their straight rear legs, which give them a pleasingly square "cow" shape, can actually make it difficult for them to walk. Artificially bred tom turkeys can be so massive, they are unable to climb atop females and impregnate them. In order to do their job, the poor things need to be lifted up by amused farmhands.

Such animals and plants have been bred, not for characters that would aid in their survival in a state of nature, but for characters that people find useful. For those of us who shop too extravagantly at the new genetic mall, the results are likely to be just as grotesque, even disastrous. Clearly we are not yet wise enough to be able to imitate natural selection in all its aspects (and we may never be).

And, just to remind you how inextricably natural and artificial selection are intermixed: the people who succumb to the blandishments of this dazzling array of genetic "fixes" will not be a random sample from our gene pool. It seems likely that one result will be selection for genes that contribute to caution, and against genes that contribute to reckless behavior!

GENETIC EVOLUTION VERSUS CULTURAL EVOLUTION— A FALSE DICHOTOMY

The changes that we have wreaked on domestic animals and plants have generally damaged their ability to survive in the wild. But what about the things we have done to ourselves? There is no doubt that we have much reduced our own ability to survive under natural conditions. Our remote ancestors such as the Australopithecines, and our closer relatives such as *Homo habilis* and *Homo erectus*, lived during times of great and continual danger, threatened by animal predators, starvation, disease, and tribal warfare. The selection that acted on them was omnipresent and fierce. We, on the other hand, are largely protected from such severities and are able to survive and reproduce without hindrance. Our cultural prowess has helped us to modify virtually every part of the planet's surface and has enabled a substantial fraction of us to live long, disease-free, enriched, and stimulating lives.

As a result, we have become very well adapted to our cultural milieu—but remarkably ill adapted to the natural world. This shift has been going on for a long time. Even hunter-gatherers who live in the warm and abundant tropics today need at least a modest assortment of artifacts in order to survive. Most of us need a lot more than that.

Just think of all the things we must take along in order to spend a week in even as relatively friendly a natural environment as California's Sierra Nevada mountains—a region that famed outdoorsman John Muir called the Gentle Wilderness. Some of the artifacts we may take are fripperies, like freeze-dried food and perhaps Nintendo games. But others are brand-new necessities, such as filters for removing dangerous pathogens like *Giardia* from the waters of streams that used to be pure before legions of backpackers polluted them. Muir himself, without the aid of fripperies, could wander off for months into those same mountains and manage very well, but he still required boots, clothing, a poncho, a knife, a gun, fire-making equipment, and so on. And to keep himself in touch with the human world, he took along books to read and notebooks to write in.

The selection we have imposed on ourselves has been, in many ways, as artificial as the selection we have imposed on domesticated animals and plants. Strip our culture away, and all of us—except perhaps a few dedicated and well-practiced survivalists—would likely perish. We have gotten ourselves into this situation, and the human gene pool has changed greatly as a result.

Because of this triumph of culture over nature, one might think, it is now our culture and not our genes that will primarily determine what will become of our descendants.

If true, this transition would mark an enormous shift in human history. It would mean a complete change in the influences that operate on us, with predominantly biological forces being replaced by predominantly cultural ones. We will see, however, that it is not true. The powerful effects of our culture have, if anything, accelerated our biological evolution.

This acceleration began very early. Our remote ancestors lived far more intimately with the savage world of nature than we do, yet they were remarkably successful creatures. They were already able, through their social activities, to diminish the more obvious rigors of natural selection. Yet their evolution did not slow—rather it accelerated. This seeming paradox provides clues to the ways in which selection is currently acting on us.

Genes and culture have been intertwined for a long time. The appearance of crude stone tools in the fossil record marks the first real indication of culture as we understand it. In their earliest manifestations these tools were simply stones that had been partially shaped to produce a sharp cutting edge at one end. Detective work has shown that they really were the creations of our hominid ancestors. For example, they are rarely found in isolation. Sometimes they are discovered in such large numbers that our ancestors must have come together in groups over long periods to make and use them. They also often have a different composition from unmodified rocks found in the same area, showing that they were carried by their makers over substantial distances.

The time span over which our ancestors can be shown to have made stone tools is continually being lengthened. Tools dated to 2.5 million years ago have been discovered in southern Ethiopia, in Israel, and in Pakistan. Even at this early stage in their history, our tool-making ancestors were peripatetic and adventurous, traveling beyond their African homeland far sooner than had been previously supposed.

Perhaps most surprisingly, the tools in these old deposits were already relatively sophisticated. Discoveries of even older stone tools are therefore sure to be made sooner or later. Even though 2.5 million years ago seems extremely ancient, we have yet to uncover the true beginnings of human tool-making.

The bones of our ancestors are anything but rarities in the fossil record, and if abundance can be considered a measure of success, then our ancestors were remarkably successful. Even before tools appeared in the fossil record, social interactions of some kind, however primitive, must have contributed to their achievements.

Toolmaking is only part of this story of accelerating evolutionary change, however. By 2.5 million years ago, for example, our Australopithecine ancestors had already been walking upright for at least a million years—though perhaps not with quite the same panache as modern humans. And there is certainly no evidence that the invention of tools halted their subsequent physical evolution, for afterward our ancestors continued to undergo many other dramatic changes in brain and body.

If culture did not halt our genetic evolution back then, why must we suppose that it has done so now?

REVEALING OUR EVOLUTIONARY POTENTIAL

We will soon be able to modify our own bodies and even our own genes more dramatically than we have modified those of domesticated animals and plants. But even in this brave new world, the things that we are able to do to ourselves will still be dictated by our evolutionary histories and the capabilities that we have acquired through natural selection.

Take brain size. The size of the human brain does not seem to be entirely governed by some innate mechanism that tells it to stop growing at a certain point. Rather, its final size, at least in part, is dictated by the rate of maturation of the skull in which the brain is imprisoned.

One baby out of every three thousand is born with a defect that causes one or more of the sutures of the skull to close too soon. This premature cranial synostosis distorts the shape of the head and puts great pressure on the brain as it grows. The problem can be corrected surgically, by preventing the suture from closing until after the brain has matured. As a result of this drastic intervention, the disfigurement and mental retardation that would otherwise result can be ameliorated and in a good many cases prevented entirely.

Cranial synostosis brings the growth of the brain to a premature halt by a combination of increased intracranial pressure and decreased blood supply. My colleague Kurt Benirschke recently pointed out to me that we simply do not know the constraints that a normally developing skull puts on the size of the brain. Suppose that, in some nightmare surgeon-dominated world of the future, it becomes common practice to perform craniotomies on normal babies. How large would their brains grow, once the barriers were removed? What effect, if any, would this have on their intelligence?

Of course, this change would be surgical rather than evolutionary. In the absence of further surgical intervention, the children of these deliberately altered people would have brains of a normal size. A true evolutionary change would take place only if larger brains were to prove advantageous to our survival in the world of the future, selecting for genes that delay the closure of skull sutures (among many others) so that the change would become a permanent part of our species's gene pool.

Benirschke's brain example may seem far-fetched, but few of us in the developed world now get through life without surgical interventions of some sort. Sometimes we owe our continued existence to them.

These interventions are often dictated by our evolutionary history. For example, one consequence of our upright posture is that our teeth are crowded toward the backs of our jaws. Most readers of this book have probably had some or all of their wisdom teeth surgically removed, either to make more room for the others or because they were perversely growing into the teeth next to them.

The collision of evolutionary forces that has led to this particular problem must have been dramatic. Dental surgery is a recent invention, going back at the most a few thousand years. During much of this time the methods that were used to treat tooth problems, while often effective, do not bear thinking about. A Roman skull has recently been found in which an artificial iron tooth had been inserted into the upper jaw—where it seems to have functioned quite well for some years before the patient died, apparently of an unrelated cause. A few years ago I watched a traveling dentist set up temporary shop in the public square of a village in Morocco's Anti-Atlas Mountains. As dentists in that part of the world had done for centuries, he extracted teeth from his surprisingly docile patients with the aid of a fearful set of rusty forceps and a large mallet and chisel. But whenever the wisdom teeth of our ancestors became impacted, the results must have been disastrous, for such primitive surgery could not have corrected that problem. The advantages of upright posture must have been overwhelming indeed, since they outweighed the negative consequences of such severe and often life-threatening dental problems.

The difference between the natural and the artificial worlds is not completely clear-cut, however. If the surgeon's knife can increase brain size—and perhaps even intelligence—by a relatively simple operation, then our brains already possess a potential that is normally masked by the opposing process of skull maturation. This potential, even though almost never realized under ordinary circumstances, may be an evolved property of the genes that control the development of the brain itself.

We are not likely to be alone in possessing this potential: other organisms must exhibit a similar property. Suppose that an operation were carried out on the skull sutures of a chimpanzee baby, so as to stretch out the period of maturation of its brain. The operation would almost certainly not result in a chimpanzee with a human level of intelligence, but it is not beyond possibility that its intelligence might be measurably increased.

Such a potential would have developed over a long period of time. Suppose that the brains of our distant mammalian or reptilian ancestors had sometimes had the opportunity to grow to unusual sizes, perhaps because those ancestral creatures happened to carry mutant genes that delayed the development of their skulls. If this genetic accident resulted in an increase in brain power, then genes that delayed skull maturation might have been selected for. So would genes that helped to adapt the brains to their new roomier homes. Thus, in the future world that we have just envisioned, the surgeon's knife would simply be exposing an evolutionary potential that has been revealed many times before, albeit under different circumstances, during our long evolutionary history.

The same thing applies to brain biochemistry. As I write, "smart" drugs are becoming the rage. Many of us are dosing ourselves with extracts of the leaves of the living fossil tree *Ginkgo biloba*, or ingesting huge quantities of rather scary biochemicals such as acetyl carnitine that are supposed to increase the activities of the energy-producing mitochondria in our cells. These substances, it is claimed, will make us think better—and indeed there is some evidence that ginkgo extract actually slows down the rate of onset of Alzheimer's disease.

These "smart" drugs may or may not work on people with normally functioning brains—there is as yet no evidence one way or the other. But the public's desire for a pill that will make one think better is so palpable that when drugs that really do work—as is inevitable—are found, the market for them will doubtless be huge. Already memory-enhancing drugs are in the testing stage at a number of pharmaceutical companies, and even more remarkable advances are on the way. A mutation that enhances long-term memory and decreases training time has recently been found in the fruit fly, and the hunt is on for the equivalent gene in humans.

In yet another variant of this brave new world, children could be given drugs that would enhance their musical or mathematical or artistic abilities. These enhancements, however, will be possible only because the genes that these drugs affect are already there—all our brains have these capabilities built in as a result of our evolutionary history. It seems likely that these drugs will have untoward and unexpected side effects, particularly at first, and that disasters will be as numerous as success stories. But we will learn from our mistakes and get better at such manipulations. (In Chapter 14 I will explore some of these incredible scenarios and their likely evolutionary consequences.)

HOW THE ENVIRONMENT CAN ENHANCE
GENETIC POTENTIAL

Even after the dust settles and we get really expert at altering and enhancing our brain function, it will be impossible to divorce the environmental changes we make from evolutionary changes that are happening at the same time. The more we know about altering our brains, the more we will force biological evolution, because genes and environment always work together.

Selection cannot act on genetic differences if those differences are not in some fashion made manifest in the bodies or minds of the people who carry the genes. As we learn more about our genes, we are discovering more and more situations in which genetic effects results from our modification of the environment.

For example, some 300 million people worldwide, primarily males of African or Mediterranean ancestry, carry a sex-linked condition called G6PD-deficiency. This condition protects against malaria, but if there is no malaria in the environment, the carriers of the gene suffer no untoward effects—that is, unless they are incautious enough to eat broad beans or breathe the fumes of mothballs! The result of these apparently innocent activities can be a life-threatening anemia. This condition is revealed only when the carriers of the gene are exposed to this particular narrow range of environments, some of which—like the mothballs—are of our own making.

Another pervasive environmental factor, which we have also generated ourselves, is smoking. A lengthening list of genes is being discovered that, in their carriers, enhance the deleterious effects of smoking. Smoking is a stupid thing to do, but for these people it is particularly stupid—the genes they carry would normally predispose them to emphysema or lung cancer only slightly, but smoking greatly enhances their risk.

Our livers are responsible for breaking down and rendering harmless all sorts of unpleasant chemicals in the environment, and they do so by manufacturing many different enzymes specific for the destruction of various classes of such chemicals. A liver enzyme called paraoxonase, for example, is manufactured by a gene that comes in a wide variety of alleles.

One of these alleles, found in about a third of Asians and ten percent of Caucasians, turns out to protect against the effects of organophosphate pesticides. These chemicals, vanishingly rare in our past but now essential to twentieth-century agriculture, are currently responsible for three million cases of severe poisoning and 220,000 deaths worldwide every year. Another allele of the gene gives little protection against these chemicals but protects against nerve gases such as sarin. Both of these alleles, of course, appeared in the population long before either kind of chemical became common. Whatever their original function, it is apparent that these alleles are going to play a much larger role in our adaptation to the horrendous world of chemicals that awaits us.

Organophosphate pesticides will be common in that world. We can only hope that disasters like the release of sarin into the Tokyo subway system by the Aum Shinrikyo cult in 1995 will not also become common.

The selective pressures that originally caused these paraoxonase alleles to be retained at substantial frequencies in the human population still remain a mystery. The presence of the allele that protects against sarin is particularly baffling, because it has recently been shown, in a study carried out in Japan, to predispose diabetics to heart attacks. Nonetheless, whatever their original function, our recent manipulations of the environment have made these alleles far more important than they were in previous generations.

The rule that genes can be selected for or against only if their effects are made visible also applies to genes that affect brain function. Brain-enhancing chemicals will be part of the environment of the future. Different people will, because of their genes, respond to some of these chemicals far more readily than others. Suppose we find ways to artificially enhance such mental differences between people, and this in turn has an effect on whether they reproduce. Then such manipulations are likely to accelerate the process of natural selection as it acts on these differences. Even as these soon-to-be-invented biochemical manipulations extend our capabilities, they will magnify the effects of natural selection.

It is not impossible to imagine the following scenario. Suppose that somebody with a potential talent is unable, through a variety of environmental circumstances, to realize that talent. Discouraged and feeling useless, he or she might turn instead to a high-risk life and suffer an early death before reproducing. But if, through some kind of biochemical or other environmental enhancement, that individual instead achieves early success, it might put his or her life on a totally different and far more fortunate track. We all know of cases in which a chance encounter with a dedicated teacher changes a pupil's career. But the teacher, in order to work that magic, must realize that the talent is there and waiting to be nurtured. Perhaps "smart" pills will help this process. Stranger things have happened.

We must always remember, as we move into a world in which surgery and biochemistry offer new ways to explore the potential of our bodies and minds, that this potential is itself a product of evolution. Realizing that potential is something that our current highly challenging environment can do very well, and it seems likely that this trend will continue.

In the middle of the eighteenth century, Thomas Gray was inspired to write a poem by some anonymous graves he found in a country churchyard. In his elegy he described the graves' occupants as being like flowers born to blush unseen and waste their sweetness on the desert air. At the end of the twentieth century, entirely because of enhanced genotype-environment interactions, far more such human flowers should be able to flourish and contribute to our cultural richness.

The converse is true as well. People who are ill adapted to the modern world will be less likely to have children—not because their own survival is directly threatened, but because of psychological pressures that they would not have encountered a century or two ago. Much of what is happening to our species at the moment is being driven by these relatively invisible psychological factors. Their effects are not as obvious as the effects of predation, starvation, and disease, but they are nonetheless potent. And at the moment, because our physical environment is not as severe as it once was, these psychological pressures are becoming more important.

As I explore with you these interwoven evolutionary processes, the important theme that will emerge is that evolutionary change can happen in many different ways. There is much more to evolution than selection for big brains and hairless bodies.

FEAR OF NATURAL SELECTION

In addition to the claim that natural selection has ceased, another enormous objection has been raised to the idea that we are continuing to undergo biological evolution. This objection emerges from a fundamental misunderstanding of a new and important idea in human affairs: that of social equality.

During the last three centuries, the wondrous Enlightenment notion that human beings are all, at least potentially, socially equal has come ever closer to realization. It has liberated an increasing fraction of us from the bondage of superstition, and from servitude to kings and tyrants. Throughout much of the developed world, societies have arisen that are predicated on this concept. If we are all socially equal, then it is self-evident that we should all be provided with an equal opportunity to make ourselves into something greater.

But the idea of social equality would seem to be in direct collision with the idea of genetic determinism. How can we construct a socially equal society if we are prisoners of our genes? How can our new and hard-won social freedoms liberate us from the genetic tyranny of natural selection?

The proponents of social equality justifiably fear that if nature is really more important than nurture, so that the differences between people are genetic rather than cultural, then our characteristics and fates must be unavoidably written in our DNA. If we are prisoners of our genes, then even though we have managed to overthrow the human tyrants of the past, we have simply exchanged one sort of bondage for another.

Because these new bonds are built into our bodies and minds, they will be far more difficult to break. And it is only a short and easy step from the conclusion that there are large genetic differences among individuals to the inference that there is a biological basis for race and class prejudice.

This concern is important. Bitter experience has taught us that the idea that we are genetically unequal, as it is usually presented, is incompatible with the idea that we are socially equal. Distressingly, when a scientist or other authority figure points out that people can be divided by some criterion into superior and inferior groups, it immediately gives the imprimatur of science to those who would maintain social inequalities.

The latest manifestation of this destructive effect is a book by Richard Herrnstein and Charles Murray called *The Bell Curve*. Utterly uninformed by any evolutionary perspective, the authors present at length a great deal of evidence for the existence of genetic inequalities among individuals. Some of this evidence is convincing, some much less so.

At the end, triumphantly waving their collection of factoids, they attempt to use them to justify and reinforce the social inequalities that have shaped our society. After all, they point out, these inequalities certainly exist. Our genes are in the process of shaping our society, and they produce an intellectual caste system in which those at the top are the natural and inevitable overlords of those at the bottom.

I find the Herrnstein and Murray approach abhorrent, because they take the undeniable fact that humans differ genetically and turn it to base and short-term political ends, writing off much of our species as unsalvageable.

In Chapter 13 I will deal with the issue of nature and nurture from the perspective of an evolutionary biologist. The evidence is overwhelming that, because of the genes they carry, some people are better at certain tasks than others. But what does this fact actually mean, in terms of its immediate impact on society as well as its long-term evolutionary effects? Not, we will discover, what Herrnstein and Murray thought that it means at all.

The famous bell curve of IQ scores is a far more complex phenomenon than is generally supposed. A dramatic increase in certain kinds of IQ scores, called the Flynn effect, has taken place over the last few decades in industrialized societies. This change is not evolutionary—it is happening far too quickly for that—but growing evidence suggests both a genetic and an environmental basis for this phenomenon.

The environment plays a larger role in such things as intelligence than had been supposed just a few years ago. It is now becoming apparent that it is impossible to establish a hierarchy of intelligence. Not only is intelligence manifested in many different ways, as psychologist Howard Gardner has pointed out, but the genes that contribute to it are constantly shifting and recombining from one generation to the next. This means that even though you may be particularly smart—and insufferably smug about it to boot—there is no way of predicting how intelligent, or in what fashion, your grandchildren will be.

The only constant in this bewildering story is selection for a diversity of intellectual capacities. The genes that contribute to these capacities are part of the rich tapestry of our evolutionary heritage. Follow the threads for even a few generations, and you will find that everybody contributes to the tapestry and that it is impossible to decide which of the threads is most important.

IQ and other human abilities are labile and easily modified, and it is the height of cruel foolishness to write off any group of humans. All of us have untapped abilities and carry important genes essential to the future survival of our species. It is the diversity of our gene pool as a whole that holds the key to our future evolution, not some particular set of genes that form the basis for an intellectual elite.

THE FUTURE

In Chapter 14 I will gaze, with some trepidation, into my crystal ball and speculate about what is likely to happen to our species in the future—provided of course that our story is not brought to a close by the sudden arrival of an asteroid. Particularly in this era of swift cultural change, prognostications are nearly always rendered irrelevant by events. Nonetheless, I will try my hand at examining the potential that is held out by the astonishing efflorescence of science that has taken place during the last two centuries and that will continue in the centuries to come.

The looming crises that confront our species are primarily of our own making. The terrible fires in Southeast Asia are only the latest in a series of environmental wake-up calls. The realization is growing that our current population size cannot be sustained without severe damage to the planet. War and disease are unlikely to control our population—we will have to do it ourselves, by modifying our behavior. There is no way we can begin to reduce our population over the next century or two without selection playing some kind of role, and I predict that this selection will primarily be at the behavioral level.

On the other hand, as far as the increased diversity of our species is concerned, the sky is no longer the limit. If life is found elsewhere in the solar system or beyond, some of us will surely visit and eventually colonize those places. As we spread to worlds that are very different from our own, the consequences for our evolution will be at least as profound as when our remote ancestors first ventured out of Africa.

Will people living on other planets evolve into new species? Given sufficient isolation from the rest of us, we have no reason to suppose otherwise. Indeed, challenged by very different environments, they will probably become new species much more quickly than the millions of years that were required to engender the timid beginnings of speciation in the orangutans of Borneo and Sumatra.

In short, the powers of natural selection that Darwin was the first to understand will certainly continue to shape our species. Here it is our task to try to glimpse the accelerating ways in which biological and cultural evolution will reinforce each other in the future, and to understand how this mutual interaction will allow us to survive the evolutionary challenges that we will face as we begin our spread through the universe.

THREE

◈

Living at the Edge of Space

*Those who sweat get frostbite easily, but I never sweat
when I am climbing. . . . Sometimes it has been suggested
that I have "three lungs" because I have so little trouble
at great heights. At this I laugh with my two mouths.
But I think it is perhaps true that I am more adapted
to heights than most men; that I was born not only in,
but for the mountains. I climb with rhythm, and it is a
natural thing for me. My hands, even in warm weather,
are usually cold, and doctors have told me that my
heartbeat is quite slow. . . . On a recent tour of India,
with the heat and the crowds, I became more sick than I
have ever been in my life on a mountain.*

TENZING NORGAY, *Tiger of the Snows* (1955)

Even though we have yet to visit other planets, some of us have spread
into environments that are almost as alien, with profound evolutionary
consequences. Think of these migrations as a kind of practice run for
our future attempts to colonize Mars or the planets of a nearby star.

The world has four roofs. One is in East Africa and encompasses
the highlands of Ethiopia; our ancestors have lived there for millions
of years. A second, which was colonized by humans only in recent historical times, is the Colorado Plateau. A third—far higher than either
of these and inhabited for perhaps ten thousand years—is the high
Andes, extending along much of the western rim of South America.

55

But the roof of the world that everybody thinks of first is the Tibetan plateau. This region encompasses some 800,000 square miles, an area more than twice as large as the high Andes. The plateau has been pushed up by a slow, grinding, but irresistible collision between two great pieces of the Earth's crust: the tectonic plate that carries the Indian subcontinent, and the immense mainland of Asia itself.

As India has plowed into Asia's southern rim during the course of the last fifty million years, the most obvious result of this collision has been the Himalayas. These mountains, however, form only the south-ernmost part of a massive pile of ancient seabed material that has been shoved ahead of the Indian plate, like snow in front of some vast snow-plow. Squeezed into accordion pleats by the force of the collision, this material has been formed into range after range of peaks. Each spring fierce rivers wear away the rocks of the high-altitude valleys between the peaks, where in spite of everything a little life manages to cling.

The nearest I have come to this roof of the world was a visit to the desert town of Dunhuang in Gansu Province of western China. Dunhaung is located one hundred kilometers from the Tibetan Plateau's northern edge, squarely at a junction of the Great Silk Road. People have lived there for a long time. The famous Buddhist caves at nearby Mogao have yielded, among other treasures, the earliest known printed book, the famous *Diamond Sutra*, which can be dated from clues in the text to A.D. 868.* But these are by no means the first traces of human occupation. In the nearby Taklimakan Desert, Chinese and European archaeologists have found more than a hundred well-preserved mummies of a Caucasian people who lived in the area four thousand years ago. The origin and fate of these people, so different in appearance and culture from the present-day inhabitants, is a complete mystery.

The immediate countryside around Dunhuang is relatively flat, invaded in places by rolling sand dunes from the desert to the south. Beyond the desert is a range of six-thousand-meter peaks known as the Denghe Nanshan, which stand out vividly whenever the air is clear. Even from so far away, these immense mountains dominate the land-scape. The sky to the south of them is pale with reflected snow from range after range of still mightier mountains. And it is there that some

---

* Earlier printed books are known, in particular books from Korea that prob-ably date from before A.D. 750, but the *Diamond Sutra* is the first printed text that can be dated with precision.

members of our species who have migrated to the edge of space can be found.

We know little about the history of human penetration of the Tibetan plateau. Perhaps a Shangri-La or two is still hidden somewhere among those remote peaks and valleys. Whether there is or not, people have dwelt in this most extreme of lands since long before written records. The severity of their lives has, until recently, been counterbalanced by their remoteness, which has made them relatively safe from invaders.

Among the diverse inhabitants of the plateau, the Sherpas of Nepal are the people who have interacted the most with the outside world. They are not native to Nepal, but migrated there during the seventeenth century from southern Tibet. Sherpas are brilliant mountaineers and make up an essential part of any climbing expedition in the Himalayas. Their mountaineering skills and ability to carry heavy loads at high altitudes have made such expeditions possible and have enabled numerous well-heeled Manhattanites to stagger to the tops of the world's highest mountains. The Sherpa Tenzing Norgay was the second person to stand on the top of Everest, courteously dropping a few feet behind Edmund Hillary during the final scramble to the summit.

The Sherpas and other indigenous peoples of the plateau have adapted remarkably to the need to live their entire lives at around four thousand meters, where each lungful of air has only a third as many molecules of oxygen as at sea level. But has this adaptation been a truly evolutionary one? Do the Sherpas actually carry genes, or combinations of genes, that are different from those carried by people who live at sea level? If so, then in the course of acquiring these genes, they have undergone a true evolutionary change.

This question is difficult to answer because it is hard to distinguish between genetic adaptations and those that are merely physiological. Any human body has the capacity to adapt fairly rapidly to extreme conditions. At high altitudes the number of red blood corpuscles in our blood can rise substantially, and over time even our lung capacity can increase somewhat. Further, any of us can soften the impact of the cold simply by bundling up warmly, by not exposing our extremities more than necessary, and so on. Are the Sherpas really different genetically from the rest of us? Or do they have these physiological adaptations simply because they have lived their entire lives, beginning at the moment of conception, at high altitudes?

The theory of evolution predicts that any population that lives under such extreme conditions for many generations should become better and better adapted; the poorly adapted members of the population will die or fail to reproduce. In order to see whether natural selection has altered the gene pool of the Sherpas, we would like to have at least a rough idea of how long their ancestors have lived under these conditions. The longer this period has been, the greater the likelihood that selection has been able to change their gene pools.

We cannot be very precise about this estimate, because uncertainties abound. Very little genetic information about Tibetan populations is available, and the human fossil record in this region ranges from sparse to nonexistent. The Tibetans have a strong genetic and physical resemblance to the peoples who currently inhabit the steppes to the north of the plateau, in the area around Dunhuang and elsewhere. That region is likely to be the primary origin of the peoples of Tibet.

But if the origin of the Tibetans is shrouded in mystery, so is the origin of their ancestors. Those puzzling Caucasians of four thousand years ago, so different from the present-day peoples of the northern steppes, are unlikely to have been the first arrivals in the area. Indeed, the first arrivals on the steppes, and perhaps even the first to venture into the Tibetan highlands, might not even have been members of our own species.

Early hominids,* in particular our immediate forerunner *Homo erectus*, likely inhabited Central Asia for a very long time. Although fossil hominid sites in Central Asia are few and far between, two different very ancient sites have been discovered that (very loosely) bracket the Tibetan Plateau. Fragmentary remains of *Homo erectus* from both of these sites have been dated to almost two million years ago. One of the sites is far to the west of Tibet, in the Caucasus Mountains; the other is located to its east, in the upper valley of the Yangtze.

---

* Hominids include our own species and our very closest relatives, all of which are extinct—the Australopithecines, *Homo erectus*, the Neandertals, and so on. Hominoids, on the other hand, are the result of a wider cast of the taxonomic net and include our closest living relatives—the chimpanzees, bonobos, gorillas and orangutans.

Such old sites have not yet been found in Central Asia. But if bands of *H. erectus* did trudge across the vast windy space that separates these two known sites, it hardly seems likely that they did so only once. Much more probably, during this immense span of time, they migrated back and forth many times across what would eventually become western China and Kirgiziya. In doing so, they would have anticipated Marco Polo and the other much later travelers belonging to our own species, for they would have been following the route of the Silk Road—long before there was any silk or presumably even any concept of a road.

Of course, since no fossil record of such migrations exists, all this is sheer speculation. We also have no evidence that tribes of *H. erectus* ventured into the mountains that they would have glimpsed far to the south of their migratory path. If they had done so, they would probably have taken the most accessible routes, from the northeast and the southeast, in order to avoid crossing the vast Gobi Desert. But we do have evidence that, about half a million years ago, they were living close to the plateau itself. The evidence comes from a discovery made in a cave near the village of Yuanmou, not far to the southeast of the plateau's edge between Chengdu and Kunming. The scattering of teeth and tools found there belonged to *H. erectus* and have been tentatively dated to 500,000–600,000 years ago, though they may be younger.

It may not be coincidental that these traces of *H. erectus* happen to have been found near a well-worn trading route into Tibet that has been used extensively during historic times. There is no doubt that this difficult but negotiable route leading to the high plateau would have been open to these predecessors of modern humans. I find it hard to imagine that some of them, impelled by curiosity or the need to escape from enemies, did not venture into the mountains.

If *H. erectus* reached high altitudes and stayed there for any time, they would have become adapted to the conditions there. Then when they were displaced—and perhaps driven to extinction—by later arrivals, those adaptations may have disappeared with them. But perhaps not. *

---

* I will do my best to ignore the persistent rumors of the yeti, or Abominable Snowman. This mysterious hairy creature figures in elaborate legends told by the Sherpas, who claim that it dines on fungi found at high altitude. Footprints and supposed samples of yeti hair have all, when carefully investigated, been traced to other animals. Still, one cannot help but wonder what the origins of the legend might be!

The much more recent migrations through this area of our own species, *Homo sapiens*, are also mostly shrouded in mystery. For centuries the plateau has lured traders who have entered through routes that snake over the towering passes and lead into Tibet proper. Up until the time of the Chinese Revolution, endless files of Chinese peasants, bent almost double under eighty-kilogram bales of tea leaves, toiled up from Chengdu into the southeastern part of the plateau, in order to slake the Tibetans' great thirst for tea. Nineteenth- and twentieth-century travelers tell of endless lines of porters stumbling up slopes that would challenge a trained mountaineer, numb and empty-faced because they were unable to withstand the cold and fatigue without the aid of opium.

But these traders left few records. Essentially nothing is known of Tibet's history prior to the seventh century A.D., although Paleolithic tools dated to about fifty thousand years ago have been found on the northern part of the plateau. So we have no idea when the earliest modern humans might have arrived, whether they found the plateau already inhabited, and what might have happened to any earlier inhabitants. But we do know that, in spite of the strong physical and genetic resemblance of the present-day Tibetans to the peoples who inhabit the steppes to the north, they did not come from that area alone. The physical resemblance does not extend to the languages they speak, which have an affinity with those of the Burmese, to the southeast. This discordance between appearance and language suggests that the peoples of the plateau have had many long-term cultural contacts with a variety of peoples who lived at lower altitudes, and that the Tibetans must be an amalgam of many different migrations.

Thus it seems reasonable to suppose that, along with new languages and trade goods, new genes have been repeatedly infused into the gene pool of the Tibetans. The genes could have come from the tribes of the northern steppes, from peoples who lived in the ever-more-populous river valleys to the east that would eventually become China, and from the diverse inhabitants of the foothills of the Himalayas. However, if the genes already carried by the earlier settlers that aided high-altitude survival were sufficiently advantageous, they might have remained in the gene pool and spread in numbers even in the face of an influx of new genes from the outside.

While we cannot identify precisely the period of years that the Tibetans have had to adapt to high altitude, it is certainly in the thousands and probably in the tens of thousands. If *H. erectus* really did leave some genes behind in Tibet before vanishing, there is a slim possibility that this adaptive period might even have extended over hundreds of thousands of years!

To get a clearer idea of the impact that this long period of adaptation has had on the Tibetans, we can compare them with lowlanders who have lived at high altitudes for only a matter of months or years. Although lowlanders such as the Han Chinese settlers tend to be much less well adapted, it can always be argued that this is simply because they were not born and brought up under such extreme conditions. If a lowlander couple were to migrate to the Tibetan plateau and have a baby, then by the time that baby grows up, it might be just as well adapted as the Tibetans who have lived there for generations.

We do not know if this is the case. The Han Chinese who now live in Lhasa and elsewhere in Tibet do so for only part of the year, and pregnant Chinese women almost always descend to lower altitudes to give birth. So it is difficult to attribute any differences between Tibetans and these lowlanders to a genetic cause.

One way to disentangle true genetic change from mere physiological adaptation would be to examine another group, who have lived at altitudes as high as the Tibetans but for fewer generations. If the evolutionary scenario is correct, these people should have adapted to their environment less thoroughly than the Tibetans, since there would have been less time for new genes and new combinations of genes to appear in their gene pool. Luckily, such a group can be found. They inhabit the second greatest roof of the world.

BEFORE THE INCAS

The human history of that other roof, the great cordillera of the Andes, has almost certainly been briefer than that of the Tibetan Plateau. The Andes were also formed by a collision of tectonic plates, though not as massive as the one that took place between India and Asia. Here the encounter was between unequals. As the thin oceanic plates of the eastern Pacific encountered the massive plate that makes up the western edge of the South American continent, they were forced to bend down, pushing up the coastal rim in the process. This relatively mild encounter has not crushed and folded the Amazon basin, which lies to the east of the collision zone. Instead, it formed the Andes, an immensely tall but relatively narrow range of mountains: the world's second tallest after the Himalayas.

The ancestors of the Quechua and Aymara Indians who inhabit the high valleys of the Andes must have made their way up into these regions a relatively short time ago, probably five or six thousand years. For one thing, human occupation in the South American mountains was dependent on the development of crops that could support them. Further, because of the shorter distances involved, the Andean altiplano was far more accessible to lowlanders than the huge and remote Tibetan Plateau. Throughout the history of the Inca empire, and undoubtedly in earlier times, there was much migration from low to high altitudes and back.

Nonetheless, during this short history, these peoples have managed astonishing things. Their accomplishments were not, as is generally supposed, all due to the Incas. The Inca empire did its best to foster this notion by carefully destroying any records of the achievements of earlier civilizations. But we know now that they were responsible only for putting some finishing touches on a vast network of roads and bridges that stretched from northern Ecuador to a point south of present-day Santiago in Chile, extending eastward into Bolivia and Argentina. This network, one of the most remarkable construction feats in the preindustrial world, was actually the result of the efforts of many different cultures.

Along these roads, in the days of the Inca empire, relays of specially trained runners called *chasqui* could carry verbal messages, or messages coded in a knotted string, from Quito to Cuzco in five days. For long stretches the *chasqui* had to run at altitudes well above four thousand meters.

The problems faced by the first colonizers of the Andean altiplano, like those that faced the first venturers into Tibet, were more challenging than a simple lack of oxygen. For most of the year, nighttime temperatures in the Andes fall below freezing. The extreme dryness of the air and the lowered vapor pressure suck moisture from the body fairly quickly. The blood literally thickens, and the brain is deprived of essential oxygen, leading to the blinding headaches, dizziness, and sleepiness—and sometimes the edema and internal bleeding—of mountain sickness.

When I arrived for the first time in Cuzco, the magnificent ancient capital of the Incas, I was advised by friends to drink as much liquid as I could manage for at least the first twenty-four hours. By following this good advice, I was able to avoid the worst of the altitude effects that often incapacitate unprepared tourists. But the first arrivals in these high desert valleys had no such warning. Their children were particularly at risk.

Life has always been severe in this region. Studies of mummified human remains from several different early cultures, ranging from 1300 B.C. to A.D. 1400, show that fifteen percent of the women had died in childbirth or immediately post-partum. Even today, among the Indians of remote areas in southern Peru, disease and the effects of the rigorous environment kill three hundred out of every thousand babies during the first year of life.

The Aymara and Quechua have developed ingenious methods for preserving the lives of their fragile children. The youngest babies are wrapped in layer after layer of cloth, rendering them immobile. This conserves heat, moisture, and the child's energy. The child in its bundle is then placed in a carrying cloth or *manta*. The amount of oxygen reaching its lungs is reduced by all the layers, but because it cannot move very much, any damage from anoxia is presumably minimized. The baby is kept tightly wrapped for the first three months, but as it grows the wrappings are gradually loosened.

How much selection for sheer survival must have taken place before this effective method of protection was invented? Cumulatively, it must have been very great. Before infant mortality was reduced to its present, albeit still brutally high, level, uncounted tiny bodies were buried or interred in the stone mausoleums on the cliffs surrounding the altiplano valleys. Not all these deaths had a genetic effect on the population, but the fraction that did, over the millennia, had the cumulative result of increasing the likelihood of survival of each generation.

It is striking and perhaps significant that Tibetan mothers swaddle their babies as well, but the babies' heads are kept freer of clothing so that they can breathe more easily. Does this mean that a Tibetan baby is better able to withstand the fierce high-altitude conditions than a baby in the altiplano? As we will see, there is evidence that it is.

A RUSH OF BLOOD

Mark Twain remarked that even though history never repeats itself, it does tend to rhyme. So does evolution. Selection for different genes and combinations of genes must have occurred in the two adventurous peoples who colonized the Tibetan plateau and the Andean altiplano. Although the Andeans have adapted less well, both groups have acquired the ability to live under these extreme conditions.

So far, scientists have made only limited comparative genetic investigations of these two groups. They have yet to track down any specific genetic changes that have contributed to their adaptation. But it is almost certainly only a matter of time before some are found.

Perhaps the most telling indication of true genetic differences between the two populations comes from an examination of the weights of their babies at birth. In virtually all parts of the world, high altitude has a strong and significant negative effect on birth weight, even when babies are carried to normal term. The effect is detectable even when the differences in altitude are relatively small. On average, for every thousand meters of altitude, the birth weight of full-term babies decreases by about one hundred grams.

This pattern turns out to hold true for the Andean Indians: their babies weigh about four hundred grams less than babies born at sea level—for the metrically challenged, this translates into a difference of a pound or so. Such a large reduction in birth weight must contribute to the high infant mortality. Babies of acclimatized parents weigh a little more than unacclimatized babies born at the same altitude, but not much.

This pattern emphatically does not hold true for Tibetan babies. In a notable exception to the worldwide trend, their birth weight at term is the same as that of babies born at sea level.

These high birth weights are not some special property of conditions on the Tibetan plateau. Chinese mothers who have lived in Lhasa throughout their pregnancies obey the birth-weight-altitude law. Even though they usually descend to lower altitudes to give birth, they have smaller babies than do Chinese mothers who live at sea level.

How do Tibetan mothers manage to nourish their babies so efficiently at high altitude? To find out, Lorna Moore of the University of Colorado visited Tibet a number of times from 1983 to 1992. Although recent political upheavals have now made further work difficult, during that window of opportunity Moore and her coworkers carried out a number of important experiments on Tibetan volunteers (see Figure 3–1). Using as controls some Han Chinese living in Lhasa who had spent long periods of time at high altitude, they found that, in vivid contrast to the Han women, Tibetan women excel at supplying oxygen to their babies during pregnancy.

Using Doppler ultrasound, it was possible to measure the rate of flow in the arteries that supply blood to the lungs and to the uterus. In most people low levels of oxygen cause the pulmonary arteries, which supply blood to the lungs, to constrict, which is exactly the reverse of what they should do. This turns out to be a reflex, an echo of the moment when we are born. As a fetus grows in the womb, the pulmonary arteries are small and little blood flows to the lungs, but at

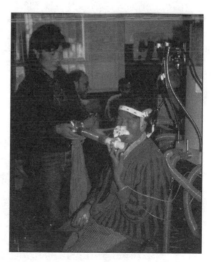

Figure 3–1. The two smiling women on the left were photographed in the Lhasa Bharkor market. They come from Kham, in eastern Tibet. The pregnant woman on the right is having her tidal volume and oxygen use measured. Pictures courtesy of Lorna Moore.

the moment of birth oxygen rushes into the lungs and the pulmonary arteries expand. We are now cursed with the reverse of this reflex— when oxygen drops, our pulmonary arteries contract in a physiological memory of fetal life. Moore and her coworkers found that this does not happen in Tibetans—soon after birth, they lose that dangerous reflex.

The rate of blood flow in the uterine arteries of the Tibetan women is very high for a different reason—their arteries are unusually large. These two factors combined mean that Tibetan women retain high rates of flow under hypoxic conditions.

After birth, other unusual adaptations are exhibited by the Tibetan babies themselves. During the period shortly after birth, Moore found that Tibetan babies had ten percent more oxygen in their arterial blood than babies born to Han Chinese parents who had lived at the same altitude. This difference persisted and actually grew more pronounced during the first weeks of life. Right from birth the newborn Tibetan babies were able to extract more oxygen from the thin air that surrounded them.

THE BODY'S PERCEPTION OF THE HIGH-ALTITUDE WORLD

The fact that Tibetan mothers can reverse the effects of high altitude and give birth to babies of normal weight, while Andean mothers cannot, is the most powerful argument for real genetic differences between the two populations. But Moore and her coworkers discovered other things about their physiologies that also undoubtedly have a genetic component. Although these phenomena are complex and interrelated, they show that Tibetans have by far the best set of adaptations to high altitude of any human group so far examined—including even the Tibetans' neighbors, the high-altitude peoples who live in Ladakh at the northern tip of India. The people of Ladakh, like those of the Andes, ventured to these extreme altitudes more recently than the people of Tibet.

Both the similarities and the differences are informative. The lungs of both Tibetans and Andean natives have more of the tiny thin-walled sacs called alveoli, in which the blood is brought in close contact with air. But such adaptations to high altitude life are to be expected— even lowlanders can eventually develop them.

Another obvious accommodation to high altitudes is shown by the Andeans: They have more red blood cells, and higher levels of hemoglobin, than lowlanders. Lowlanders, too, tend to increase the number of red cells in their blood as they become acclimatized to high-altitude life. Remarkably, however, Tibetans show neither of these adaptations. In fact, they have slightly fewer red blood cells, and slightly less hemoglobin, than people living at sea level. This is not due to an increase in blood volume, which is the same as that of altitude-adapted Han. Further, when Tibetans climb to even more extreme altitudes than those at which they normally live, the hemoglobin in their blood does not increase to the same extent as it does in the blood of lowlanders when they become adapted to the same altitude. Recently suggestions have been made that an allele among Tibetans may control the rate of saturation of hemoglobin with oxygen, though hard biochemical evidence is as yet lacking. Oddly, the statistical analysis suggests that all Tibetans do not have this allele.

In the Tibetans increased blood flow is enough to overcome their deficiency in red blood cells. And their thinner blood has other consequences, the most important of which is protection against high-altitude pulmonary edema.

This severe illness constitutes the greatest danger to adult survival. When lowlanders climb quickly to high altitudes, their arterial pressure shoots up, especially the pressure in the pulmonary arteries. In mountaineers and others under severe stress, fluid is sometimes actually forced out of the circulation and into the lungs.

More insidious is chronic mountain sickness, in which the increased pressure forces blood out of the capillary beds into the tissues. A great variety of symptoms can result, some taking years to develop, including bleeding under the skin and gastrointestinal hemorrhage. The symptoms of the disease are obvious even to the casual observer: the faces of the victims are a darkened mahoganylike color, stained with the blood that has become trapped.

This slowly developing disease is common in people who have lived at high altitudes for long periods, or even for many generations. Surprisingly, it often develops in Peruvian highlanders, who one might have thought would have evolved resistance.

Chronic mountain sickness is also widespread among the Han Chinese living in Lhasa, who make it worse by incessant smoking. But it is almost never seen among Tibetans, even those who smoke heavily. Their cheeks remain pink and healthy without the telltale broken vessels or dark engorgement suffered by the lowland Chinese.

Resistance to the effects of dehydration may be the key to resistance to mountain sickness. Recall the remark of Tenzing Norgay's that I quoted at the beginning of this chapter. Tibetans may lose less moisture from perspiration, although this has not been systematically studied. In addition, the fact that Tibetans have a relatively small number of red blood cells means that their blood should not be as affected by dehydration. Even when they do lose fluids through exertion at high altitude, their blood will retain its normal low viscosity for longer. This too has yet to be studied in detail.

Yet another Tibetan adaptation, almost invisible but very effective, has to do with the way their bodies perceive the high-altitude world. They respond differently to the stresses of living at the edge of space.

When people who have spent their lives at sea level are first transported to a high altitude, their rate of breathing does the reverse of what one might expect—it actually slows down. This lowers the amount of oxygen in the blood and increases the amount of carbon dioxide, which can worsen the symptoms of mountain sickness. With time, as these newcomers become acclimatized, their breathing rates return to normal levels. Remarkably, the Quechua of the Andes, even after many generations at high altitude, always show a pattern of slow breathing, even though they are presumably thoroughly acclimatized. This may help to explain why chronic mountain sickness is so prevalent among them. They show what seems to be an inappropriate physiological response to their extreme conditions.

If you expect the Tibetans to show a different response from the Andeans, you will not be disappointed. At high altitude Tibetans breathe at the same rate as acclimatized newcomers. They do this because their bodies are sensitive to changes that are largely invisible to the rest of us.

In the laboratory it is possible to manipulate the amount of oxygen in the air supply that an experimental subject is breathing through a mask, and to do so in such a way that the subject gets no hint of what is going on. When oxygen is lowered to about two-thirds of normal, Tibetans react very differently from either Quechua or lowlanders. Their heartbeat rises rapidly from seventy to a hundred beats a minute.

The Quechua show a much smaller response, and lowlanders show almost no change at all. It takes a further substantial reduction in oxygen before their hearts begin to respond.

The bodies of the Tibetans, it appears, are far better than those of the rest of us at measuring oxygen levels. Their ability to sense conditions on the edge of space has been extended by the process of evolution, in just the direction that is needed to enhance their survival.

THE COEVOLUTION OF HUMANS AND YAKS

Tibetans are not alone in their superb adaptation to the high-altitude world. They share many of these adaptations with yaks, the high-altitude cattle of the Tibetan plateau. Yaks, too, have large and unusually thin-walled pulmonary arteries (Figure 3–2). This means that, under hypoxic conditions, the hearts of both Tibetans and their cattle do not have to work as hard as the hearts of lowland humans or lowland cattle to achieve the same results.

Just when the yaks first came to occupy the high-altitude plateau of Tibet is as much of a puzzle as the history of the first human occupation of the area. We do know that yaks have been plentiful in Tibet for a long time, long enough for people to have invented all sorts of uses for them. They pull plows in the high-altitude fields and carry huge loads over passes as high as six thousand meters. One recent traveler described "their great lungs inhaling and exhaling in puffs like a blowing locomotive." Indeed, they are so thoroughly adapted to high altitude that they tend not to do well below about three thousand meters.

In Tibet it is not sensible to waste anything. Every part of each yak is utilized. The hide has many uses, from clothing to coverings for the boats in which the Tibetans navigate the high lakes. The long and massive horns have been ground into powder for medication and have even been employed as building materials—in Lhasa, crushed yak horn is often used to reinforce masonry, and intact horns are inserted at intervals along the resulting walls for decoration. Yak oil is essential for lamps, and the butter is used to flavor tea, which the Tibetans drink in immense quantities to counteract the severe liquid loss at high altitude. Slabs of yak butter are often carved into intricate sculptures. So honored is the yak that two stuffed carcasses hang from

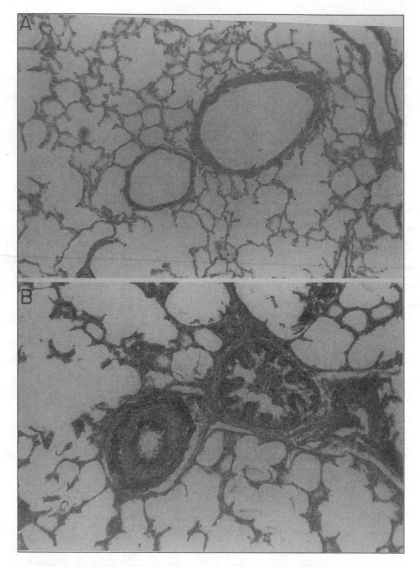

Figure 3–2. Cross-section of yak (top) and lowland calf pulmonary arteries (bottom) from Kurt Stenmark. In each case, the artery on the left is associated with an air passage on the right. Notice how thin-walled both the artery and its associated air passage are in the yak.

the ceiling of the great monastery that overlooks the remote town of Gyantse, southwest of Lhasa. Perhaps they are a relic of the time before Buddhism when the Tibetans worshipped a pantheon that included yak-headed gods.

Their long coarse hair gives yaks a very different appearance from lowland cattle. The differences extend to their internal organs as well. Their lungs are enormous—three times as heavy as the lungs of the cattle that live in nearby Inner Mongolia, which is a far greater difference than that seen between the lungs of human Tibetans and lowlanders. And, like human Tibetans, yaks have surprisingly little hemoglobin in their blood—less than in lowland cattle.

In spite of these and many other differences, however, one can think of yaks, without much loss of precision, as simply cattle with lots of hair. Zoologists have put them into a separate species (currently most authorities call them *Bos grunniens*), but their gene pool is not clearly separated from that of cattle. Because purebred yaks tend to be ill-tempered and balky beasts, for centuries the Tibetans have hybridized them with lowland cattle in order to produce animals that are a little more manageable. These hybrids turn out to be perfectly healthy, although there have been reports that they are not fully fertile.

All this hybridization has, however, made the task of disentangling the relationship between yaks and domestic cattle difficult. The best way to determine just how genetically separate they are is to compare the DNA of the two species. But it is difficult to find genetically pure yaks. Many animals exhibited as yaks in zoos are actually mixtures. Rather than venturing into the fastnesses of Tibet, scientists have understandably tended to obtain blood from these zoo animals, so they may have been misled into concluding that yaks and cattle are more closely related than they really are.

Still, the general position of yaks on the family tree of cattle and their relatives is undisputed: all the molecular studies show that yaks are by far the closest relatives of domestic cattle that have yet been found. They are much closer to cattle than the Indian gaur and the Javanese banteng, both of which have been suggested as ancestors of domestic cattle breeds. But exactly how much closer they are will remain problematical until this matter of genetic admixture has been straightened out. We may never know, for after centuries or millennia of forced interbreeding with cattle and escape of hybrids into the wild, it is possible that truly pure wild yaks no longer exist.

Further, artificial selection may have been imposed on top of their natural adaptation. The original wild yaks, grazing peacefully on their high-altitude grasslands, probably had no need for such giant lungs and extreme physiology. But when people began capturing them and forcing them to carry immense burdens over the high passes and pull heavy primitive plows through stony soil in the thin air, the yaks best able to carry out these grueling tasks must have been selected.

Selection, whether natural or artificial, is much more likely to proceed rapidly when a population is full of genetic variation. The numerous hybridizations of yaks with domestic cattle could have provided some of this variation, resulting in a kind of evolutionary kick-start.

Humans, may have benefited from the same sort of kick-start. As we have seen, the introduction of genes from lowland human populations into the Tibetan highlands may have helped provide evolutionary fuel for their adaptation to extreme conditions.

There is no doubt that selection is continuing in humans as well as yaks. While infant mortality is slowly decreasing in both Nepal and the Peruvian altiplano (we currently have only fragmentary information from Tibet itself), the rigors of high-altitude life remain severe. The Tibetans, the Quechua, the people of Ladakh, the yaks, and all the other human groups and human-associated species living at the edge of space will continue to evolve better mechanisms to survive these conditions for as long as they continue to inhabit this extreme environment.

It may not be long before some of the genes involved in high-altitude adaptation are tracked down. It recently has been found that one of the two common alleles in the English population for a gene controlling blood pressure is more frequent in people who are able to perform severe physical labor. It is particularly frequent in mountain climbers. Is this allele also frequent in the Tibetan population and less so in the Andean? Or do the Tibetans have yet another allele of this gene that adapts them even better to high-altitude living? The answers will soon be forthcoming.

This is the kind of natural selection that we can easily understand: severe conditions that select for new adaptations absolutely necessary for survival. But what about the rest of us? We are not faced daily with lung-freezing cold or a paucity of oxygen. Still, our lives are not risk-free. Many of us are faced with a different set of dangerous problems that require continuing adaptation. We are only beginning to see just how pervasive selective pressures are, and how difficult or impossible it will be to free ourselves entirely from their rigors. We have not, as many assume, overcome these selective pressures—far from it. Some of them are pressures connected with disease, which are far more pervasive and subtle than we had thought even a few years ago. The next chapter provides a glimpse of some of the difficulties that we still face in our fight for survival against the multitude of organisms that prey on us, and how we have evolved and continue to evolve in response.

FOUR

◈

Besieged by Invisible Armies

[T]he struggle against disease, and particularly infectious
disease, has been a very influential evolutionary agent,
and . . . some of its results have been rather unlike those
of the struggle against natural forces, hunger, and
predators, or with members of the same species.

J.B.S. HALDANE, *Disease and Evolution* (1949)

In 1946 my colleague Jon Singer arrived as a young postdoc in the Caltech laboratory of chemist Linus Pauling. One of his first tasks was to help in what was then a very tricky problem: to detect a difference between two almost identical proteins.

The project had its genesis in an accidental encounter. A year earlier Pauling had been invited to join a committee that would eventually recommend, among other things, an enormous infusion of resources into health sciences after the war. The eventual result was the emergence of a great research empire, the National Institutes of Health, that has dominated biomedical research in the postwar years.

In the train on his way to the committee's first meeting in Chicago, Pauling bumped into another committee member, geneticist William Castle. In the course of their conversation he learned about a strange disease, found among African-Americans, called sickle cell anemia. Castle's description of the symptoms of the disease immediately fascinated Pauling.

Although reports of what was probably sickle cell anemia have been traced to as long ago as 1870, the first clear evidence for the disease was found in 1910. In that year James Herrick, a Chicago doctor, noticed an odd phenomenon in the blood of Walter Noel, a black dentistry student from Grenada. Noel's normally round red blood cells would suddenly distort into a sickled shape while they were being examined under the microscope. Herrick and others soon found that many other people of African ancestry had the same puzzling trait.

On the train Pauling's attention was caught when Castle told him why the sickling took place. If a thin film of blood sits for some time on a microscope slide, trapped under a cover slip, the cells in the blood will use up the oxygen. It is only then that the sickling happens, accompanied by an increase in the polarization of light as it passes through the cells. What particularly intrigued Pauling was that as soon as oxygen is reintroduced, the cells quickly snap back into their normal shape and the polarization disappears.

Pauling realized that this visible phenomenon must be due to an invisible molecular event. Since red blood cells are little more than bags of hemoglobin, what was happening to the cells was almost certainly due to some property of the homogeneous population of protein molecules that they contained.

The most likely explanation seemed to be that, in the absence of oxygen, the hemoglobin molecules became linked together in a reversible fashion. These linked protein molecules would polarize light that passed through their fiberlike chains and would actually distort the shape of the cell itself. Then, when oxygen was reintroduced, the chains would break up and allow the normal shape of the cell to reassert itself. Pauling was familiar with such reversible reactions from laboratory experiments. If his idea was right, then the sickling must be a clear signature of a change at the molecular level.

He realized further that people with sickle cell anemia should be carriers of a mutant gene, specifying a mutant form of the hemoglobin protein. Since the dogma current at the time was that genes themselves consist of proteins—only a few heterodox researchers were beginning to turn their attention to DNA—Pauling had no clear idea of the nature of the mutation itself. But he thought that it might be possible to approach the problem by seeing if the hemoglobin protein in sickle cell patients really was different from the normal protein.

Another arrival in Pauling's lab just after the war was a brand-new M.D. named Harvey Itano. In common with most other Japanese-Americans, Itano had begun the war years branded as an enemy alien. He was actually being sent with his family to an internment camp when he received a letter telling him he had been accepted at St. Louis University Medical School. A few months later, through the efforts of many friends, he was permitted to leave the camp and go to St. Louis. But his love was science, and he had no interest in practicing medicine after getting his degree.

As soon as Itano arrived at Pauling's lab, Pauling asked him to look at the sickle cell problem, suggested some physical chemistry tricks that Itano might use, and then promptly disappeared to take up a temporary fellowship in England.

Itano tried Pauling's ideas but got nowhere. In talking over his difficulties with Singer, it seemed to him that the overall electrical charge of the molecule might be a better measure than the ones Pauling had suggested. Luckily, Caltech had a massive apparatus, invented and built in Sweden, for carrying out this sort of experiment. Singer remarked to me with much amusement that he quickly became essential to the project, since he was the only one in the lab who was strong enough to lift the enormous apparatus and lower it into the vast buffer-filled holding tank.

After much experimentation, they homed in on just the right pH of buffer that would distinguish clearly between the electrical charges on the normal and mutant proteins. At a pH very close to neutral, normal hemoglobin migrated toward the positive pole, while hemoglobin from patients with sickle cell anemia migrated toward the negative pole.

They went a step further and examined hemoglobin from relatives of sickle cell anemics, who had no sign of the disease but whose blood cells nonetheless sickled when the oxygen tension was reduced to very low levels. The hemoglobin of these relatives turned out to be made up of a mixture of the two kinds of molecule. This meant that these relatives must be genetic heterozygotes, who had inherited an allele for the normal protein from one parent and an allele for the sickle cell protein from the other. People with the more severe sickle cell disease were homozygotes, who by ill luck had inherited a mutant allele from both parents.

Since the heterozygous relatives did not show the disease, possession of one of these alleles was not enough to produce any medical problems. Possession of two, on the other hand, could be very dangerous. Sickle cell anemia, as Pauling had originally suspected, could truly be called a molecular disease. And it was this memorable phrase that Pauling, on his return from England, scrawled as a title across the top of the paper that Singer had drafted.

A few years later, in 1953, James Watson and Francis Crick announced their famous double-helix model for the structure of DNA. They realized that complementary sequences of bases must be strung along both halves of the helix. Their model explained how genes could be replicated and passed from one generation to the next—each chain of the double helix could serve as a template for a new complementary copy of itself, resulting in two double helices where there had been only one before. But Watson and Crick also realized that each half-helix of the DNA was made up of a long string of bases. Proteins were already known to be long strings of amino acids; both gene and protein could be read like sentences, and the first must code for the second. The language of the DNA was translated directly into the language of the protein. Changes in the sequence of bases along the DNA molecule must be the ultimate cause of changes in the amino acids of proteins.

This reasoning led Crick to suggest to protein chemist Vernon Ingram that he take a careful look at the normal and sickle cell hemoglobin proteins. Crick predicted that the difference between them would turn out to be a small one, and indeed Ingram found that the two molecules were identical in all but one of their 146 amino acids.

Once the genetic code embodied in the DNA itself was worked out, it could be seen precisely how a single change in the DNA could lead to the small change that Ingram had detected in the protein. Tiny as it was, this change could nonetheless have severe effects on people unlucky enough to carry two mutant alleles. All the many billions of copies of the hemoglobin molecule in their bodies would be manufactured with the defect, which would multiply the effect of the mutation enormously.

The discovery by Itano, Singer, and Pauling had huge consequences. Not only did it shed much light on the molecular basis of sickle cell anemia, it helped point the way to an entire new scientific field, that of molecular biology.

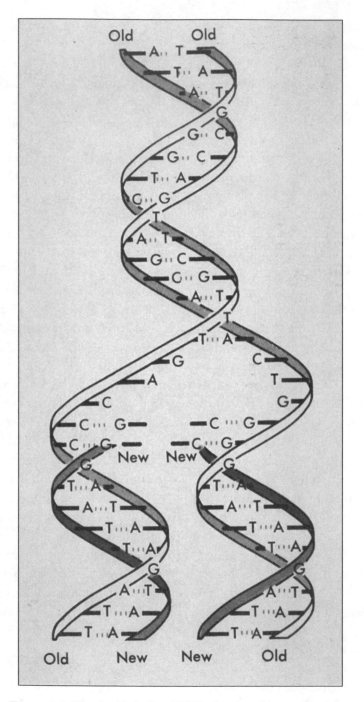

Figure 4–1. The double helix of DNA, showing the complementary bases, the way the code is carried by the molecule, and the synthesis of new complementary strands.

But their finding had still not explained why the mutant gene is so common in the African-American population, and why its effects are so obvious that it had come to Pauling's attention in the first place. We now know the answer to this question: various kinds of complex, rapid, and continuing evolution explain a great deal about the sickle cell and many other puzzling alleles. These evolutionary processes have been accelerated by our cultural inventions, particularly the invention of agriculture. Had we stayed in the forests like our orangutan and chimpanzee relatives, none of them would have happened.

A CLASSIC CASE OF SELECTION

Sickle cell anemia is a serious disease. It is found in one of every four hundred African-American babies in the United States. As the children with the disease grow up, they are susceptible to a series of sudden hemolytic crises, in the course of which large numbers of their red cells are destroyed. These crises can be brought on by something as small as a bout of unusual exertion or even a minor viral illness. They trigger a great variety of symptoms—enormous swelling of the spleen, swelling of the hands and feet, bone necrosis, and sometimes detachment of the retinas. The symptoms tend not to be obvious during the first few months of life, but the cumulative effects of the repeated crises soon build up.

In the 1970s infant mortality from the disease was fifteen percent during the first year of life. Now that the generally white medical establishment has finally realized the seriousness of the disease, this rate has gone down substantially—a 1995 study reported that only twenty of 694 homozygous children had died over a ten-year period, even though there was much serious sickle cell-related illness among the group.

Although some belated progress has been made in palliation of the disease, there has been no dramatic breakthrough in finding a cure. But this is probably only a matter of time. Eventually gene therapy—in which the gene itself is modified or replaced—will hold out hope for a lucky few.

The wide prevalence of the allele for sickle cell anemia is something of a surprise, in view of how harmful the disease is. Of the many different genetic traits that can damage their carriers severely, sickle cell is among the commonest. But the disease itself is only the tip of the iceberg. Heterozygous carriers of sickle cell trait are far more common than people who actually have the disease. Most heterozygotes marry homozygotes for the nonmutant allele, because they outnumber them in the population, and such marriages do not produce children with the disease. Even if two heterozygous people do happen to marry, only one quarter of their children will have the disease. This means that the heterozygous carriers in turn greatly outnumber the homozygotes who have sickle cell anemia. Carriers are in fact forty times as common as mutant homozygotes.

A five percent frequency of the mutant allele—about what is seen in the African-American population—means that about one person in every ten of the population is a heterozygous carrier. Most of the mutant sickle cell alleles in the population lurk almost invisibly in these heterozygotes, and they may be passed on in the heterozygous state for many generations before they come together by chance in an unlucky homozygote.

Even heterozygosity for this mutant allele has its costs. Under extreme conditions possessors of a single mutant allele can suffer a sickle cell crisis. This was first noticed during the Second World War, when African-American soldiers who were being transported in unpressurized planes would sometimes and unexpectedly suffer hemolytic crises at high altitude. More recently, sickle cell trait might have been implicated in a case of sudden death of an African-American athlete. Because this condition is not usually noticed on autopsy, there are likely to have been other cases (although the presence of the allele in one dose does not seem to influence the choice of an athletic career). These harmful effects, uncommon as they are, make it even more puzzling that the mutant alleles are so numerous.

Most of the ancestors of the American children with the disease were originally taken by force from West Africa, where the allele is even more common. In many West African tribes it is found at an astonishing frequency—twenty percent or even higher. Half the members of such tribes are heterozygous carriers of the gene, and five percent of babies are born with the disease—twenty times the rate in the United States. To make matters worse, in some tribes of equatorial Africa there is another allele of the same gene, called hemoglobin C. In combination with the sickle cell allele, it too can also result in serious illness. (This combination occasionally turns up in the United States as well.)

Even with advancing medical knowledge, the devastation caused by sickle cell anemia is still vast. About 150,000 babies are born with the disease every year in sub-Saharan Africa, and even those who have access to adequate medical care face enormous risk from recurrent health crises. Until recently, half the affected children died during their first two years, and very few of them survived to adulthood—though that picture is beginning to change, particularly in the larger cities.

So why, we are driven to ask, in the face of this horrendous illness and mortality, is this damaging allele still found in these populations in such numbers? One would surely expect it to disappear quickly, or at least to fall to very low levels, as a result of natural selection.

In 1949, the same year that Pauling and his group published their paper on the molecular nature of the disease, the eminent geneticist J.B.S. Haldane made a rather hesitant suggestion. Perhaps, he thought, the allele is so prevalent in Africa because the heterozygotes are protected against a blood disease, such as malaria. If this protection were to be sufficiently strong, Haldane pointed out, a kind of genetic balance of death would be set up. People homozygous for the normal allele would be killed by malaria in large numbers, as they presumably always have been. And people homozygous for the sickle cell allele would also die in large numbers—not because of malaria, but because of the effects of the mutant allele that they carry in double dose.

But the story would be different with the heterozygotes. Some of them would die as well—life in West Africa has never been easy. But as long as they did better than the other two groups, then the balance that Haldane suggested would emerge. Even though death will take its toll of all three genotypes every generation, the heterozygotes will be the winners in this grim sweepstakes.

The victory of the heterozygotes will only be temporary, however. Each generation, heterozygous people will meet, marry each other, and have babies. One quarter of these babies will be homozygous for the normal allele and thus at great risk of dying from malaria, and one quarter of them will be homozygous for the terrible sickle cell allele. Only half their children will be heterozygotes and therefore protected. This means that, even though heterozygotes have the highest fitness, they cannot take over the entire population. Even if malaria were so severe that only heterozygotes survived, homozygotes of both types would inevitably appear in the next generation, and the grim toll of death and disease would continue. Figure 4–2 shows how this works.

This kind of balance, known as a balanced polymorphism, has another dramatic genetic consequence. It gives a huge advantage to sickle cell mutations as soon as they first appear. Consider a population that lives in a malarious region, in which all the individuals happen to carry only normal alleles. In such a population malaria will be rampant and highly destructive. Then, if a sickle cell mutation arises and is not lost by chance during the first generation or two, all the copies of the mutation that are passed down to subsequent generations will have an immediate advantage.

The result is that the mutant allele will spread with tremendous speed through the population. It will continue to rise in frequency until the advantage it gives to the heterozygotes is eventually counterbalanced

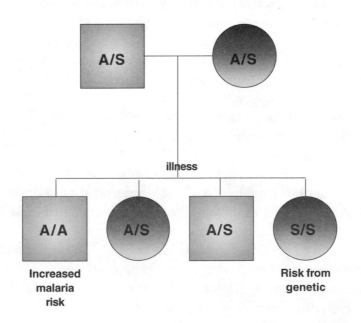

Figure 4–2. Even if two people heterozygous for the sickle cell allele marry and have many children, only half the children will be heterozygous. A quarter will be homozygous normal, and at greater risk from malaria, and a quarter will be homozygous for the mutant allele and will suffer from sickle cell anemia.

by the cumulative disadvantage of the growing numbers of homozygotes. Thus, once a sickle cell mutation arises anywhere in a malarious region, the genetic composition of the population will change swiftly and a balanced polymorphism will emerge.

E. A. Beet, a colonial medical officer working during the mid-1940s in what was then northern Rhodesia, found some evidence for this phenomenon. He noticed that malaria patients who had the sickle cell trait were less heavily infected with the malaria parasite than those who did not. It took decades of further accumulated data to convince the skeptics, but it is now generally agreed that Haldane's inspired guess was indeed right. Heterozygotes for the sickle cell allele really are protected against malaria. But this protection turns out, rather surprisingly, to be a clumsy and ill-constructed shield against the disease.

Contrary to an often-repeated myth, heterozygotes are not protected against getting the disease in the first place. Because the disease is spread by mosquitoes, such protection would imply that carriers of the mutant alleles could somehow fend off the mosquitoes, or could manage to destroy the parasites as soon as the mosquitoes injected them into their bloodstream. They can do neither of these things. Instead, they are protected simply because the severity of the infection is reduced. Even though this protection might seem a case of too little too late, it can often make the difference between life and death.

The gene's effects are exerted most strongly in the bloodstream of the malaria victim. When a malaria parasite enters a red blood cell, the cell is doomed—the parasite's progeny will multiply in the cell, consuming its hemoglobin as a food source. Finally, the cell will burst open and the progeny of the original invading parasite will be released in a jostling crowd, ready to infect other red cells in a kind of biological chain reaction. It is this simultaneous bursting of millions of red cells that produces the fierce fevers of malaria. The icy chills that follow cause the victim to tremble violently, giving the disease its older English name of the ague.

As if this were not enough, the parasites of some types of malaria cause even more damage, starting from the moment they invade the red cells. They change the infected cells' outer membranes, causing them to stick to nearby uninfected cells. As these clumps of red cells accrete, they take on a flowerlike shape known as a rosette.

Rosettes, when they are forced through capillaries by circulatory pressure, can clog them up. The more rosettes, the greater the danger. This helps to explain why the parasite *Plasmodium falciparum*, which produces rosettes, causes the severest kind of malaria. This species multiplies to enormous numbers in the blood, reaching levels several times as high as the other (and somewhat milder) human malarias. The result can be damage to the brain or the kidneys.

The situation is different in people with sickle cell trait. When their red cells are infected by parasites, the oxygen inside the cells is used up as the parasites grow and multiply. In the bloodstream of sickle cell heterozygotes this triggers the sickling reaction, just as it does on a microscope slide. Once the cell is distorted in this way, it is less able to form rosettes with nearby cells. The result is less damage to the capillaries.

An added bonus is that the sickled cell is more likely to be removed from the circulation before the parasites can mature. Our spleen serves as a filter, selectively removing defective red blood cells. The unsickled cells are round, smooth, and flexible, so that they can slip unharmed through the many complex passages inside the spleen. The irregular sickled cells get stuck and are soon destroyed by white blood cells. This filtration has the effect of reducing the total numbers of parasites in the blood, but at the cost of a dangerously swollen spleen.

In spite of all these benefits, there is a point beyond which the sickle cell gene can no longer provide even this crude and clumsy protection. A really severe malaria infection can cause too many red cells to sickle. If the infection is overwhelming, even heterozygous individuals can succumb to the effects of the disease.

THE AGRICULTURAL REVOLUTION AND MALARIA— LIFE AND DEATH GO HAND IN HAND

Sickle cell anemia is such a well-known story in biological circles that, whenever I bring it up in one of my classes, my students' immediate reaction is, "Oh, no! Not this again!" They have already heard about the disease in class after class and assume that by this time there is nothing left to learn about it.

They are wrong, as I gently assure them. The more the story is examined, the more complex it becomes. Our genetic defenses against malaria are influencing the future evolution of our species in an astonishing variety of ways—and perhaps may be affecting the future evolution of the malaria parasites that prey on us as well.

The most important actors in this evolutionary tale are the mosquitoes that spread the disease. People who have done fieldwork in the forests of central Africa find that, provided the forests have not been modified by human activity, their dark and sheltered floors are surprisingly mosquito free. The mosquitoes that are found in the forests tend to live high up in the tree canopy, where they prey on birds, snakes, and arboreal mammals.

The forest mosquitoes themselves belong to a great many different species, each of which is fairly rare, and relatively few of them are inclined to prey on humans. Such diversity among mosquito species is typical of the great variety of animals and plants that are usually found in any undisturbed forest ecosystem, especially in the tropics.

Fields and farms, in contrast, with their plentiful pools of water and disturbed vegetation, swarm with mosquitoes. In equatorial Africa agriculture has had particularly disastrous effects on the environment. The soil in this region tends to be impoverished and filled with iron salts, so that when vegetation is stripped away, it quickly hardens into a red cementlike substance known as laterite. Any ruts and depressions in the ground are made permanent by this crust and tend to fill rapidly with water, providing places for mosquitoes to breed. This problem has been compounded in recent decades by the accumulating detritus of civilization. Abandoned tires, bottles, and plastic containers, filling up with rainwater, provide even more breeding places.

As a result the mosquito population explodes—but not all of the mosquito species found in the region take part in the explosion. Throughout much of equatorial Africa, the new mosquito universe is dominated by a formerly rare species called *Anopheles gambiae*. Currently, this species is responsible for spreading three-quarters of the malaria in Africa. It is particularly common in small villages and outlying farms, wherever rainfall is high. Every inhabitant of such a village may be bitten by malaria-infected mosquitoes several times a night.

This change has been relatively recent and has been particularly marked during the last three or four thousand years. When agriculture, and later the technology of ironworking, were introduced into central Africa by waves of Bantu-speaking peoples who migrated down from the north, the immediate result was widespread deforestation. This increased the incidence of malaria, as the mosquito population altered.

An even more recent and better-documented upsurge in the disease has taken place in the highlands of Madagascar. During the eighteenth century European travelers to this island remarked with surprise on the great contrast between the fever-ridden coasts and the thickly forested, malaria-free highlands. But around the year 1800, following the introduction of rice cultivation to the highlands, this distinction rapidly disappeared. Just as on the mainland of Africa, as soon as the ancient forests began to be destroyed, a formerly rare mosquito species exploded in numbers. This species, *Anopheles funestus*, quickly became the primary carrier of the disease in these formerly healthy parts of the island.

These dramatic, human-caused ecological changes have, like so many human-caused changes, accelerated our own evolution as well. Along with the spread of agriculture and the huge migrations of peoples, the sickle cell allele has risen in frequency. Anthropologist Frank Livingstone has found that the history of the region can be traced in its genetics. For example, tribes speaking the Kwa subfamily of languages have lived in what is now Liberia and the Ivory Coast for a long time, since before the migrations from the north. Near the coast, these tribes have low or even zero frequencies of the sickle cell gene. But the gene is far more common in the eastern regions, where they have come into contact with other invading peoples.

These invasions of the Kwa territory by outsiders have been dated, though with some uncertainty, to the seventeenth century. This means that the gene has had perhaps twenty generations to reach the high levels that are currently found in the contact region. Livingstone suspects that, nearer the coast, the gene has been introduced so recently that it has not yet come into balance between selection for the heterozygotes and selection against the homozygotes.

A ZOO OF NEW ALLELES

The sickle cell mutation is, as we have seen, anything but an unalloyed blessing. It causes a terrible disease in some of its carriers even as it protects other carriers against a different terrible disease. This seems like a very poor bargain. Why has evolution not replaced it with some other allele that confers protection against malaria without entailing such a severe cost?

The sickle cell mutation is only a single mutational step away from the normal allele, and at present in each generation some one hundred brand-new sickle cell mutations are appearing somewhere in the human population. If such a single-step mutation, readily arising somewhere in the population, confers some limited amount of protection, it will immediately be selected for. More "sophisticated" mutations that confer better protection without such damaging side effects are likely to require more than one mutational step, perhaps occurring in the same gene or perhaps in different genes. These multiple mutations need more time to appear, since the probability that they will turn up at all is much lower. Only later, perhaps thousands of years later, will such far less probable combinations of mutations arise. Thus, the apparently clumsy evolutionary response of sickle cell anemia is a function simply of probabilities, not of some evolutionary ineptitude on our part. Given enough time, evolution should be able to solve the problem.

The sophisticated mutations have not yet appeared. Instead, a wide variety of other clumsy mutations have appeared at a variety of genes, some just as harmful as the sickle cell gene. They have been driven up in frequency, in Africa and elsewhere, by the relentless selective pressure of malaria. This zoo of mutations includes other amino acid changes that affect the same hemoglobin gene. They are called hemoglobin C, hemoglobin E, and various other designations. Some of these mutations were originally found by Harvey Itano, who has devoted his life to understanding the many things that can happen to this hemoglobin gene. In many cases we still do not understand what their protective effect might be.

A rather different kind of mutation, widespread in Mediterranean countries, turns the hemoglobin gene off completely. Homozygotes for this gene suffer from a very severe anemia known as thalassemia (from the Greek *thalassa*, "sea," since it is so common among the people who live on the malarious Mediterranean coasts). The people of Greece, southern Italy, and especially coastal Sardinia have high frequencies of this gene.

The parallel with the sickle cell story is striking. Thalassemia, too, is held in the population as a balanced polymorphism, and the protection against malaria that it confers on heterozygotes is equally clumsy. Turning the gene off does not make all the hemoglobin in the red blood cells disappear, since other hemoglobin genes are still active. But it does damage the hemoglobin molecule itself, which tends to shorten the lives of infected red cells so that they are destroyed early. Once again the protection comes at the cost of severe anemia and swollen spleens.

In Chapter 2 we met the G6PD gene, which has dangerous effects if its carriers eat broad beans. Hundreds of different mutations of this gene, which codes for an enzyme that is very important in red blood cells, have been found throughout the Mediterranean and parts of Africa. The mutant alleles are particularly common among the little villages of Greece. Sometimes the inhabitants of villages only a few kilometers apart can carry quite different mutations. Again, the damaged genes shorten the lives of red blood cells and diminish the effects of malarial infection.

In New Guinea and Micronesia a completely different kind of mutation has appeared that also protects against malaria. It changes the properties of the membrane that surrounds the red cell, altering its shape from a circle to an ellipse. This change makes the cell more resistant to the entry of the malaria parasite. Because the mutation actually increases the resistance to the parasite itself, it would appear at first sight to confer protection that is less clumsy and damaging than sickle cell or thalassemia. Unfortunately, these elliptical cells are not as flexible as normal cells, so the spleen detects them as abnormal. Homozygotes for this gene can also suffer from anemia.

This list is not exhaustive by any means. It is striking that all these mutations appear to be half-baked and jury-rigged. They are exactly the kind of "easy" mutations that one would expect to be selected for immediately in a population that has been subject to a brand-new and severe selective pressure. And indeed, there is genetic evidence that many of this zoo of new alleles have arisen in the last few thousand years—which is just the period during which we human beings have had a substantial impact on our environment.

MILD OLD DISEASES AND SAVAGE NEW ONES

In a perverse way these various protective mutations may actually be making things worse for us, at least in the short term. In a kind of evolutionary warfare, every genetic move that we make is countered by a move made by the parasites. Although this situation is temporary, it has contributed substantially to our rapid evolution.

The received wisdom (to which, by and large, I subscribe) is that long-continued host-parasite relationships will eventually lead to milder forms of a disease, because hosts and their parasites will co-evolve. Strains of parasites less damaging to their hosts will be selected for, since this lengthens the period of time during which the parasites can be passed on to new hosts before the old hosts die. Strains of hosts more resistant to their parasites will also be selected for, which will make it more difficult for the parasites to spread. When this happens, fewer parasites can survive in any individual host. But although the chance that the parasites can spread to new hosts during any given time period may be decreased, the very mildness of the disease lengthens the window of opportunity during which the parasites can spread.

Epidemiologist Paul Ewald of Amherst College argues that this pattern does not always hold true. Under some conditions highly pathogenic strains of disease organisms can survive or even thrive, so that these strains will continue to be selected for. Suppose a pathogen spreads easily: Then, even if it kills its host quickly, it can go on spreading so long as there is a plentiful supply of other hosts nearby.

I agree that this can happen—the history of our battle with diseases is rife with examples of selection for high virulence. Perhaps the most extreme of these was the terrible Black Death. Many other diseases have swept through our species during historical times with disastrous results. But this apparent plethora of extremely virulent diseases is an artifact, the result of our understandable tendency to concentrate our attention on our own diseases. When we look at the natural world at large, highly virulent diseases are not so common. Animal and plant plagues, while not unknown, appear to be rare in undisturbed ecosystems.

The reason the situation is very different for us is that we are the planet's preeminent disturbers of ecosystems. An ecological disturbance, such as the introduction of agriculture to central Africa, can lead to the temporary success of a poorly adapted pathogen. The pathogen can survive even if it causes a severe disease in its human hosts, because the hosts are now so numerous. As a result of our efforts to change the world, we (and our domesticated animals and plants) have suffered dreadfully from diseases.

Our rapid cultural change has driven not only our own evolution but the evolution of our parasites as well. This becomes vividly apparent when we compare the malarias that we suffer from with the malarias that afflict our far more slowly evolving relatives, the chimpanzees.

Like humans, chimpanzees have at least four different malarias, but they cause much milder forms of the disease. Zookeepers who have examined chimpanzees freshly captured from the wild have often found malaria parasites in their bloodstreams, but the chimps do not seem incommoded by their presence.

The nearest known relative of our own *falciparum* malaria (and not a very close one at that) is a chimpanzee malaria parasite called *Plasmodium reichenowi*. DNA evidence suggests that that the *falciparum* and *reichenowi* plasmodia have each had long histories of association with their unwilling hosts.

If malaria in humans is such an old disease, and chimpanzee malaria is so mild in its effects, then the conventional wisdom would predict that *falciparum* should also be mild in us. Instead, it is vicious. A likely explanation for this contradiction is that our disease has not always been so nasty.

Before the arrival of agriculture, *falciparum* parasites must have already infected the peoples living in the central African forests. We do not know how serious a disease they caused, and we do not know which species of mosquitoes were responsible for spreading the parasites through those sparsely populated and heavily forested regions. But it seems likely that they were not the same mosquitoes as the ones that spread the disease today.

Perhaps the unknown mosquito vectors that once carried the parasites from human to human in the forest have disappeared or become confined to the few remaining undisturbed regions. In order to survive this sudden change, the parasites had to adapt to one or more of the relatively small number of mosquito species that are able to thrive in open and disturbed country. It would be highly unlikely that these mosquitoes would be particularly good at transmitting the parasites.

Indeed, from the parasite's point of view, today's mosquitoes do a rather mediocre job. *Falciparum* malaria is transmitted poorly by its most numerous host, *Anopheles gambiae*. Only a small fraction of the mosquitoes of this species that drink *falciparum*-infested blood manage to transmit the parasites to new human hosts. Further, as was recently shown by Jakob Koella of the University of Åarhus in Denmark, infected mosquitoes live a shorter time than uninfected ones. Even though infection does stimulate the mosquitoes to bite more victims, they are on the whole poorly adapted to the parasites.

These days, therefore, the parasites must multiply in swarms in the blood of their victims. Otherwise, their inept mosquito vectors would not be able to pick them up in numbers sufficient to ensure that some of them will be passed on to other human hosts.

The risk to the parasites of this uncontrolled multiplication can be high, for they might kill their current hosts from cerebral malaria or other severe symptoms before they manage to spread further. But the benefit also seems to be substantial, since their increased virulence allows them to spread through the crowded host populations in which they now find themselves.

We have no direct proof of a recent increase in the level of virulence in the *falciparum* parasite. Nor have we any firm information about what *falciparum* malaria might have been like before the advent of agriculture. But it is striking that *reichenowi*, the chimpanzee relative of *falciparum*, is so mild by comparison. Moreover, additional genetic evidence shows that chimpanzees, unlike humans, have not been forced by ecological upheaval to fight an escalating evolutionary war against their parasites. No equivalents of the human sickle cell and thalassemia genes have been found in chimpanzee populations; such clumsy, half-baked protections have simply not arisen there.

The transmission process, too, is likely to be very different in chimpanzees. Nothing is currently known about the mosquitoes that transmit *Plasmodium reichenowi* from one chimpanzee to another. But when these mosquitoes are found and tested, I confidently predict that they will turn out to be excellent carriers of the parasite. Even though they must be rarer than the clouds of *Anopheles gambiae* that swarm in the fields and villages where most humans now live, they still manage to spread *reichenowi* among small and scattered bands of chimpanzees. The *reichenowi* parasite must be superbly adapted, not only to chimpanzees but also to the mosquitoes that carry it readily from one animal to another.

We should not be surprised, therefore, that the disease seems to have so little effect on the chimpanzees that acquire it. The chimpanzees and their malarias have reached an evolutionary compromise, in which ready transmission has been accompanied by lowered virulence.

Every year brings a new discovery about how clever chimpanzees are, but they are not yet clever enough to invent agriculture. They have lived in forested regions in central Africa for a very long time, perhaps for several million years. During all that time, both they and their environment have remained relatively unchanged. They have thus had plenty of time to reach such accommodations with their parasites. (They have lots of other parasites, by the way, in addition to those that cause malaria.) But recent human-caused ecological disturbances are already taking their toll. As Jane Goodall observed at her site at Gombe Reserve in East Africa, chimpanzees have proved to be tragically susceptible to several newly introduced human diseases, notably polio. Through no fault of their own, like so many other endangered species, chimpanzees are succumbing to human alterations in their environment.

KEEPING UP WITH THE PATHOGENS

While the sickle cell allele and other mutations provide limited protection to the human populations that carry them, they make it more difficult for the malaria parasites to spread. In turn, this drives the parasites to make an evolutionary response.

Some West African populations, as we have seen, have a twenty-five percent frequency of the sickle cell gene, and almost half the people are heterozygous carriers. The parasites that happen to invade heterozygous carriers are immediately put at a disadvantage, compared with the luckier parasites whose hosts do not carry the sickle cell allele.

The disadvantaged parasites will be able to survive and spread only if they have mutated to higher-than-average virulence, allowing them to overwhelm their victims' defenses. This increased virulence will raise the genetic stakes, selecting for even more half-baked, partially protective mutant genes like sickle cell in the host population. And this will in turn select for even more virulent strains of parasite. Thus the spread of agriculture is not the only reason for increased parasite virulence: the genetic response of the human population has contributed to the problem as well.

Recent work has shown that the alleles are doing even more damage to the human population. *Anopheles* mosquitoes that have fed on the blood of malaria-infected sickle cell carriers are more likely to spread the disease than those that have fed on malaria-infected blood that has normal hemoglobin. In sickle cell heterozygotes the parasites must multiply to huge numbers if they are to be picked up by the mosquitoes, but if they are picked up they are more likely to spread. This is another way in which the sickle cell allele may do more damage than it prevents.

Leigh Van Valen of the University of Chicago has described such evolutionary races in terms of Lewis Carroll's Red Queen. In *Through the Looking Glass* the Red Queen remarks that in her land (where everything is reversed) everybody has to run as quickly as possible just to stay in the same place. Van Valen points out that very often in the past we have had to run furiously in an evolutionary sense simply to keep up with our pathogens.

If we invent some technological means to conquer malaria, the terms of this race will change completely. But suppose that we do not: How might we and our parasites break free of this vicious tit-for-tat? One obvious escape route is for the parasites to become better adapted to their mosquito hosts. Doing so would relieve some of the pressure, halting the mindless escalation in parasite numbers. This already seems to have happened with the other human malarias, which have been shown to survive better in their mosquito vectors than the *falciparum* parasite can manage to do.

Another escape route might open up if a new mutation appears in the human population that confers strong protection against *falciparum* malaria without the terrible cost of sickle cell. Just such a mutation, which bestows essentially complete protection against a less dangerous form of malaria called *vivax*, has already appeared in central Africa. People homozygous for this mutation, called Duffy-negative, lack a protein on the surface of their red cells that aids the entry of the *vivax* parasite. In contrast to sickle cell homozygotes, the homozygotes for Duffy-negative seem to suffer no ill effects—people can, it seems, get along quite nicely without this particular protein.*

Since homozygotes for Duffy-negative are not obviously harmed, no balanced polymorphism slows the spread of the allele, with the result that it has spread through much of central Africa. The incidence

---

* What does the Duffy-negative protein do when the *vivax* parasite is not binding to it? Nobody is quite sure, but we do know that it resembles proteins that are involved in the inflammation response.

of *vivax* malaria has been driven down in these regions, but this does not mean that the Red Queen race between ourselves and the *vivax* parasite has been halted. It is only a matter of time before a mutant arises in the parasite population that will allow the parasites to enter the blood cells by a new route, which will circumvent the Duffy-negative protection and start the whole grim cycle all over again.

Over the last few thousand years, the invention of agriculture has dramatically accelerated our evolution. But the Red Queen race against malaria that has formed such an important part of this acceleration may be coming to an end. When a good immunization against malaria is developed—as it surely will be soon—then the many genetic changes in our species that currently provide clumsy and dangerous half-protections against the disease may all be rendered moot.

Before we assume that the conquest of malaria and other diseases will halt all such evolution, however, we must remember that these dangerous diseases are sure to be replaced by others.

THUNDERING HERDS OF RED QUEENS

Malaria is not the only story of this kind. As a species, we find ourselves in the middle of dozens or perhaps hundreds of such evolutionary Red Queen races that involve many different diseases. Let me give just a few examples.

AIDS has spread from Africa to many other parts of the world in a mere two decades. Although it appears to be a very new disease in our species, having recently made the jump from other African primates to us, evidence is growing that it is not unique. It seems that we have been caught up for a long time in a Red Queen race with similar viruses—a race that is just as complex as the race against the malaria parasites.

One reason that AIDS infections cannot currently be cured is that the viruses are able to incorporate themselves directly into our chromosomes and lurk there for years. The RNA of the AIDS virus is converted to DNA, then slips in among our own genes. This process reverses the normal flow of genetic information in the cell, which goes from DNA to RNA. It is because of this property that such viruses are called retroviruses.

Although this ability makes the AIDS virus almost impossible to eradicate from an infected host, it also provides us with a clue to the prevalence of retroviruses in our own past. Thousands of bits of "fossil" virus DNA lie in our chromosomes, consisting of the much-damaged remnants of old retroviruses. They constitute strong evidence that we have met retroviruses very much like the AIDS virus in the past.

Moreover, we probably waged a Red Queen race with them. Stephen O'Brien of the National Cancer Institute has recently found that people of European ancestry carry an allele conferring strong protection against the commonest AIDS virus. This protection works in a way that is very similar to the protection that the Duffy-negative genotype provides against *vivax* malaria. People who are homozygous for the resistance gene lack a protein that facilitates the entry of the virus into cells.

About one percent of the population is homozygous for this protective allele, and almost all of this one percent are completely resistant to the virus. Far more people—about twenty percent—are heterozygous; they are also somewhat protected against the consequences of infection. Unlike homozygotes, heterozygotes can be invaded by the virus, but the invasion seems to be more difficult. Thus, although infected heterozygous people will eventually develop full-blown AIDS and succumb to an unstoppable tide of opportunistic infections, the course of the disease will be significantly slowed.

O'Brien speculates that an earlier retrovirus, now vanished, drove the resistance allele up to such high frequencies in Europe. Or perhaps the selective pressure that drove the allele up was due to a completely different disease that just happened to use the same entry mechanism into cells. The disease might have vanished, or it might still be among us. Whatever the disease was or is, it seems to have been most active in Europe, since the resistance allele is found there much more commonly than in the Middle East and India.

O'Brien suspects that the earlier disease was probably different from AIDS in one respect. The protective allele (and a protective allele of another gene that has been discovered more recently) are common among Europeans but appear to be rare in Africa. The European disease, whatever it was, probably did not originate in Africa. If it had, then the protective alleles would have been selected for there as well. It appears to be a coincidence that the same alleles confer protection against the new threat posed by the AIDS virus.

HIV resistance is also seen in Africa, but its genetics is still mysterious. There, too, a minority of people have been found who cannot be infected by the AIDS virus, but the gene or genes responsible have not yet been tracked down.

Does this mean that the AIDS virus has been loose among humans in Africa for longer than we thought? Or that other viruses or bacteria, perhaps similar to those in Europe, have been involved in selection for these unknown alleles? Even in Europe the resistance alleles that have been found so far account for only a minority of the people who are resistant to the AIDS virus, which means that many other genes must be involved. If O'Brien is right, then AIDS is only one of a large set of similar diseases that we have battled against in the past. And very recently, he and a large group of colleagues have traced one of the HIV-resistance alleles to an origination time of roughly 700 years ago. Could the Black Death have been the selective agent that drove up the allele in frequency? Experiments are underway to test this possibility.

AIDS-like viruses and malaria are just some of the diseases to which our species has responded genetically. We seem capable of carrying on such genetic wars on many fronts simultaneously. Other fascinating relationships between genes and diseases are coming to light, particularly now that it is possible to carry out experiments that were undreamed of just a few years ago. Consider the debilitating, and ultimately fatal, disease known as cystic fibrosis (CF for short).

CF results from a fundamental difficulty with chloride transport across cell membranes, which triggers a variety of symptoms. The most serious, and the one from which the disease gets its name, is the formation of destructive cysts in the pancreas. In the days before CF patients could receive the missing pancreatic digestive enzymes, these cysts were inevitably fatal.

If enzyme supplements have lengthened the life spans of victims of the disease, they have also provided time for new and more insidious symptoms to appear. The worst of these is gummy mucus that forms in the lungs, gradually blocking off air passages and allowing bacteria to multiply. Eventually, and inevitably, these repeated bacterial infections destroy the lungs.

About one European baby in 2,500 is born with CF, which means that about one person in twenty in the European population is a heterozygous carrier of the mutation. CF is so common among

Europeans that it is sometimes called the Caucasian sickle cell anemia, though there seems to be no connection between the disease and malaria.

CF can be caused by many different mutational changes in the gene. One such altered allele accounts for about a third of the cases, and seems to have been in the European population for at least ten thousand years. Perhaps some kind of heterozygote advantage kept it and other CF alleles there. But against what diseases do CF alleles protect?

The suspected villains are not exotic tropical infections but rather are common diarrheas, the chief cause of mortality among children and the elderly in less developed parts of the world. The worst such diarrhea, without doubt, is the terrible scourge of cholera, which has caused worldwide pandemics since the eighteenth century and still breaks out periodically.

These diarrheal diseases result from a sudden rush of body fluids into the intestine. Bacterial or viral toxins trigger the rush by changing the electrolyte balance of nerve cells in the walls of the intestine which control the fluid transfer. It has been suggested that nerve cells in the intestines of CF heterozygotes, with their defect in chloride transport, are less sensitive to the triggering signal.

It would hardly be possible to infect carriers of the CF gene with a disease such as cholera for research purposes. Luckily, molecular biologists have recently provided some remarkable and suggestive evidence by using mice.

A mouse model for CF has now been made by destroying the mouse equivalent of the human CF gene and replacing it with normal or mutant human genes. Mice that are homozygous and heterozygous for the mutant CF allele can now be bred in any numbers in the laboratory. The homozygous mice are not a perfect model, since their lives are too brief for them to develop more than the earliest stages of the disease. The heterozygous mice, like heterozygous humans, show no obvious symptoms. But these mutant strains can be used in many experiments.

Sherif Gabriel and his colleagues at the University of North Carolina examined the sensitivity of these mouse strains to the toxin produced by the cholera bacillus. Mice that did not carry any human CF alleles quickly succumbed to the toxin's effects, but mice that were homozygous for the human CF mutation were completely resistant—their intestines functioned normally. Heterozygotes, although affected by the toxin, did not become as ill as normal mice. Just as with heterozygotes for the sickle cell and AIDS-resistance alleles, the heterozygous CF mice were not protected against getting the disease in the first place but seemed protected against its severest effects.

Cholera toxin was the obvious choice for these initial experiments, because its effects are so powerful. But the cholera bacillus, which only appeared in Europe in the early nineteenth century, is only one of a large army of bacteria and viruses that cause diarrhea around the world. In the bad old days in Europe before there were any notions of public health, such diseases must have been even more powerful selective agents than they are today. Very recently, researchers in England and the United States have shown that the typhoid bacillus attaches to the CF protein as it enters intestinal cells—typhoid is therefore probably one of the diarrheal diseases that helped to select for the CF allele in Europe.

Problems remain with the theory connecting CF with diarrhea. For example, why are CF mutant alleles not common in the tropics, where diarrhea is even more widespread than it is in the temperate zones? One would expect the alleles to be at particularly high frequencies in Bengal, where most of the great cholera pandemics of the last three centuries have originated. There, cholera has devastated the population for centuries and even millennia—providing a very long span of time for selection to have driven up the CF allele frequencies. Yet CF mutant genes do not seem particularly common in that part of India.

Paul Quinton, a physiologist at the University of California at Riverside, has an ingenious explanation for this discrepancy. One striking symptom of CF is a high concentration of salt in the sweat. Indeed, a "sweat test" for salt is commonly used to detect the onset of the disease in children. Quinton suggests that in hot countries heterozygotes for CF might have been selected against because they would lose dangerous amounts of salt through their sweat. This theory makes a lot of sense—we take salt for granted now, but for most of human history and for most groups of people, it has been in desperately short supply. Such selection, since it would operate each hot season, would outweigh any advantage that CF's resistance to diarrheal diseases might confer. CF heterozygotes who happened to live in a cold region like Europe would not suffer so much salt loss, and their resistance to diarrheal diseases could then become important.

Quinton's theory is ingenious, but there are still difficulties. For example, it is not obvious why CF mutations are very rare among the Chinese, many of whom live in a relatively cool climate. In spite of these puzzles, however, evidence is growing that CF, like sickle cell and thalassemia, is held at high frequency because it forms a balanced polymorphism.

ECOLOGICAL AND GENETIC UPHEAVAL

As our environment changes, the situations that have led to these various balanced polymorphisms are likely to disappear. In the next century we will likely stamp out many of the most contagious and frightening human diseases, just as we have already conquered or nearly conquered smallpox, chicken pox, and polio. But four great factors remain that will lead to further disease-driven evolution.

The first and perhaps most important factor is the continuing global ecological upheaval that we are causing. In the next hundred years most of the natural world will have either disappeared entirely or become drastically simplified. In the introduction to this book we glimpsed some of these changes taking place in Borneo. Ecological changes in central Africa have resulted in the emergence of AIDS viruses. During our lifetimes and those of our children, more such epidemiological emergencies are likely to come thick and fast. Newly resurgent dengue hemorrhagic fever and yellow fever are recent examples.

Second, most of the disease-caused mortality in the world today is the result of common respiratory and diarrheal illnesses that chiefly affect the very young and the very old. These diseases are most prevalent in the developing world, and it will take a very long time before improved sanitation and nutrition finally conquer them.

Third, our technology is also producing many environmentally driven Red Queen races. The indiscriminate use of antibiotics and antimalarials is producing new generations of resistant pathogens. Some people will be better able to survive these resistant strains than others, and our gene pool will change as a result.

Fourth, even in the developed world pathogens play a larger role in disease than we had imagined until recently. Protozoa, bacteria, and viruses that were formerly unknown, or were formerly supposed to be benign simply because they are found everywhere, are now suspected to make a contribution to many different illnesses that were previously attributed to poor diet, unwise living habits, or psychological factors. Even though most of these pathogens may not cause obvious and dramatic symptoms, they nonetheless affect our health and reproductive abilities.

One recent example that has gained worldwide attention is the bacterium *Helicobacter pylori*, which is now known to be responsible for many stomach and duodenal ulcers. Ulcers may not at first glance seem life-threatening, but in the underdeveloped world they often provide a route for systemic bacterial infection. Throughout much of the world *Helicobacter* is now being revealed as a very dangerous organism indeed.

It is also now being discovered that even heart attacks are influenced by bacterial and viral infections. Although heart disease is of course correlated with such factors as fatty diet and lack of exercise, these are not the whole story. There are persistent and growing reports that the formation of arterial plaques may be triggered by the primitive but widespread bacterium *Chlamydia pneumoniae* and by the equally ubiquitous cytomegalovirus. In fact, the rise in heart attacks since the beginning of this century may not even be due to our current terrible diet after all—diets in the eighteenth and nineteenth centuries tended to be even fattier and less balanced! Some of the rise may be the result of crowding and of ecological changes that have allowed these plaque-triggering organisms to become more widespread in our population.

Very recently a group of Swiss researchers discovered that if they infected mice with a specific mouse retrovirus, the infection could trigger the subsequent destruction of pancreatic islet cells, producing diabetic symptoms. In humans this type of diabetes has many of the properties of a so-called autoimmune disease, in which we make destructive antibodies against our own pancreatic cells. But the trigger that first stimulates our immune system to produce these antibodies remains elusive. It now seems possible that the trigger in humans will also turn out to be a viral infection. The reason the connection between infection and diabetes was not found earlier might be that the virus tends to come and go early in life, long before damage to the pancreas begins to develop.

A variety of puzzling illnesses may also be associated with infectious agents, including autoimmune diseases like multiple sclerosis and perhaps even such hard-to-define conditions as chronic fatigue syndrome (CFS). The jury is still out on whether this syndrome is real or entirely psychosomatic, but many people with CFS are convinced that their suffering has a physical cause. Some studies have suggested there may be as many as two hundred cases of the syndrome for every hundred thousand people, though it is difficult to obtain reliable information about such an amorphous illness. Complicating things further, it has been found that simply telling a patient that he or she might have CFS causes the patient to feel worse. The act of putting a name to the disease can make it seem more real. Several viruses have been implicated in CFS, ranging from bornaviruses through various herpes viruses to cytomegalovirus, but the smoking gun has yet to be found. Indeed, the presence of a virus alone does not mean that it causes the syndrome.

Such a syndrome, so difficult to diagnose and variable in its effects, might have an almost invisible selective effect. Let us suppose that at least some cases of CFS turn out to be due to the spread of pathogenic agents. As the disease drains away energy and initiative, does it make infected people less likely to reproduce? If it does, then such selection could be powerful and yet at the same time virtually undetectable.

Other plagues, not necessarily associated with disease organisms, are becoming more important in our lives—ironically, just as the role of extremely severe infectious diseases would appear to be diminishing in the developed world.

Addictions to tobacco, alcohol, and drugs are obvious examples. It can be argued that mortality from tobacco happens too late in life to have much impact on fitness in an evolutionary sense. But like CFS, tobacco-related illnesses can affect people at every age and may also affect their reproductive abilities. The high rate of mortality and severe injury among young people that result from some of these addictions must also be having an effect, not just on the victims but on the gene pool of our species. The related plague of violence, which in the United States and in many developing countries has been carried to an extreme, must also be having an impact. It too is a disease, though its causes are more psychological than physical. Death by gunshot or a drug overdose can be just as effective selective agents as death by infectious disease.

As we conquer some of our obvious disease enemies, the mix of those to which we are still subjected will change. The effects of disease-caused selection may become less obvious with time, but they are sure to continue. And the frequencies of many different alleles in our gene pools will shift in compensation, as we run these many subtle but still threatening Red Queen races.

When we first think about diseases and evolution, it seems obvious that the selective pressures involved ought to involve life-or-death situations, nature red in tooth and claw. But the ramifications of even such dramatic types of selection can be very widespread and subtle. Further, the terms of many of these evolutionary battles have recently changed. As we have "subdued" the natural world, some of our pathogens have become even more dangerous. There is no escaping from the consequences of such natural selection—Darwin is, even from beyond the grave, continuing to change our lives.

FIVE

❖

Perils of the Civil Service

The perfect bureaucrat everywhere is the man who
manages to make no decisions and escape all
responsibility.

BROOKS ATKINSON, *Once Around the Sun* (1951)

Sometimes selective pressures can arise as a result of our daily interactions with other members of our society. As our social environment
increases in complexity, these selective pressures will become stronger. In this chapter and the next, we will turn from the physical sources
of selective pressures to the far more subtle psychological sources.

The effects of psychological stress on health are sometimes profound, but few studies have examined the long-term consequences of
such selection on our species. In some cases, however, we can get a
glimpse of the selective pressures involved.

The University College School of Public Health stands on a quiet
London side street, not far from the great complex of university and
medical research buildings centered on Gower Street. Here, in a special
suite of rooms, epidemiologist Michael Marmot and his colleagues have
been examining some of these subtle psychological interactions, in particular how position in the social hierarchy can affect a person's health.

At the nearby School of Tropical Medicine and Hygiene, doctors and epidemiologists wrestle daily with the problems of malaria
and schistosomiasis in far-flung parts of the world. Marmot's work
deals with far less threatening but nonetheless real situations that are

much closer to home. His group has, over a span of three decades, carried out two massive studies of health in the British civil service. During a recent visit I talked with his group about the evolutionary implications of the studies, which have become known as Whitehall I and II.

Whitehall is the ancient street that runs between Charing Cross and the Cenotaph; it passes through the very center of English government. The whole area is filled with government buildings, each with its own maze of offices. Here tourists who have strayed off the beaten track find themselves in a world of pin-striped bureaucrats, clad in discreetly tailored suits and dark overcoats and carrying tightly furled umbrellas. Sometimes top functionaries can be glimpsed as they are whisked from one important meeting to another in black Bentleys and Jaguars.

Surely, one would think, this is an unlikely place to detect the ravages of natural selection. Nonetheless, Marmot and his colleagues may have found distinct traces of it at work among these hurrying civil servants.

One of the many cultural differences between England and the United States is that Americans tend to worry about societal inequalities that have resulted from racial prejudice, while the English worry about those that have been generated by the class system. These two causes of inequality are entangled in both societies, but sometimes a story emerges in which the effects of one or the other stand out in vivid relief. The Whitehall studies are particularly striking examples.

THE MINISTRY OF FEAR

The British civil service, like that of many other countries, is a hierarchical system in which pay and responsibility are carefully graded. At the top are the permanent secretaries and undersecretaries, who serve under an ever-shifting collection of politically appointed cabinet ministers and provide the government's continuity (and inertia—you may remember Sir Humphrey, the permanent secretary in the wonderful television series *Yes, Minister*, who used every art at his command to ensure that nothing ever got done).

In the middle regions of the hierarchy are vast numbers of functionaries, along with their secretaries and clerks, and at the bottom are the people who sort mail, carry messages, and perform other simple tasks.

None of the work is physically demanding, job turnover is very low, and every employee has access to roughly the same level of health care through the National Health Service (though most of those at the top supplement it with their own private doctors or additional health plans). The salary differential is steep, though not as precipitous as at a private company—the permanent secretaries at the top make about twenty times as much as the employees at the bottom.

The first Whitehall study dealt only with males, but the second was expanded to include women. In both studies the civil servants who staffed twenty different government departments were divided into six categories depending on their positions in the hierarchy. They were asked to fill out detailed questionnaires about their health, job satisfaction, financial situation, and degree of familial and social support. In addition they were given electrocardiograms, their blood pressures were measured, and they were subjected to a variety of other medical tests.

During the period 1967–69, data were collected from the first cohort of 17,530 men. Then the cohort was followed for the subsequent ten years, in order to see what would happen to them. Data began to be collected from the second cohort of 6,900 men and 3,414 women almost twenty years later, during 1985–88. The second cohort is still being followed, and so far the patterns that are emerging in the males are remarkably similar to the patterns seen in the first cohort. The inclusion of women, however, has added a whole new dimension. It was this new data that has revealed the possible evolutionary implications of the second Whitehall study.

The most dramatic pattern to emerge from both studies is that, once age is factored out, position in the hierarchy is strongly correlated with the likelihood of mortality. With each step down the hierarchy, the likelihood of death increases. When the bottom and the top categories are compared, the differences are surprisingly large: a civil servant in the bottommost category is three times as likely to die during a given period as one of the same age who happens to occupy a position at the top.

This greater mortality is correlated with a number of obvious health-related factors. In both studies, with each step down in the hierarchy, smoking becomes commoner, as does obesity. Diet tends to be less healthy at the bottom end, and both men and women are less likely to exercise regularly. Signs of early heart disease are more common in the lowest groups.

These factors are not the whole explanation for the increased mortality, however. This is because the physical risk factors tend to be more concentrated at the bottom of the hierarchy, while the risk of mortality increases with each step down.

For example, when the Whitehall II cohort of civil servants was tested during 1985–88, detailed ECG measurements turned up indications of incipient ischemic heart disease—the effects of narrowing of the coronary arteries—in about five to six percent of the people in each of the top five categories. But the signs of ischemia jumped to 10.5 percent among those in the sixth and lowest job category.

In contrast, when the top administrative category is compared with the category of professionals and executives just below it, even this small step down the hierarchy results in a highly significant increase in mortality. People in this slightly lower-ranked group, all of whom are still at the managerial level, are about one and two-thirds times as likely to die (again, once age is factored out) as those in the top group.

Marmot discovered that mortality rates seemed linked less closely to physical risk factors than to the mindset of the study participants. When the members of the Whitehall II group were asked to self-rate their health, the ratings were again correlated linearly with position in the hierarchy, rather than reflecting the health risk factor. The proportion of those rating their own health as average or below rose steadily from 15.3 percent in the highest job category to 33.7 percent in the lowest.

It would seem that, particularly among the highest categories on the scale, the differences in mortality are not due to physical risk factors—which leaves psychological factors. Even an artificial construct like the civil service system can apparently create its own selective effects.

More recently, Marmot and his collaborators carried out two studies that reinforced their conclusion. In the first they concentrated on one of the twenty government departments, singling it out because it had just become widely known that the department was scheduled to be privatized. This traumatic news caused the incidence of self-reported health problems to shoot up immediately, particularly among the men, although during the same period these measures showed no change in the other nineteen departments.

In the second study they examined death rates over the long term. They returned to the first Whitehall cohort, for which statistics had been collected in the 1960s, and examined what had happened to its members after they retired. Of course, mortality kept going up as the cohort got older. But the researchers wanted to find out how well the job category that a civil servant had occupied before retirement predicted the likelihood of death during the study. What happened, in short, once the job-related pressure was off?

They found that hierarchical position was still a predictor, but only about half as good as it had been among the civil servants who had met their demise while still in harness. After the psychosocial stress had eased, the gaps in mortality between the higher and lower job categories tended to close.

Unfortunately, this study was confounded by the probability that the people who had survived to retirement, regardless of their hierarchical level, were those who by definition were less likely to succumb to work- and status-related pressures. Yet the weight of evidence is that the psychological pressures are real and affect some people disproportionately. Another piece of evidence came from cases of coronary heart disease among the Whitehall II cohort, cases that had turned up since the first statistics were collected on the cohort. These new cases, Marmot's group found, were most common among those civil servants who felt they had little or no control over their jobs, regardless of which job category they occupied.

MENTAL HEALTH AND EVOLUTION

The health implications of the Whitehall studies and others like them are far-reaching. They are unusual among sociological studies in that they clearly factor out and demonstrate the importance of psychological influences on physical health.

Earlier studies had shown clearly that in most societies position in the socioeconomic hierarchy has a huge influence on health, but it had proved difficult to disentangle the psychological and physical effects of being low on the socioeconomic scale. The Whitehall studies have been able to remove at least some of the influence of physical factors. They show that, at least in the upper echelons of the Whitehall social hierarchy, position has a greater impact on health and survival than mere income. The far less visible, and thus far less easily measured, psychological selective pressures are created by the society of which the individual is a part and are perceived by individuals in many subtle ways. Some of these ways, it seems, can have a detrimental effect on health.

To be sure, these psychological pressures act in tandem with physical factors, and the effects of the two may often be multiplied together. While it used to be fashionable to attribute heart attacks to diet or to a hard-driven, overachieving "type A" personality, we now know that pathogenic organisms may also play a role. To make the story even more complicated, it is now becoming apparent that a remarkably good predictor of the likelihood of a heart attack is a person's mental health. Twenty to thirty percent of people who have suffered myocardial infarcts, and an astonishing forty percent of people with coronary heart disease, are clinically depressed. The problem is not simply that people who know they have a heart problem tend to be gloomy about it. Prospective studies have shown that, even prior to a heart attack, the presence of a major depressive disorder is a powerful predictor that the attack will take place.

When depression is combined with all other known predictive factors—ischemia, smoking, blood pressure, cholesterol levels, lifestyle—its presence accounts for a full forty percent of the predictive ability of this entire suite of factors. We must be cautious about attributing too much to the power of the mind, for the state of psychological health might have been affected by the early developing symptoms of heart disease, or some effect of the infectious disease that could have triggered the condition. But a growing body of research shows that state of the mind has a dramatic effect on physical health.

Even if, in the future, it is discovered that heart disease really is primarily caused by infectious agents, the role of clinical depression will still have to be explained. Why should depressed people be so much more susceptible to the damaging effects of these infections than people with sunnier dispositions? And what roles do genetic differences play? The interplay between visible and invisible selection, between the influences of the body and the mind, will provide fodder for scientists and sociologists for decades to come.

Adding yet another layer of complexity, the effect of these psychological factors may vary dramatically from one society to another.

Some fascinating patterns, for example, as Marmot and I discussed, have recently been observed in Japan.

In Japanese society, even though social hierarchies are extremely rigid, the differences in mortality from the highest to the lowest socioeconomic classes are reportedly very small. Indeed, they seem to be the smallest that have been measured in any industrialized society. Marmot's group is currently collaborating with Japanese sociologists to find out whether these differences in mortality are really as small as these early studies indicate.

If the early studies are borne out, the reason may be that the Japanese tend to be more accepting of class differences and take their place in the social hierarchy more willingly. The British, on the other hand, might be straining more to fight free of the chains imposed by their class structure. But if the early studies prove to be wrong, and pronounced differences in mortality—or some other indicator of stress—are discovered at the two ends of the Japanese socioeconomic scale, then the health of the Japanese, too, may turn out to be deeply affected by their position in their society's hierarchy. It also seems likely that, as younger Japanese begin to rebel against their rigid society, these stresses will increase. Socioeconomic status may not exert a differential stress, but the health effects of being out of step in Japanese society can be profound—although the gap is now closing somewhat, in the middle part of this century unmarried Japanese had life spans fifteen years shorter than married Japanese!

WHICH CIVIL SERVANTS WILL SURVIVE?

The civil servants of Whitehall, stressful though their lives may sometimes be, are not exactly dropping like flies in the corridors of power. The dry numbers summarized in these studies are difficult to reconcile with some nightmare vision of natural selection at work—Mother Nature grimly wielding her incarnadined teeth and claws while bowler-hatted civil servants scurry for cover.

Any selection that is detected by these studies would seem to be very weak. Most of the differential mortality and illness found among the Whitehall civil servants occurred in midlife, which means that it took place near or after the end of their reproductive period. But what matters most in an evolutionary sense is whether and how often the individuals occupying this social hierarchy reproduced, and what happened to their children in turn. Until recently, the Whitehall studies gave us no such information; they did not measure the effects that hierarchical position might have had earlier in life, when the civil servants were just starting out in their careers and their marriages. Did position have any effect on reproduction, or on success in parenting? And to what extent might such effects be passed from one generation to the next?

The inclusion of women in the second study should eventually allow some of these effects to be measured. A variety of gynecological problems, such as severe effects of menstruation, already appear to be strongly correlated with hierarchical position, although these results have yet to be published in detail. Again, differences are seen even between the first and second civil service categories. Do such problems, and others yet to be discovered, have an effect on reproductive fitness? The answer is still concealed in the dry lists of numbers waiting to be analyzed in Marmot's laboratory.

Even though any evolutionary connection remains elusive, the Whitehall studies have already shown that the industrialized societies of the latter part of the twentieth century are not quite as benign as they seem on the surface. Although the consequences of differences among civil service grades in Whitehall are mild, they may not be perceived as mild by people who have known nothing else. People who have relatively little else to worry about may well magnify the importance of tiny differences in status.

This tale of the civil service raises another question. It is quite clear that the Tibetans and the Quechua are being selected for the ability to survive at high altitude, and it is clear that Central Africans are being selected for resistance to malaria and other diseases. But what traits are currently being selected for or against among the civil servants of Whitehall? Are they being selected for their ability to reconcile themselves to their place in the hierarchy, so that they don't worry themselves into a state of ill health? Are genes that have a detrimental influence on mental health being selected against? Probably both, but another mental factor may be subject to selection as well.

One of the more intriguing questions on the Whitehall questionnaires was "Do you think you can reduce your risk of having a heart attack?" Seventy-five percent of the males in the highest category answered yes, while only fifty percent of those in the lowest category did. Those at the upper levels were more likely to avoid behaviors that they had learned would increase their chances of ill health, just as they were less likely to feel that their lives were buffeted by forces beyond their control. I suspect that one thing that is being selected for, even in the calm purlieus of Whitehall, is the ability to take advantage of new information and turn it to one's own benefit.

Such an amorphous ability is far less easy to define than the ability to survive at high altitude. And we have absolutely no idea of the extent to which genes might contribute to it. But throughout our history as a species, we have been subject to greater and greater onslaughts of information of all kinds. Those best able to handle this tide have benefited from it. And one day, perhaps soon, we will understand the nature of the genes that have made this possible.

HIERARCHIES IN THE JUNGLES BEYOND WHITEHALL

Although the current selection acting on the Whitehall civil servants is weak, it gives us a glimpse of the complex ways in which selection might still operate on our species.

Primate studies have already demonstrated that position in a hierarchy can have a dramatic effect on levels of stress. Studies on the baboons of Amboseli Park in Kenya by Robert Sapolsky and his co-workers show that in stable baboon hierarchies the lowest-ranked males show very high levels of endocrine indicators of stress, while the highest-ranked males have much lower levels of these indicators. But during times of upheaval, even the highest-ranking members suddenly shoot up on the stress scale. A threat to one's position in a hierarchy can be extremely stressful. It is fascinating to learn from Marmot's study of the ministry that was about to be privatized that similar effects can also be seen, however dimly, in our own species.

In addition, the baboon studies tell us something about the role of variation in a population. Sapolsky and his students found that in spite of their position, some low-level males did not show high levels of stress indicators. These males would often be able subsequently to move up in the hierarchy. They were also, it turned out, the ones most adept at stealing opportunities to copulate with females, despite the vigilance of the dominant males. These findings suggest that the ability to withstand the stress of occupying a low position on the hierarchy can increase reproductive success.

While these low-ranking males may have boosted their reproductive success by a small amount because they were alert to fleeting opportunities, recent genetic studies have shown that the high-ranking baboon males, during the fairly brief time they occupy the top ranks, reproduce most effectively. At least among baboon males, hierarchical position really does have an impact on fitness. Does this pattern extend to females as well?

Such correlations among females have been frustratingly difficult to measure, in baboons or in other animals. Although females' reproductive success is much easier to measure than that of males, the place of females in a group's hierarchy is very difficult to determine. The reason is that they usually do not show the clearly aggressive behavior of the males and tend to spend far more time alone. But despite these difficulties intriguing evidence has recently emerged that for female chimpanzees reproductive success is indeed correlated with their hierarchical position. Jane Goodall and her collaborators have patiently collected the data over a long period of time from the female members of chimpanzee groups living in Gombe National Park in Tanzania.

The researchers could find only one signal that allowed them to determine the females' social status. This is a sound called a "pant-grunt" that a submissive chimpanzee makes after a dominant one has behaved aggressively. One female would often be observed to pant-grunt at another, even though to a human observer the second female did not seem to have done anything particularly aggressive. On the assumption that an aggressive interaction had really taken place, and that this interaction, although invisible to humans, could be perceived by the lower-ranked chimpanzees, the researchers decided to use the sound as a marker.

With astonishing foresight, they had already collected just the data they needed to test this possibility. Between 1970 and 1992 they had painstakingly recorded hundreds of interactions between pairs of females. They had tallied the numbers of pant-grunts and kept track of which female was the grunter and which the gruntee. Now they were able to use this record of "invisible aggression" to unravel the female hierarchy.

Once they worked out the hierarchy, much other data fell into place. They discovered that females at the top of the hierarchy tend to have more babies, and that their babies not only have a higher chance of survival but actually mature more quickly. Thus, even though the hierarchical interactions among the females were difficult to detect, they were very important to reproductive success. And in the evolutionary long term, this is all that matters.

A final question, and a very important one, is not answered by these studies: Even if reproductive success is correlated with hierarchical position, there is no guarantee that a propensity for success will be passed down to succeeding generations. For such an evolutionary change to take place, genes that increase reproductive success must be selected for. No studies have yet shown direct evidence of such an effect.

There are, however, some hints. Frans de Waal, who has spent a lifetime observing primate behavior, found in a rhesus monkey colony in Wisconsin that mothers act in such a way as to increase the likelihood that their offspring will rise in the hierarchy. He observed a behavior called a double-hold, in which a mother enfolds her own baby and an unrelated baby together in her arms. Almost always the unrelated baby belongs to a mother with a higher rank than the double-holding mother. Similar behavior is not unknown among humans—mothers get upset when their children play with children from the wrong side of the tracks. While genes for social climbing per se are unlikely to exist, the fact that there is a hierarchy to be climbed and that there are some apes—and humans—who attempt to climb it, suggests a genetic benefit to such climbing.

It is a long way from the wilds of Gombe, or even a Wisconsin primate colony, to the calm precincts of Whitehall. Yet the social forces acting on Jane Goodall's chimpanzees and Frans de Waal's monkeys seem not so far removed from those that impinge on British civil servants. It is ironic that we know less about the effects of hierarchical position on humans than their effects on chimpanzees! Until we learn more, it would be foolish for us to rule out the possibility that natural selection is still taking place among the civil servants of Whitehall or indeed among the members of any human social hierarchy.

SIX

◈

Farewell to the Master Race

*I do detest people who consider themselves superior. It
makes things so difficult for those of us who really are!*
HYACINTH BUCKET, *Keeping Up Appearances*

The Whitehall studies focus on death and illness—events that we can
measure fairly easily, because they happen before our eyes. But what
of the events that do not happen at all? How, for example, can we
measure the impact on the gene pool of a child who is never born?
Just as factors that change the death rate can have an effect on evolution, so can factors that change the birth rate.

THE CHORUS OF UNBORN CHILDREN

One hallmark of primitive societies, remarked on by many anthropologists, is that almost all their members marry and have children.
Many of these children die from disease, deliberate infanticide, starvation, or accidents, but in general virtually all those who survive have
at least a chance to pass on their genes to the next generation. Even
so, the variation in reproductive success from one human group to
another can be remarkable. Among tribes in the Congo (formerly
Zaire), the incidence of women without surviving children ranges from
six percent among the Ngbaka to sixty-five percent among the Mbelo.
It is unclear how much of this variation is due to disease, but it is likely

to play a substantial role: Sexually transmitted diseases, as well as malaria and trypanosomiasis, can have large effects on fertility as well as on the survival of children.

The numbers of children can vary dramatically not only from tribe to tribe but also from family to family. Extreme variation of this type can have large genetic consequences. In one study human geneticist James Neel of the University of Michigan collected data on numbers of children among families of a South American Indian group. These people, the Yanomama, live in a scattering of small villages in a heavily forested area that spans the border between Venezuela and Brazil. Their remarkable social structure was revealed to the world in a memorable book by Napoleon Chagnon, *The Fierce People*.

Neel found that even though most of the adult members of the four villages that he examined had managed to reproduce, the numbers of their children varied wildly. Further, the likelihood of survival also varied greatly from family to family. While on average only about half the children survived to reproductive age, some families were far luckier than others. Using a computer model because he did not have long-term data, he extrapolated these numbers for a further generation, to see how many grandchildren the various members of the villages would be expected to have. He was able to predict that for fifty-five percent of a typical cohort of eighty-six males in a village, none of their grandchildren would survive to adulthood. In effect, the genes of over half the cohort would disappear in the space of two generations. At least some of the genes of the other forty-five percent would survive, however, and one of the males in the cohort would probably end up with about thirty-two surviving grandchildren. His genes would almost all be passed on, and some of them would be represented many times in the population.

This real-world situation is very different from a situation governed by random chance. When I revisited Neel's data, I calculated what might have happened if the likelihood of having children were distributed equally among all males. I assumed that, just as in the real tribes, mortality among the children was fifty percent; but I specified that the deaths took place at random. Assuming that the population size remained constant from one generation to the next, I found that forty percent of the men would have no surviving grandchildren—a slightly smaller fraction than Neel predicted. But the largest number of grandchildren to be expected for any of the men would be about ten, far fewer than Neel's maximum of thirty-two.

In the real world, a lucky few of the Yanomama males will pass on their genes to subsequent generations in disproportionately large numbers. In the artificial world of my calculations, in contrast, no such extremely superfecund minority would exist.

This difference between Neel's numbers and random expectation arises because, for a variety of very nonrandom reasons, some people in a Yanomama group will be more fertile than others, and the children in some families will be more likely to survive than those in others. From Neel's data alone we cannot tell how much of these differences are due to environmental factors and how much to genetic differences among individuals. But even assuming only a small genetic contribution, the great differences in fertility and survival among the Yanomama will unquestionably have a substantial impact on their gene pool. In essence, Neel was able to show that in each generation some genes manage to pick their way through the perils that are faced by these tribespeople much more readily than others.

All these genetic events are taking place in little villages on the banks of isolated jungle rivers. The overall effect on the Yanomama of the success of some genes and the failure of others might have remained slight as long as they stayed isolated from the rest of the world. Neel has estimated that the Yanomama and their antecedents have lived in this region for a long time, perhaps a large fraction of the time since humans first arrived in South America. A time traveler able to visit a Yanomama village of a thousand years ago would probably not see many differences from the villages of today, and certainly the appearance of the people would be little different.

Now, as the twentieth century is beginning to impinge on their ancient way of life, that world is vanishing. Many Yanomama have died from introduced diseases, and others have moved away to the cities and shantytowns that are springing up everywhere in the Amazon basin. As a recognizable people, they will be gone in an evolutionary moment, even though some of their genes will undoubtedly persist. And as the genes of the Yanomama are catapulted into a new environmental context, they will survive or disappear according to very different rules.

Theirs is only a tiny part of the enormous genetic upheaval that is taking place throughout the world. The effects of all these selective processes, especially the effects of children who were never born, can have dramatic and sometimes unexpected consequences on this larger canvas. The gene pools of entire continents of people are being altered.

THE CHANGING FACE OF EUROPE

Two kinds of genetic change are taking place globally, one obvious and one far less so. The obvious change is the rapid mixing of gene pools that have remained separate for millennia. The results are seen most dramatically in people who live in major cities throughout the world. The London that my parents knew, for example, has undergone enormous alteration in the space of two generations, and the rate of change is accelerating.

The second kind of change is one that receives its impetus from this mixing of different gene pools. As we saw in Chapter 1, mixing can alter allele frequencies and produce new gene combinations. But real evolutionary change will only happen when these new gene combinations are sorted out by natural selection, so that the frequencies of the alleles in these new mixed populations begin to shift.

Europe provides some particularly dramatic examples of both kinds of change, and of how traumatic the changes can be. In recent history, various groups who imagined themselves to be indigenous and permanent have had great difficulty coming to terms with the fact that, as genetically separated populations, they are doomed to extinction.

Two generations ago the gene pools of Europe were the jealous preserves of xenophobic demagogues. Attempts to "purify" them led to some of the most ghastly events in the history of our species. At the present time this "purification" process, with all its attendant hatred and misery, continues in various sad fragments of the former Yugoslavia. But despite the efforts of Hitler and his successor Karadzic, and in one of the great ironies of history, the current populations of Europe will soon be altered beyond recognition.

New mixed populations can swamp and overwhelm old ones for many reasons. New technologies, such as agricultural innovation or rapid transportation, may drive the process. In a collision of cultures the vibrant new will tend to outstrip the fossilized and hidebound old. Psychological factors can also play a central role. When sociologists have examined the effects of mental attitude on demographic trends, they have discovered them to be profound. And demographic trends are, after all, the very stuff of the evolutionary process.

One important finding that has emerged from such studies is that social attitudes can have a substantial effect on birth rates. In an astonishing reversal of historical trends, the native populations of many European countries are currently declining in numbers. In European Russia, from 1989 to 1993, the birth rate dropped by thirty-five percent, and in eastern Germany during the same period, the rate plunged an astounding sixty percent. On average, if a population is to remain at steady state, each woman must have about 2.2 children. In eastern Germany, assuming current trends are not reversed, the average number of children born to each woman during her lifetime will be less than one.

At the same time death rates have soared in Eastern Europe, putting this region at odds with every other part of the world. Even in Africa, in spite of warfare and economic and epidemiological disasters, death rates have by and large fallen.

The role of psychology in this trend is illustrated vividly by the gap between the life expectancies of Russian men and women. Female babies born in Russia at the present time are expected to live 73.2 years, compared with an expectation for white female babies in the United States of 79.6 years. But male babies born in Russia will only live 58 years, while white male babies in the United States live an average of 73.4 years. The gap between Russian men and women is more than twice as large as the gap between American men and women, and indeed Russian men live about as long as men in Liberia.

Why the huge difference between Russians and Americans? Many causes have been suggested, from the declining quality of health care in the former Soviet Union to the effects of widespread and extreme environmental pollution. The disproportionate effect on Russian males can in part be traced to high levels of binge drinking and smoking. But there is another important cause, a psychological one. A combination of stress, discouragement, and helplessness—of the sort that seems to affect the lower echelons of Whitehall civil servants but greatly magnified—becomes clearly visible when we examine these Russian numbers more carefully.

Although I compared the Russian life-expectancy statistics with those for white Americans, it might be more informative to compare them with the statistics for black Americans. Life expectancy for black females born in the United States in 1995 is 74.0 years, but for black males it is only 65.4 years. These numbers, and the huge spread between them, are remarkably similar to the equivalent numbers for whites living in Russia.

The psychological situations faced by these two superficially very different populations may have similarities as well. In the United States the self-image of black males is under severe assault, chiefly because of the pervasiveness of racial discrimination. The self-image of Russian males is under equally severe assault, as they find that their society is no longer either stable or a superpower but has been demoted to a marginal third-world entity. Females, for a variety of reasons, seem to be less severely affected by such assaults on self-image.

These birth- and death-rate trends have already had an enormous impact on the Russian population, which in 1995 alone actually decreased by 900,000. If the trends continue, then by the year 2030 the population of Russia will, in spite of increasing immigration from other republics, have fallen from 148 million to 123 million. And the ethnic mix will have changed dramatically.

Eastern Europe and Russia present us with a gigantic experiment in which human misery and a decline in fertility are obviously correlated, although just how much the first causes the second remains uncertain. The effects of psychological attitudes on reproduction are notoriously difficult to measure. It has been repeatedly suggested, by analogy with John Calhoun's famous experiments on rats in the early 1960s, that stress due to crowding should decrease birth rates in people. Psychologist Jonathan Freedman, in summarizing many studies that measured the effects of crowding on the attitudes of individuals, found that these effects can range from positive to very negative depending on the design of the experiment. But one generalization does emerge: If people have the perception that their surroundings are both crowded and unpleasant, they acquire strikingly negative attitudes toward life in general.

In another few decades a sunnier economic outlook in eastern Germany and Russia, and the disappearance of the older generation that had been used to life under Communism, will likely change things again. Birth rates will rise, although they are unlikely to reach the levels prevalent before the collapse of Communism.

The demographic upheavals in Eastern Europe are an extreme manifestation of a swift and continuing demographic transition throughout the continent. Italy, in spite of the pronatalist stand of the Catholic Church, now has the lowest birth rate in Western Europe. At 1.3 children per woman, it is far below the 2.2 required for replacement. Prosperous and politically stable western Germany has a rate that is almost as low. The united Germany is now predicted, if current trends remain unchanged, to experience a population decline from 80 million today to 55 million by 2050. This prediction has surprised demographers, who have assumed that as prosperity and education levels increase, populations tend to come into equilibrium rather than plunge.

Even the general public seems to have expected an equilibrium, not a decline. In 1947, 73 percent of French people surveyed thought that the population of France should increase, while 22 percent thought it should stay the same and only one percent thought it should decrease. By 1974 the upbeat pronatalists had declined to only 23 percent, those who thought the population should stay the same had risen to 63 percent, and those who thought their numbers should decrease had risen to 10 percent. The decline in birth rates has actually gone far beyond the expectations of the people experiencing the decline.

Demographic transitions, in which declining birth rates follow upon declining death rates, are now taking place in many parts of the world. Eastern Europe's transitions are particularly striking, since death rates are rising as birth rates are falling. Before these calamitous changes, the most rapid demographic transitions for which we have records took place in Korea and Taiwan. In Korea, from 1960 to 1974, the crude birth rate was cut almost in half, while similar declines were seen in Taiwan.

The reasons for these declines are numerous, but the transitions undergone by these countries are very different from the widespread economic upheaval and discouragement that have had such an impact on Eastern Europe. Korea and Taiwan appear to have reached a kind of societal consensus, for all classes and educational levels have been affected by the decline.

This pattern is very different from that seen in less-developed countries that have less economic equality. There any decline in birth rate tends to be concentrated in the well-off or educated classes. In sum, the reasons for birth declines are as numerous as social structures and local histories.

Worldwide, the demographic change over the last few decades has been dramatic. Population growth has a kind of inertia—the huge numbers of people currently moving into their reproductive years ensure that world population will continue to grow overall well into the next century. The peak numbers that will be reached, however, are continually being revised downward. In 1960 the median United Nations estimate for global population by 2050 was twelve billion. Now the median 1997 prediction is that it will reach a mere 9.1 billion by the middle of the next century, and this estimate seems certain to drop further. This number is still huge and frightening, but it is three billion less than was predicted less than forty years ago.

Most tellingly, by the year 2015, at least two-thirds of the world's population will live in countries in which the number of children per woman has dropped below the replacement level of 2.2. Malthusian terrors may still await us, but as a species we seem to be responding with commendable rapidity to the new pressures of an overcrowded world.

These precipitous drops in birth rate have caught everyone by surprise—or almost everyone. In 1952 Charles Galton Darwin, a grandson of Charles Darwin, wrote a book entitled *The Next Million Years*, in which he predicted that in the future people will move from the developing world into the first world, as a result of the population vacuum caused by the demographic transition there. His prediction now appears to be coming true in Europe. In spite of intense xenophobic reactions in many countries, migration from the less developed world into Europe is increasing rapidly. Almost ten percent of the population of Germany is now foreign-born.

Such changes are happening far more quickly in some cities and countries than in others. London and Paris will soon be as polyglot as Los Angeles, but Madrid and Stockholm are likely to lag much further behind. Even the melting pot is defined differently in different places. I recently met an Irish archaeologist and over dinner regaled him with tales of this European transformation. "I know just what you mean!" he exclaimed. "One of my coworkers back in Dublin has just married"—and here he paused for effect—"a Frenchwoman!"

AN EVOLUTIONARY EYE-BLINK

My collaborator Pascal Gagneux, a student of chimpanzee behavior, tells me that when he was a child in Switzerland, all his relatives and indeed everyone he knew was white. Now, each time he revisits his native country, he finds that the ethnic mix has changed dramatically. So exotic is this mixture becoming that children of Tibetan refugees are now joining the Swiss army in substantial numbers. As in most of the rest of Europe, the population of Switzerland is altering at unprecedented speed.

A few centuries from now—an eye-blink in evolutionary terms—the peoples who occupy the European peninsula will have undergone a huge change. We can bid farewell, at least in their present forms, to the "master races"—and the new, dynamic mixtures of peoples emerging throughout the planet will likely not mourn their passing.

Indeed, "master races" seem to be on the decline everywhere around the world. In South Africa in 1951, people officially classified as white by the apartheid regime made up a fifth of the total population outside the "black homelands." By the year 2020 they will make up one-eleventh of the total. This demographic shift helped bring about the demise of that unlamented regime.

Other groups that have acquired economic power and high levels of education, and thus constitute what we can call "master races," are declining almost as rapidly. Sometimes, by an accident of history or geography, these groups are diminishing in numbers but immigrants are not yet altering their gene pool. The Japanese are a notable example.

Japanese birth rates began to decline in the 1920s, then shot up just before and during World War II, because a government desperate for soldiers instituted an unrelenting pronatalist campaign, including severe penalties for abortions. In 1947 the Japanese lifetime birthrate was 4.5 children per woman. But the decline soon resumed, aided by the legalization of abortion in 1949, and by 1957 it had plunged to 2.0. Starting in 1974, a further decline began, and Japanese birthrates are now well below replacement.

Dramatic demographic shifts have accentuated the effects of the decline. The Japanese countryside has emptied out as people have moved in huge numbers to the cities. There, overcrowding makes it virtually impossible to have large numbers of children. More than twenty desperate towns and villages, faced with the prospect of becoming ghosts of their former selves, have posted rewards for large families. The richest prizes are being offered by the government of Hachijojima, an island southeast of Tokyo: a graduated scale ranging from half a million yen for a third child up to a dazzling 2 million (about $20,000) for a fifth. During the first year after the reward system was begun, five couples made claims—not enough, however, to have much of a demographic impact.

The Japanese have one of the world's longest life expectancies and have reduced infant mortality to the lowest rate found anywhere in the world. Married women still have an average of about two children, but they are getting married in smaller and smaller numbers. In 1975, 21 percent of women between the ages of 25 and 29 had never been married. By 1990 that proportion had doubled, to 40 percent. Women complain bitterly that men are too exhausted by their endless workdays to manage the additional strenuous activity of baby-making. And once the babies have arrived, men do nothing to help their wives, even though most of the wives are also holding down demanding jobs.

If these trends continue, a population vacuum will emerge in Japan. Despite the isolation of Japanese society, it will be necessary for industries to begin importing guest workers. Then, as has already happened in Europe, the transformation of the gene pool in Japan will begin in earnest.

In my own state of California, people who call themselves white on census forms will become a minority by the year 2000, and the Census Bureau estimates conservatively that whites will be in a minority in the entire United States by 2050. My own hunch is that these demographic trends will accelerate, as they are doing in Europe, and whites will be demoted to minority status much more quickly than that.

A good deal of the demographic change will be driven by intermarriage. My own grandchildren, when (as I devoutly hope) they eventually make their appearance, will be one quarter northern European, one quarter Chinese, and the rest a mix of southern European and Peruvian Indian. In the United States the number of children classified as being of more than one race has risen from fewer than half a million to more than two million from 1970 to 1993. The rate of racial mixing continues to accelerate.

Extrapolating demographic trends over the long term is dangerous in the extreme. Declining population sizes have as many different causes as there are populations undergoing decline. Moreover, these trends will often be locally reversed, at least temporarily. But two important points can be made that have implications for the future evolution of our species.

First, the average member of our species a century or two from now will be heterozygous at more different genes than the average person is today. Isolated islands of relative genetic homogeneity that have built up over tens or hundreds of thousands of years will have disappeared.

Second, even though recognizable members of the various "master races" will be far less common than they are today, their genes will not disappear as well. Rather, their genetic legacy will still be there, combining and recombining in ways that the fierce xenophobes of the past could never have imagined.

Is all this change evolutionary in nature? You bet it is—rapid and fundamental changes in the gene pool do not require terrible plagues or the slaughter of war.

Our current population size is unsustainable—a millennium from now, the world population will doubtless be far smaller than it is today. Much—and I hope most—of this shrinkage will be the result of individual decisions to have fewer children or none at all. Although we cannot predict what the effect on the gene pool of the survivors will be, it is likely to be profound.

In this first part of this book, we have looked at the wide range of evolutionary pressures acting on us, some obvious and some more subtle. Physical factors in the environment, such as disease, are playing a smaller role, though they are not declining as much as optimistic scientists had suggested a few years ago. Social and psychological pressures are increasing, and their nature is changing rapidly.

Some authorities contend that the benefits conferred by our complex culture have somehow brought our pell-mell evolution to a halt. If they are right, it would be an event unparalleled in our long history. If the evolutionary brakes have been applied so strongly, the smell of burning rubber should be everywhere!

PART II

Our Stormy Evolutionary History

SEVEN

❖

The Road We Did Not Take

As recently as 35,000 years ago western Europe was
still occupied by Neandertals, primitive beings for
whom art and progress scarcely existed.

JARED DIAMOND, "The Great Leap Forward" (1989)

The rolling brown and olive-green hills of the northern Spanish province of Castile and Léon have seen a great deal of history. In medieval
times millions of pilgrims tramped the length of the province on their
way to the great cathedral town of Santiago de Compostela. The pilgrims believed that, in the ninth century, Saint James himself had arisen
from his grave and beaten back the Moors in a great battle. During
succeeding centuries the pilgrimage route to the site was marked with
thriving towns and astonishing cathedrals, standing as testaments to
the overwhelming force of faith.

But the history of the region dates back much, much further. These
same hills have witnessed the triumphs and tragedies of an ancient group
of western European people who had evolved their own unique characteristics. Their story was played out over a very long span of time,
starting with their arrival in the region well over a million years ago and
ending with their final disappearance less than thirty thousand years ago.

The remains of the last people of this lineage were first discov-
ered a century and a half ago. We call them the Neandertals, after the
Neander valley in Germany, where traces of them were first found.
The Neandertals' forebears were ancient tribes whose real names have

long been lost to history but whom we now call pre-Neandertals. Some of the most vivid evidence for their history has been found among the Spanish hills.

New fossil finds are casting fresh light on this remarkable story, which I recount here for three reasons.

First, these finds show that the roots of our own species go a long way back in time, much further back than most people think. Human beings did not suddenly appear on the planet a few tens of thousands of years ago, in the form of modern people, different from anybody who had gone before. We did not suddenly acquire the ability to decorate the cave walls with pictures and patterns and make elegant artifacts of wood and bone. Our emergence, like the parallel emergence of the Neandertals, took far longer.

Second, the same strong forces that have driven our evolution also drove the evolution of the Neandertals. We are not unique. Both we and they were caught up in a period of ever-accelerating evolution.

Third, and recent news stories to the contrary, the genetic differences between us and the Neandertals are not profound. Genetic differences among present-day human groups are actually very slight, primarily involving different frequencies of alleles of various genes that all human groups hold in common. We saw in Chapter 1 that if populations differ not in their genes but in the frequencies of alleles of their genes, the effects can be substantial. Selective pressures or chance events can easily change such populations by shifting the frequencies of the alleles. The Neandertals, while somewhat different from us, were not so far away as to have a different collection of genes; rather, they were primarily distinguishable from us by different allele frequencies.

Many anthropologists ignore this evidence and continue to consider Neandertals to be not quite human. But I think it is scientifically far more valid to hold up their story as a mirror of our own, and to learn as much as we can from it.

The emerging story of the pre-Neandertals of northern Spain was uncovered through the patience and determination of the Spanish archaeologists Emiliano Aguirre, Eudald Carbonell, Juan Luis Arsuaga, and their colleagues, who persisted for years in excavating sites that everyone else told them were hopelessly unpromising.

The sites lie quite near the ancient cathedral town of Burgos, with its charming gardenlike moat and immense carved city gate. To the east of Burgos the undistinguished village of Atapuerca has given its name to a low range of limestone hills. The limestone was originally laid down in a warm tropical sea during the last part of the age of dinosaurs. Much later, water seeping from above carved out a vast honeycomb of caves.

Very little of this region has been excavated by archaeologists, so it is particularly striking that two out of the three sites that have been examined intensively have yielded important human fossils. For such fossils to be so plentiful, the whole region must once have supported large populations.

Superficially, the world inhabited by these people would not have been much different from what we see today. Grazing and farming have now transformed the nearby lowlands, but the hills are still covered with scrub oak and wild rosebushes. Grasshoppers make a deafening noise. Pollen grains found in the digs suggest that the vegetation has not changed greatly over the last million years—the peoples who lived here during all that time probably looked out over a very similar landscape, sometimes a little more heavily forested, sometimes a little less. The big difference was in the animals. In earlier times many different animals flourished here, including elephants, boars, cave bears, and rhinoceros; deer flourished in the region, making it ideal for hunting. The last of the remaining animals were hunted to near-extinction by the nobles of Castile during medieval times.

A RAILWAY TO NOWHERE

During the latter part of the last century a deep and narrow railway cutting known as the Gran Dolina was sliced through the nearby hills. The railway was used only briefly before it was abandoned. Today it is one of the most dramatic archaeological sites in the area.

At several points in the cutting, the workers had exposed old caves, actually pockets in the limestone that, over the millennia, had been completely filled by detritus washed in from above. The detritus is made up of compressed orange earth, some of which spilled out into the cutting.

The archaeologists excavated down through these layers, looking for information about the hundreds of millennia during which the caverns had slowly filled up. They excavated two of these caverns, each filled with material some eighteen meters thick. The first, discouragingly, yielded nothing. The second also seemed empty, except for a few possible stone tools. But, in the early 1990s, the excavators took a chance. Leaping ahead of their normally methodical investigation, they tunneled a one-meter-square hole down toward the bottom of the deposit. In 1994, their gamble paid off, as they discovered stone tools and fragmentary human remains at the nine-meter level.

Dating the find might have been a problem, but luckily it lay beneath a layer of deposit that marks a sudden reversal of the earth's magnetic field. The switch is one of many that have taken place periodically throughout our planet's history. Traces of this particular reversal, known as Matuyama-Brunhes, have been found in many other places, allowing the event to be dated with precision at 780,000 years before the present. The Gran Dolina bones and artifacts are therefore older than that, perhaps substantially older. Indeed, they are the oldest fossil human bones, by at least three hundred thousand years, that have ever been found in Europe.

At the Natural History Museum in Madrid, José Bermudez de Castro showed me the finds. The stone tools are very primitive, of a type known as Oldowan. The bones themselves are few, including some isolated teeth, a fragment of an upper jaw, and some bits of lower mandible that include molars. They are, however, remarkably well preserved—it is still possible to see the gum line on some of the teeth.

The most important find is a piece of facial bone, including the lower rim of an eye socket and enough of the upper jaw to determine that it belonged to a child. The angle of the facial bone below the eye socket is surprisingly vertical, and as a result the child who owned this face has been put by its discoverers into a separate human species, *Homo antecessor.* This new species has been proposed to be an ancestor of both Neandertals and modern humans. But the child's modern-looking face could also simply be the result of the fact that the skull is immature—even the skull of an immature chimpanzee has much in common with that of a human child. My own guess is that this child, living the better part of a million years ago in these remote hills, was already following the separate evolutionary path that led not to modern humans but to Neandertals.

The fact that so few artifacts and remains have been found after so much excavation suggests that the people of the area probably did not spend much time in the caves but rather lived and hunted out in the open. Little can be inferred about how the child's remains, along with a few fragments of an adult cranium, came to be there. Cut marks on the bones suggest that they might have been butchered and eaten, and their remains thrown into the cave afterward.

The discoveries in the Gran Dolina have already pushed the first appearance of hominids in Europe much further back in time than might have been imagined just a few short years ago. In addition to this discovery at the nine-meter level, at least two deeper layers—which have yet to be investigated thoroughly—are known to contain stone tools tentatively dated to well over a million years ago. It is likely that further excavations will yield even older traces of human activity.

Hominids first ventured out of Africa into Asia, and perhaps into Eastern Europe, at least 2.5 million years ago. It seems unlikely that it took them a million years to travel as far west as the Sierra de Atapuerca. Either at this site or elsewhere in this honeycomb of caves, bones will certainly be discovered that will help fill in that million-year gap

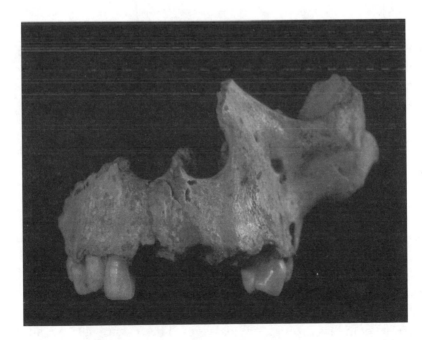

Figure 7–1. The face of the Gran Dolina child. In spite of its great age, it is surprisingly narrow and vertical.

between the first ancient peoples to leave Africa and the people of the Gran Dolina.

These excavations provide only part of the complex story of early human history in Europe. There is much evidence that hominids occupied other parts of the continent, beginning at least half a million years ago.

These other finds, like those at Gran Dolina, are fragmentary, consisting of isolated jaws, crania, and leg bones. They have turned up at sites scattered over a huge area that extends from the Pyrenees to southern England and northern Germany and east into Central Europe and Greece. The finds are puzzlingly variable, showing some features that are Neandertal-like, others that are more primitive, and still others that defy categorization altogether. Yet their variability may be no more significant than the variability found among human populations today. Study of these early remains, like that of the Gran Dolina people, has been hampered by their very fragmentary nature. For example, one species, *Homo heidelbergensis*, which probably lived half a million years ago, has been named entirely on the basis of a single lower jaw.

The fragmentary nature of these finds has been the curse of paleontologists, who have been forced to infer complex evolutionary stories from a few bits of bone. For this reason a second great discovery that has emerged from the caves of the Sierra de Atapuerca is dazzling. It allows archaeologists to examine ancient Europeans in enough detail to get a good idea of what they must have been like as a group, not as a few broken bits of one or two individuals.

THE CAVERN OF BONES

A few hundred meters from the Gran Dolina railway cut, a winding path leads to a large cave entrance under an overhanging vertical cliff of limestone. Until it was recently closed off by a metal gate, generations of explorers had found bones of cave bears in abundance in the thick wet mud deposits on the cave's floor.

The cave consists of a series of large chambers, separated by narrow scrambles through tiny passages. They eventually lead, after some five hundred meters, to a heart-stopping drop-off. At the bottom of this hole, thirteen meters deep, is a cramped, claustrophobic, sloping chamber with a floor only a few meters square. Over the years many cave explorers climbed down into this dark hole, trampling the floor and leaving garbage behind. It became known as the Sima de los Huesos, the cavern of bones, from the bear bones that were found there.

One day in 1976 Trinidad Torres, a professor of mining, and a spelunker named Carlos Puch had descended together into this uttermost cave in search of cave bear bones. Suddenly Puch made a remarkable discovery—in the midst of the cave bear bones was a human mandible! As soon as the two emerged, they took their find to Emiliano Aguirre, an expert on fossils, who realized its importance.

Two years were to pass before Aguirre managed to get a team together to explore the cave further. They soon found that excavation was a daunting task—part of the cave's cramped floor was covered with huge limestone blocks that had to be removed. This task, and the excavation of the overburden of trampled and damaged sediments, took several years. The excavators had to fill backpacks with rock, then climb and crawl through the narrow passages back to the surface. A couple of these grueling trips a day were the most that could be managed. So deep was the cave that if more than two or three people worked there simultaneously, they soon began to run out of oxygen and were forced to climb out.

Finally, after they removed three tons of material, undisturbed sediments were revealed. Excitement was growing, for sieving of the overburden had already revealed many fragments of human bones. The degree of preservation was extraordinary—even some tiny bones of the middle ear had been found. This little cave clearly contained many disarticulated, badly fragmented, but complete skeletons.

Some of the undisturbed sediments were found under a layer of stalagmitic limestone, which could be dated at about 300,000 years old. The bones are therefore at least that old, which makes them approximately half as old as the fragments that would later be found in the excavations of the nearby Gran Dolina.

It is now estimated that the complete skeletons in the Sima de los Huesos number more than a hundred and that it will take decades to finish the excavation. The way they are piled together in a tangled heap indicates that these people perished over a very short span of time. How did they come to be there, so far below ground, in such a tiny place?

We simply do not know. Perhaps they died elsewhere and their bodies were thrown down that thirteen-meter shaft. Or perhaps they wandered in through a lower entrance that has since disappeared— such an entrance may have existed at the time. If they did, a terrible fate awaited them. They might have been overcome by lack of oxygen, as nearly happened to some of the later excavators. They might have been trapped behind a rockfall. Or they might have been herded or thrown in there by some conquering tribe and left to die, a grim early case of genocide lost to history. Clues to what happened to them may eventually be found among the undisturbed layers.

Once those undisturbed layers were reached, the pace of excavation slowed, for the position of every bone had to be measured carefully. To make things more difficult, the bones had been wet for hundreds of thousands of years and were now so soft and fragile that a touch could damage them. Whole chunks of the floor of the cave had to be dug out and carefully dried in order to allow the bones to harden. A shaft was drilled down from the surface, through which material could be hauled up. As a bonus, precious oxygen could now percolate down the shaft, so the excavators far below no longer had to put their brain cells at risk from repeated oxygen deprivation.

Finally the years of work paid off. On July 7, 1992, a complete cranium was found, and others were soon discovered nearby. Freeing the blocks containing the crania and bringing them to the surface took two weeks of exquisite effort. The skulls could not be matched to any of the lower jaws that had been brought up earlier, although a year later a jaw was found nearby that matched the largest and best-preserved skull.

Over the years I have seen many different fossil finds from sites in Africa, Europe, and Asia, but nothing prepared me for the impact of the bones of the Sima de los Huesos. Their sheer abundance is staggering. With a flourish José Bermudez de Castro opened a drawer in a cabinet at the museum to reveal a dozen superbly preserved lower jaws. At the nearby Universidad Complutense, Juan Luis Arsuaga showed me the skulls, which were also remarkably well preserved. One showed signs of a massive upper jaw infection, which might have been the cause of death.

Both of the best-preserved skulls are adult, but their crania are very different. One had a large brain, comparable in size to those of modern humans, though not quite as large as those of the much later Neandertals. The other had a much smaller brain, so small that it actually fell within the range of our even earlier ancestor *Homo erectus.* If so huge a variation in brain size had been found in different excavations, the fossils could easily have been attributed to different species. We know, however, that they belonged to the same tribe. It is likely that, as the excavation proceeds, even more variation will be discovered. Even though this variation may seem extreme, a similar range of variation sometimes appears within human groups today.

Such variation cannot be investigated among any of the other early European fossil finds, since they usually consist of a few isolated bones. Indeed, so abundant are the fossils of the Sima de los Huesos that Arsuaga and his coworkers have been able to show that the differences in body size between males and females are about the same as in present-day human populations. These people were like us in many ways.

But not in all ways. Like the skulls of the much later Neandertals and unlike our own, the skulls in the Cavern of Bones have prominent browridges and rather forward-thrusting faces. These features were, however, not as pronounced as they were later to become in the Neandertals, especially those that lived in the west of Europe, nor were the bones of their skeletons as heavy as those of the Neandertals. (See Figure 7–2.)

The differences in size do not extend to the jaws and teeth, which are remarkably uniform. The mandibles are massive and very like those of Neandertals—and very like that of Heidelberg man. Neandertals, like modern Eskimos, used their front teeth as pliers, gripping objects with such force that the front teeth wore down more quickly than the molars. The people of the Sima de los Huesos were doing the same, for without exception the front teeth of their jaws are badly worn down while the crowns of the molars are relatively intact.

The skulls of the Sima de los Huesos differ from those of the western Neandertals in one striking respect, however. Neandertal skulls, as Jean-Jacques Hublin had earlier shown me in a basement room of the Musée de l'Homme in Paris, have a rounded, swollen appearance, almost as if the back of the head has been blown up like a balloon. Such an extreme "occipital boss" is seen in no other hominid fossils from any other era. It is certainly not present in the skulls of the Sima de los Huesos. Rather, when observed from the rear, those

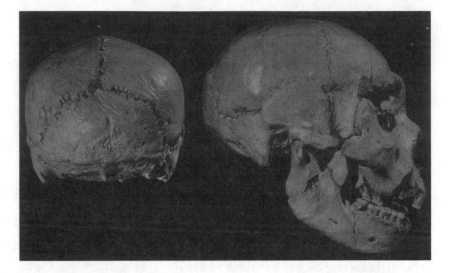

Figure 7–2. The best preserved Sima de los Huesos skull, from the rear and from the side. Note the pentagonal shape of the skull from the rear.

skulls have a pentagonal appearance, shaped by the clearly visible angles of the parietal and temporal bones on their sides. When viewed from that vantage, apart from their low cranial vault they could be mistaken for modern human skulls. (See Figures 7–2 and 7–3.)

Nevertheless, the bones that Arsuaga and Bermudez de Castro showed me convinced me that the people of the Sima de los Huesos were firmly in the lineage that eventually led to the western Neandertals. The only important differences between them were the enlarged and swollen braincase at the rear of the Neandertals' skulls, and the fact that the Neandertals tended to have more robust skeletons. The pattern of tooth wear, the shape of the jaws, and the size and shape of the projecting browridges at Sima de los Huesos were all clearly like those of the much later Neandertals.

A SEPARATE LINEAGE

Thanks to these Spanish discoveries, our picture of the Neandertal lineage is growing more complete. The Neandertals' history has been dramatically lengthened and has been revealed to be more complex than scientists had previously imagined.

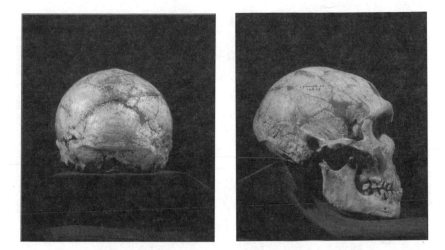

Figure 7–3. On the left, a rear view, and on the right, a side view of a typical late Neandertal skull, from La Ferrassie. Note the swollen appearance of the skull from the rear, very different from that of the skulls found in the Sima de los Huesos.

The child of the Gran Dolina probably lived relatively near the base of the branch that led to the western Neandertals. It must have been the descendant of some group of people who, more than a million years ago, wandered as far west as it is possible to go in Southern Europe. There they became isolated from the evolutionary events happening elsewhere, and started on their own separate evolutionary path.

Three or four hundred thousand years passed between the time of that child and the time of the people of the Sima de los Huesos. These later people, like other peoples living in Western Europe at the same time, had large brains. They probably had an advanced technology, although no tools were discovered with the bones: the evidence comes from a remarkable archaeological find much farther to the north.

This stunning discovery was made in early 1997, by Hartmut Thieme of the Institute for Historic Site Preservation in Hanover, Germany. For years he and his coworkers had been monitoring excavations at an open-cast brown coal mine near the city. The gigantic digging machines used at the mine often turn up important sites and artifacts, and Thieme and his colleagues watch for such events. When something is found, they must react quickly in order to stop the machines and examine the site in detail. As soon as they are finished, the machines start up again.

The site that was uncovered in 1997 was particularly remarkable. It yielded three long, heavy, and painstakingly carved wooden spears, along with a throwing-stick and a variety of stone tools. These wooden artifacts would normally have long since rotted away, but they had been preserved in peat. Numerous bone fragments were found nearby, including the remains of ten horses with cut-marks on their bones, showing that the site had been used for the butchery of animals. Other signs of human activity, including a primitive hearth, were also uncovered.

These finds would have been unremarkable if they had only been a few thousand years old, but to everybody's surprise they were dated very clearly to about four hundred thousand years ago. This astonishing date makes the spears the oldest wooden artifacts ever found. The people who made them were approximate contemporaries of the people of Sima de los Huesos.

This discovery triples the period of time for which we have evidence of systematic hunting among our ancestors or near-ancestors. The spears are not simply pointed sticks but are very sophisticated, indicating a long spear-making tradition that went back even further in time. They were undoubtedly meant to be thrown and were quite capable of killing big game. Their points had been carved from the base of the sapling's trunk, where the wood is hardest. And just as with a modern javelin, they are carefully balanced so that their centers of gravity are about a third of the way back from the point. They provide unequivocal evidence that at least some of the people of western Europe were technologically well advanced at the time the tribe at Sima de los Huesos met its fate.

We are constrained to infer the level of any really ancient culture from the few artifacts that happen to have been preserved. Stone tools survive very well, while wooden implements do not. It now appears that their primitive stone tools may have fooled us into selling the ancestors of the Neandertals short. These people were much smarter, for much longer, than we had imagined. We will probably not find advanced wood-based technology in the Atapuerca cave deposits, but the Sima de los Huesos people may very well have used such weapons to hunt in the surrounding hills and plains.

These lucky finds from 300,000 to 400,000 years ago give us remarkable clues about these peoples. But a frustrating gap appears in the fossil record, from the time of the Sima de los Huesos people down to about 100,000 years ago. Only then do we find the first evidence for true Neandertals.

The most extreme types of Neandertal remains have been found in caves ranging from southern Spain to northern Germany. Less extreme types have turned up throughout Eastern Europe, and as far east as Israel, the Crimea, and Iran. Erik Trinkaus, an expert on the later Neandertals, suspects that these less extreme types were already mixing genetically with other types of humans who were more like ourselves. The western Neandertals, however, because they lived in such remote and isolated regions, remained separate.

ANCIENT DNA

The great age of the western Neandertal lineage has recently been demonstrated directly, through a daring exploit by Matthias Krings, a graduate student at the University of Munich. Krings disobeyed the first rule followed by students who want to get a Ph.D: He embarked on a project that had very little chance of succeeding.

Was it, Krings wondered, possible to get intact DNA out of a Neandertal bone and sequence it? Svante Pääbo, his adviser, was the world's leading expert on extracting old DNA from museum specimens and fossils, but he was very doubtful that it would work.

The DNA that Krings set out to find is carried in structures inside our cells known as mitochondria. These structures, tiny but absolutely essential, are the factories that produce the huge amounts of energy-rich compounds that we need for our growth and development.

Mitochondria are not an integral part of our cells but are actually the remote descendants of creatures that were originally quite alien to us. Their presence in such intimate association with our own bodies is the result of a remarkable evolutionary event that took place among some of our single-celled ancestors about two billion years ago.

At that time a species of parasitic bacterium attacked and entered our ancestors' cells—probably producing a nasty disease in the process. Over time, however, just as can happen with present-day diseases, host and pathogen gradually reached an accommodation. The result was beneficial to both sides: the bacteria found a safe place to live, passing from one generation to the next inside their hosts' cells, while at the same time they brought a great gift.

Unlike the cells they were attacking, the bacteria were able to utilize their food fully, extracting most of the available energy from their diet. They did so by using oxygen to burn the food in a highly controlled way. When they had evolved to the point of invading the cells of their hosts without killing them, the hosts also began to bene-fit from all this new energy, taking evolutionary pathways that had previously been closed to them. They could, for example, grow in size and complexity, and eventually some of them could become clever multicellular organisms like us.

We think of ourselves as oxygen breathers, but in fact we are not. Without our mitochondria, we would die instantly. That is why cya-nide is fatal—it destroys this essential function of our mitochondria, while leaving the rest of our bodies quite unharmed. (Bacteria that do not breathe oxygen are unaffected by cyanide; some of them actually use it as an energy source.)

By now, the mitochondria have long since degenerated into nothing more than tiny energy factories, and most of their genes have been lost or transferred to the chromosomes of their hosts. But for reasons that are not understood, a few genes have been left behind, carried on a tiny bit of DNA.

Mitochondria are not transmitted by sperm, which means that they can be passed down only from mother to daughter through the egg. Nor do they seem to recombine with each other, with the result that each mitochondrial chromosome sends intact copies of itself to the next generation. But these chromosomes, like any other pieces of DNA, do acquire mutations. Over time they gradually diverge from each other, like sentences copied by scribes who make the occasional mistake. Krings realized that if he could isolate Neandertal DNA and compare it with that of present-day humans, it would tell him a great deal about the differences between the Neandertal lineage and ours.

As Krings knew, mitochondrial DNA has one other overwhelm-ing advantage. Most of our genes are present in two copies in each cell, one from our mothers and one from our fathers. But mitochon-drial genes are present in hundreds or even thousands of copies, because there are many mitochondria in each cell and each mitochondrion may have many chromosomes. Because the DNA in fossil bones breaks down over time, it is much easier to find a given piece of ancient mitochondrial DNA that is still intact than it is to find a given piece of DNA from the nuclear chromosomes.

Krings began by examining the very first Neandertal ever discovered, a partial skeleton found by quarry workers in the Neander valley in 1856. The bones had been buried in freezing cold mud for thirty thousand years. Then they had been transferred to a museum display, where they had sat quietly—except of course for the innumerable times during which they had been brought out and handled by scientists and the simply curious. Tests showed that their protein had suffered very little damage, so perhaps some of the DNA had survived intact as well.

If any of the original DNA were still left in the remains of the bone cells, Krings knew, it would probably have been broken up into fairly short pieces after all this time. But the many people who had handled the bones must have left bits of their DNA behind, and this DNA would be in much better shape. Somehow this contaminating DNA had to be distinguished from the ancient DNA of the Neandertal.

To make things harder, most of the ancient pieces of DNA were likely to have been chemically modified in all sorts of ways, making it impossible to resurrect them. So Krings used a molecular trick called PCR* to amplify the DNA, beginning with a short section about one hundred bases long. He found to his delight that while some of the amplifications of this short section resembled modern humans and were therefore probably contaminants, most of them did not.

Were these new bits of DNA really a molecular echo from the distant past? Patiently, Krings used PCR to amplify short sequences

---

* PCR, polymerase chain reaction, allows an experimenter to make huge numbers of copies of a short piece of DNA. It requires that something be known about that piece, however. If the experimenter knows the sequence of two short regions on either side of the piece, he or she can use an enzyme to fill in the intervening region. The reaction is arranged in such a way that, each time this is done, the number of copies of the intervening region doubles. After twenty or thirty cycles there are billions or copies—enough to use standard laboratory techniques to determine the sequence of the intervening region.

that overlapped the first. These pieces could be joined together into a longer fragment, eventually almost four hundred bases long. A variety of control experiments convinced him that this really was thirty-thousand-year-old DNA. To clinch the matter, he sent his material to another laboratory in the United States, where Anne Stone, an expert on Native American mitochondrial DNA, was able to amplify the same sequence.

The sequence of bases in the Neandertal DNA (for such it almost certainly is) shows distinct differences from the equivalent region in modern human DNA. The lineage that led to this western Neandertal really had been separate from us for hundreds of thousands of years, just as the fossil discoveries in Atapuerca had suggested.

More Neandertal DNA sequences will have to be studied—particularly from the Neandertals of Central Europe and the Middle East—if the true place of the Neandertals in human evolution is to be determined. But don't try this at home—Krings very sensibly added a warning at the end of his paper, pointing out that the Neander valley skeleton is particularly well preserved. Most Neandertal bones are unlikely to yield any DNA, unless tests to determine the extent of molecular damage are carried out first. If the drills and saws of molecular paleontologists are not used with caution, irreplaceable Neandertal remains may end up looking like Swiss cheese without telling us anything new about their DNA.

One other intriguing feature of the Neandertal sequence emerged from Krings's study. Although it is indeed very different from our DNA, some mitochondrial DNA sequences from modern humans are substantially different from each other as well. Some sequences taken from African pygmies and Australian aborigines, for example, are about two-thirds as different from each other as human sequences are from the Neandertals. If we knew very little about our own DNA but had just by chance obtained a few sequences from very isolated human groups, then the Neandertal DNA might not seem particularly unusual.

The Neandertals were far, far closer to us than our nearest living relatives, the chimpanzees. Some of them may have been even closer to us than the morphologically extreme member of its race that Krings examined. After all, the bones of that Neandertal were found in the isolated Neander valley, far from Eastern Europe and the Middle East where the real evolutionary action was taking place.

THE FINAL DAYS OF THE NEANDERTALS

In Western Europe, probably between about 35,000 and 28,000 years ago, the ancient western Neandertal lineage came to an end. Their disappearance coincided with the arrival in Southern Europe of the Cro-Magnon people and a number of other groups, who were virtually indistinguishable from modern humans and had far more sophisticated cultures. Most people have assumed that the Cro-Magnons were simply too clever for the Neandertals and were able to drive them swiftly and mercilessly to extinction.

Yet there is more to the story. The Neandertals were hardly pushovers—the first appearance of the Cro-Magnons and the final disappearance of the Neandertals were separated by at least seven thousand years, a period of time substantially longer than the entire history of recorded civilization.

It was presumed until recently, in the absence of evidence, that the Neandertals were savages indeed and knew nothing of the decorative arts or what we might suppose were the niceties of civilization. While some of their remains show signs that they had been buried, the burials seemed to have been nothing more than simple interments. The bodies were neither accompanied by artifacts nor covered with the pigments used by the Cro-Magnons and other later peoples. But excavations of Neandertal remains from a cave in the Crimea show that spring flowers had been buried along with the bodies, so that these people had invented at least some elements of ritual burial. And some of the skeletons show signs of severe injuries that had subsequently healed, indicating that the victims must have been nursed through their trauma by other members of the tribe.

Recent important discoveries suggest that toward the end of their career, the Neandertals might have progressed considerably in their technology, though it is not yet clear whether this happened because of contact with the Cro-Magnons and other more advanced peoples or whether they accomplished these advances without outside help.

In early 1996 Eric Boëda and a group of French and Syrian colleagues working at a remote site in western Syria discovered stone tools, at least forty thousand years old, that had traces of black pitch adhering to them. Careful chemical analysis showed that the pitch must have been heated and melted in a fire. Probably the pitch had been used to bond the tools to wooden handles. Were the makers of these sophisticated tools Neandertals? Boëda thinks so, since he recently found a Neandertal skull in the same deposit.

Until this discovery, the oldest tools showing traces of pitch as an adhesive were a mere ten thousand years old. By an odd coincidence those tools had also been found in the same area, which raises the intriguing possibility of cultural continuity between the Neandertal makers of the older tools and the more modern people who made the more recent tools.

Perhaps the most remarkable discovery of an advanced Neandertal artifact was made in the summer of 1996 by a team headed by Ivan Turk of the Slovenian Academy of Sciences. His team was excavating deposits in a cave in hilly country to the west of the city of Ljubljana, near the current border between Italy and the former Yugoslavia. There were many signs that the cave had been occupied by Neandertals, including some typical stone tools buried in the cave floor. But along with these tools the excavators also found the remains of an astonishingly advanced artifact. It was a piece of the femur of a juvenile cave bear in which four neat circular holes had been punched, all along the same side. It seemed likely to them that this fragment formed part of a flute.

If it was a flute, then it is the oldest ever discovered. It has been dated using electron spin resonance measurements of cave bear teeth that had been found in the same layer; the results show that it was made, and perhaps played, somewhere between 40,000 and 80,000 years ago. It must already have been part of a long musical tradition, for the punched holes are precisely placed and beautifully round (Figure 7–4). Their position suggests that the flute could have been used to play a diatonic scale, with full and half tones.*

This remarkable object implies a maker who had a deep understanding of how to coax a variety of sounds from a hollow tube. The Neandertals might not have made it themselves; they may have obtained it through trade or warfare with its real makers, perhaps a more

---

* It has recently been suggested that the bone was not part of a flute at all, and that the holes were punched by animal teeth. The edges of the holes show no smoothing of the type that would be expected if the instrument had been used. This flute-versus-natural-object argument is likely to continue, and its final outcome is uncertain. But the holes are remarkably round and regular in their placement. The flute may have been broken in its manufacture before it had a chance to be used.

Figure 7–4. A possible Neandertal flute on the left, and on the right, a closeup of the strikingly round yet puzzlingly unworn holes in the bone.

advanced tribe of early Cro-Magnons living in the vicinity (but for whose existence there is no evidence). But if the Neandertals were as brutish as they are supposed to be, why would they have wanted such an object?

The more we learn about Neandertals, the more like us they become. If they really could use a flute to make music, they must also have invented complex ritual occasions for doing so. Any impartial observer would agree that this constitutes fully human behavior. Indeed, it seems less than kind of us to relegate these flute-playing hominids to a subhuman status—particularly when most of *us* are not able to play flutes!

The western Neandertals, in spite of their isolation, showed remarkable adaptability right up to the end. They seem to have invented their own moderately advanced culture of stone tools and implements. It has been named the Chatelperronian and was quite distinct from the somewhat more complex Aurignacian culture of the Cro-Magnons who replaced them. We will probably never know whether this burst of invention at the end of their career was stimulated by the arrival of the Cro-Magnons, or whether they would have done it on their own.

They may have gone even further than these suggestions imply. Some years ago a collection of tools and decorative objects was found in the caves of Arcy-sur-Cure, near Auxerre, in the Nivernais region south of Paris. The artifacts include shells and other objects with holes drilled in them, so that they could be dangled on thongs. They have been dated with some assurance to 35,000 years ago and are among the very earliest such objects found anywhere.

The Arcy-sur-Cure finds are very typical of objects made by Cro-Magnon people. But were the people who made them, or at least possessed them, really Cro-Magnon? Tantalizing bits of skull and other bones were found along with the artifacts, but these bits were so fragmentary that it seemed at first that they could tell nothing about the people who lived in the caves.

Then in 1994 a group led by Jean-Jacques Hublin of Paris's Musée de l'Homme employed a CAT scanner to look inside some of the skull fragments and visualize the semicircular canals of the middle ear. For some reason that has yet to be discovered, the shapes of these canals in Neandertals and modern humans, including Cro-Magnons, are clearly different. The CAT scans showed unequivocally that the bone fragments were Neandertal, not Cro-Magnon.

It now seems that Neandertals occupied the cave of Arcy-sur-Cure and may have been the owners—if not the makers—of those advanced artifacts. Possible scenarios abound. Did they steal the objects, or acquire them through trade? Did they learn how to make them by imitation? Alternatively, was the cave really occupied by Cro-Magnons, who were in the habit of hunting Neandertals, bringing their heads back, and decorating these gruesome trophies with necklaces and earrings? The most daring possibility of all: Did the Neandertals invent these objects on their own—and then teach the Cro-Magnons how to make them? Whatever the explanation, western Neandertals and Cro-Magnons apparently came into contact, perhaps repeatedly, and some technology could have been transferred.

The fate of the western Neandertals is made more poignant by these new discoveries. Were they driven utterly to extinction? No

DNA sequences resembling that found by Matthias Krings have yet been found among living humans, so there is no direct evidence that the Neandertals left behind any genes when they disappeared just under thirty thousand years ago. But even though their incipient speciation seems to have been aborted, perhaps they did manage to exchange at least a few genes with the Cro-Magnons. If so, then a Neandertal-like sequence may be found at some future time in a modern human, providing direct evidence of such an exchange.

Further, as Erik Trinkaus points out, when some of the Neandertals escaped from their isolation in the west, they seem to have made repeated contact with more culturally advanced humans in middle Europe and the Middle East. If gene exchanges between these two groups did not take place, then it may be our loss.

The Neandertals lived a life of unimaginable rigor, able to survive and even thrive during the great climatic swings of the most recent ice ages. Yet in spite of all these physical challenges, their brains grew larger and their culture became more complex over time, just as ours did. In their lineage, as in ours, cultural evolution did not preclude physical evolution. Acquiring more elaborate culture did not slow down their physical evolution—if anything, it seems to have accelerated it.

Modern humans, too, were pushed in roughly the same direction. Why? Not because we have different genes from the Neandertals, but because when modern humans met Neandertals, both groups had roughly the same set of alleles, though in different proportions. Remember that Neandertals are not much farther from us genetically than the maximum distance between living human groups. Just as with human groups living today, the differences between the Cro-Magnons and the Neandertals must have been primarily allele frequency differences, not the presence of an allele in one group and its absence in the other.

These allele frequency differences were expressed in many ways. Some of them had to do with brain size increases.

The pressures to increase brain size must have been very strong, powerful enough to have found a way around equally strong evolutionary restraints. The shapes of the fronts of the Neandertals' skull were constrained by the requirement for strong jaw muscles, particularly the temporalis muscle that attaches to the side of the skull and passes beneath the cheekbone to attach to the jaw.

You can feel this muscle by touching the area above your temple and clenching and unclenching your jaw. In modern humans it is a feeble thing, but in the Neandertals it was far larger. The projecting browridge above the eye sockets and the sloping forehead, which to us gives the Neandertal skull such a primitive appearance, were needed to provide both room and a substantial attachment point for this muscle. The front part of our own skulls has grown larger, resulting in our near-vertical foreheads and a reduction in the thickness of the temporalis muscle sheet. This option for increasing brain size may not have been possible for the Neandertals, since it would have weakened their jaw muscle too much.

We do not know the diet of the Neandertals or how, if at all, they prepared their food. We do know that present-day Eskimos, who probably live under very similar conditions, use their teeth as tools to make many different items of clothing and artifacts, and it seems likely that the Neandertals did the same. They are, perhaps, a vivid example of the old saw that you are what you eat.

Given these constraints, which prevented their foreheads from becoming more vertical as their brains expanded, how could they have acquired larger brains? The only way, it would seem, was by expanding the skull to the rear. The occipital boss, so pronounced in late western Neandertals, is the result of this final expansion. It had not yet appeared by the time of the people of Sima de los Huesos.

In view of the sometimes substantial variation in skull shape among living human groups, it is not difficult to see how the different shapes of our skulls and those of the Neandertals could have been produced, not by the appearance of new genes, but simply by shifts in the frequencies of preexisting alleles.

LADINS AND NEANDERTALS

The tale of the western Neandertals casts light on our own recent and continuing evolution. Although they lived in a kind of parallel universe, following a different evolutionary path from ours, in the end they seem to have become remarkably like us. Then they either disappeared entirely or managed to contribute some of their genes to our gene pool.

But what do we really mean by "contribute to a gene pool"? The phrase is glibly tossed off by generations of geneticists and anthropologists whenever they talk about mixtures of human groups, but without specifics the term has little meaning. To make it more concrete, we must recall that the blending taking place among different human gene pools may be having a substantial effect on our evolutionary potential. This will, however, only be the case if the various gene pools that are undergoing the blending are substantially different from each other.

When population geneticists have looked at samplings of the genes from different human groups, they have been able to find very few real differences in alleles between groups. Most human groups have roughly the same collection of blood group alleles, enzyme alleles, and so on. The differences lie primarily in the frequencies of these alleles.

Further, the total amount of allelic diversity within a single human group is far greater than the differences between separate groups. According to an estimate made by geneticist Richard Lewontin, variation within groups accounts for at least ninety percent of the total variation in our species.

Sometimes we can see in our own species milder examples of what happened to the Neandertals. Geneticists have recently examined a people who live in the valleys of the Dolomite Mountains in northeast Italy and speak a language called Ladin. The origin of their language is a mystery, yet the people who speak it seem no different in appearance from their neighbors who speak other languages. In addition to speaking their unusual language, some of them carry unusual mitochondrial chromosomes, as genetically distant from those of their neighbors as the mitochondrial chromosomes of Africans. Like the origin of their language, the origin of their mitochondria is a mystery.

Their story illustrates the very different way in which mitochondrial and nuclear chromosomes are inherited. Some of the ancestors of the Ladins may once have been very different in physical appearance from the surrounding tribes, but over time, as they married across tribal boundaries, their nuclear genes shuffled and recombined. The result was that any physical differences between them and their neighbors disappeared. Their mitochondrial DNA, however, did not recombine, and a few copies of it came down a separate lineage as an echo of their distant past. Their language, too, has remained separate, and while it has undoubtedly changed greatly from the ancestral language of the Ladin people, it has retained its unique identity.

The Ladins have contributed to the cultural diversity of the northern Italians and must have added genetic diversity as well. But except for those telltale mitochondrial chromosomes, no sign of their genetic contribution persists.

To find out whether the same process happened to the Neandertals, we need more evidence from their DNA—not just from their mitochondrial chromosomes but from their nuclear genes as well. Will their nuclear genes, as I strongly suspect, turn out to be very like our own? Did they, too, have ABO blood group alleles, and Duffy-a and -b alleles, and perhaps the same set of dopamine receptor alleles, only in a slightly different mix? Because copies of nuclear genes are likely to be very rare in Neandertal bones, finding them is probably beyond the reach of present-day technology. Yet I suspect it will only be a matter of time before the DNA hunters are successful. And when these genes are found, I predict they will show that we are closer to the Neandertals than mitochondrial DNA alone would suggest.

Now let us leave the remarkably parallel story of the Neandertals and turn to the question of how our own species has reached its current state. As we will see in the next chapter, the evolutionary histories of chimpanzees and gorillas have been very different from our own. If chimpanzees from everywhere in their natural range—which extends across the entire width of tropical Africa—were suddenly to be mixed together in a huge involuntary gene-blending experiment of the type that our species is currently undergoing, the effect on their physical appearance and behavior would be slight and perhaps even unnoticeable. This implies that the differences between human populations must somehow be greater than the differences between chimpanzee populations—even though it turns out evidence from mitochondrial chromosomes seems to point in the opposite direction. New genetic advances means that we will soon be able to glimpse just how our evolution took place and why it has been so similar to that of the Neandertals and yet so different from the evolutionary history of the great apes.

EIGHT

❖

Why Are We Such Evolutionary Speed Demons?

*The most casual student of animal history is struck by
the fact that while most phyletic lines evolve regularly
at rates more or less comparable to those of their allies,
here and there appear some lines that seem to have
evolved with altogether exceptional rapidity.*

GEORGE GAYLORD SIMPSON, *Tempo and Mode in
Evolution* (1944)

CHIMPANZEES ALL THE WAY BACK

DNA evidence tells us that the evolutionary lineage that led to the gorillas split off from ours perhaps seven to nine million years ago; the lineage that led to the chimpanzees departed along its own path between five and six million years ago. During much of that time the evolutionary stage on which these apes trod was a narrow one, for they have apparently always been confined to the forests of sub-Saharan Africa. Our own ancestors, in contrast, beginning at an unknown point somewhere in Africa, spread first through the entire African continent and later through all of Eurasia.

On the basis of this evidence alone, we might expect that the great apes have evolved less than ourselves. They have not been such brave explorers of new environments. Like the orangutans of Borneo and Sumatra, relatively unchanged for at least two million years, the chimpanzees and gorillas may be stuck in a kind of evolutionary time warp.

Because chimpanzees and gorillas are clearly different from each other, they can hardly have been in complete evolutionary stasis. But, at least over the last few million years, they do not seem to have undergone the enormous changes that can be seen in our own lineage.

I can make this statement because of the remarkably complete fossil record that we have for our species. Although paleontologists are always complaining about how fragmentary it is—and indeed, vast areas of the world and vast spans of time are still empty of traces of our ancestors—the record is now sufficiently good that we can get a fairly clear idea of what happened. Perhaps the most striking finding is that the further back in time we go, the more chimpanzeelike our ancestors become.

This allows us to make a prediction: The evolutionary history of chimpanzees for the last five million years should be, by comparison to our own, supremely uninteresting—chimpanzees all the way back.

The strongest evidence for the chimpanzeelike nature of our own ancestors was found in 1992 by Gen Suwa of the University of Tokyo, along with the Berkeley paleoanthropologist Tim White and an international team of field-workers. As these patient anthropologists were carefully quartering a stretch of eroding ground at a desolate site in the hot, dry Afar region of Ethiopia, Suwa found a single molar tooth peeking out from the clay.

He and the others then began to comb every inch of the surrounding area, soon finding bits of skull and arm bone nearby. The most exciting discovery was made by Alemayehu Asfaw, an Ethiopian team member. It was a fragment of the lower jaw of a young child, with the deciduous teeth still in place. To the excited anthropologists, this fossil seemed clearly different from anything previously found in the area. In particular, it was different from the jaws and teeth of the famous Lucy skeleton that had been discovered, two decades earlier and not very far away, by Donald Johanson. For one thing, these hominids lived about 4.4 million years ago—almost a million years earlier than Lucy.

The teeth are very like those of chimpanzees, more so than those of any of our later ancestors in the fossil record. The relatively thin enamel was much more typical of chimpanzees and other apes than of humans and was only half as thick as Lucy's.

By sheer luck, White's group also found a bit of the base of a skull. It included part of the foramen magnum, the opening through which the spinal cord passes. This very diagnostic feature can tell us a great deal about a hominid's habitual posture. The heads of modern humans are balanced on top of the spinal column, and our foramen magnum is more or less at the center of the base of the skull. But in the ancient hominid from the Afar, the opening was much farther back, toward the rear of the skull, almost as far back as in present-day chimpanzees.

In something of an excess of taxonomic zeal, White and his coworkers named their find *Ardipithecus ramidus*, thus creating for it a new genus as well as a new species. They were following that old tradition among fossil hunters: discoverers of new and exciting fossils have always tended to give them new names, even though it is not really possible to tell from the bits of bone themselves how much their owners differed genetically from similar creatures or even from ourselves.

Since this first discovery, more *Ardipithecus* fragments have been found. While these latest discoveries show that it was probably able to stand upright, its posture was definitely not the same as ours. Standing fully erect would have been as tiring for these hominids as it is for present-day chimpanzees, since their heads projected so far forward. They must have spent only rather short periods fully upright.

In the first paper published about this find, White and his colleagues admitted that one would be hard put to distinguish *Ardipithecus* from a chimpanzee. There were only some slight differences in the teeth, the small shift in position of the foramen magnum, and some minor additional features of the skull.

The resemblance to chimpanzees extends beyond skeletal similarities, to embrace the world in which *Ardipithecus* lived. These hominids, like the majority of present-day chimpanzees, spent their lives in fairly dense forest. The nature of their habitat can be gathered from a variety of bits of evidence. Many animal bones have been discovered in the same slopes as the *Ardipithecus* fragments, including those of numerous colobus monkeys. Today such animals are typical of dense closed-canopy forest, and their ancestors most likely lived in a similar environment. It was a setting quite different from the open and more fragmented woodlands inhabited by the later hominids in our own lineage, such as Lucy.

These findings mesh with what we know about *Ardipithecus*, particularly the thinness of the enamel on their teeth. In the forest their diet would have been rich in fruit and tender leaves. Only later, when their descendants had moved out into more open country, did their diet change to include more difficult-to-chew foodstuffs. Lucy's thicker tooth enamel was the evolutionary result of that change.

Ardipithecus is the most chimpanzeelike of our forebears to have been discovered, and it also happens to be the oldest. But the overall resemblance to chimpanzees persisted for a long time in our lineage. In 1924, when Raymond Dart found the first skull of our strikingly ape-like ancestors in material excavated from a South African quarry, he named it *Australopithecus*, "southern ape." The date of this young child's skull is still a bit uncertain, but it is probably about two million years old—less than half as ancient as *Ardipithecus*. It too, however, is so distinctly chimpanzeelike that the British scientific establishment initially dismissed it as an ancestor not of humans but of apes.

A few years ago, at the South African site of Sterkfontein, paleoanthropologist Ron Clarke demonstrated the similarities to me. Side by side, on a rough wooden table, he placed a cast of a 2.5 million-year-old *Australopithecus* skull and a skull from a modern chimpanzee. I was struck, just as those establishment authorities of the 1920s had been, far more by their overall resemblance than by their differences, which were primarily in the teeth. And this resemblence had persisted even though the *Australopithecus* skull had undergone two additional million years of evolutionary change since the time of *Ardipithecus*.

Of course, those remote ancestors of ours must have been different from chimpanzees. Chimpanzees, too, have undergone five or so million years of separate evolution, and their remote ancestors must have been different from the chimpanzees that we see today. But the differences must have been relatively minor.

Paleoanthropologists have probably not yet found fossils that date all the way back to the human-chimpanzee branch point. Through a geological accident, fossil-rich deposits that date to the period just before *Ardipithecus* have been hard to find in East Africa. But some are being discovered, and as paleoanthropologists scan these deposits they will continue to close the gap between our ancestors and those of the chimpanzees. One day soon a lucky fossil-hunter will find pieces of a fossil that is very close to the common ancestor. The bones will probably be discovered projecting inconspicuously from a baked and eroding hillside somewhere in Ethiopia, but they might turn up anywhere in Africa—even in West Africa, which until some recent discoveries was thought to be empty of hominid fossils.

It is safe to predict that these ancestors will turn out to have been even more apelike than *Ardipithecus* and to have had chimpanzee-size brains, chimpanzeelike dentition, and many other chimpanzeelike features. They were probably able to walk upright only for short periods and lived very like present-day chimpanzees, banding together into small groups and roaming a fairly dense forest. They probably lived five million or more years ago. If so, this date will match extremely well with dates that have been inferred from DNA and from protein molecules.

Such ancestors, when they are discovered, will also be the first true ancestors of chimpanzees to be discovered in the fossil record, despite the better part of a century of African fossil-hunting. This dearth is not for lack of looking—any bone that a paleontologist finds that might have belonged to an ape or a monkey will receive just as much scrutiny as one that might have belonged to our own ancestors.

The usual explanation given for the lack of fossil remains of chimpanzees or gorillas is that these apes have always lived in tropical forests. As soon as they died, the parts of their bodies that were not eaten by scavengers must have been destroyed by the plentiful insects and bacteria on the forest floor. This argument, however, has recently lost much of its force. Even though *Ardipithecus*, our own oldest ancestor, lived in quite dense forest, a substantial number of fossils of this very chimpanzeelike creature were preserved.

Further, floods or volcanic eruptions occasionally killed our own ancestors and covered their bodies before they could decay. Such accidents appear to have been responsible for a number of famous fossil finds, such as the First Family, a large collection of three-million-year-old Australopithecine bones discovered by Donald Johanson in Ethiopia. The ancestors of chimpanzees and gorillas lived in volcanic areas for just as many millions of years as our own ancestors, and it seems odd that there were no opportunities for such scientifically happy (though personally distressing) accidents to have happened to them as well.

Other kinds of fossils giving clues to our ancestors, traces that are much more ephemeral than bones, have also been preserved. In 1976 paleontologists in a group led by Mary Leakey were working at Laetoli, a site not far from the famous Olduvai Gorge, when they made one of the most important discoveries in all of paleoanthropology. Hardened into the rock, and barely visible unless the light slanted in just the right direction, were sets of footprints from a variety of animals. Excavation revealed that some of these footprints belonged to two, or possibly three, Australopithecines.

As these hominids trudged across a field of ash that had been freshly deposited by the eruption of a nearby volcano, they left their footprints behind. Then, through a further lucky coincidence, rain fell briefly, just enough to harden the ash but not enough to destroy the fresh prints. This astonishing find provided proof positive that by 3.5 million years ago Australopithecines were already able to walk upright.

Why did such a lucky accident never befall the footprints of chimpanzee or gorilla ancestors? The most likely explanation is that their ancestors were never as numerous or as widespread as the human ancestors. Although fossils of the ancestors of these great apes or of their footprints must be somewhere out in the endless baking gullies of East Africa, perhaps exposed on the surface at this very moment, they are so rare that no one has yet stumbled on them. Whatever first great adventure took *Ardipithecus* on the new path that eventually led to us, the apes had no part in it.

A QUICK OVERVIEW OF HUMAN PREHISTORY

Our own pell-mell evolutionary story has been very different from those of the chimpanzees and gorillas. Many different evolutionary changes, at many different times, have happened that resulted in our big-brained, clever-handed, highly social, language-speaking selves. The ability to stand upright seems to date back at least four million years.* Hands that, like our own, had extreme touch sensitivity and

---

* Upright posture may not be unique to our own lineage. An ape that lived ten million years ago on Sardinia, *Oreopithecus bambolii*, seems to have acquired similar capabilities, perhaps independently.

great manipulative skills evolved 2.5 million years ago. Their appearance coincides with the first stone tools in the fossil record. Substantial increases in brain size and in stature took place about two million years ago. Fire was probably first used at around that time.

The appearance of language is more problematical, since it has proved impossible to trace the ancestral tongues of present-day languages back more than ten thousand years, and even these efforts have been dogged by controversy. But our brains are so hard-wired for the acquisition of language that this skill must have had its genesis a long time ago. It is unimaginable that the spear-throwing people who lived in Germany 400,000 years ago were incapable of communicating verbally with each other during the course of their hunts.

Although writers on the subject tend to point to one or another of these events, particularly the acquisition of language, as the watershed event in the appearance of modern humans, they are all part of a remarkably seamless movement toward creatures like ourselves. We cannot single out one event that catapulted us into "full" humanity.

Further, we have only a few glimpses of our ancestors' diversity. If we knew how genetically variable they were, we could say something about their evolutionary potential, since diversity is a powerful resource for continuing evolution. But because very few fossils can be assigned to a given time period, we do not know the full range of types of people who lived at various points. The people of the Sima de los Huesos are an exception, and even they represent only one tribe. They had a great deal of variation in brain size, while at the same time their jaws and teeth were remarkably uniform. Such glimpses of the variation among our ancestors are unusual. Yet it is just this kind of variation that has powered our rapid evolution.

It is certain that there was enormous diversity among the hominids of about two million years ago, although we do not know how much there was among the subset who were our direct ancestors. Discoveries in East Africa, particularly by Louis and Mary Leakey and their son Richard, have showed that around that time an extremely varied collection of hominids roamed the plains and patchy forests. Small-brained Australopithecines were present in large numbers; they had, by this time, evolved beyond Lucy and the older and more chimpanzeelike *Ardipithecus*. They stood more nearly upright, and their hands were very similar to ours. Randall Susman of the State

University of New York at Stony Brook, who has examined many of these hand bones, strongly suspects that their owners were capable of using tools.

By two million years ago the Australopithecines had split into at least two lineages, one of which was robust and had huge teeth and jaws, and another that was relatively more light-boned or gracile (though compared with us fragile creatures, these gracile Australopithecines were pretty robust as well). (See Figure 8–1.) It is not easy to determine what place any of these Australopithecines have occupied in our own ancestry, though the light-boned ones are generally agreed to have been more like us. To make things more confusing, the distinction between the robust and the gracile Australopithecines has blurred a bit in recent years—fossils showing features that are intermediate between them have turned up in South Africa.

It was at about this time that the much larger-brained *Homo habilis* appeared on the scene. They are the first creatures to be sufficiently like ourselves that general scientific consensus has permitted them to join the genus *Homo*. All anthropologists agree that a major evolutionary step was necessary for them to do so, although it is not clear just how quickly that step took place, or whether it happened all at

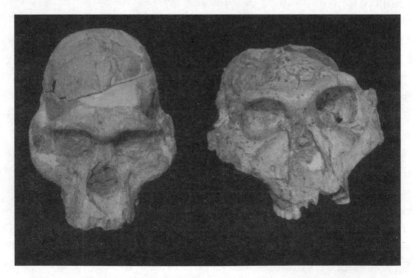

Figure 8–1. Skulls of a gracile Australopithecine ("Mrs. Ples," Sts5) on the left, and of a far more massive robust Australopithecine (SK48) on the right.

once or in piecemeal fashion. An increase in brain size was certainly an important part of this change, but not the only one. *Homo habilis*, like the Australopithecines, was short in stature; an increase in body size to something more like that of modern humans seems to have come much later than these first increases in brain size.

When paleoanthropologists penetrate this far back into the past, they simply do not have enough information to follow the stories in detail. As nearly as we can tell, over a span of perhaps half a million years the brain sizes of the hominids now grouped together as *H. habilis* increased by about fifty percent over those of the Australopithecines. This increase is certainly striking, but it must be remembered that it is based on a mere six skulls—the only ones that are sufficiently complete that the sizes of the brain cavity can be measured.

The most famous of these skulls, known as KNM-ER 1470, was found by Bernard Ngeneo, an associate of Richard Leakey, in 1972. It is relatively well preserved and is shown on the left of Figure 8–2. Its owner lived almost two million years ago, in quite open country near a lakeshore in what is now northern Kenya. The skull has space for a very large brain, with a volume of perhaps 750 cubic centimeters. Here *large* is a relative term, since its volume was equal only to two cans of soda, a little more than half the volume of our own brains. It was, nonetheless, more than half again as large as the brain of an average Australopithecine.

Had the braincase been destroyed while the rest of the skull remained intact, this find would almost certainly have been classified as an Australopithecine. Like an Australopithecine's, the face is very prognathous or forward-thrusting and very broad. KNM-ER 1470 seems to have been classified as *Homo habilis* primarily by virtue of its mighty brain.

Some of the other *Homo habilis* skulls are very different in size from KNM-ER 1470. Although they vary greatly in their completeness and their overall appearance, it has nonetheless been possible to make an approximate estimate of their brain sizes. While some were large, others were much smaller and actually appear to have overlapped those of the Australopithecines.

So if their brains were so small, why have these creatures been grouped with *Homo* rather than *Australopithecus*? One important reason is that their faces tend to be narrower and more vertical—less

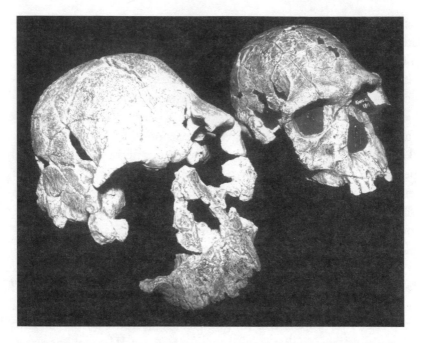

Figure 8–2. The larger, more fragmented, but clearly more robust KNM-ER 1470 on the left, and the smaller KNM-ER 1813 with its narrower and more vertical face on the right.

prognathous or forward-thrusting—than the more chimplike faces of the Australopithecines. (See the skull named KNM-ER 1813 in Figure 8–2.) The evolution of a relatively narrow and vertical face, which we think of as typically human, is surprisingly separable from the evolution of increasing brain size. Anthropologists appear to be quite accommodating about the characteristics that a hominid need display in order to be admitted to the select company of earliest *Homo*. Either a large brain or a vertical face will do.

The skulls of *H. habilis* pose difficult problems for the scientists who have to sort them out. One of them, known as KNM-ER 1805, has such an extravagant mix of Australopithecine and more human characteristics that it seems to be a blend of the two. The existence of this skull raises an important question: Were there, as most authorities on the subject have concluded, really several different early species of *Homo*? Or could they all have been one species, made up of individuals that varied greatly among themselves, so that KNM-ER 1805 is an example of this mixture of characteristics? What was the real relationship of this mix of hominids to the Australopithecines that some of them resemble so closely?

While this problem may seem highly technical, it actually has a direct bearing on our own recent history, since similar questions have been raised about the evolutionary events that lead directly to *Homo sapiens*. If multiple species of *Homo* existed two million years ago, then all but one of them must have died out, since by definition the various species would have been unable to exchange genes once they had parted genetic company. But did they? What really happened? Did ecological upheavals provide multiple opportunities for gene exchange, of the sort now happening between the Bornean and Sumatran orangutans and that possibly took place between Neandertals and modern humans?

THE BEGINNINGS OF LANGUAGE

The brains of *H. habilis* have long since disappeared, but it is possible to learn something about them from the impressions they left behind on the interiors of the skulls. Careful study by Ralph Holloway, Phillip Tobias, and others has shown that the brains of the Australopithecines really do seem to have differed from those of the best-preserved and largest-brained *H. habilis*.

The differences were more profound than mere size. They were concentrated in the front and side regions of the brain, which in modern humans are involved with the production and understanding of speech. These regions appear to have undergone a great enlargement in *H. habilis* compared with the Australopithecines. So, apparently, did the mysterious cerebellum or "little brain," nestled near the brain stem. Recently, functional magnetic resonance imaging (MRI) scans on volunteers have shown that this primitive brain structure is very active during speech. Until the advent of MRI, it had been thought that the cerebellum was concerned only with fundamental abilities like balance, but it too has taken part in our rapid recent evolution and has acquired important new functions.

Did all these changes in its brain mean that *H. habilis* could talk? Probably nowhere near as readily or as incessantly as we do, but it could doubtless use its brain in ways that Australopithecines could not. Though its language skills were probably extremely limited, *H. habilis* was probably capable, to some degree, of applying labels to things and processes in the world around it, and of communicating these labels to other members of its tribe.

There is likely to have been another difference between *H. habilis* and its ancestors. When researchers attempt to teach chimpanzees sign language, or the use of symbols as communication tools, it takes endless persuasion and reinforcement to produce results that are rather marginal and difficult to interpret. One of the most obvious hallmarks of our own species is the way in which young children acquire language rapidly, in an eager, unforced, and enthusiastic way. If *H. habilis* were able to speak in some fashion, as the shape of its brain seems to indicate, then the joy of communication must have been evolving at the same time as that capability. This enormous step on the road to humanity, like all the others, was a complicated one and did not happen instantaneously.

PEOPLE MORE LIKE US

This plethora of African hominids was eventually supplanted by a more advanced, and to all appearances even more successful, hominid known as *Homo erectus*. The new hominid was almost certainly a direct ancestor of ours. Its earliest fossil remains date from almost two million years ago in both Africa and Asia, but it probably made its actual debut about half a million years earlier. *H. erectus* was, so far as we know, our first ancestor to leave Africa and travel through the Old World.

The earliest complete African skeleton of this hominid happens to have been superbly well preserved. Australopithecines and *H. habilis* tended to be small in stature, but this adolescent boy was very tall and probably would have reached six feet at maturity. His brain, too, was larger than that of even the brainiest *H. habilis*, approaching nine hundred cubic centimeters for the first time.

The triumph of *Homo erectus*, while it seems eventually to have been complete, was nonetheless anything but instantaneous. The gracile or light-boned Australopithecines did vanish at around the time that *H. erectus* appeared, an event that has haunting resemblances to the much later replacement of Neandertals by Cro-Magnons. This earlier replacement process, however, took as long as half a million years, not a mere seven thousand.

The robust Australopithecines managed to persist in various parts of southern and eastern Africa until less than a million years ago. Because these animals shared much of the eastern region of Africa with *H. erectus* for a million years or more, this suggests that the two hominids had such divergent ways of life that they were able to coexist without coming into mortal conflict.

Because no Australopithecine fossils have been found outside Africa, we are fairly confident that it was *H. erectus*, not the Australopithecines, who first migrated through the Middle East and into Asia about 2.5 million years ago. At that time the Pleistocene ice ages still lay in the future and the climate was relatively mild, so that no climatic barrier stopped them from venturing into Europe or further into Asia and probably back again. Even though we have found only Australopithecines and *H. habilis* from that time in Africa, early representatives of *H. erectus* were probably there too.

The first indication of these early population movements comes from some remarkable new discoveries of ancient stone tools in Israel by Avraham Ronen and his colleagues at Haifa University, and in Pakistan by Robin Dennell of the University of Sheffield. In both cases the tools have been dated quite firmly at about 2.5 million years ago— a remarkable finding which makes them as old as any that have been found in Africa. As soon as our *H. erectus* ancestors invented such tools, it seems, they used them to conquer new worlds. It has generally been assumed that stone tools were first used in Africa, but these new finds raise the possibility that they were invented somewhere else and taken back to Africa by *H. erectus* as they moved back and forth over a wide swath of territory.

The regions into which *H. erectus* ventured were very different climatically from the deserts and exhausted badlands that make up so much of today's Middle East and Southwest Asia. Sometimes one can obtain a vivid glimpse of that ancient world.

In 1994 a group from the Georgian Academy of Sciences began to excavate some old building foundations in the city of Dmanisi, near Tbilisi in Georgia. In the floors of these ancient basements they found deep pits that had been dug by the medieval inhabitants for grain storage. After the debris that filled these pits had been carefully removed, the scientists found that the pits themselves could be used as probes into far deeper strata. The walls of the pits first yielded ice age animal bones. Then, some distance below them, the archaeologists discovered a hominid jawbone, some stone artifacts, and some earlier animal bones that had been missed or ignored by the original medieval diggers.

The jaw shows clear similarities—notably, its lack of a well-developed chin—to *H. erectus*. It has now been dated to between 1.6 and 1.8 million years ago. The animal bones are evidence that the Georgia of those days was a very different place, for among them are bones of giraffes and other warm-weather animals. The mild climatic conditions meant that the animals that are now confined to the East African savannas were far more widespread in those days.

The Israeli site of 'Ubeidiyah, in the Jordan valley some four kilometers south of the Sea of Galilee, also lay in this salubrious region. Excavations starting in 1960 have turned up a large number of remains of animals dated to about 1.4 million years ago. Here too there were giraffes, elephants, and antelope, showing that the climate and range of fauna were very similar to those of East Africa today. Early humans lived in the Jordan valley as well—the archaeologists have discovered primitive stone tools and some very fragmentary bones that almost certainly belonged to *H. erectus*.

Surely, one might think, the more adventurous among those bands of *H. erectus* would soon have taken the path leading through such a pleasant valley directly into Europe. It seems unlikely that they dithered at 'Ubeidiyah for four hundred millennia or so before finally venturing west. More probably we have simply not yet found the earliest traces of hominids in Europe itself. Some of those yet-to-be-discovered early migrations eventually gave rise to the people of the Gran Dolina and the tribe whose bones were preserved in the Sima de los Huesos.

PREDECESSORS OF MARCO POLO

We are not sure where and when people closer to us than *H. erectus* first appeared, but some quite modern-looking skulls, dating from about 100,000 years ago, have been discovered at a variety of sites in southern and eastern Africa and in the Middle East. They have been given a variety of names and are generally referred to as archaic or early modern *Homo sapiens*. These people, with their large brains, quite vertical faces, and high foreheads, were not, however, quite like us.

They were more robust, their teeth were larger, and they had primitive-looking characteristics like pronounced browridges. It appears that even our immediate ancestors, close though they were to us, continued to evolve over the last 100,000 years.

Moreover, these brainy ancestors of ours were not instantaneously all-conquering. In the Middle East they lived in the same region and at roughly the same time as other groups who had more Neandertal characteristics—and they may, though the evidence is only suggestive, have interbred with them.

It took the Cro-Magnons seven thousand years to replace the western Neandertals. Now evidence indicates that, at the other end of Eurasia, our ancestors shared the world of *H. erectus* for an even longer period. The overlap of *H. erectus* with people like us did not last for half a million years, like the overlap of *Homo* and the gracile Australopithecines, but it did last for much longer than anybody had imagined until recently. Evidence for this astonishing persistence of *H. erectus*, in spite of what must have been innumerable contacts with our own species, was found recently on the island of Java.

Of all the Indonesian islands, Java is the most heavily populated, and its landscape has been greatly modified by human activity. The countryside is now cultivated so intensively that terraces have been built right to the tops of quite high mountains. The farmers who work them must spend most of each day climbing up to them and back. Despite such efforts the island is unable to feed its burgeoning population and is a net rice importer.

During its less crowded past, however, Java was a rich prize. Here Buddhist, Hindu, and Mogul empires rose, collided, and fell. They left behind the remains of hundreds of temples, particularly in the central highlands. Earthquakes and volcanic eruptions have shaken most of these ancient structures into rubble and covered them with layers of ash. Only a tiny fraction of them have recently been excavated and rebuilt.

Even though the oldest temples are relatively recent, dating to about the seventh century A.D., their builders are unknown. Our ignorance of Java's more ancient history is even more profound. Were we able to read it, however, that history would surely be long, eventful, and highly informative about human evolution.

During the last 1.64 million years, whenever the world went through a period of glaciation and the sea level dropped, Java became part of a much larger landmass. This huge peninsula extended down from Southeast Asia, encompassing most of the islands of Indonesia and the shallow Sunda continental shelf on which they rest. There were often times during the last part of this period when modern humans were able to migrate south across this shelf from what is now Indochina.

Some of these waves of peoples penetrated as far as New Guinea and Australia, and there is growing evidence that they first arrived in those regions a long time ago. The human history of Australia, in particular, may be older than scientists had supposed until recently. Starting quite suddenly, about 140,000 years ago, great fires swept again and again through huge areas of that continent. They transformed the landscape, destroying dense acacia forest and replacing it with open grasslands and scattered eucalyptus groves. There is a strong possibility that these fires were set by humans. There is far less controversy over other and firmer evidence that people arrived in northern Australia at least sixty thousand years ago.

The migrants must have passed over the Sunda landmass, including Java, on their way to Australia and points east, yet on Java itself there are no fossils or artifacts that can be traced to migrants from this period. The earliest remains of modern humans found on the island date to a mere ten thousand years ago, although a modern skull that might be fifty thousand years old has been discovered on Borneo to the north.

The only Javanese fossil bones that predate this relatively recent find are not modern at all. And rather than a record of migration and change, the story they tell appears to be one of virtual stasis.

These ancient peoples lived on or near the Solo River, which springs from the central Sangiran Plateau, flows muddily across most of the width of the central part of the island, and eventually wends its sluggish way to the north coast. While the river has been a feature of the landscape for a long time, it has often changed its course, carving out a wide floodplain in the process. Here and there on this plain, the floods that shaped it have left traces, in the form of the bones of drowned animals and hominids.

The remains tended to collect at bends where the onrushing flow of the river slowed briefly, allowing them to settle to the muddy bottom and be buried. After the river shifted its bed, these old accumulation sites were sometimes left high and dry. It was detritus from these ancient disasters that in 1891 provided Eugène Dubois, a Dutch doctor, with the first clues to these ancient people.

Dubois put the hominid fossils that he discovered, chiefly crania, into the genus *Pithecanthropus*, because they seemed extremely primitive and ancient—a true missing link between apes and men. It now seems quite certain that these hominids were not unique but were really representatives of *H. erectus*.

Subsequent excavations by the Dutch occupiers of Java turned up even more of these fossils, including one enormous treasure trove of hominid and animal bones found near the village of Ngandong in 1931. Unfortunately, except for a few skullcaps, this entire collection of 25,000 bones, including who knows what treasures, has now disappeared.

The dating of these Javanese specimens, however, has turned out to be extraordinarily difficult. As we saw, nothing—not even massive stone temples—lasts long on Java. Its landscape has been continually rearranged by volcanic eruptions, floods, and earthquakes, which makes its geology much harder to understand than the more stable geology of East Africa. Moreover, the exact provenance of many of the fossils is often lost, because farmers who find bones eroding out of the banks of soft earth immediately pull them out of their geological context and sell them.

This unfortunate process is continuing today. A sad little museum at Sangiran, not far from the banks of the river, enshrines a few extremely sketchy exhibits—not, luckily, including any of the precious crania. Every day, at the base of the steps leading to the museum, jostling hawkers try to sell bits of monkey skull and suspiciously symmetrical stone tools to a scattering of gullible tourists. The real fossils, I was told, are sold to private collectors and disappear.

Carl Swisher, from the Geochronology Center in Berkeley, California, is a pioneer in using the extraordinarily precise argon-argon method of isotopic dating. He has been able to obtain approximate dates for the skulls using individual microscopic crystals of mineral obtained from strata that were probably laid down at the same time. Over the last several years, employing this and other methods, he has obtained a dramatic set of dates that have utterly changed our concept of the history of *H. erectus* on Java.

His oldest date, announced in 1995, comes from strata associated with a juvenile skull that had been found near the town of Mojokerto, in a river valley to the east of the Solo River. Astonishingly, the strata and hence the skull could be as old as 1.8 million years. This date would make it quite as old as the very oldest *H. erectus* yet found in East Africa.

Did tribes of *H. erectus* shake the African dust from their cracked and calloused feet almost as soon as they evolved and sprint the entire length of Asia, ending up in Java? Unlikely—if Swisher's date holds up, it will simply be another piece of evidence that each type of hominid probably evolved long before the time of the earliest fossil of that type that we have yet been lucky enough to discover. The migrations from Africa to Java (and perhaps vice versa) were probably more leisurely.

Yet Swisher's remarkable date increases the mystery surrounding the origin of *H. erectus*. Where—and when—did these people first appear? Perhaps the most intriguing aspect of Swisher's find is that it raises the possibility that *H. erectus* might have appeared first in Asia, more than two million years ago, possibly from some peripatetic group of *H. habilis*, then migrated to Africa!

A year after this discovery, Swisher and his colleagues announced a new set of possible dates, from a different fossil collection, that illuminated the very end of Java Man's long and strangely static history on the island. The dates were obtained from the Ngandong collection of skullcaps excavated by the Dutch in 1931, the remnants of that huge lost treasure trove of hominid and animal fossils. The skullcaps were a great puzzle—over the years they had been assigned different dates by different workers, some of them unnervingly recent.

The Ngandong material is indeed too young to be dated by the argon-argon method, but luckily Swisher could use other techniques that allow bones to be dated directly. The Javanese scientists in charge of the fossils were understandably reluctant to allow him to use bits of the skullcaps themselves, but he did manage to obtain some animal teeth from (he hoped) the same strata as the original collection of bones.

The dates he determined from the animal teeth were indeed very recent—somewhere between 27,000 and 53,000 years old. If these numbers are correct, they show that *H. erectus* managed to survive on Java for almost two million years, finally going extinct at a time so recent that it was only a geological instant away from the present.

Remarkably, Swisher's date places the *H. erectus* extinction at roughly the same time as the disappearance of the western Neandertals at the other end of Eurasia.

These numbers were greeted with disbelief by much of the rest of the archaeological community. Swisher's dates imply that if the first modern humans really did come through Java on their way east 140,000 or more years ago, they must have been passing through *H. erectus* country for well over 100,000 years. It seems astonishing that these two types of people could have ignored each other for so long.

Assuming that Swisher's very old and very young dates are correct, what kind of history can we reconstruct for the *H. erectus* of Java? Were the people of the Solo River somehow sequestered from all the other currents of human evolution? Did they, like the orangutans that also lived on Java during that time, live in jungles so remote that the more modern humans migrating through the area seldom interacted with them? If so, then they would have had to conceal themselves for tens of thousands of years. This behavior would seem to be remarkably timid for hominids whose ancestors had ventured across the whole width of Asia almost two million years earlier. And, if they were so rare and timid, how did they manage to be so abundant in the Javanese fossil record, where no sign of modern humans appears until ten thousand years ago?

Perhaps our assumption of numerous migration routes through Java is incorrect, and the island—however lush, tempting, and richly forested—was really a backwater where *H. erectus* could preserve its primitive way of life undisturbed—rather as we currently suppose all of Europe must have been for the Neandertals.

Another possibility is that, once *H. erectus* reached the verdant forests of Java, they took up a simple and self-sustaining forest life and remained invisible to later modern human migrants. We know nothing about that mode of life, for we know far less about Javanese *H. erectus* than we do about the Neandertals of Europe and the Middle East. Few stone tools have been found in Java. We do not know how advanced (or primitive) the culture of the last *H. erectus* who lived at Ngandong might have been, or how much it had changed during the previous two million or so years. All that information was washed away by the flooding Solo River.

Very recent evidence suggests that *H. erectus* was both more advanced and more peripatetic than has been previously thought. To the east of Java, a string of islands stretches away toward New Guinea. Dominated by towering volcanoes, they are now fairly dry but must have been far lusher in the past. Two of them, Komodo and Rinca, are still home to the gigantic Komodo dragon, a lizard once found throughout the chain but which disappeared from the larger islands the better part of a million years ago.

Because these islands have always been fragmented by narrow ocean channels, even during the height of glaciation, it has been assumed that modern humans were the only ones smart enough to be able to raft across from one island to the other. Yet in the 1960s Theodor Verhoeven, a Dutch missionary, reported the discovery of stone tools on the island of Flores that were about 750,000 years old. His claim was ignored, but it has now been substantiated through careful redating by a group of Australian scientists—in fact, the tools now appear to be almost a million years old. *H. erectus*, it seems, was not simply sitting around in the deep forests while dramatic evolutionary events were going on elsewhere. These hominids may even have been responsible for the disappearance of the Komodo dragons, along with giant tortoises and small relatives of elephants, that lived on Flores at the time.

We can tell one more thing from these fossils. During this huge span of time the Javanese *H. erectus* were not entirely static in their physical attributes: there was a small but significant increase in brain size from the earliest fossils of Trinil and Mojokerto to the last of them at Ngandong. Indeed, the last survivors of *H. erectus* have brain volumes of as much as 1,250 cubic centimeters, which quite comfortably overlaps the low end of brain sizes in modern humans.

Further, they may not have spent all their time hiding in the jungle. One tantalizing hint that some of them were still wanderers is the discovery of ten-thousand-year-old skulls in Australia that have characteristics very like those of *H. erectus*, even though their brain sizes are larger. These skulls, oddly, are more recent than some skulls of more modern humans found in other Australian excavations. This has suggested, to anthropologist Alan Thorne and others, that several groups with clearly different physical appearances colonized Australia over a long span of time. Somehow these tribes retained their distinctive characteristics until remarkably recently—just as *H. erectus* seems to have done on Java.*

The reader bewildered by all of this is not alone. These new results have overturned most of our assumptions about the prehistory of Southeast Asia, raising far more questions than they have answered.

It does seem safe to predict, in the light of our growing knowledge of the Neandertals, that Javanese *H. erectus* were far more advanced and had a far more complicated culture than the apparent absence of artifacts would suggest. The banks of the Solo River, and the eroding gullies and slopes of the Sangiran Plateau from which it flows, are certainly filled with undiscovered deposits. Modern excavation techniques should be able to tell us far more about these people. It is essential that the farmers of Sangiran be educated to leave bones and tools where they discover them. Perhaps this can be done by a system of rewards. And in order to make even more sense out of this story, other parts of Indonesia must also be explored. No traces of fossil hominids have been found on the vast island of Sumatra to the north of Java, for example, but they must be there somewhere.

Did the rich evolutionary ferment that led to our species bubble as briskly on the islands of Southeast Asia as it did in East Africa and Europe? There seems every reason to suppose it did. Indeed, the Neandertal story sketched in Chapter 7 is not unique. Consider the pattern. In Africa *H. habilis* coexisted with and eventually replaced the gracile Australopithecines. Later *H. erectus* coexisted with and eventually replaced *H. habilis*. In Java modern humans coexisted with and eventually replaced *H. erectus*. During each of these long periods of coexistence, gene exchanges between the different groups may have helped to power further evolutionary transformations. While it is likely that we differ from *H. erectus* by more unique alleles than we differ from Neandertals, the great majority of the differences between our gene pool and theirs were probably still differences in allele frequency rather than in alleles. These groups and our ancestors shared a diverse, complex gene pool with immense capabilities for further evolution.

* This idea, too, has generated much controversy. The skulls of these Australians might simply have been deformed because their heads had been bound as babies.

THE DIORAMA EFFECT

The fossil record offers only a glimpse of this evolutionary ferment. Between roughly 2.5 million and two million years ago, the savannas and gallery forests of East Africa were the scene of particularly dramatic evolutionary change in our lineage. All during those five thousand centuries, a heterogeneous collection of *H. habilis*, probably along with early representatives of *H. erectus*, were sharing a savage and dangerous world with an equally heterogeneous collection of Australopithecines.

We do not know whether these various hominids lived cheek by jowl or were separated by long distances, though fossils of the Australopithecines have been found over a wide geographic area. But all these groups were surely split into many tribes, perhaps hundreds or even thousands of them. Those tribes must have interacted with each other in innumerable ways, ranging from peaceful coexistence through cautious trade to vicious all-out war. If *H. habilis* and the various Australopithecines really were separate species, the beginnings of this separation may have involved such fierce tribal interactions. Many other nascent separations must have been aborted before they went so far: two separate tribes must often have joined up again, and one tribe or alliance of tribes must often have wiped out another.

To get some idea of the sheer scale of all this, it is reasonable to compare it to something closer to our own time. Consider the recorded tribal histories of the Celtic peoples of Wales, Scotland, and Ireland and of the Saxons of England during the Middle Ages. That period was marked by innumerable fierce external and internal wars, triggered by endless battles over succession to leadership within the various groups and marked by an ever-changing mosaic of alliances and enmities among the rulers who managed to survive. More recently the hatreds of these groups have lessened somewhat, and their gene pools have even begun to blend, but under different circumstances their fragmentation might have increased instead. Extremist groups among them—Sinn Fein, Orangemen, Welsh and Scottish nationalists—are still determined to help accelerate such incipient speciation.

Now imagine such a bloody history lasting not five hundred years but five hundred thousand, and played out over an entire continent. Such an imaginative leap gives us a dim idea of the selective pressures to which our ancestors must have been subjected during the critical half million years of genus *Homo's* emergence. What an immense amount of history must have taken place!

We are forced to imagine virtually all of this history, since the only bits of evidence we have are a few fairly complete skulls, a bucketful of assorted bone fragments, and some tantalizing collections of stone tools. But just because we have no direct evidence for it does not mean it did not happen.

The bones themselves hint that things could be rather unpleasant. Recently, prehistorians have begun to revise their assessment about the role of cannibalism, or at least of savage burnings and hackings, among our hominid ancestors. While it was politically correct during the 1970s to presume that savages were peaceful and noble, the discovery of what appear to be burned, split, and butchered human bones at many places in both the Old and New Worlds has changed the assumptions. (The cuts found on the Gran Dolina bone fragments from the hills near Burgos are just one example.) Such unspeakable practices might have played a much larger role in our past than we have hitherto dared to admit.

I suspect we have difficulty admitting to such things because of what I like to call the Diorama Effect. All those museum dioramas that show happy hominids gathered around their campfires, chipping intently away at stone tools, have lulled us into a sense of the peacefulness of the past. Even scientists fall into the trap of supposing that the past was virtually static, that the various hominids of Africa led separate, essentially event-free lives over thousands of generations, slowly evolving or not evolving as the case may be.

This Diorama Effect has to some degree brainwashed all of us, reinforcing our disinclination to admit just how brawling, ugly, dangerous, and complex were the realities that faced our distant ancestors. Here scientists must defer to novelists, who are well aware of the savageries buried in even the most apparently peaceful breasts. Some of them, like Björn Kurtén, have done an excellent job of counteracting the Diorama Effect.

Most of the events that have driven our evolution have been lost to history. But powerful forces are still influencing us, ever more strongly. It took half a million years for the genus *Homo* to evolve from the Australopithecines. Can we doubt that half a million years from now our species, if it survives, will have undergone even more dramatic changes?

NINE

◈

Bottlenecks and Selective Sweeps

When Adam delved and Eve span,
Who was then the gentle man?

JOHN BALL, inciting the Peasants' Revolt (1381)

The dizzying history sketched so briefly in Chapter 8 marks us out as evolutionary speed demons par excellence. Although we have concentrated on the fossil record, traces of all this speed have been left, not only in our bones, but also in our molecules. Recently my colleagues and I have begun to look more intensively than ever before at human DNA molecules and those of our nearest primate relatives, to find out why we differ so dramatically from them. The most vivid of these molecular stories has emerged from work with our nearest relatives, the chimpanzees.

Pascal Gagneux has spent much of his scientific life investigating chimpanzees, both in the wild and in the laboratory. Deep in the forests of the Côte d'Ivoire in West Africa, he and his mentor Christophe Boesch have observed chimpanzees exhibiting a remarkably advanced set of tool-use behaviors. At certain times of the year when there are sudden windfalls of nuts, the chimpanzees will gather in boisterous groups to break the nuts open by pounding them with stones or short, heavy bits of branch. At full spate the sound echoes through the forest like a cheerful carpenter's shop. Young chimpanzees slowly learn, over a period of three to five years, to join this rewarding activity by watching their parents and other elders. Remarkably, the same tools

174

are often used by generations of chimpanzees. And the behavior is probably cultural, for it seems to be confined to this one forested region.

Now, employing techniques of molecular biology, Gagneux has been able to detect other patterns of behavior that are normally hidden from human observers.

Chimpanzee groups in this West African forest are relatively small, each consisting of twenty to a hundred members. Were such groups to remain genetically isolated from each other for many generations, they would begin to suffer from the effects of inbreeding. This can be avoided, or at least postponed, if genes can be exchanged between groups. Even before Gagneux began his investigations, it was known that such exchanges can happen, since young females will often leave a group and join another some distance away. He was soon able to determine yet another, far less obvious, mechanism of gene exchange.

He studied one group of about fifty chimpanzees intensively and identified them all at the DNA level. This was done very elegantly, simply by collecting hair left behind in the nests made from tree boughs that the chimps built each night.*

Only a tiny amount of DNA can be extracted from the root bulb of a single hair, but it is enough for the PCR technique to work its magic. (You will remember that PCR is so powerful, it can amplify the vanishingly small remnants of thirty-thousand-year-old DNA from a Neandertal. Getting enough DNA from the root bulb of a recently plucked chimpanzee hair is child's play by comparison!) For his family studies Gagneux used bits of DNA from the cell nucleus that have a different kind of inheritance from mitochondrial DNA. Like most of our other genes, these pieces of DNA are inherited from both parents. Because of this, he could determine the paternity of each offspring. This determination would have been impossible by simple observation, since receptive females normally mate with several males.

To his surprise, he found that new patterns of DNA continually appeared among the babies born in the group. The babies carried half the genes of their mothers, of course, but the new DNA patterns from their fathers were different from those carried by any of the males

---

* Such hair collection is harder than it sounds. Gagneux had to climb high into the trees, negotiating a gauntlet of sweat bees, wasps' nests, and biting ants, in order to get those tufts of hair.

that he had typed. There was only one possible explanation: Females, even those who seemed well integrated into the group's social hierarchy, must often be sneaking away into the woods for brief trysts with males of other groups in nearby territories. Although he was never able to observe this behavior directly, the story the genes told was unequivocal.

He presented this remarkable insight into chimpanzee mating behavior at a small conference at the end of 1995. Those of us in the audience nodded our heads wisely and exclaimed over what an interesting way this was to avoid inbreeding. How naïve and unworldly we all were! For when he later published his work in the journal *Nature*, it was not the long-term genetic consequences of this behavior that excited headline writers in newspapers around the world, but the behavior itself—the deliciously naughty "adultery" of the chimpanzee females.

The inconvenient fact that chimpanzees know nothing of marriage vows—let alone the guilt associated with breaking them—played no role in the brief media frenzy that followed. Nor did anyone seem to make much of the fact that if the females were committing adultery, the males were perfectly happy to take part in this wicked behavior as well. It takes two to tryst, after all.

Despite this story's soap-opera aspects, in the long term the prevention of inbreeding is most important to the survival of the chimpanzees. The genetic exchange allows chimpanzees living in the wild to be remarkably free of any harmful genetic effects that might otherwise accumulate as a result of their small group sizes. This may help to explain a puzzle that we explored earlier: Even though chimpanzees seem always to have been so rare that no fossils of them have been found, they have somehow managed to persist for millions of years.

The protection conferred by this outbreeding behavior may not last much longer. Chimpanzees' habitats are being destroyed rapidly, and surviving groups are being dispersed ever more thinly as we humans invade their territory—often killing them for meat or to protect cocoa crops against raiding. This invasion will make it more and more difficult for surviving chimpanzee females to bring new genes into their groups, or to find new groups into which to migrate. Human intervention is likely to do to the chimpanzees in a few brief centuries what millions of years of small population size could not do: bring about the demise of all of their wild populations.

FLIRTING WITH EXTINCTION

The relatively plentiful fossil record of our own species suggests that human population sizes tended to be large and that our ancestors did not have quite the same imperative to avoid the effects of inbreeding. While people do of course mate outside their immediate social groups, hunter-gatherer societies do not seem to show such strong outbreeding behavior as chimpanzees.

The pattern of our genes, however, tells us that for some brief moments we too may have flirted with extinction. Had those moments been prolonged, we might really have gone extinct.

Humans and chimpanzees are still so closely related that both species have essentially the same number of genes. But small DNA differences have accumulated between virtually all of our genes and the equivalent genes of chimpanzees. We do not yet know which of these genes help to produce such a strong urge to outbreed in chimpanzees, or what differences between those genes and ours might contribute to a less overwhelming urge in us. But while we wait for these genes to be discovered, we can in the meantime detect some other important genetic differences that can tell us a great deal about our different evolutionary histories.

For some years many groups of scientists around the world have been examining in detail the variation in a small piece of mitochondrial DNA that is shared by humans and apes. (It is the same piece, by the way, that was recovered from that Neandertal who lived long ago in the Neander valley.) The pattern of variation in this fragment provides, among other things, clues to the last time when our own species may have had a really serious brush with extinction.

We can glimpse the history of a human or ape population by determining how many of these genetic differences have accumulated in the population. If a great deal of such variation has accumulated, the population must have been quite large far back into the past—because mitochondrial variants that appear by mutation can accumulate relatively undisturbed in a large population. Fewer of these variants will be lost by chance, as would have happened if the population were small.

Small population size, by contrast, is the enemy of genetic variation. It is striking that chimpanzees have gone to such extraordinary lengths to increase their population size through outbreeding. Nor are chimpanzees alone in this respect. Other species of animals and plants have also acquired a great variety of outbreeding mechanisms. Indeed, the prevalence of outbreeding in the natural world has led many evolutionists to suspect that genetic variation must in itself be valuable.

If a population is large and yet has little variation, the most reasonable explanation is that something happened in the past to reduce the population's size. Such an event is what geneticists call a *size bottleneck*.

The term is quite descriptive. A population that has been large for some time may occasionally undergo a temporary reduction in numbers—perhaps as a result of ecological change, disease, or the migration of a small part of the population into a new territory. If such a reduction is then followed by a recovery period, during which the population once again achieves large numbers, then no visible signs of the bottleneck would remain. But the effects of the bottleneck would still be detectable at the gene level, for the population would have less genetic variation than its current large size would lead us to expect. We can infer that such bottleneck events must have happened to animals as diverse as cheetahs and elephant seals, because these species currently have unusually small amounts of genetic variation.

Just as an X-ray of the body can reveal an old lesion that has healed without an external trace, a look at the genetic diversity of a population can speak volumes about what has happened to that population in the past. Examining DNA variation can be a kind of genetic X-ray, showing signs of such traumatic events in a population.

Figure 9–1 shows the family tree that Pascal Gagneux and I constructed, using almost twelve hundred different mitochondrial sequences from humans, chimpanzees, bonobos, and gorillas. Like an X-ray, it presents in vivid form the remarkable differences between the amounts of diversity in humans and in our close relatives. Some of these populations, including ours, show evidence of fairly recent "lesions," and some do not.

The DNA samples were gathered from lowland and mountain gorillas by Karen Garner and Oliver Ryder of the San Diego Zoo; from East and Central African chimpanzees by Phillip Morin of the Sequana Corporation and Tony Goldberg of Harvard University; from western African chimpanzees by Pascal Gagneux and Rosalind Alp; and from the isolated Central African bonobos (pygmy chimpanzees)

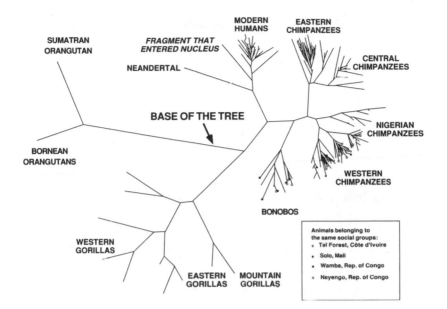

Figure 9–1. A family tree of 1,158 related pieces of DNA, taken from the mitochondrial chromosomes of our own species and our African and Asian ape relatives. The base of the tree is shown with an arrow. I have done a little editing of the data, to get rid of some noise.

by Ulrike Gerloff and her colleagues of the University of Munich. The human samples were gathered by many different scientists, most notably Linda Vigilant of Pennsylvania State University.

While they were collecting the data, the individual researchers had tended to concentrate on their particular species. But as Gagneux and I built the tree from all the sequences, some remarkable patterns emerged. We were able to see our own evolution very clearly in the wider context of the evolution of our close relatives.

We constructed the tree using an ingenious computer method invented by Naruya Saitou and Masatoshi Nei. In the tree, distantly related sequences are separated by long branches, while closely related ones form little twiglets at the end of a branch.

We can think of this tree as like a tree in a forest, with its trunk rooted firmly in the past and its branches pointed skyward toward the present. The tip of each twiglet represents an individual who is alive at the present time. The tree in the figure may be somewhat confusing, for its resemblance to a real tree seems to be minimal. But the branches were too numerous for us to draw them in a typical treelike shape viewed from the side. As in a real tree, some of the branches would have obscured, or been obscured by, other branches.

To avoid this problem, we have drawn the tree as a kind of star, with the branches radiating out from a common point (marked with an arrow). This shape makes sense if you imagine yourself suspended in space over the tree, looking down at it. From such a vantage point, all the branches radiating out from the base are clearly distinguishable.

The base of the tree clearly divides the orangutans from all the other groups. The branches of the rest of the species all tend to cluster on the other side of the tree, away from the orangutans. This is what one might expect, because by many measures orangutans are only distantly related to the other species on the tree. In effect, the orangutan branches have grown away from one side of the base, while those of the various other species have grown away in the other direction.

SQUEEZING THROUGH THE BOTTLENECK

At first the tree seems very complicated, but certain features emerge. Let us begin with our own species.

The human branch of this tree is straight and, except for the Neandertal sequence and another sequence that escaped into the nucleus from the mitochondrial chromosome a million years ago, is unadorned with twigs until the very end. This small and dense bundle of twigs, made up of more than eight hundred different sequences, contains representatives from many different human tribes and races, including a large number of African tribes, as well as Asians, Europeans, American Indians, Australian Aborigines, and New Guinea tribespeople. Although not all human groups are represented in the data, enough are present to give a good idea of the extent of human variation. It is unlikely that this tuft will change much, aside from getting even denser as more human groups are added in the future.

This tiny, dense human tuft stands in vivid contrast to the long, sprawling branches that lead to the gorillas. Soon after leaving the base, the main gorilla branch fans out immediately into a number of very long subbranches. Here and there, at the tips of the subbranches, are little tufts of twigs. The gorilla sequences within each of these tufts resemble each other very closely, but the sequences on the tips of the different long subbranches are only very distantly related. So far as their mitochondrial DNA is concerned, gorillas are enormously more diverse than humans.

Gorillas are found over a relatively restricted geographic region. Indeed, the very dangerous population decline that they are currently suffering is made even more serious by the fact that they were never common in the first place. Yet their genetic diversity may be even greater than we see here. The sequences shown in the figure are—because of the tremendous logistical problems involved in their collection—a rather poor sampling even of the few remaining wild gorilla populations.

Our guess is that gorillas have lived in isolated bands for very long periods of time, and with little genetic exchange among them. They really have flirted with extinction and have done so for millions of years. Like the chimpanzees, they have undergone little evolutionary change for a very long time. Since the gorillas at the ends of these long branches are similar to each other, most of the evolution that resulted in their gorilla-ness must have occurred during the time that passed between the base of the tree and the beginning of the divergence of the gorilla subbranches. These long subbranches tell us that gorillas have not done much evolving for at least the last four or five million years.

The bonobos are another isolated and highly threatened group of animals with a highly restricted geographic range, inhabiting only the forests of Central Africa south of the Zaire River. For decades these chimplike animals were occasionally captured and brought to zoos in Europe and America. There they were simply put in cages and allowed to fraternize with chimpanzees that had come from other parts of Africa. By the 1920s, however, careful anatomical and behavioral studies had shown that they really were different from chimps. Now they have been put into a separate species, *Pan paniscus*, which is clearly different from the true chimpanzees, *Pan troglodytes*. Their common name, *bonobo*, is an apparently African word of unknown origin.

Frans de Waal of the Yerkes Primate Center in Georgia, who has studied the behavior of bonobos intensively, recently summarized studies showing that they are significantly milder-mannered than *Pan troglodytes*. Some of them also seem more adept at learning words than the average *troglodytes*, though they may not be quite as good at tool use. Their sexual behavior is complex as well—they often use a wide variety of sexual encounters to smooth social interactions, and unlike *troglodytes* they often couple sexually face to face.

In spite of all these differences, however, bonobos are not some kind of chimpanzee–human hybrid (as has actually been suggested). They are really very similar to chimpanzees. This is obvious from their general appearance and behavior, and it has been confirmed from studies of the sequences of their DNA molecules and of the shape and number of their chromosomes. So few of them remain now, from such a restricted region, that their branch may be the only survivor of several bonobo branches that formerly existed. In the unlikely event that they are left alone in their forest and are allowed to migrate to other regions, they may regain that lost genetic diversity, though they will probably continue to be bonobos or something very like them.

The chimpanzees in the tree show yet another kind of pattern. They are currently far more widely distributed geographically than the gorillas or bonobos, and are found across a great swath of Western, Central, and East Africa. Although the sprawling chimpanzee branches in the tree reflect this dispersal, their branches are no more diverged than those of the gorillas. The mitochondrial sequences of the eastern, central, and western chimpanzees are clearly separated from each other; the western chimpanzees themselves are divided into complex subbranches. The ends of some of the subbranches have fairly dense tufts of twiglets. One particularly dense tuft, which looks very much like the one on the end of the human branch, is seen in the eastern chimpanzees. Most of the chimpanzees that contributed their DNA to this tuft live in Tanzania and include those that Jane Goodall has studied for three decades in the Gombe Reserve.

These various groups of chimpanzees are quite similar in appearance and behavior. They have all (after many taxonomic misadventures) now been put into the same species, *Pan troglodytes*. Although *P. troglodytes* has in turn been divided into three different subspecies, these groups are based primarily on geographic location and not on any consistent physically distinguishing characteristics. If we were to put an assortment of these chimpanzees into a police lineup, as we imagined ourselves doing in the Introduction with the orangutans, not even experts would be able to determine their origin with any consistency. To tell them apart, it is necessary to examine their genes.

The chimpanzee–bonobo split in the tree probably happened about three million years ago, and the various major branches of the chimpanzees parted from each other at least two million years ago. Thus the eastern, central, and western chimpanzees went their separate ways not very long after the chimpanzee–bonobo split. It astonished us, when we first saw this huge sprawl, to realize that these very different DNA sequences belong to members of a single species.

The little hollow squares in the figure, scattered across the western chimpanzee subbranches, represent members of the little group of chimpanzees that Pascal Gagneux studied in the Ivory Coast forest. Even this tiny group contains a huge diversity of different mitochondrial types, which have been separated for well over a million years. The mitochondrial chromosomes have recently been brought together in this single band in the forest, adding to its genetic diversity, by the laudable and untiring efforts of the females who migrated into the group. Even within Gagneux's little group, the genetic differences from chimpanzee to chimpanzee can be far greater than those that separate humans living in the farthest-flung parts of the old world.

The tree illustrates the enormous power of the outbreeding behavior practiced by female chimpanzees. Through their labors they have managed to bring together, in one tiny group in one part of a forest, mitochondrial lineages that long predate the appearance of *Homo sapiens*. Such dramatic genetic mixing has not happened with the gorillas, and it has certainly not happened with humans.

The most likely explanation, though not the only one, is that our own ancestors were squeezed through a size bottleneck, while the chimpanzees as an entire species were not. When this bottleneck took place, and whether there might have been a series of bottlenecks, remains to be discovered. But that something unusual happened to our ancestors seems unquestionable.

We are not unique in experiencing a bottleneck. Look again at that little tuft of sequences representing the eastern chimpanzees: it is almost indistinguishable from our own tuft. These eastern chimpanzees, too, may have gone through a size bottleneck.

Our best guess is that, some hundreds of thousands of years ago, a few chimpanzees managed to travel from Central Africa across the western part of the Great Rift Valley, possibly almost impassable in those days, into what is now Tanzania. There they quickly established a new population. As they passed through this size bottleneck, they lost genetic variation, but as the new population became established, they slowly gained it back. Nothing much else happened, however. The scenery in their new territory, with its vast valleys and mighty volcanoes, was certainly spectacular, but chimpanzees tend to pay little attention to scenery. For them, little had changed. Chimpanzees they were, and chimpanzees they remained.

Our little tuft of twigs, in contrast, stands alone. The subbranches that led to the various Australopithecines, to *Homo habilis* and *Homo erectus*, and most recently to the Neandertals, have all disappeared, and at this writing only one Neandertal sequence and an ancient sequence that managed to escape into the nucleus are left. If we did go through a size bottleneck, then our lineage managed to survive when all those others did not. And unlike the various chimpanzee subbranches that have lasted down quite nicely to the present time, the hominids at the tips of those lost subbranches were clearly different from us. Unlike the chimpanzee branches, our lineage has undergone enormous evolutionary changes and many of our close evolutionary companions did not make it.

SWEEPING CHANGE

Although our ancestors probably experienced a size bottleneck, we cannot now tell its cause. It may have been some chance event, perhaps a migration into a new territory, as with the East African chimpanzees; or it may have reflected some desperate time in our past, when a few of our ancestors were forced to adapt to new environmental circumstances. Such rapid adaptation would have drawn on the genetic variation present in the population in those days and could have led to huge changes in our gene pool.

A possible consequence of such an environmental change would be what population geneticists call a selective sweep. In such a sweep an advantageous allele spreads rapidly through a population and replaces the existing alleles.

Suppose that, hundreds of thousands of years ago, a group of our ancestors found themselves in a new environment that required rapid adaptation. Formerly rare alleles of many different genes would now be catapulted into prominence. The individuals who possessed these alleles would leave many progeny.

Over the next few generations these alleles would sweep through the population. If the sweep were rapid enough, even an evolutionarily irrelevant piece of DNA—for example, one of the population's mitochondrial chromosomes—might be carried along for the ride. It would replace the other mitochondrial chromosomes in the population, even though it had little, if anything, to do with adaptation to the new environment.

If the new environmental stress resulted in reduced population size, selective sweeps would become more probable. Formerly rare alleles can sweep more quickly if the population is small, and in order for a selective sweep to be really effective, speed is of the essence. If the sweep does not happen quickly, the genes being dragged along with the selected alleles will have time to lose their association with them. They will tend to be unmounted from their positions nearby on the chromosomes, through the processes of genetic recombination and assortment.

We do know one thing about any size bottleneck, with or without an accompanying selective sweep, that might have happened among our ancestors. It is difficult to see how it could have taken place if they had already dispersed themselves over a wide area. Suppose they had spread through Africa but had not yet fanned out across the Old World. Africa is a huge and complex place, with many different geographically dispersed habitats. Once hominids had spread through several of them, no single disaster or other event could have affected all of them simultaneously.

The bottleneck must have affected some hominid group that inhabited a single small region, almost certainly somewhere in Africa. Afterwards the survivors must have spread through Africa, displacing and perhaps driving to extinction all the other hominids that lived there. Then they broke free into Asia and Europe, perhaps in their bloodthirsty way to do the same thing there. The mitochondrial sequences show some signs of a second bottleneck that accompanied the migration from Africa into the rest of the Old World, but it is not as "clean"—there are genetic traces of subsequent migrations back and forth among African, European, and Asian populations.

ALL ABOUT EVE

The bottleneck question—whether one actually happened—is inextricably linked with a great scientific debate that is currently going on about human origins. This debate has to do with the age of a particular lady—something that is of course generally not discussed in polite society. Scientists, however, are not inhibited by such restraints and have spent years of intense argument over the age of the lady in question. She has become known as the mitochondrial Eve.

The Eve was first named a decade ago by Jim Wainscoat of Oxford University. He was careful to define who she was. Because mitochondrial chromosomes are passed down the female lineage without recombination, they must have had an ultimate source. That source was the mitochondrial Eve. She carried a particular mitochondrial chromosome from which all the mitochondrial chromosomes carried by humans living at the present time are descended.

But she was not—repeat, not—the equivalent of the Eve in the Bible, a woman from whom we are all descended! I emphasize this point because, while Wainscoat was careful not to make this mistake, both scientists and science writers in the decade since have repeatedly stated that the mitochondrial Eve, like the biblical Eve, was our sole ancestress. Wainscoat realized that the mitochondrial Eve had companions, and that both she and her companions bequeathed many other genes to us. Those genes were on nuclear chromosomes and not on the small mitochondrial chromosome. Although she was indeed the ultimate origin of all the mitochondrial chromosomes in our species, she played only a minor role in passing down the overwhelming majority of our genes.

The debate about where and when Eve lived is ongoing and inconclusive, but we can say some things here.

We know, from Krings's Neandertal DNA sequence, that Eve lived sometime after the split between humans and Neandertals. Further, if a bottleneck or sweep took place, then it must have done so at or after the time when the Eve lived.

We also know that Eve was probably not particularly remarkable. The fact that she carried the ancestor of all our mitochondrial chromosomes was after all a statistical accident, because of the way these chromosomes are inherited. She might have lived at some interesting time in our past, such as a period during which a bottleneck took place, but the chances are slim. After all, she was just one of our ancestors, most of whom lived in times that, although fraught with many dangers, were (luckily for them) not unusually so!

Mathematical theory and computer simulations show that the Eve could easily have lived long before a size bottleneck. She might, for example, have lived in a small tribe that managed to retain, for thousands of generations after she died, a few different mitochondrial chromosomes. Then, after hundreds of thousands of years, all but the descendants of her chromosome were lost, and the tribe fanned out in a sudden population expansion that permitted more chromosomal types to accumulate. There is no way to distinguish this possibility from the possibility that the Eve lived near the time of the bottleneck.

Thus the time when Eve herself lived cannot be settled at the moment. Most calculations have placed her at a date of between one and two hundred thousand years, at about the time of the appearance of early modern humans in southern Africa and the Middle East. But even a small change in the assumptions underlying these calculations will produce a much older date. By making some changes that I thought were justified, I recently concluded that she might have lived half a million or more years ago, and a few other scientists have arrived at similar figures.

All these are estimates of the median age of Eve, the most probable age. To add even more to the uncertainty, the errors on these estimates, particularly at the upper boundary, are so immense that we cannot be confident about the figure to within a few hundred thousand years! Eve, like any proper lady, remains coy about her age.

Nonetheless, combining evidence from molecules with evidence from bones makes things a bit clearer. The molecules tell us that we have far less mitochondrial DNA variation than chimpanzees do. So far, they are silent about when, how, and why we lost any earlier variation. The bones, on the other hand, tell us that many dramatic things really did happen in the course of our evolution, and we must always keep them in mind.

It would certainly be a mistake to attribute too much of our evolution to bottlenecks. Modern humans differ so much from our ancestors in so many ways that our species must have been shaped by far more than a single event such as a size bottleneck, with or without a selective sweep.

Popular science books tend to attribute the entire process of evolution to a single phenomenon—waves of cosmic radiation or giant asteroids from outer space, for example. But such an approach tells only part of the story. Although bottlenecks, and whatever caused them, may easily have been important, the genes that managed to squeeze through them were even more so. It is these genes that made us what we are and that set us on our new evolutionary path, so different from the evolutionary stasis of chimpanzees, gorillas, and orangutans.

AN EVOLUTIONARY TIME-WARP

We can tell many other things from these little bits of mitochondrial DNA. One of them gives at least a partial answer to the question: were the chimpanzees really chimpanzees all the way back? Have chimpanzees really changed hardly at all, even though during all those millions of years our own ancestors were undergoing one dramatic transformation after another? The fact that our own remote ancestors were very chimpanzeelike does not guarantee that the ancestors of chimpanzees were chimpanzeelike as well—they might have been quite different. Remember that we have no fossil evidence about those ancestors.

The bonobos provide us with the evolutionary yardstick that we need to answer this question. The common ancestor of the chimpanzees and bonobos must have closely resembled both species. If we knew when that common ancestor lived, then we could determine the minimum amount of time during which chimpanzees have been chimpanzees rather than something else. If the two species split recently, then we would have to guess what their common ancestors might have looked like during the millions of years before the split took place. But if they split much longer ago, closer to the human–chimpanzee divergence, then both the bonobo and common chimpanzee lineages must have been chimpanzeelike for all that time.

We can get a rough idea of the date of this common ancestor by comparing two numbers. The first is the amount of DNA divergence that has accumulated between the chimpanzees and bonobos, and the second is the amount that has accumulated between chimpanzees (or bonobos) and humans. After some statistical correction the ratio of these numbers shows that the chimp–bonobo split occurred about sixty percent of the way back to the human–chimpanzee split. This means that they probably diverged about three million years ago.

During those three million years many different evolutionary events must have taken place in these diverging chimpanzee and bonobo lineages. Yet none of them were so dramatic as to efface the essential chimpanzee-ness of either lineage. Indeed, as I mentioned earlier, before and even after the differences between chimpanzees and bonobos became known, they had been mixed together in zoo cages. Some of them readily mated with each other and produced hybrid offspring. The handful of these hybrids that we know about seem to be healthy, and there is no reason to suppose that they would be unable to have babies of their own. As with the hybrid orangs, however, it is unclear how well they might do if they were reintroduced into the wild, or whether subsequent generations will continue to be normal.

Although these different evolutionary lineages have now been brought together again in zoos, they pursued their separate courses for an even longer period than the Bornean and Sumatran orangutans. Indeed, at the time that the chimpanzee and bonobo lineages separated, our own African ancestors were all Australopithecines. Even *Homo habilis* had not yet appeared. It is remarkable that, even after all this time, gene flow can still take place between these species.

The ranges of chimpanzees and bonobos currently do not overlap; but what might happen if they were to meet in the wild? Would they mate, would they fight, or would they simply ignore one another? If it turns out that they do not mate in the wild, then they will have proceeded farther along the speciation process than the zoo hybridizations would indicate.

Let me emphasize once again that during that same three million years, our own ancestors changed dramatically. They acquired enormous brains, a truly upright posture, flexible and sensitive hands, the ability to construct elaborate cultures, the physical and mental equipment to be able to communicate through complex languages, and many other things.

In other words, since the time of the split between us and the chimpanzee–bonobo lineage, we have changed and they have not. The contrast is a vivid illustration of the uniqueness of our own evolutionary history.

THE TWO FACES OF SELECTION

As if these huge differences between ourselves and the chimpanzees were not enough, mitochondrial DNA provides still further clues to the uniqueness of our history. The clues, which seem at first to be paradoxical, arise from the nature of this region of the mitochondrial DNA itself.

The piece of DNA that Pascal Gagneux and I used to build the tree is not a gene in the usual sense. The part of the mitochondrial chromosome from which it is taken does not code for proteins. Instead, it is a necessary structural part of the chromosome, where the enzymes that duplicate the mitochondrial chromosome first attach themselves to the DNA. The reason that this particular piece of the chromosome has been used in evolutionary studies is a purely utilitarian one—it tends to accumulate mutations at a very high rate. Some of these mutations survive, and the resulting rapid evolution means that even closely related people or closely related apes tend to fall on separate little twiglets of the evolutionary tree.

The exact sequences of such pieces of DNA appear not to matter very much, since this type of DNA changes so quickly. But it does have a pleasing property that lends itself to mathematical investigation. Because the DNA is free to accumulate mutational changes, the rate at which it diverges from its relatives should be quite constant. Such a piece of DNA should act like a little molecular clock, ticking off the millennia in a regular fashion. So when we measure the amount of divergence that has accumulated between such sequences in different evolutionary lineages, we should be able to calculate the time at which the divergence between the lineages took place.

This supposed property is what allows us to measure the age of the mitochondrial Eve. But Pascal Gagneux, Tony Goldberg, and I found that while such clocklike behavior seems to be true for some kinds of DNA change, it is dramatically not true for one kind of change in one part of this mitochondrial region. There is statistical evidence that since the very beginning of the lineage that led to our species, right from the time when *Ardipithecus* lived in ancient Ethiopia, the kinds of changes that make substantial differences to the structure of this piece of DNA have been selected against, and selected against very strongly. Whatever its cause, this selection has not happened, or at least not as much, in our great ape relatives. Once again we are unique, but this time our uniqueness lies in our resistance to anything but small changes.

I have been emphasizing up to this point that our own evolution has been proceeding pell-mell, and that much more has happened to us than to our nearest relatives, the chimpanzees. But selection has two faces: it can accelerate some kinds of change and retard others. In the evolution of our bodies and minds, we have been revolutionaries. But in this bit of our mitochondrial DNA, it seems, we have been extreme conservatives.

Figure 9–2 shows this contrast clearly. This pair of trees, much simpler than the earlier tree, shows the number of changes in this piece of DNA that have happened in us and in our nearest living relatives. Because these trees are so simple, I have been able to make them look more like real trees, with our common ancestor at the base and with branches rising through time to the present.

The "transition tree" at the top shows changes of a type that have small effect on the DNA. These changes, technically known as transitions, occur when one DNA base has been substituted for another with similar chemical properties. Summing these changes gives the result that we, the bonobos, and the various branches of the chimpanzees are all about equally far from the gorillas. Since the time of that ancient split between the gorilla lineage and the rest of us, the transition clock has ticked along quite evenly.

This is emphatically not the case with the lower "transversion tree," which shows another type of change that has accumulated incrementally over the same time. These changes, called transversions, have a larger effect on the DNA. (The diagram to the right in the figure shows the difference between transitional and transversional changes.) In transversional changes, humans have lagged far behind. The lag has affected our entire branch, apparently right from the beginning.

This lag is continuing right down to the present. Some pairs of sequences taken from different people are separated by many transitional differences, but these same pairs show far fewer transversional differences than one would expect. At this very moment in our own species, transversional changes in this bit of DNA are being selected against.

The transversion slowdown seems confined to this one bit of DNA. The piece that lies right next to it does not show the effect— in this nearby region, transversions have accumulated as briskly down the human lineage as they have down all the others.

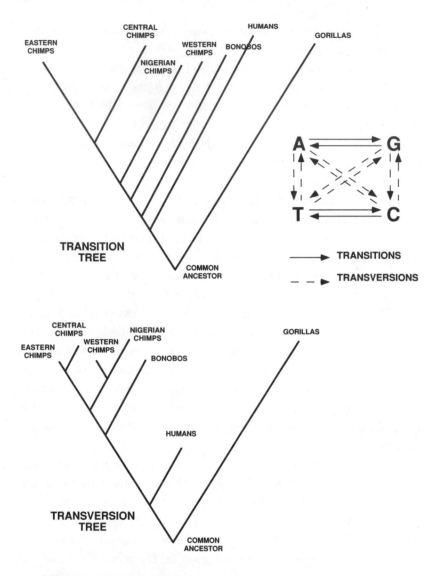

Figure 9–2. Trees showing how the transversion clock has slowed down for humans. The small diagram shows the difference between transitional and transversional changes in the DNA. The DNA bases A and G (adenine and guamine) are *purines*, larger than the *pyrimidines* T and C (thymine and cytosine). A change from a large base to a small one or vice versa appears to have a greater impact on the function of this DNA region than a change from large to large or from small to small.

What can be going on? Our current guess is that most transversion changes may damage the function of this important bit of DNA slightly and that this has been enough to select against these mutations. Many changes of the transversion type might interfere with the binding of enzymes to the DNA. This hypothesis is testable these days, and we are currently exploring ways to do so at the molecular level.

Mitochondria are thus very important to our survival, and not just any old mitochondria: human beings need tip-top mitochondria. The requirements for apes, it would seem, are a bit less stringent.

Throughout our ancestry, from *Ardipithecus* on, it seems to have been essential for our mitochondria to function at their very best. Conservative selection, which tends to weed out any changes that would do even the slightest harm, has been necessary to keep them at a peak level of functionality.

Several genetic diseases have now been traced to specific mutations in the mitochondrial chromosome. One of them is known by the mouth-filling name of Leber's hereditary optic neuropathy. As people with this condition grow older, a blind spot appears in their fovea, the highly acute central region of the visual field. The nerve cells in this region, which are among the most metabolically active in the entire body, begin to die because their mitochondria are not able to function at the extreme levels that are required. Mitochondria that have the same defect can continue to play their less demanding roles in other cells, including nerve cells, with no difficulty. If a fovea is to function properly, however, tip-top mitochondria appear to be needed. And if the requirement is to evolve big-brained, clever people like us, the demands on the mitochondria are likely to be stringent as well.

Why has it been more necessary for human beings than for chimpanzees or bonobos to preserve every ounce of the biochemical powers of their mitochondria? The big difference between us and the apes is the sheer size of our brains. Our brains have insatiable appetites—sixty percent of the energy consumed by a young child is funneled into brain metabolism and growth. Highly conservative selection, weeding out mitochondria with the slightest defect, appears to have been essential in order to ensure a plentiful energy supply to power revolutionary changes in our bodies and our brains.

Recently Jean-Jacques Hublin and his collaborators in Paris have obtained more information about the intensity of the selection that has shaped our brains. Their evidence comes from recent and fossil human skulls and from the skulls of chimpanzees. Using CAT scans, they measured the size of the sinuses through which the carotid arteries supply blood to the brain. The relationship between blood flow and brain size is a very direct one in humans and our ancestors; it is much less obvious in chimpanzees. If only the size of the sinuses is known, it is possible to predict very accurately the size of the brain in humans, but it is far less possible to do so in chimpanzees.

Human beings seem to be pushing the evolutionary envelope: if there is more blood flow, we can develop bigger brains. In chimpanzees, even if the blood flow is substantial, a bigger brain does not necessarily result.

The various phenomena that we have talked about here are all connected. The preservation of the base sequence of that small piece of mitochondrial DNA, and the strong relationship between blood flow and brain size, may be two aspects of the same thing. The overwhelming evolutionary imperative that has led to our large brains seems to have done much to shape both our genes and our bodies!

THE REST OF OUR GENES

This little bit of mitochondrial DNA has told us some amazing stories. It has illuminated our own history and shown us how different that history has been from that of our ape relatives. But this piece of DNA is very short, only four hundred or so bases long, which means that it makes up a mere one-hundred-thousandth of one percent of all the DNA in our cells. And as we have seen, it is rather unusual.

What have the rest of our genes been up to? They must have equally fascinating stories to tell. We are now beginning, dimly, to see what those stories might be.

TEN

◈

Sticking Out Like Cyrano's Nose

My nose is huge! . . . [L]et me inform you that I am
proud of such an appendage, since a big nose is the
proper sign of a friendly, good, courteous, witty, liberal
and brave man, such as I am.

A REMARK OF CYRANO DE BERGERAC, EMBROIDERED ON BY

EDMOND ROSTAND (1897)

THE WORLD OF DNA

Any DNA molecule can be represented as a string of bases of four different types, arranged like letters in a sentence. Over time DNA molecules tend to diverge from each other, as has happened with the mitochondrial DNA of humans, chimpanzees, and gorillas. They will diverge because they can become different from each other in so many more ways than they can become similar again. This property allows evolutionary trees to be constructed from DNA sequences.

Consider any sentence, even a short one, such as:

To be or not to be, that is the question.

Suppose this sentence is copied repeatedly by illiterate scribes, who occasionally and at random substitute one letter for another. As time goes on, the sentences will diverge more and more from each other. I wrote a little computer program to act as such a clumsy scribe. After ten random mutations, the sentence looked like this:*

Td be vr nov to be, zhat im dhe quqstion.

After a hundred, it looked like this:

Fe st dp byy ar ta, tlxu xx nnj qmcpjztu.

Then I ran the program again, starting with a different random number. After ten mutations, I got:

Tt bl or not to bi, that it hhe question.

And after a hundred, I got:

Tw kl or gaa rp iz, awtp an asy uqadpwbi.

After only ten mutations in each line, it is still possible to trace out the familial resemblance. But after a hundred mutations, neither of the highly mutated sentences shows much resemblance either to each other or to the original sentence (though the word *or* was spared in the second computer run). Any further mutations would simply, in Churchill's phrase, make the rubble bounce. But I could have run the program any number of times and obtained a different nonsense sentence each time. DNA sequences are capable of diverging in many different directions.

An evolutionary tree built from DNA sequences has another pleasing feature. DNA molecules will continue to diverge even if the visible characters for which they code happen to be converging. Convergence of visible characters can often happen in the course of evolution. Consider the Ainu, an aboriginal group of people who are now reduced to living on the northern Japanese island of Hokkaido but who used to live throughout the Japanese archipelago. In appearance they differ greatly from the subsequent invaders of these islands—they have prominent facial features, an abundance of wavy brown hair, much body hair, and no epicanthic fold on their upper eyelid. Indeed, they are so Caucasian in appearance that anthropologists at first classified them in that group.

---

* I specified that if a mutation happened in a space between words, it left the space unchanged, to retain some recognizability in the sentence.

They are not Caucasian, however. Genetic studies show that the alleles of various genes that they carry are quite typical of the surrounding northern Asian populations. The Caucasian-like characteristics of their physical appearance seem to have arisen independently, though why they did so remains a mystery. (It should be pointed out, in order to avoid Eurocentrism, that the reason Caucasians themselves look the way they do also remains a mystery!)

This example illustrates vividly the dangers of making a tree using physical characteristics. If the characteristics have converged from independent origins, one can easily be fooled about their degree of relationship.

Because of DNA's property of divergence, evolutionary trees based on it tend to be more accurate, and more reflective of underlying processes, than traditional trees based on visible characters. Often it is easier to see the "true" evolutionary relationships among such species in a DNA tree than it is in a tree based on visible characters.

A typical gene tree for ourselves and our close relatives appears in Figure 10–1. It looks very much like the transition tree in Figure 9–2,

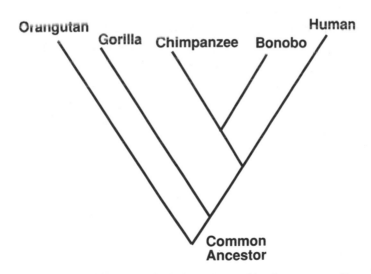

Figure 10–1. This tree was built from the combined sequences of four genes shared by humans and the great apes. The common ancestor occupies the bottom of the tree. It shows nothing unusual about humans—in this particular evolutionary sweepstakes, we have remained part of the pack.

in which the transitions seem to behave in clocklike fashion. There is no sign of anything unusual about the human branch; it is slightly longer than the others, but no more than one would expect by chance.

Surely, one would think, if human evolution really has taken off like a rocket, this fact should somehow be obvious in a DNA tree. Yet when a DNA tree is constructed, no matter which genes are used, the branch on which we are perched always blends nicely into the tree. Our branch is about as long as the branches of our nearest relatives. We really have to look very closely, as we did with the transversion tree in Chapter 9, to find any differences.

THE WORLD OF VISIBLE CHARACTERS

Once a month or so for the last two years, a group of scientists from the San Diego area have come together to discuss human evolution. This discussion group was the brainchild of my colleagues Ajit Varki and Rusty Gage. Members of the group hail from many parts of the huge biomedical research establishment that has grown up around the University of California, San Diego, campus, including the Salk Institute, the Scripps Research Institute, the Scripps Institution of Oceanography, and the San Diego Zoo. Our discussions have ranged widely, with intense debates about fossils, genes, primate behavior, and many other factors that have contributed to our evolution.

Already a number of collaborations are emerging from these discussions. One of them is a project that Ajit Varki and I have begun, to get a handle on a longstanding and extremely puzzling paradox about human evolution. Reduced to its essence, the paradox is this: The rates of evolution of our physical and mental attributes and of our genes simply do not seem to match. Our physical evolution seems to proceed much faster than our genetic evolution.

Varki and I realized that, unless we could explain this paradox, we could not explain human evolution.

In Chapter 9 I traced the story of how, from the time of *Ardipithecus* to the present, one very small but important piece of our DNA has slowed down in its evolution. But this slowing must surely be unusual. The many changes in our physical and mental capabilities during all this time strongly suggest that the rate of evolution of other pieces of our DNA must have sped up rather than slowed down.

Whatever genes control these characteristics must have changed at an accelerating pace in order to account for all the rapid evolution that is so plain in our fossil record.

Something else truly remarkable has happened to us as we diverged from the apes: Compared with those close relatives of ours, we can do so many different things! Alleles of various genes must have been selected for that contributed to that increase in the range of our capabilities. What might these genes be like, and indeed would we even recognize them if we came across them?

The heart of the paradox is this: When, as in Figure 10–1, we compare ourselves with our closest relatives at the DNA level, we simply cannot detect any sign of all this evolutionary ferment. We find no trace of this breakneck evolutionary pace in our genes—at least, not in the relatively small sample of genes that have so far been looked at in detail.

But in the visible world, the world of the phenotype, evidence for rapid change is abundant. Varki and I tried to build evolutionary trees that had nothing to do with DNA but rather were based on physical and mental resemblances. Trees of this kind hearken back to the early days of evolutionary studies, the dark ages when nothing was known about molecules. Darwin himself built such trees, and the relationships among living organisms that became apparent from them helped to convince him and other scientists of the reality of evolution.

The only difference between our trees and the ones that Darwin and his contemporaries constructed was that we built ours using up-to-date methods. These methods allowed us to see a very striking relationship indeed.

For years, Varki had been putting together a list of the many differences and similarities between modern humans and the great apes. The full list is enormous and highly elaborated. He found that the number of characteristics by which we differ markedly from our close relatives is astonishing.

He and I began our project by choosing random sets of the characters from his list and constructing a tree. Like a DNA tree, this tree was arranged so that closely related organisms tended to be grouped together.

To build this tree, we picked twenty-two physical and behavioral characters, including the ability to speak, the ability to recognize

oneself in a mirror, the ability to walk bipedally rather than quadru-
pedally, the degree of difference in body size between the sexes, the
difference in size between the two halves of the brain, the amount of
thumb mobility, the period over which the young are nursed, the
amount of skeletal muscle strength, the presence of breasts in a
nonlactating female, and a number of others. We assigned a number
to each characteristic on a scale of one to ten.

We found that while humans and apes sometimes had similar
numbers for a given characteristic, usually they were very different.
These differences make the astonishingly rapid pace of evolutionary
change in our species glaringly obvious. As you can see in Figure 10–2,
our own branch sticks out from among the branches of our close relatives like Cyrano's nose.

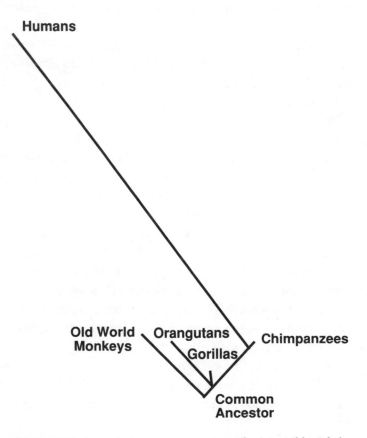

Figure 10–2. An evolutionary tree constructed using visible or behavioral characters. The base of the tree represents the common ancestor.

Note that the apes and monkeys differ among themselves in these characters. As one might expect, our closest relatives, the chimpanzees, are the nearest to us on the tree. But as we broke free of the pack, we have left even the chimpanzees far behind.

Sometimes, as in language ability, humans score high relative to apes and monkeys. Sometimes, as in skeletal muscle strength, we score low; the muscles of chimpanzees are three times as strong as ours. For most of these characters, however, we are clearly different from our close relatives, and when they are summed, the differences become enormous.

This tree is the mirror image of the DNA tree at the end of Chapter 9, in which humans lagged behind the apes in transversional DNA changes. A comparison of the two trees demonstrates very clearly the enormous disparity between the small amount of change that has happened to at least some bits of our DNA and the huge amount of change that has happened to us physically and mentally.

JUST HOW UNIQUE ARE WE?

Ajit Varki and I are hardly the first to have noticed that humans are so dramatically different from other animals, though so far as we know ours is the first attempt to put hard numbers on the differences. So great and obvious is this gap that for millennia people took for granted that our species resulted from some special creation event.

The special creation idea was sorely battered a century and a half ago, when Darwin used his theory of evolution to bridge the immense gulf between ourselves and the other animals. With one flash of insight, he showed that we really are part of the rest of the living world.

But old ideas die hard. Many people, including many scientists, felt strongly that Darwin's theory did not explain everything. He did not, it seemed to them, completely close the gap between ourselves and the rest of the animal world. Darwin's contemporary Alfred Russel Wallace, who had independently stumbled on the essential idea of natural selection, could not believe that human evolution was entirely the result of such an automatic process. In Wallace's eyes, the gap between ourselves and the other animals was still there, and it was as inexplicable as ever. He felt that some other, perhaps supernatural force must have been involved. Yet he could not explain the nature of that supernatural force, nor why we as a species were singled out for its benefits.*

In this book, as I hope you will have noticed, we nowhere invoke supernatural forces to explain human evolution. It can be explained by well-understood and perfectly ordinary evolutionary processes. What is unique is the speed and the power with which these processes have acted on us.

Long after Darwin, and after most scientists had abandoned special creation as an explanation, other aftereffects of the assumption of human uniqueness persisted. During the early part of this century paleontologists who attempted to reconstruct the history of our species were hampered because the fossil record they had access to was far more fragmentary than the one we know today. As a result they, too, were led astray by our obvious uniqueness.

The conventional view at the time, espoused most strongly by the great paleontologist Henry Fairfield Osborn, was that the human lineage and those of the great apes must have diverged very early. Osborn surmised that we probably originated in Asia, and that we and the apes must have embarked on separate evolutionary paths some thirty million years ago. This huge span of time was surely necessary in order to explain how we have become so different from our relatives.

The idea of an Asian origin was soon abandoned, but the idea of an early split was widely accepted by anthropologists right down to the 1960s, when it was overthrown with the advent of the science of molecular evolution. The chemists Linus Pauling and Emil Zuckerkandl of Stanford University were the first to suggest that molecules have evolved at a surprisingly constant rate and can be used as an evolutionary clock.

A molecular clock, it was soon realized, could be used to date the divergence between humans and chimpanzees. If the difference between the proteins of humans and chimpanzees was large compared to the divergence among mammals as a whole, then the split must have happened a long time ago. But if the difference was small, then humans and chimpanzees must have had a recent common ancestor.

---

* Wallace's idea was resurrected by Arthur C. Clarke and Stanley Kubrick in the film *2001*. The mysterious black obelisk that altered the course of human evolution in that film is nothing but Wallace's supernatural force, brought up to date and attributed to a galactic civilization rather than a deity.

Vincent Sarich and the late Allan Wilson of the University of California, Berkeley, were the first to carry out these measurements. Their data showed quite clearly that humans and chimpanzees have pursued separate evolutionary paths not for thirty million years but for a trifling five or six million. Their daring claim, at first resisted furiously by anthropologists, has now been proven correct by measurements on a wide variety of molecules in many different laboratories, and it has been confirmed, as we have seen, by new discoveries in the fossil record.

Once again, a worldview has been shattered, this time that of the anthropologists who had speculated about human origins. How much intellectual and scientific history is encapsulated in that little evolutionary tree of Figure 10–2! The immense physical and behavioral gap between ourselves and the apes seems to have led us badly astray at least twice.

Does the gap itself really exist? We are, after all, a terribly self-important species; perhaps we have simply exaggerated small differences in order to make ourselves seem unique. This possibility of bias, conscious or unconscious, poses a real problem for evolutionists. For just this reason many of them have never been satisfied with evolutionary trees of the kind that Ajit Varki and I constructed that are based on visible characters.

An enormous amount of subjectivity, after all, is involved in creating such trees. Why, for example, should we have picked one particular subset of visible characters out of the thousands by which humans differ from the apes? Did we load the dice too much by picking characters that are obviously different? Did we simply find what we expected to find?

We think not. Varki's list is made up of physical and behavioral characteristics by which we often, though not always, differ from the apes. In the majority of cases the variation from one ape species to another or from ape to monkey is dwarfed by the differences between these primates and humans. These differences are particularly striking for behavioral characteristics such as language and tool-making. Unfortunately these are just the characteristics that are most subjective and thus most prone to bias.

To get around this problem, we constructed another tree using only directly measurable physical characters and avoiding behavioral ones. Even after restricting ourselves in this fashion, however, our species still stuck out like that famous nose.

Bias is impossible to avoid completely, though I suspect that it is built into the way that any intelligent creature would look at itself and its near relatives. The composition of Varki's list was driven by the fact that humans happen to be available for comparison. I can illustrate this point with a thought experiment.

Suppose that our world is visited by a saucerful of utterly alien biologists from another planet. These aliens have a completely different body plan from our own—they resemble, say, intelligent insects. On their arrival they discover colonies of great apes and Old World monkeys, but for some reason they find no humans. (Highly unlikely, but since this is a thought experiment, I can postulate anything I please.)

The aliens study these primates by classifying them according to their visible or behavioral characteristics. As they do so, it will probably never occur to them to employ such features as bipedal gait, fine motor skills, or the ability to communicate through language. Such abilities will be only dimly or fleetingly apparent among these apes and monkeys. As a result, their list of characters will probably end up being much shorter, and far less varied, than Varki's list.

But if these aliens had been able to observe humans along with apes and monkeys, their list of characters would have been much longer. The reason is that, in the course of our own evolution, humans have taken to an extreme many of the slight and evanescent abilities exhibited by the other primates. At the end of such study, the aliens would almost certainly reach the same conclusions as ours. In any physical- and behavioral-character tree that they might build, the human branch would stick out dramatically.

BARKING UP THE WRONG EVOLUTIONARY TREE

The writer and cartoonist James Thurber loved to draw pictures of bloodhounds. His friend Hendrik van Loon, enraptured by the pictures, actually went out and purchased a bloodhound. But van Loon found the dog something of a disappointment. He later told Thurber, "That dog didn't care a damn about where I was. All he was interested in was how I got there."*

* "How to Name a Dog," *The Beast in Me and Other Animals* (New York: Harcourt, Brace, 1948).

Evolutionists who construct DNA trees are like van Loon's bloodhound. Such trees do indeed provide an overall picture of the shape of evolution and show us the proper evolutionary relationships among groups of species. But these trees cannot distinguish important changes from unimportant ones. The old-fashioned trees that are based on physical characters might be better for making such a distinction.

At the DNA level, as far as a tree-builder is concerned, all changes are given equal weight, even though most of them have little or no impact on the individuals carrying them. Unbeknownst to the tree-builder, only a tiny subset of these changes are the really important ones.

The consequences of tracking down this subset will be immense, for they are the very characters, at the DNA level, that contribute to the essence of humanness—or of chimpanzeeness. When we find them, we can measure accurately how these changes are continuing to accumulate among humans or chimpanzees at present.

When we reach that point we will not be confined to odd little bits of DNA like that mitochondrial sequence we spent so much time on in Chapter 9. We will be looking at the very stuff of our evolution itself. And we will finally be able to provide a firm answer to the central question posed by this book: Why are we evolving so quickly?

IMPORTANT GENETIC CHANGES

Sorting out the important from the unimportant changes in our DNA will soon be possible. The Human Genome Project, that vast scientific effort to sequence all our genes, is currently well on its way to realization. One of the many exciting prospects it opens up is that, over the next few decades, we will be able to compare all our DNA with that of chimpanzees and gorillas and thereby track down the many differences.

The substantial work that has been done so far has already revealed some surprising things about those differences. One is that humans are much closer to our near relatives than we might think, or than even sophisticated measurements tell us. It is often stated that we share 98 percent (or sometimes 99 percent) of our genes with chimpanzees. This statement is not right, and the error arises from confusing two rather different things, DNA sequences and genes.

The 98 percent figure really means that, when human DNA is matched with the corresponding stretch of DNA from a chimpanzee, we share 98 percent of the bases. But this is not the same as sharing 98 percent of our genes. Consider the gene at which sickle cell mutations arise, the beta hemoglobin gene. We have one of these genes, and so do chimpanzees. Even though the genes differ at a few bases, they are the same gene. Thus, by the criterion of whether both we and the chimpanzees have a beta hemoglobin gene, we are a hundred percent identical!

This is not surprising. Humans have been separated from chimpanzees for about five million years, but this period is less than 0.2 percent of the total span of 3.8 billion years that life has existed on the planet. By this measure, we should be more than 99.8 percent identical with chimpanzees—and indeed, so far as the number and types of our genes are concerned, we seem to be about this close.

The beta hemoglobin gene is just one of the hundred thousand genes we share with chimpanzees. In fact, despite some quite strenuous efforts, researchers have not been able to find any really "new" genes in humans—that is to say, any human genes that are completely different from those of chimpanzees. It appears that there has simply not been enough time, during our relatively brief separate evolutionary paths, for entirely new genes to evolve. We share not 98 percent but essentially a hundred percent of our genes with our close relatives.

To be sure, virtually all our genes are a little bit different from those of chimpanzees. In a few cases we have more copies of a gene than chimpanzees do, and there are a few cases in which chimpanzees have more. All our genes depart from those of chimpanzees by a few DNA bases here and there—a few letters in the sentences that spell out the genes' meanings. As a result, they are slightly different models of the same thing, rather like furniture that comes from different historical periods. A Louis XIV and a Louis XVI chair are both chairs, but they are distinctly different from each other. Yet most of these differences have little effect on their function—someone may perch with equal discomfort on both.

Like chairs from different historical periods, the same genes from different organisms also turn out to be quite equivalent. In some remarkable "gene swaps" made by molecular biologists over the past few years, genes taken from humans have been found to work very nicely in amphibians and even in fruit flies and bacteria.

But in order to work properly the genes must be brought under the control of the regulatory systems that are already present in those very different organisms. It is here that the differences between humans and chimpanzees chiefly reside.

In a royal court the regulation of chairs is at least as important as their function. The factotum who decided how many chairs should be placed in which rooms of Versailles had a great influence on the palace's livability. Similarly, the regulation of genes can be just as important as their function. Some regions of DNA that lie close to genes influence when and where in the body the genes are expressed. These regions have a great influence on the sort of organism that results.

A small change in such a regulatory region is quite enough to turn a gene on or off. More subtly, a change may delay or speed up the gene's timing during critical stages of development, or alter its expression in different tissues.

Many examples of such changes, some with huge effects, have now been found. In a fruit fly, for instance, the insertion of a small piece of DNA near one important set of genes is enough to give the fly an extra pair of wings. Humans all make large quantities of an enzyme called tyrosinase, which manufactures the dark pigment called melanin that colors our skin, our hair, and the irises of our eyes. When the tyrosinase enzyme itself is first manufactured in melanocyte cells, a regulatory region of the DNA decides whether it will be active or inactive. In some of us the enzyme is produced in its active form, so that we make large quantities of melanin. But somewhere in the ancestry of Europeans and Asians, mutations arose in this regulatory region that caused most of the enzyme to be made in an inactive form.

People who carry these mutations manufacture very little melanin. If the enzyme present in the cells of Europeans and Asians had been made in the active form, the result would be copious quantities of melanin, producing skins and hair as dark as those of Africans and Melanesians.

This small genetic abnormality, found in the "superior" white races, has had vast and terrible consequences for our species. Yet if such a change were to be examined solely at the DNA level, without reference to its effects on the body, it would be indistinguishable from the overwhelming number of other genetic changes that are scattered everywhere in our genes and that have far smaller effects or no effects at all. This is why evolutionary trees that are built using DNA sequences, like that in Figure 10–1, look so different from the physical character tree in Figure 10–2. The really important changes are certainly there, and most of them have important effects on gene regulation, but they are buried in the noise.

A small number of such highly significant regulatory bits of our genomes must have been changing at top speed over the last few million years. We suspect, without direct proof as yet, that fewer such changes have happened in the chimpanzee lineage. Thus, even though at first sight the evidence from DNA trees might seem to argue against it, we really must be evolutionary speed demons. The trick will be to track down the changes that demonstrate this.

The small subset of genes that are really responsible for our evolution will not continue to elude us for long. In Part III of the book we will look for clues to their nature, clues that we can glean from our remarkable behavioral diversity and our growing knowledge of brain function. We will see how changes in our environment can actually accelerate the evolution of our behavioral diversity. And we will explore scientists' first tentative attempts to track down these genetic changes. Finally, we will look at what all this growing understanding of our evolutionary heritage might mean for our long-term future.

PART III

Selection for Diversity

ELEVEN

◈

Going to Extremes

*At this campus, twelve Nobel Prize winners have
taught or studied from nine different countries. A half
century from now, when your own grandchildren are in
college, there will be no majority race in America. . . .
As you have shown us today, our diversity will enrich
our lives in nonmaterial ways, deepening our understanding of human nature and human differences,
making our communities more exciting, more enjoy-
able, more meaningful.*

PRESIDENT BILL CLINTON,
UC San Diego commencement address (June 14, 1997)

One Saturday in 1997 my wife and I visited the X-games in San Diego. X stands for "extreme," and the games certainly lived up to their
name. This twice-yearly event, started in 1994 by the cable television
channel ESPN, celebrates newly invented competitive events. They
involve brand-new skills, such as in-line skating and bungee-jumping,
that did not exist a few brief years ago.

The hastily erected X-games park was filled with thirty thousand bronzed and healthy young people, watching a range of contests
that their parents would have found unimaginable. The games fall into
eleven categories, made up of twenty-seven different disciplines. While
most of the 450 athletes in attendance were male, about twenty percent were female and this fraction is growing rapidly. So are the

rewards—top athletes in the most challenging X-game sports are now earning more money through product endorsements than their Olympic counterparts. The athletes themselves are an unusual mix and include a number of professionals such as doctors and lawyers.

ESPN has managed to change the habits of the American television-watching public. When X-games are on television, they sometimes outdraw more conventional sports like football and baseball.

The most extreme of the extreme was the snowboarding competition. Although snowboarding is now an Olympic sport, these summer games took place before the Olympian transformation. Here the snowboarders used a hundred-foot tower, built approximately in the shape of a squat, wide ski-jump. The tower was festooned with cables along which the automated ESPN cameras could scoot, allowing them to follow almost the same trajectories as the athletes and catch every bit of the action.

That morning, in cheeky defiance of nature, workers had coated the surface of the plastic-covered jump with a layer of snow. By the time the event started, the snow was already melting briskly in the southern California sun. Between jumps, game attendants with shovels crawled here and there over the slope, trying to break up the worst of the icy patches.

Balanced on snowboards, which are essentially oversize skateboards without the wheels, jumpers hurtled down the upper slope. Sometimes they would use the board as a snowplow to brake their descent; sometimes they turned it parallel to the slope in order to hurtle even more rapidly toward the jumping-off point. Once in the air, they showed how far they had diverged from the staid ski-jumpers who formed part of the distant ancestry of this fearsome sport. They began to somersault, or to whirl around like tops. Then they divided into two groups. The first group met the ground in a less-than-graceful collision, cartwheeling to an eventual halt in a confusion of arms and legs. The second group, as casual as could be, finished their last spin or somersault exactly on time, landed with their snowboards in exactly the right position, and coasted to an elegant stop over the bumpy remnants of the snow.

The reader will not be surprised to learn that, confronted with a spectacle such as this, I immediately wondered about its evolutionary implications for our species. They are, I hope you will agree, profound.

SELECTION FOR DIVERSE CAPABILITIES

Ecologists divide species of animals and plants into specialists and generalists. Specialists, like the gorillas we examined in Chapter 8, are confined to narrow ecological niches. It is difficult to imagine that, if they were accidentally to be transported to another continent or even to another part of Africa, they would multiply uncontrollably and produce a plague of gorillas. But many other animals and plants—generalists—have done just that: starlings and house sparrows imported from Europe to America, rabbits brought from Europe to Australia, kudzu that has spread from Japan to remote parts of the world such as the Galapagos archipelago. Humans are perhaps the most successful generalists of all, and the world is currently suffering from a plague of humans.

Part of our success is due to our ability to modify our behaviors in the face of new environmental challenges, and to build on a knowledge base in order to do so. Snowboarding is a madly logical blending of skateboarding and ski-jumping, but the idea for it would have come only to somebody who was familiar with both sports. Such a sport would never have occurred to a Cro-Magnon, even though, judging by brain size, he or she had essentially the same mental capacities as a modern human.

In range and variety, the basic skills needed to survive in the Cro-Magnon world were not much different from those needed to survive in the modern world. It is a challenge to make useful tools from stone and bone. Hunting skills are complex and subtle, as are the abilities and knowledge that one needs to gather wild plants for food and medicine. The skills we currently need, while different, may actually be less difficult than those required of a Cro-Magnon. Being able to read at a minimal level, to do one's income taxes, and to drive a car are probably less demanding than being able to track and kill a deer in a winter forest.

But our world differs dramatically from the Cro-Magnon world in one important aspect: its sheer bewildering variety. Moving beyond the realm of what we *must* do to the realm of what we *can* do, we are confronted with choices and opportunities that grow with every passing year. And yet, no matter how outré the challenges may be, some of us will rise to meet them.

If a group of Cro-Magnon babies were suddenly transported to the present time, they could learn how to take advantage of many of the opportunities offered by our complex world. They would probably not exhibit the full range of possible behaviors—the chances are slight that such a small group could produce concert musicians, top-flight engineers, brilliant mathematicians, and expert snowboarders. But they might easily produce a star in one of these categories, and many of them would certainly prove moderately skilled in others.

All evolutionists would agree with this scenario, but they have always been puzzled by how Cro-Magnons—or indeed any primitive people—might have acquired such capabilities if their ancestors had never been exposed to such challenges. How could the genes for these capabilities have been selected for before the selective pressures existed?

The solution to this problem will give us important clues to how our brains have evolved and are continuing to evolve. But as yet we have a limited understanding about how our brains function. Some distressingly simple-minded explanations have been proposed: geneticists, for example, have attempted to build models of how our genetics might have been influenced by our culture and vice versa. They begin by postulating a straightforward one-to-one connection between genes and behaviors.

One of the most detailed of such models was made by Charles Lumsden and E. O. Wilson in their 1981 book *Genes, Minds, and Culture*. They suggested that particular features of culture, features that they called *culturgens*, might increase the survival of those members of the population who carry particular allelic forms of genes.*

Simple-minded though this approach is, on its face it seems not entirely unreasonable. The model is, after all, based on well-established evolutionary theory and on much observational data from animals and plants. Often there is a straightforward connection between a gene and its environment. We saw some vivid examples of this in Chapter 4 when we looked at the selective response of humans to malaria, which resulted in rapid selection for alleles such as sickle cell that confer at least some resistance to the disease.

Another clear example is one that we all remember from our high school days, the story of the moth *Biston betularia* and how it changed

---

* They were remarkably coy about defining culturgens, giving only an extremely general definition and virtually no examples beyond such vague things as "avoidance of incest."

as a result of the English industrial revolution. Effluent from factories killed the lichens that covered the tree trunks on which the moths habitually rested. The bare trunks were darkened still further with a coating of soot. Within a few years light-colored moths were replaced by dark-colored ones. But the environmental change did not cause the genetic change. Even in the days before industrial pollution, moths that were dark-colored had always been arising by mutation. The difference was that these dark moths had suddenly become advantageous.

If evolution worked entirely in ways similar to this moth example, then we might well suppose that particular forms of genes could be selected for by particular features of our culture as they were invented. But the moth example actually illustrates only one very simple kind of evolution. We have already explored some much more complicated evolutionary stories, in which the connection between the selective pressure and the result is nowhere near as obvious. Evolution works on the brain in ways that are very complicated indeed.

Geneticists wedded to the idea that each gene has its function tend to have a hard time coming to grips with this, but ordinary genetic rules do not apply in the world of the mind. For one thing, few if any genes are directly connected to culturgens. Specific genes for specific behaviors—like playing the piano, or playing the stock market—do not exist.

While this observation may seem trivial, it has not prevented geneticists from continuing to search for such genes. At various times over the last few years, groups of scientists have announced the discovery of genes that are claimed to explain a great variety of behaviors. Lately, the triumph of genetics spearheaded by the Human Genome Project has increased the number of such claims. But most of these claims, on closer examination, have melted away like the snow at the X-games. When we examine the actual details of the searches, we can begin to see why this has happened. And, as we proceed, we will begin to uncover clues to why this is so, and why the evolution of the brain has been so unusual.

HITTING THE BRAIN WITH A HAMMER

An obvious place to look for genes that influence behaviors is in people who are mentally retarded, since they show a loss of abilities and greatly altered behaviors. One might think that if we could pin down which genes cause retardation, we would know the genes for these abilities and behaviors.

Many people incarcerated in homes for the mentally retarded are there because of genetic damage. In most cases the damage is localized to a single gene, or a small region of a chromosome carrying a few adjacent genes, or a single chromosome. It is easy for genetic recombination to get rid of the damaged gene or chromosome—about half of all children born to people who are severely mentally retarded are of perfectly normal intelligence. Because many of these genetically normal children have extremely difficult childhoods, this fraction will certainly turn out to be higher as social workers carry out more early interventions. If retardation were due to some generalized damage that spreads throughout the chromosomes, then all or most of these children would be abnormal.

In general, the causes of retardation are not problems with the genes that control specific brain capabilities. Instead, the damage is caused by genes that affect the overall development of the brain or its overall function.

The common genetic causes of really severe brain malfunction can be counted on the fingers of one hand, although any one cause may claim hundreds of thousands of victims. For example, among all cases of severe mental retardation, regardless of whether they are caused by genes or by some environmental or developmental accident, fully a third are in fact genetic and can be traced to one particular genetic abnormality.

This abnormality is Down's syndrome. It results when, by accident, a fertilized egg ends up with an extra chromosome 21, so that each cell carries three instead of two copies of all the genes on that chromosome. Even though all these genes are likely to be perfectly functional, too many copies of them can be profoundly damaging.

Down's syndrome affects development in many ways. At least some of the associated mental retardation has been traced not to an abnormality of the brain itself but instead to a tendency for the pulmonary arteries, which supply blood to the lungs, to become blocked by excess tissue growth. Unimpeded blood flow is crucial to a developing fetus—so much so that a high rate of blood flow has been selected for in Tibetan mothers. The brain of a Down's baby is affected by this excess tissue growth taking place elsewhere in its body.

Although the brain of a Down's syndrome child is potentially normal, it cannot reach its potential because of a genetic disturbance that has nothing whatsoever to do with the brain's function per se. Later in life, other accumulating abnormalities, some with a striking resemblance to Alzheimer's disease, cause further damage, but the early arterial blockage triggered by the genetic abnormality helps to set the pattern of mental retardation.

A substantial number of non-Down's cases of mental retardation are due to a genetic condition called fragile X syndrome, in which a mutation causes an abnormal lengthening of a part of the DNA itself. This first mutation increases the likelihood of subsequent mutations, so that the lengthening can actually grow worse in succeeding generations. The syndrome becomes progressively more severe as it passes down through a family.

Lengthening this DNA turns off a nearby gene that seems to be an important part of the pathway by which information passes from the cell's nucleus to the rest of the cell. This obstruction appears to damage the metabolically active cells of the brain more than it damages the cells of the rest of the body. Unlike Down's, the abnormality does lie in the brain itself, but it is a generalized abnormality that causes generalized brain damage. In effect, the brain has been hit with a hammer.

Molecular biologists are now probing other less common genetic defects that lead to mental retardation. One of the most remarkable of these is Williams syndrome, a rare condition that happens about once in every twenty thousand births. It produces a striking set of symptoms.

People with Williams syndrome are frail and suffer from heart problems. Their faces are thin and their chins are pointed, giving them an appearance described as "pixielike." The irises of their eyes show a lacy texture. And they suffer from varying degrees of mental retardation, from slight to profound. Yet their retardation is odd and strangely selective.

Victims of Williams syndrome are cheerful, outgoing, and gregarious and often have remarkable musical talent. They love to tell complicated and highly imaginative stories and use an extensive spoken vocabulary. A Williams child might, when asked to name animals, come up with an amazing collection of real, mythical, and extinct beasts ranging from apatosaurs through unicorns to zebras.

Many Williams children imagine that they might become writers someday, even though most of them can only read or write to a limited degree. All of them show another strange deficit—they are unable to put together a pattern out of its component pieces. A drawing of a bicycle made by a child with Williams will consist of highly distorted bits of bicycle joined together in a seemingly random fashion.

This pattern, in which some mental abilities are apparently spared while others are damaged, has led to tremendous interest in the syndrome. Is Williams, which is clearly genetic, caused by damage to genes that code for some brain functions but have little influence on others? Are there, in short, genes for these specific behaviors?

A large group of researchers at the University of Utah and five other institutions have now untangled some of the genetics of Williams syndrome. The story is one of genes that interact with each other in dramatic ways, and sometimes do seem to affect specific behaviors.

The syndrome results from a tiny deletion of part of chromosome 7. The size of the deletion varies among different people with Williams syndrome. It is sometimes substantial enough to remove a dozen genes and occasionally small enough to remove only two. People heterozygous for this deletion have just one copy of each of the genes in that region, a copy that was carried on the undeleted chromosome that they inherited from their other parent. Note that this is the reverse of the situation with Down's syndrome—in Down's, the problems arise from having too many genes, while in Williams the problems come from having too few.

One gene that always seems to be included in the Williams deletion codes for a protein called elastin. This protein contributes to the elasticity and strength of arterial walls; its deficiency may explain the circulatory problems and the strange lacy appearance of the irises so typical of the syndrome. But by itself it does not explain the unique pattern of mental retardation, for some people have only one copy of the elastin gene but are missing none of the nearby genes. Even though they have the same circulatory problems, most of them are mentally quite normal.

It is the lack of the other genes that seems to contribute to the mental deficiencies. The Utah researchers examined carriers of very small deletions, who are missing one elastin gene and one other nearby gene. These people are not mentally retarded, though they have the physical appearance and circulatory problems typical of the syndrome. But they do express very strongly one behavioral characteristic of Williams: the inability to draw coherent pictures.

This second gene, rather than the elastin gene, is most likely the one that influences brain function. Unlike the elastin gene, it is turned on in various parts of the brain. It codes for a type of protein that is able to modify other proteins in the cell chemically, but nothing else is known about it. Why it affects brain function remains a mystery. Nonetheless, having too little of the protein must lead, through who knows how many developmental steps, to this remarkably selective mental deficit.

The gene is unlikely to act alone. To begin with, it may interact with the elastin gene. People who have two copies of the brain protein gene but are missing one copy of the elastin gene seem, for the most part, to be quite normal. But what about people who have the reverse situation—two copies of the elastin gene but only one copy of the brain protein gene? Team member Michael Frangiskakis tells me that they have looked for such people but without success. They may remain undiscovered among the many people who have slight learning deficits, or they may be completely asymptomatic, concealed somewhere in the population at large.

This story resembles the story of Leber's optic neuropathy that we encountered in Chapter 9. In that disease the harmful effects of the mitochondrial mutation are expressed only under the severest conditions, while the same slightly damaged mitochondria are able to function normally in most of the cells of the body. Similarly, people with Williams syndrome make enough of the brain protein in most of their nerve cells, but during critical times in development, some cells may be particularly liable to damage because they do not make quite enough of the protein.

Is this second gene a gene for a specific behavior? Perhaps, but far more likely, it is interacting with many other genes. The brain is too complicated an organ, the result of too many complicated gene–gene and gene–environment interactions, for specific functions to be controlled by single genes.

Many other rare types of mental retardation, with a wide variety of symptoms and names like Angelman, Langer-Giedon, and Prader-Willi, have been traced to little deletions on various chromosomes.

Some of these deletions remove a single gene, but most, like Williams, remove more than one. They cause a great range of symptoms, and one day, as with Williams, these symptoms will be traced to deficits of specific proteins. But these proteins are not likely to be the ones that control these behavioral deficits—instead, I would predict that when the proteins are made in reduced amounts, they will be found to damage complex functions of the brain that are each the sum total of many different genes acting together.

These various syndromes show that in order for one or two genetic changes to affect mental function, the changes must be severe enough to hit the brain with the biological equivalent of a hammer. Gene hunters can track down genes with such large effects easily, but tracing the many different genes that together produced these complex brain functions in the first place is far harder. The gene for piano-playing will continue to elude us.

TWELVE

⬥

How Brain Function Evolves

No one, I presume, doubts that the large proportion
which the size of man's brain bears to his body,
compared to the same proportion in the gorilla and
orang, is closely connected with his higher mental
powers. . . . On the other hand, no one supposes that the
intellect of any two animals or of any two men can be
accurately gauged by the cubic contents of their skulls.
CHARLES DARWIN, The Descent of Man (1871)

As we saw in the last chapter, there seems to be no obvious connection between specific genes and specific behavior patterns. But without such a connection, how could our brains, with their wide variety of abilities, possibly have evolved? Various alleles of genes that regulate brain development must have some connection with the environment that selects for them, for otherwise natural selection would have no effect. The connection is indeed there, but it turns out to be a subtle one. Genes and environment interact in such a complicated way that the whole process resembles the complex weave of a tapestry. Although it is difficult to follow the individual threads, in this chapter we will try.

THREADS OF THE GENE–ENVIRONMENT TAPESTRY

People with the condition called narcolepsy exhibit an odd and life-threatening behavior: They fall asleep suddenly and inappropriately.

An old joke has it that it is normal for an audience at a lecture to fall asleep, but you know you are in the presence of narcolepsy when the *speaker* falls asleep!

Jokes aside, narcolepsy is a very serious condition. Among Caucasians there happens to be a direct connection between this behavior and a particular gene, one of the strongest that has yet been discovered. Virtually all Caucasians who exhibit narcolepsy carry a particular allele in one part of a complex collection of genes called HLA, which is found on chromosome 6.

Merely having the allele does not sentence the carrier to narcolepsy, however. While almost all Caucasian narcoleptics have the allele, only a small fraction of those with the allele are narcoleptic. The allele is found in twenty-one percent of the Caucasian population, but we do not see one in every five of this group suddenly nodding off; only a third of one percent of those who carry the allele actually show the condition. A few narcolepts do not carry the allele, yet develop the disease nonetheless.

Having the allele, then, does not appear to determine one's fate with certainty but constitutes only one step on the road to narcolepsy. The story, when it is finally unraveled (which will likely take decades), will certainly turn out to be very complicated. But it is possible to speculate a little. The gene is located in the HLA complex, which is known to influence disease resistance and susceptibility; this suggests that some childhood episode of infectious disease might act as a trigger. Such an environmental event would presumably be fairly rare—perhaps an unusual disease or a particularly severe bout of some common disease. Those of us who do not carry the allele might easily have had the disease during childhood, but we would have recovered without long-term effects.

Is the HLA allele, then, the gene for narcolepsy? No, although in some as-yet-to-be-determined way it does contribute to the condition. The damage it does to the brain is not the result of the activity of the gene itself, but probably happens because carriers of the gene are more susceptible to some environmental stress; without that stress, nothing would happen.

Genes that supposedly contribute to behaviors in some fashion have become the new darlings of the media. Television and newspapers have recently trumpeted the discovery of genes "for" alcoholism, hyperactivity, and criminal behavior, among others—even a putative gene "for" divorce!

Undoubtedly some genes affect behavior. But in general they are not genes "for" specific behaviors. Consider the newly discovered gene "for" attention deficit hyperactivity disorder (ADHD), which recently received worldwide publicity. Children with ADHD have great difficulty concentrating on schoolwork and cannot sit still for long periods. Usually, but not always, they outgrow these symptoms. Ritalin is the medication of choice for ADHD, and more than a million children throughout the United States are currently on this drug.

Such wholesale prescribing may in part be a result of the way ADHD is diagnosed. The standard diagnostic test for ADHD lists fourteen different criteria, and a child with a high score in eight or more is considered to have hit the jackpot, so to speak. But a majority of male children in the population score high on at least a few of the criteria (females tend not to show such high scores), which means that ADHD is really something of an artificial construct.

The syndrome lies at one end of a continuum of behaviors. Drawing the line between ADHD and behaviors that are not quite ADHD is thus very difficult, and highly subjective. Because the drug is prescribed in such wholesale fashion, I suspect that all children in literature would have ended up on Ritalin if they were real and alive today, with the possible exception of Little Lord Fauntleroy.

The connections between specific genes and ADHD are tenuous but intriguing. It has been known for a long time that some hyperactive people simply do not respond to thyroid hormone treatment; in a few cases this refractoriness runs in families. Starting in the mid-1980s, work in many laboratories led to the discovery that the defect in these families lies in a gene that codes for a receptor protein that binds to thyroid hormone. The altered protein is unable to bind the hormone properly.

In 1993 a group at the National Institutes of Health examined eighteen of these families and found that over half the children with a defective gene fit the criteria for ADHD. Further, half the adults carrying the defective gene probably had ADHD when they were children. By comparison, thirteen percent of family members without the mutant gene either exhibited ADHD symptoms or had the condition when they were children, which is slightly but not significantly higher than the frequency of the condition in the general population.

The Associated Press headlined its story about this work "Defective Gene Produces Hyperactivity." This is, to say the least, overstating the case. The gene "for" ADHD actually has much in common with the gene "for" narcolepsy. A number of family members who carry the defective gene show few if any symptoms, and a number of those who do not carry the defect show many symptoms. Probably the defective gene is able to *affect* the level of hyperactivity, but it hardly *produces* the behavior.

The discovery that a defect in this one gene has a statistically significant impact on attention span and activity levels is nonetheless exciting, and much research is currently being directed toward the whole complex issue of thyroid hormone activity and its influence on these behaviors. But the discovery of this defect does not constitute the discovery of a specific "gene for hyperactivity." Rather, it tells us that a specific defect in the mechanism of binding of thyroid hormone to cells may exaggerate certain behaviors.

The thyroid hormone gene is not alone in its influence on ADHD: three different genes that are involved in binding or transmitting the important neurotransmitter dopamine have been discovered to be polymorphic in the human population. This is important because sensitivity to dopamine, and a similar compound serotonin, have been found to be correlated with a whole web of behaviors. Some of the various alleles of these genes have slight but significant associations with ADHD, including Tourette syndrome, stuttering, obsessive-compulsive disorder, substance abuse, and other psychiatric ailments.

It must be emphasized that in none of these cases is the correlation anywhere near perfect. These genes are a long way from directly affecting the mechanisms, whatever they are, that define this wide assortment of behaviors.

Yet these findings are unquestionably of enormous importance. As a result of these and similar discoveries, psychiatry is rapidly moving into a new and exciting phase that holds promise for specific palliatives and even cures for some extreme and damaging behaviors. But as exciting as these discoveries are, we are only just beginning to penetrate the first of many levels of brain function. It is as if, in the course of investigating how an automobile works, we discovered that increasing the richness of the fuel vapor entering the engine causes it to speed up. While this finding may suggest many other experiments, it tells us nothing about the engine itself.

Let us take a moment to define our terms. In studies such as these, the word *behavior* has a very specific meaning. When authors speak of "behaviors," they do not mean some aspect of intellectual function, but rather some thoughtless, driven, or inappropriate activity. A person who exhibits one or more such "behaviors" may at the same time be able to function intellectually at a high level.

Consider drug addiction. People who become drug addicts appear often to be driven toward that addiction by biochemical imbalances that produce emotional dysfunction, rather than intellectual deficits. We define *drug addiction* as a behavior, but it has nothing to do with the higher functions of the brain. Better to call it, as many psychiatrists and psychologists do, a behavior pattern, with the implication that it is caused by some global influence which distorts the function of the brain. A gene that influences such a behavior pattern is a long way from a gene for piano-playing.

Many such behavior patterns are unacceptable in our highly structured society; hence all of the crude chemical attempts to correct them. In some cases, our remarkable new armamentarium of drugs actually helps us succeed in modifying the behavior patterns so that they can conform more closely to societal norms. Children successfully treated with Ritalin and the newer generation of ADHD drugs are able to acquire skills that permit them to survive more effectively in our society.

As that society changes, acceptable behavior patterns change as well. These continually shifting societal parameters are a moving target for the selective pressures that act on us. We are also adding new and unexpected selection pressures: a child who cannot respond to Ritalin may grow up, carry out some act that is unacceptable in our current society, and end up incarcerated—and effectively removed from the gene pool—as a result. What is being selected against here: genes that predispose a child to an unacceptable behavior pattern, or genes that make a child unable to respond to the drug of the moment? It is impossible to tell—this thread of the tapestry cannot be disentangled from the others.

BENDING THE TWIG

The genetic connections to these behavior patterns are still tenuous and volatile. Yesterday's highly touted gene "for" this or that behavior pattern often turns out, after further work, to be an artifact, usually because subsequent studies are unable to replicate the original observation. This has not been the case for the associations of ADHD and other behavior patterns with alleles of dopamine receptor and transporter genes, as weak as these associations usually are; many of these findings have now been independently replicated in other laboratories. But a number of claims that genes have been discovered "for" such conditions as bipolar depression and schizophrenia are now known to have been premature.

Retractions of highly publicized discoveries rarely appear in newspapers or on television, so the public gains the impression that scientists are adding to an ever-growing list of genes "for" this or that behavior. In fact, they are doing nothing of the sort. What they are doing is finding a tentative list of genes that affect brain function, on some global or slightly more specific biochemical level. At the same time, they are discovering that these genes are influenced tremendously by environmental factors. Environmental effects are often so strong that they can cause the supposed effects of a gene to appear in one study and disappear in the next. If we are to begin to understand how genes for enhanced brain function can be selected for, we must understand the role of the environment, since it can both mask and enhance a genetic connection.

The impact of the environment is often overwhelming. Poor nutrition and disease can affect our mental capabilities, sometimes more than we suspect. One of many vivid examples comes from the annals of mental illness.

Schizophrenia afflicts, at one time or another, about one percent of the population, a fraction that is remarkably constant throughout the world and even across cultures. The disease unquestionably has a genetic component, even though the genes responsible have so far evaded some fairly intensive searches.

One strong piece of evidence for this genetic component comes from twin studies. In identical twins it has been repeatedly observed that if one twin is afflicted with schizophrenia, the other often develops the disease as well. The similarities may be even more striking, for sometimes the time of onset and the range and types of symptoms are very similar in both twins.

But not all affected pairs of identical twins show such a pattern. Sometimes only one twin develops schizophrenia and sometime they both do. The fraction of affected twins in which both develop the condition is called the concordance rate. If this always happens, the concordance rate will be a hundred percent. In fact, the rate for schizophrenia is only about half that. Half of the identical twin pairs are discordant for the disease.

This pattern is remarkably universal. A twin concordance rate well below a hundred percent is also seen in other major types of mental illness, although in each case the concordance rate is much higher than would be expected by chance. But if identical twins have identical genes, and genes are controlling this condition, shouldn't the concordance rates be a hundred percent? Why are there so many pairs of identical twins in which one twin develops a condition and the other does not?

The usual explanation given is that some traumatic event—an emotional crisis or a disease—must trigger the mental illness, and that about half the time it happens to one twin and not to the other. No candidates for such hypothetical events have yet been found, but a striking and little-noticed effect shown by these twins may provide a clue. Among discordant twin pairs, the twin who develops schizophrenia is almost always the lower birth-weight twin!

This remarkable observation, mentioned casually in a few papers on the subject, seems to me to hold great promise for understanding the causes of the disease. It may mean that schizophrenia can be triggered by something as simple as poor nutrition in utero. Or it may mean that the lower birth-weight twin is more susceptible to a disease of early childhood that brings on the mental illness, sometimes decades later. The genes give rise to the twig, but the environment must bend it.

If the reason for this birth-weight effect can be discovered, many cases of schizophrenia may be prevented simply by improving maternal nutrition during pregnancy. Or perhaps scientists will discover some disease or diseases that spur the condition and that preferentially affect the less well-nourished of the twin pair, perhaps shortly after birth. In sum, we would be well advised to investigate the roles of both early nutrition and childhood disease in the development of schizophrenia. The short-term payoff from these workaday studies may be far larger than the discovery of some of the many genes involved.

The interaction of schizophrenia and the environment raises a larger question: Do nutrition and disease have subtle effects on the mental capabilities of the rest of us? We will see in a moment that this may be so. The evolutionary consequences may be startling.

The bottom line of our search so far for the genes for behavior is that we are indeed beginning to discover genes that contribute, however subtly, to behavior patterns as we have narrowly defined them. But genes that contribute to behaviors in the more commonly understood sense continue to be as elusive as ever. We have found no genes that contribute to musical talent, to mathematical ability, or to other complex behaviors. We have certainly found no direct gene-culturgen relationships of the type suggested by Lumsden and Wilson.

We can now dismiss the paradox that there would have been no way to select for genes for piano-playing before the invention of the piano. This naïve view must be replaced by a much more subtle one, in which nature selects for complex sets of genes that interact in complex ways. An evolutionary process, involving many genes, has increased our brain capacity and flexibility over a broad front. This selection explains how we can turn our brains to a variety of tasks—like snowboarding—of which our forebears never dreamed.

Further, our brains have evolved into even better instruments than we suppose. It is often suggested, on the basis of rather specious arguments, that we use only ten percent of our brains. In fact, functional MRI scans show that we use a good deal more than that. But for a variety of environmental reasons, we do not do so very efficiently. Were we able to use more of our brain capacity, it would actually accelerate our evolution, because adaptive evolutionary change can take place only if physical or mental differences exist on which natural selection can act. If our environment reveals previously hidden genetic variation in our gene pool, then the effect of natural selection can become greater. The progressive unveiling of this variation, which is taking place as each succeeding human generation faces a more complex environment, can have unexpected results.

THE FLYNN EFFECT

Even those alien biologists from another planet whom we hypothesized in Chapter 10 would be forced to admit that the human brain is the most remarkable product of evolution to be found among the

Earth's living organisms. Although our culture has been shaped largely through the activities of people whose mental abilities are far above the ordinary, we cannot yet begin to define what those abilities are, or to understand what genetic differences might exist among us and how they have led to the occasional appearance of such outstanding people.

Despite the profundity of our ignorance, however, some interesting patterns do emerge. In the course of our evolution, our brains seem to have acquired capabilities that are more varied than even our current highly stimulating environment can elicit. The X-games of the future will involve feats of skill, balance, and coordination that we cannot begin to imagine, yet some of our children will be able to accomplish them with a flourish. Future societies will be more complex and challenging than any we have yet seen. Some of our descendants will thrive on the challenges, while others will be bewildered and discouraged. As we have already seen, such discouragement is likely to affect the gene pool, if people become so depressed and dysfunctional that they are unable to have children.

Our brains are capable of a wide variety of tasks. Sometimes they are damaged by biochemical and developmental hammers, padded or otherwise, but if they are not, they can manage very nicely. Such a basic or Model T brain, as we might call it, is perfectly capable of being tweaked, by training and by its inbuilt capabilities, in a great variety of directions.

All of us possess that basic brain. The various combinations of bells and whistles, the various factory options that each of us has in addition, are invisible even to the trained eye. Were we to line up the brains of an Einstein, a Mozart, and a Michael Jordan on a dissecting table, they would appear indistinguishable—mere "lumps of porridge," in the memorable phrase of psychologist Richard Gregory. While some differences must exist among them, my guess is that they are not large. The difference between a Mozart and an ordinary duffer who can just manage to pick out "Mary Had a Little Lamb" on the piano is a second-order difference.

One day soon we will be able to trace this particular difference to the number of synaptic connections between certain cells in the cortex, or to some unusual mix of neuropeptides. We will even be able to recreate the mix that ought to produce a Mozart and successfully apply it to any reasonably undamaged basic brain. Given such a biochemical and developmental boost and the right environment, any of us could then become a musical genius.*

We seem, though unconsciously, to be carrying out such manipulations already. In the early 1980s, James Flynn, a New Zealand psychologist, sent a questionnaire to researchers around the world. The researchers had been measuring intelligence test scores in their various countries over periods of years or decades. He asked a deceptively simple question: Did they see any trend? Was it true, for example, as is predicted by the prophets of genetic disaster, that IQ scores are going down because less intelligent people are having more children?

The period covered in the studies was from the 1950s to the 1980s. Flynn processed the masses of data with great care and reached a remarkable conclusion: During this time, rather than going down, IQ scores were in general going up, in some cases very far up. This increase appeared to be a very widespread phenomenon, since it was seen in all the industrialized countries of Europe, the United States, Australia, New Zealand, and Japan. The details of the pattern were puzzling: the largest gain was seen in the Netherlands, while one of the smallest was seen in nearby Great Britain.

For some subtests of the IQ tests, the average gain was as much as twenty percent. The gains were in many cases extremely significant and could not have been the result of some artifact of sampling. In the Netherlands, Norway, and France, which had universal conscription, IQ tests were given to virtually all males at age eighteen. Such data approach the statistician's ideal of measuring an entire population, rather than some sample that might or might not be randomly drawn.

The lengthy study that Flynn published in 1987 is festooned with caveats. First of all, the change can hardly have been an evolutionary one. Although evolution can be quick, it is not that quick. Probably some genetic alteration took place in the populations measured during those few brief decades, but it would primarily have been the result of mixing with immigrants—the very thing that racists predict should lead to a lowering of IQ. Flynn calculated that even the most extreme genetic scenario, in which genes play an overwhelming role in determining

---

* A tall order, probably, since the child would have to grow up with a group of people to whom music is absolutely central to existence and would in addition have to receive the undivided attention of the equivalent of a Leopold Mozart.

IQ, could have produced a change of only one percent or so each generation. Such a change, even if it were in the positive direction, could account for only a fraction of the huge increases. The effect must be due to changes in the environment.

Second, not all the components of the IQ tests showed the same increase in average score. Strangely enough, those that showed very little of the Flynn effect were the very ones that might have been expected to increase.

In recent decades succeeding cohorts of children have been exposed to more radio and television and more sophisticated schooling. One might suppose that they would therefore do better on so-called culture-loaded subtests, such as vocabulary tests that depend on special knowledge. In many cases there were increases in these subtests, but they tended to be small. Instead, the subtests that showed the largest increases were those that were perhaps the least culture-loaded.

One of these components is a test called Raven Progressive Matrices, in which the examinee must match ever more difficult patterns (Figure 12–1). While other parts of the tests are often rewritten to reflect changes in the tested population, this subtest has generally remained unaltered. It therefore serves as an excellent benchmark with which to follow a tested population over time. Flynn found that, on the whole, the young people of today are far better at solving these matching problems than their parents had been a generation earlier.

It is unlikely that these gains could have been due to better schooling. Indeed, by a variety of measures, schooling seems to have been getting less effective during the decades covered by the IQ measurements. Scores in widely applied tests of knowledge such as the American SAT and the English O- and A-levels have (until recently) been decreasing.

At the end of his paper, Flynn raised an interesting question. If one assumes that the IQ increases really do reflect an increase in intellectual ability, then why has the population at large not become measurably more brilliant with time? Surely far more geniuses should be abroad now than thirty years ago! As he looked about him, however, doubtless observing with horror the excesses and stupidities of the modern age, he failed to find evidence for such an increase.

Reading this part of his paper, one imagines that Flynn must have written it in a leather chair in his club, while harrumphing to fellow

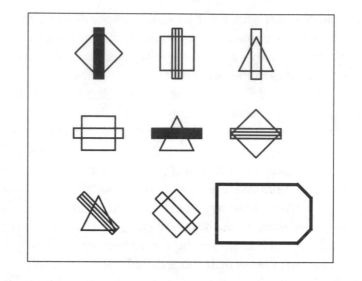

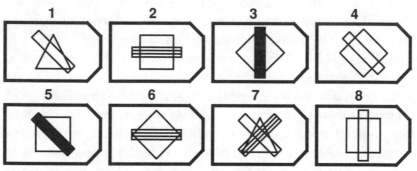

Figure 12–1. An example of Raven Progressive Matrices, not taken from an official test. The trick is to find the rules that make only one of the inserts below the figure the correct match. The answer is in the endnote for this page.

club members about the sad state of the younger generation. Nonetheless, he has a point: It is not terribly obvious that we are getting to be as brilliant as the numbers seem to indicate.

Flynn came to the tentative conclusion that the IQ subscores that he had been looking at so carefully may not have much to do with intelligence after all. Perhaps they were measuring some other factor, one that might not even be very strongly correlated with intelligence. It must be that other factor, he thought, that had changed so dramatically over the last few decades.

Perhaps he is right. But the tests do measure brain function at some level—and that function has changed in such a way as to improve our ability to do certain challenging tasks. Moreover, this effect has spilled over from the less culture-loaded to the more culture-loaded subtests, since smaller but significant gains were seen in a number of those that were more culture-loaded as well.

What this effect might be remains a mystery. Schooling, as we noted, seems to be ruled out. Three likely factors remain, all of which have the potential to influence overall brain function.

The first is television, a new factor that has become more important (and often central) to our lives during the postwar period. While it seems risible in the extreme to suppose that television as we know it might somehow be the cause of the Flynn effect, it is true that researchers at the University of Kansas have found that children who watch *Sesame Street* and its many imitators tend to perform better in school. The results of this study, unfortunately, are confounded by many other environmental factors that may be operating, notably class differences among the children. And *Sesame Street* is only a small part of the diet of television that most children watch.

But television may also have a more general effect, on speed of cognition. When people play fast-moving computer games, they quickly improve their cognitive skills. Perhaps the flashing of images in rapid succession during a television program or commercial can have a similar effect, speeding up reaction time and brain function. One certainly hopes so, and that the trillions of hours people spend in front of the tube worldwide every year are having some kind of benefit.

The second factor is nutrition. Since World War II diets in the industrialized world have improved markedly in terms of balance and completeness. During the last months of the war, the Dutch were deliberately starved by their German occupiers; after the war their diet became, to say the least, dramatically better. Flynn found that they show the largest IQ gains. On the other hand, the British, who showed the smallest overall IQ gains, were the last to benefit from this increase in the quality of nutrition. An insular nation, they have not yet completely abandoned the culinary habits that have made their cuisine a standing joke throughout the rest of the world. Having experienced a childhood of English food, I am quite willing to entertain the possibility that my intellect suffered as a result!

But the third factor, I suspect, is potentially the most important of all: the lessened influence of childhood disease. Measles, mumps, diphtheria, and whooping cough, so common only a few decades ago, may have had more of an effect on long-term brain function than we realize. Every child, as it grows up, suffers repeated bouts of illness, not all of which have an obvious cause. Some of these diseases, though slight in their immediate effects, may well leave behind a legacy of damage.

Perhaps not coincidentally, immunizations against these diseases have become more widespread and effective during the period of Flynn's study. Studies have shown that even children with moderately low birth weight suffer a higher incidence, and greater severity, of these childhood diseases than unimpaired children do. It is unclear whether the disabilities have caused the disease susceptibility or vice versa, but more detailed studies of these effects cry out to be done.

DISEASES AND THE MIND

The recent change in disease incidence has two evolutionary effects. The first is obvious: people whose brain function survives the onslaught of disease are at an advantage. The second is that as the influence of disease lessens, these genetic differences are becoming less important to survival. Intellectual challenges from our environment, not disease-based challenges, will become more important in the future.

We do not know, however, how large a role diseases still play in brain function. If this role is still substantial, then we can examine which childhood diseases might have the greatest negative impact on brain function, for example, and why. Such studies will be important for the underdeveloped world, where many of these diseases are still prevalent.

In the industrialized world we have reduced or eliminated the most obvious diseases, but these are the easy targets. The ones that remain have less blatant effects, which will make their long-term harmful influences much more difficult to study.

That such early, almost invisible, damage can indeed have long-term consequences is becoming apparent. One of the most unnerving discoveries about brain function in recent years has emerged from an extensive study of Alzheimer's disease. About ten percent of elderly people, numbering perhaps four million in the United States, are affected by this debilitating and ultimately fatal neurodegenerative condition.

In 1986 epidemiologist David Snowdon of the University of Kentucky began to recruit nuns of the School Sisters of Notre Dame for a longitudinal study of the effects of aging. He soon found that this gigantic subject was too broad to study properly, given his limited resources, so he narrowed his focus and concentrated on the effects of one particular manifestation of aging, Alzheimer's disease.

The Sisterhood is a teaching order with convents scattered throughout the East and the Midwest. Snowdon succeeded in persuading 678 of the nuns to join the study. At the time of recruitment all his subjects were seventy-five or older. They provided information about themselves and their careers and agreed to undergo repeated psychological tests to follow the effects of aging. Most important, they also agreed to donate their brains for study after death.

The brains of the nuns who have died so far have been carefully preserved and examined under the microscope for the plaques and tangles that are symptomatic of Alzheimer's. Snowdon and his fellow researchers are now able to make direct correlations between symptoms of dementia observed before death and the types of brain damage that are due to this degenerative condition.

Many important results have already emerged from the study. A particularly striking one is that plaques and tangles in the brain do not necessarily result in dementia. Several of the nuns remained alert and intensely interested in the world around them right up until their deaths, and yet their brains showed signs of advanced Alzheimer's on autopsy. It will be very interesting to learn how their mental capabilities managed to survive a disease that, in its extreme form, literally eats tiny holes in the brain. Some of us are clearly far more resistant to the effects of Alzheimer's than others.

A second observation of great interest is that the onset of Alzheimer's can be sharply accelerated by other diseases, particularly strokes. Synergistic interactions between Alzheimer's and other disease states are only beginning to be explored. Infections of the brain with herpes virus, for example, may be associated with Alzheimer's, but this virus has also been found in normal aging brains. Its presence may simply mean that it has had time to spread into previously inaccessible tissues.

The third observation is perhaps the most remarkable of all. Half a century earlier, when the nuns were in their twenties and were being recruited into the order as young novitiates, many of them wrote short biographical essays. They made these essays available to Snowdon, on the chance that their early writings might contain some hint about their future fate.

Snowdon examined two characteristics of the essays: idea density and grammatical complexity. Idea density, as its name implies, is a measure of the content of the essay, and it can be quantified rather crudely as number of ideas per ten words. Grammatical complexity measures the number and type of sentences beyond the simple declarative, such as those that have dependent clauses or employ the conditional. Using these criteria, he divided the essays into the lowest-scoring third and the highest-scoring two-thirds and asked what proportion of each group belonged to the nuns whose brains showed signs of Alzheimer's.

The results for idea density seemed to be remarkably clear. Among the twenty-five now-deceased nuns who had written the essays, the brains of ten were found to show clear signs of Alzheimer's. Nine of these ten fell in the lowest third on the idea density scale. In contrast, among the remaining fifteen who showed few or no signs of the plaques and tangles, only two fell in the lowest third. This result was quite significant statistically.

At the same time, however, Snowdon found no significant pattern in the relation between the grammatical complexity of the essays and subsequent Alzheimer's. This finding was odd, because both idea density and grammatical complexity had been shown in previous studies to be good indicators of brain function. Why should the two results differ—unless the hypothetical early stages of Alzheimer's somehow affect only idea density and not grammatical complexity?

It is of course possible that the significant effect Snowden found was simply a statistical fluke, a function of the small sample for the study. This possibility, at least, should be resolvable. At the risk of sounding a bit ghoulish, time is on Snowdon's side. As the nuns inevitably die, the numbers in his database will increase—and so will his confidence in the results.

If we assume that idea density really is negatively related to the later development of Alzheimer's, this study seems to suggest one of two things.

First, Alzheimer's may be more likely to strike those who, for whatever reason, have been exposed to few ideas and whose prose reflects this. Indeed, a number of studies have shown that Alzheimer's is more prevalent among people of lower socioeconomic status. Perhaps this is because poor people are more likely to be poorly nourished, or to be exposed to diseases. These factors could serve as triggers for events that will eventually culminate in Alzheimer's.

Such people may simply be in the wrong place at the wrong time. If disease organisms are the ultimate villains in this train of events, then by the time Alzheimer's becomes apparent, the diseases will long since have disappeared. Such situations are not unknown for other diseases—rheumatoid arthritis, for example, can be triggered by a bacterial infection, causing life-threatening symptoms many years after the infection has been cured.

Alternatively, Alzheimer's may be built into the system, so that it starts wreaking its damage on the brain even while the fetus is still developing. Anybody, depending on the luck of the genetic draw, could then develop Alzheimer's. Reinforcing this idea is the growing body of evidence that the disease may have a genetic component; a number of genes have been found that are strongly implicated in the small minority of Alzheimer's cases that tend to run in families.

In fact, disentangling these genetic and environmental alternatives is almost impossible in practice. Probably, as with so many of these situations, we are seeing a contribution from both effects. If you are unlucky enough to draw the wrong set of alleles from the genetic lottery, but then are lucky enough to grow up in an environment free of the diseases that might trigger the condition, you might easily live to a ripe old age with all your faculties intact. But even in the absence of such alleles, a sufficiently severe combination of malnutrition and disease during childhood might be enough to trigger Alzheimer's. There are likely to be no simple answers.

Assuming the results of Snowdon's study hold up, they seem to fall into the pattern that we have seen again and again: Our brains are truly remarkable organs, but our environments are filled with factors that can prevent them from achieving their full potential. Some of these environmental insults are severe and obvious, while some are much less so.

Factors such as nutrition and childhood disease may lower the brain's capabilities in subtle ways, but they also may damage the brain so slightly that it still appears to be perfectly functional. Yet people who suffer this damage may lose the intellectual edge that they need to take full advantage of the opportunities provided to them by our current highly challenging environment.

Even in the face of damage, surprisingly minor environmental interventions can help bring the brain up to a higher level of function. One of the most important studies of the effectiveness of such intervention was a carefully randomized trial reported in the *Journal of the American Medical Association* in 1990. From eight different geographic regions around the United States, a total of almost a thousand children were chosen who had been born prematurely or otherwise had low birth weight. Numerous studies had already shown that such children were at high risk for mental retardation and behavior problems later in life. Could early intervention change this prospect?

The children were divided into two groups, carefully balanced for such factors as race, gender, and mother's educational level, but otherwise randomized. Both groups received normal follow-up pediatric care, but one group participated in a program that included home visits, periodic meetings with the parents, and intensive training. This training, certainly the most important component of the intervention, was carried out at Child Development Centers specially set up for the program. For two years, starting when they reached the age of twelve months, the children were bused to the centers and given intensive learning activities by teachers. There was one teacher for every three children.

The results were dramatic. In every geographical region highly significant differences in IQ score emerged between the two groups. The children in the intensive program scored an average of fourteen points higher (which is, by the way, the average difference in adult IQ scores between blacks and whites in the United States). The gain was a little less among the children who had been the most severely underweight when they were born, but it was still highly significant. The mothers of the children in the intervention group also reported fewer behavior problems.

It seems that such intervention really can have an effect. Similar results, though less well controlled, have been reported for Head Start, a U.S. government program that employs comparable though less intense intervention efforts. Head Start has primarily concentrated on preschool children, starting at age three, though it has now—in part as a result of the *JAMA* study—been extended to younger ages.

Unfortunately, follow-up studies of Head Start children have shown that the gains from this program are often lost when the children start regular schooling. Critics of the program have used such results to suggest that Head Start has no lasting effect and should be discontinued. (See that unpleasant book *The Bell Curve* for a sampling of these arguments.) I find this suggestion astonishingly stupid. In view of the fact that the gains from Head Start are clearly measurable, it would seem much more sensible to ask how regular schooling manages to undo them!

At least some of these factors are likely to be important in the Flynn effect, which leads me to be much more sanguine about his effect than Flynn is himself. To me, the Flynn effect says that we have recently begun to do a surprisingly good job of preventing damage to our brains and enhancing their capabilities. Over the space of a few decades, we have succeeded in ameliorating and perhaps even reversing much of the early-childhood damage that in earlier generations was almost unavoidable.

How we succeeded in doing this remains a mystery, since of course we do not yet know the nature of the damage. But whatever the mechanism of the Flynn effect may be, the effect itself has been strong enough to overcome the simultaneous onslaughts of family breakup, declining school standards, and the worst damage to our intellects that our popular culture can wreak. This is, let me emphasize again, not an evolutionary change but rather a change in the environment. However, as we will see, it can result in evolutionary change.

Taken together, all these studies lead to an important conclusion: *The brains of most of us are capable of working better than we suppose.* The genes that seem to be responsible for aberrant behavior patterns are now, on careful study, resolving themselves into crude genetic defects that bludgeon the brain biochemically or developmentally, and that are often enhanced by negative environmental factors such as disease and low birth weight.

I am not—repeat, not—suggesting that all human brains are equivalent. As a result of genetic differences, the abilities of human brains surely do differ from one to another. But as long as that basic brain of which I spoke earlier is not harmed by its environment or its owner, it is quite capable of doing an excellent job.

Once we realize this, the stigma attached to not being very bright will disappear, and blame for lower intelligence will be placed where it belongs: not with the individual but with deleterious factors in the environment.

Presumably our ability to enhance brain function through improvement in our diets, the prevention of disease, and alterations in the environment has limits. But if we deliberately set out to improve brain function through more direct chemical means or even through genetic manipulations, we can perhaps go much farther. Can it be that all of us are potential Mozarts or Einsteins, simply awaiting the appropriate biochemical manipulations to bring out these hidden talents?

Probably not—from their rarity, as I suggested earlier, such remarkable talents must result from highly fortuitous combinations of genes and environment. But one prediction seems safe: The mental capabilities of our grandchildren, while they may not be in the Einstein league, will be dramatically greater, on the average, than our own.

Demand for artificial enhancement of mental capacity is already great. "Smart" pills—combinations of plant products and other chemicals that are, one hopes, relatively innocuous—are currently available in health food stores. At the moment it is difficult to separate the placebo effect from anything that these pills might do to enhance brain function.

But a new generation of real "smart pills" that work on the structure and function of the brain cells themselves will soon be available. What would happen, for example, if that mysterious second protein involved·in Williams syndrome were overexpressed in the dendrites of nerve cells, instead of being underexpressed as it is in people with the syndrome, or normally expressed as it is in most of us? Would it enable us to think in two and three dimensions with unparalleled ease? Or would it simply produce some new and damaging biochemical imbalance? We do not know, but we will probably find out soon, first in experimental animals and then—if it seems to work—inevitably in humans.

Now that we have a clearer idea of the potential resident in our brains, let us return to that deeper question: Why is that potential so great? Why can so many of us respond so effectively to the breathtaking new challenges presented by our environment, whether it be designing a Mars lander or learning how to snowboard? What genes have been, and are continuing to be, selected for?

GENETIC STUTTERS

In the course of our evolution, two things have happened to the genes that control the development of the human brain. First, the brain has become larger and more complex. In a straightforward evolutionary process, alleles of various regulatory genes that caused this enlargement must have spread through our population and become "fixed," replacing older alleles in every member of our species.

The outcome has been an increase in both the size and the capabilities of the basic or Model T brain that all of us now possess. It happened relatively quickly in evolutionary terms, building on genetic equipment that was already in place back when our brains were very much like those of chimpanzees.

Second, and more subtly, we have accumulated additional new alleles of many of these genes. We know this must have happened because the range of talents that we exhibit as a species has increased so markedly since our chimpanzeelike days, and at least some of this increase must be due to our genes. The addition of new alleles increases the diversity of a population, providing the bells and whistles or "factory options" that we talked about earlier. Because each gene may have many alleles, none of us can possess all this variation. We can have a maximum of only two allelic forms at each of our genes, regardless of how many different alleles there may be in the population at large. Other members of the population may have a quite different mix of alleles, which is one reason we are so diverse in our abilities.

Where are these alleles, and what are they like? We do not yet know, but one newly discovered category of genes provides an excellent place to look for them.

As we have seen, the relatively common type of mental retardation known as fragile X syndrome is caused by an expansion or lengthening of a short repeated bit of DNA on the X chromosome. In all of us this stretch consists of the three-base motif CGG, repeated over and over in a kind of genetic stutter. We carry it on our X chromosomes, and most of us have no more than twenty or thirty copies of the three-base motif in our repeated region.

Geneticists have given this and similar repeats various names, but one that is becoming increasingly popular is *microsatellite*. Satellite DNA is found in huge quantities in our chromosomes, consisting of quite long stretches of noncoding DNA repeated many times. It can be separated from the rest of the DNA in a centrifuge, where it forms a satellite band in the centrifuge tube. Microsatellites are far too small and far too thinly scattered around the chromosomes to be separated in this way, but the name has stuck.

Mutations are constantly appearing that, by adding or removing three-base motifs, make this bit of DNA longer or shorter. For the majority of us, it seems not to make much of a difference—this microsatellite is not even in a gene, although it happens to be very near a gene of unknown function named FMR1. But in people who have too many copies of the little motif—about a hundred or more—a runaway mutational process begins to happen. As the repeated region is passed down from one generation to the next it gets longer and longer—so long, indeed, that the two parts of the X chromosome that it connects begin to spread apart. The function of FMR1, and perhaps of many other genes nearby, is disrupted. The growing thread of out-of-control DNA connecting the two parts of the chromosome becomes so fragile that the chromosome often breaks just at this point.

Bad news indeed for the carriers of this out-of-control microsatellite. But this process is not limited to the fragile X syndrome: another microsatellite in another part of the X chromosome occasionally does the same thing, and other microsatellites on the rest of our chromosomes can also go out of control, breaking those chromosomes as well.

In addition many other, shorter microsatellite repeats are scattered throughout our chromosomes. They do not damage the chromosomes, but when they get to be more than forty or fifty motifs long they too tend to go out of control, getting longer and longer and giving rise to genetic diseases.

The striking thing is that virtually all these so-called repeat diseases affect brain function. More than a dozen of them are now known, and the list is lengthening swiftly. Some of them, like Huntington's disease, have received a great deal of publicity. Others, like myotonic dystrophy and Machado-Joseph disease, are much rarer and therefore less studied. But all have dramatic effects, and most of them result from having too many CGG or CAG repeats. Often these repeats are found inside genes, which means that the genes code for strange proteins that have long stretches made up the same amino acid—usually, but not always, glutamine.

When these distorted proteins are introduced into bacteria or into animal cells in culture, they can kill the cells, but nobody yet knows why. As you might imagine, hundreds of laboratories around the world are working on the problem.

It seems remarkable that so many of these odd and dangerous mutations affect brain function, rather than anything else. Why do some of them not affect liver function or the immune system? Perhaps so many of our genes are involved in brain function that repeat diseases are by chance more likely to affect the brain than something else. Alternatively, repeat mutations may indeed damage different organs, but because their consequences are less dramatic than the ones affecting the brain, they have not yet been discovered.

Even common mental illnesses like schizophrenia and manic depression may be influenced by these genetic stutters. One of the hallmarks of repeat diseases is that they can get worse as the distorted genes are passed down from one generation to the next and the repeat lengthens. This process has been given the name *genetic anticipation*— it is as if previous generations anticipate the more terrible consequences seen in generations yet to come. Although specific DNA repeats have yet to be traced to schizophrenia and manic depression, in some families these illnesses show signs of genetic anticipation. And recent studies have shown that unusually long repeats tend to be found among the genes of schizophrenics, though again the specific genes involved have yet to be tracked down.

Regardless of whether their association with brain function is real or a statistical accident, these microsatellites have intriguing properties. They can evolve very swiftly—and, as I have repeatedly emphasized, our large and highly flexible brains have been the result of swift evolution.

A decade ago (a lifetime in today's fast-moving scientific world) I wrote a book called *The Wisdom of the Genes* in which I predicted the existence of a class of genetic variants that can evolve swiftly—though I had no clear idea at the time of what they were like. This class, I suggested, would consist of genes that have become very good at evolving.

To make a convincing argument for such genes, I had to tread through a bit of a theoretical minefield. Obviously, neither organisms nor their genes can see into the future; genes cannot evolve quickly in order to protect some of us against an unexpected disaster, like the arrival of a giant asteroid. But over time genes can change swiftly and adapt their carriers to alterations in the environment that are more predictable—the coming and going of ice ages, the alternation of dry and wet periods in the climate, changes in the amount of sunshine that select for increases or decreases in skin pigment. If such environmental cycles are repeated often enough, they will select for genes that are capable of mutating quickly, adapting their carriers to these new but not totally unexpected environments.

I called such genes potential-generating genes, because they were capable of generating new evolutionary potential. Soon afterward one type of potential-generating gene was discovered, in an unexpected place, by several groups of workers, including Richard Moxon and his group at the University of Oxford. They found that many disease-causing bacteria carry short microsatellites, in or near a number of their genes. When these microsatellites change in length, they can turn the genes on or off.* The genes act as tiny switches, regulating the production of proteins that are found on the bacterial surface or of the toxins that the bacteria excrete.

Moxon, together with Richard Lenski of the University of Michigan, called these repeated regions contingency genes, because changes in their length could adapt the bacteria swiftly to different environmental contingencies. The striking thing was how quickly and easily these contingency genes could mutate back and forth, providing the bacteria with remarkable evolutionary flexibility.

All mutations can in theory be reversed by subsequent mutations that precisely undo the first change. But most of the time reverse mutations are rare—the chance that a given base change in a gene will be reversed is small, when you consider all the other base changes that are more likely to happen to the gene first. Microsatellites are highly unusual, because the mutation process is so reversible—they can become longer or shorter in units of one or more motifs, because the enzyme that copies the DNA is very prone to make such mistakes.

Having such contingency genes is very costly. If a bacterium must switch some genes off and others on in order to survive some new

---

* They do this by throwing the DNA of the genes in and out of reading frame. Because the information in a gene is read in groups of three bases, a repeat that consists of motifs four or five bases in length can throw the reading frame off if it becomes shorter or longer by one or more motifs. But subsequent mutations can easily put the gene back on track.

challenge from its host, then only the rare mutants that have this capability will survive. Most of the bacterial population will die. This is no problem for bacteria, which can quickly replenish their populations from a few survivors. But for more elaborate organisms such as ourselves it is a problem, which is almost certainly why such crude on-off contingency genes are not found in our own genomes.

We do, however, have many little microsatellites that are scattered among our own genes. Even though they are not serving as simple on-off switches, they may have more subtle functions. The burning question that our laboratory and many others are investigating is whether our little microsatellite stutters, too, can affect the genes that they are in or near. And if so, what might their effect be?

Moxon and I have speculated that they might be important in brain function, since so many of them, when they get out of control and become too long, can give rise to neurological conditions. Most of the time, luckily, they remain short and do not cause disease. But even in people without disease they are not all of one length—many alleles of different lengths are found at each of these potentially disease-causing microsatellites. These normal alleles do not cause disease, but do they have a function? Do they fine-tune our brains in some fashion, adding innumerable "factory options" to that basic brain?

Most of the microsatellites probably have little effect, but the list of microsatellites that have an influence is growing. The most striking example is a pair of microsatellites in a male sex hormone receptor gene. If one of the microsatellites is too long, it produces a neurological disease. But shorter alleles can have striking effects too—alleles that are not quite long enough to cause overt disease are associated with infertility, and very short alleles of both microsatellites are associated with prostate cancer.

Even if only a small fraction of the thousands of microsatellites in our genomes exert detectable effects, the number of possible combinations of options would be huge, because there are so many different-length microsatellite alleles in or near so many of our genes. Each of us is sure to have a different mix.

Since these alleles are known to mutate so swiftly, increasing and decreasing in length, the mix in our whole gene pool can change relatively quickly. Further, we know that these little stutters can come and go in the course of evolution. The microsatellites that we humans carry are sometimes in different places from the microsatellites found on the chromosomes of chimpanzees.

Exciting questions abound: Do humans have more of these little stutters than chimpanzees? If so, many new potential-generating mutations may have appeared in the course of our evolution. There are some indications, particularly from the laboratory of Simon Easteal at Australia's National University, that this may be so, although the data are difficult to interpret at the moment because most of the microsatellites have been found in humans and then looked for in chimpanzees, rather than the other way around. A concerted effort to find similar microsatellites unique to chimpanzees will provide a truer gauge of their prevalence in the two species.

Another question is whether neurological diseases caused by microsatellite repeat expansions are found uniquely in our own species or whether they can occur in chimpanzees and other apes as well. If they are unique to us, then perhaps they represent a genetic cost because our brains have evolved so quickly. Preliminary results suggest that this may be the case—several repeats that can lead to disease tend to have longer alleles in normal humans than in normal chimpanzees—though there is currently much argument over whether this, like Easteal's finding, may be a statistical artifact.

Schizophrenia, which afflicts one percent of the population, is a likely source of such genetic cost. Because schizophrenics tend to have few children, the genes involved should be removed rapidly from the population unless their damage is balanced by some benefit. Some early studies had suggested that the siblings of schizophrenics tend to be brighter than average, perhaps giving the genes a heterozygote advantage like sickle cell, but more careful work has now shown that the "unaffected" siblings of schizophrenics often show slight neurological deficits. This makes the prevalence of the disease even more puzzling.

As we saw earlier, there is growing evidence that repeat expansions may be involved in the disease. So, if repeat diseases are at work here, it is reasonable to wonder what possible advantage these repeats might have that counters the very large disadvantage of the disease. The answer may be hard to come by, since it may require a deeper understanding of how our brains work than we have at the present.

It is also important to ask whether the incidence of schizophrenia is as high in other primates as it is in us. If it is not, then the likelihood will be increased that this and other widely prevalent mental illnesses may be a cost incurred by our runaway brain evolution. But this, too, will be difficult to ascertain, because the symptoms of schizophrenia are likely to be very different in chimpanzees and gorillas. This will make the disease far more difficult to recognize in these primates— and goodness knows, schizophrenia is already hard enough to diagnose consistently in humans. Almost certainly, the answer will have to await the discovery of the genes that contribute to schizophrenia, followed by a comparison of these genes in ourselves and our close primate relatives.

HAVE WE DOMESTICATED OURSELVES?

Microsatellites can hardly be the whole explanation for the evolving diversity in the abilities of our brains. But they may help explain how we have been selected for greater and greater behavioral diversity in the past, and how we continue to be selected for even more diversity in the present.

If the growing diversity of our cultures is selecting us for genetic diversity, it is likely to be somewhat different from the diversity that our ancestors possessed. The emergence of new skills that were necessary for survival, and of continually shifting behavioral norms in our evolving societies, must have had an impact on our genes. How flexible are we, and how easily can we adapt to such shifting circumstances?

Neuroscientist Fred (Rusty) Gage of La Jolla's Salk Institute has recently reopened the old question of just how much our environment really influences our brains. His recent experiments have built on a series of remarkable studies by previous investigators, chiefly a group at Berkeley who started work in the 1950s. In the Berkeley experiments rats were raised in either stimulating or deprived environments, and the effects on their brains were measured.

The researchers found that the brains of the stimulated rats had a thicker cortex, and that the connections between their neurons were more numerous and complex. But at the time it was not technically possible to determine whether the number of neurons had increased.

In the mid-1990s Gage carried out similar experiments with mice. He used a combination of techniques to show that very young mice that are exposed to a stimulating environment gain neurons as they grow up. The increase that he measured occurred in the dentate gyrus, a part of the temporal lobe that has been implicated in memory, though many other parts of the brain may have gained as well.

In addition, and remarkably, Gage and his group showed that the stimulated mice were cleverer than the others. Mice in both groups were put into a water "maze," essentially a round tank filled with opaque water. Like their enemies the cats, mice do not like to swim, but they will do so if necessary. A glass platform was placed at a specific point below the surface of the water, undetectable to the mice until they found it with their feet and were able to stop swimming.

After repeated trials, Gage and his colleagues found that the stimulated mice discovered the platform significantly more quickly than the unstimulated mice. Even though the stimulation they had undergone earlier had had nothing to do with water mazes, they did very well at meeting this new challenge. As a result of their environmental enrichment, their brains had become better at dealing with the unexpected.

His experiments also illustrated another aspect of evolution. His mice had, in a way, actually cooperated with the researchers in the course of the experiments because of their recent evolutionary history. This remarkable fact emerged when he discussed his experiments during a meeting of our human evolution group, and it illustrates vividly a central point of this book. The ways in which the mice had responded to the new challenge of the water maze were functions of both their environment and their evolutionary history.

The mice that Gage and his group normally used were laboratory animals, bred for docility. They swam about, found the platform, and then simply stood on it, relieved that they could finally stop swimming. This enabled the experimenter to pick them up and repeat the challenge.

But when mice of a strain that had recently been captured from the wild were used, the outcome was very different. The mice swam furiously about, found the platform—and then immediately leaped from it and scurried away across the laboratory floor!

These wild mice were particularly strong and active: perhaps the laboratory mice would have done the same thing had they been physically capable. But they didn't, and the maze experiment could therefore be used to measure the improvement in their intellectual capabilities.

Indeed, Gage's entire experiment had been designed to take advantage of the capabilities of laboratory, not wild, mice. He observed that when wild mice were put in the original enriched environment, they did not react the way the laboratory mice did. The laboratory mice rushed around and explored things, while the wild mice spent very little time exploring. As soon as they determined there was no food to be had, they tended to fall asleep in a corner.

As Gage soon found, measuring the brain power of the wild mice would have required a different sort of experiment. They had greater physical resources than the laboratory mice and had no hesitation about using them. Did they get smarter in their enriched environment? Perhaps not, since they seemed to pay so little attention to it.

In miniature Gage's experiment may illustrate what has happened during our own evolution. Like his domesticated mice, we can rise to new intellectual and even physical challenges, but we are probably doing so in a different way from the routes our ancestors took. Things might have been different for our species in the past, even the recent past, before people were tamed by civilization.

Suppose I could travel back in time to the court of Attila the Hun and attempt to give him an IQ test. He would probably have my head lopped off without thinking about it, and then get on with the business of subduing his enemies and raping their women.

We have, on the whole and for better or worse, intellectualized our world. Our efforts have made that world a tamer place and have modified our behaviors accordingly. Over time, in such a diverse but less dangerous world, selection for survival may have had effects on us that are similar to the selection for docility that had taken place in Gage's domesticated mice. Hun-like behavior is less advantageous than it once was. Perhaps we, too, have been tamed.

THIRTEEN

◈

The Final Objection

*There exist no data which should lead a prudent man
to accept the hypothesis that IQ test scores are in any
degree heritable. This conclusion is so much at odds
with prevailing wisdom that it is necessary to ask, how
can so many psychologists believe the opposite?*

LEON J. KAMIN, *The Science and Politics of IQ* (1974)

In the last few chapters I emphasized the roles that genes, and their
interaction with the environment, have played in our rapid evolution.
Most of my readers will, I assume, feel quite comfortable with a viewpoint that gives weight to both factors. But some will also be seething
by this time. They will demand to know how I can suggest that genes
play any role at all.

A number of researchers, notably Leon Kamin of Princeton
University, have suggested that as far as they can see, no role for genes
need be postulated to explain differences in the ways people think.
Such genes may be there, they point out, but there is no evidence for
it. Environmental variation can explain all the observed differences.

Kamin's arguments are interesting, and he has a superb ability
to find flaws in the experimental designs of studies that psychologists
and geneticists have carried out to measure the impact of genes on
brain function. But flaws or not, one gigantic and unavoidable fact
tells us that the genes must really be there.

This fact is simple: Our brains have evolved in the past. They could not have done so unless genes played a role. Evolution does not take place unless genetic differences exist on which natural selection can act.

As the Dutch botanist Wilhelm Johannsen first showed at the turn of the century, variation that is entirely the result of the environment cannot be selected for. He examined populations of highly inbred plants that had almost no genetic variation and found that the plants varied among themselves. The variation they displayed, however, was entirely due to environmental differences that were encountered by each seedling as it grew up. When he tried to select for seed size, he found that the seeds of subsequent generations were unchanged.

Critics of a role for genes will concede that humans have certainly evolved a great deal, both physically and mentally, in the past. But now, they will claim, everything has changed. We are no longer simply the prisoners of our genes, forced onto some continuing evolutionary path by factors beyond our conscious control.

Actually, our conscious control has probably played a huge role in our physical and mental evolution throughout the process. In *The Runaway Brain* I elaborated on the old idea that much of our rapid evolution over the last few million years has been the result of our response, not to changes in the natural environment, but to modifications that we ourselves have made in the environment. This progressive modification, and the ever greater and more human-oriented complexity that resulted, selected for those of our ancestors who were best able to take advantage of these new circumstances. The upshot was that we have been continually pushed further and further away from adaptation to the natural world.

This idea dates back, like so much else, at least to Darwin, who remarked in chapter 5 of *The Descent of Man* (1871):

> Now, if some man in a tribe, more sagacious than the
> others, invented a new snare or weapon, or other means
> of attack or defense, the plainest self-interest, without
> the assistance of much reasoning power, would prompt
> the other members of the tribe to imitate him; and all
> would thus profit. . . . If the new invention were an

important one, the tribe would increase in number, spread, and supplant other tribes. In a tribe thus rendered more numerous there would always be a rather better chance of the birth of other superior and inventive members. If such men left children to inherit their mental superiority, the chance of the birth of still more ingenious members would be somewhat better. . . . Even if they left no children, the tribe would still include their blood-relations; and it has been ascertained by agriculturists that by preserving and breeding from the family of an animal, which when slaughtered was found to be valuable, the desired character has been obtained.*

In my book I took this argument a step further and suggested that such events have escalated into a runaway process. Once this brain–body–environment feedback loop was established, it proceeded at an ever-increasing pace as the complexity of culture and the opportunity for new inventions increased. Because this feedback loop involves cultural changes, and these changes have accelerated dramatically, there is no reason to suppose that selection for increased intellectual capacity in our species has slackened.

Many genes can increase brain capacity in a general way, and these genes have been selected for. Additional variant alleles, accumulating in the population over a long span of time but not possessed by all of us, have provided us with a still more elaborate array of potential behavioral options. But, as we saw in Chapter 12, so many genes are involved that a given gene has no obvious connection to a given behavior. And the environment plays an enormous role.

This argument is rather different from the hereditarian–environmentalist dispute that has disfigured so much of social science—and that is responsible for so many of the conflicts in our society at large. Even extreme partisans of these two views know that the truth lies somewhere between, or at least they pay lip service to such a compromise. But it is useful to examine the assumptions that underlie these

---

* Here Darwin also anticipated the important evolutionary idea of kin selection, in which our inherited capacities can be passed on by our relatives even if we die without issue. One stands in awe of the man, who could cram two such seminal ideas into one paragraph!

extreme views in order to see what they imply in evolutionary terms. When we do so, the fundamental illogic of both extremes becomes glaringly apparent.

The extreme environmentalist view is that we all have the same innate capacities. Only because of unfortunate genetic or environmental accidents do some of us do less well in society than others. If this extreme were true, then selection *for* increased capabilities would be impossible, since there is simply be no genetic variation for such capabilities. Moreover, the environmentalist view implies that all our evolution has taken place in the past: For reasons that environmentalists have not explained, our brains are finished pieces of work, and we can only mess them up. The only natural selection that can operate today is selection against harmful mutations, such as the deletion that causes Williams syndrome. These mutations appear occasionally, damaging the brains of their carriers by bludgeoning them biochemically or developmentally.

The extreme hereditarian view, on the other hand, is that our capabilities are dictated entirely by our genotypes. Genetic variation is sufficient to explain all the differences among people. In this view we are all the prisoners of our genes; environmental manipulation will have no effect, or at least no lasting effect. An extreme hereditarian is able to predict confidently that the gains of Head Start will be only temporary, like the muscles produced by exercise that shrink away soon after we stop.

The particularly pernicious hereditarian view espoused by the authors of our favorite stalking horse, *The Bell Curve*, makes a further assumption: that all of us can be ranked from best to worst according to our genotypes. If our evolution is to continue, in this view, then we must all be sorted out quite mercilessly according to this ranking. After all, this is how it happened back in a state of nature, when only the best-adapted members of the population were able to survive and reproduce. But because we have been incautious enough to relax the rigors of natural selection, the less well-adapted genotypes have begun to survive, to the detriment of our species.

Partisans of both extremes agree that a relaxation of natural selection would be undesirable because it would allow deleterious genes to accumulate. To a pure environmentalist, however, most people carry a single good genotype, while bad genes are rare and distressing accidents.

To a pure hereditarian, most of us are bad, and only a few noble genotypes, especially those belonging to the hereditarians themselves,
deserve to survive.

In the United States the seesaw battle between these extremes
has coincided with the alternating political ascendancy of liberal and
conservative views. The most recent triumph of the hereditarian viewpoint took place during that intellectual nadir known as the Reagan
years. But this temporary dominance of the hereditarians is already
beginning to melt away.

As we have seen, evidence is growing that crude genetic and
environmental damage causes most cases of diminished intellectual
capacity, and that without such damage anybody's brain is perfectly
capable of functioning at a high level. Further, it has now been shown
that early environmental intervention can have a dramatic effect. Even
if the effect fades away in the absence of continued stimulation, this
does not mean that the stimulation should not be carried out in the
first place, merely that it must be continued throughout life. In these
very important matters, the environmentalists are right.

The environmentalists are also winning on another front, where
a bastion of the extreme hereditarian viewpoint has recently taken a
severe battering. Over the years a number of apparently clear-cut studies seemed to show that genes play a large role in determining how
well we can think. The conclusions of these studies have now come
into question.

The idea of using twins to study the influence of genes was first
proposed by Darwin's cousin Francis Galton over a century ago. Since
Galton's time, it has been realized that the best way to do so would be
to find identical twins who were separated at birth and raised apart in
different adoptive families. Identical twins have identical genotypes,
but those pairs who are separated are subject to different environments,
presumably as different as those encountered by two unrelated children. So if twins turn out to be very similar to each other, it must be
because their genes have had a large influence, while if they are very
different from each other, their different environments must have
played the larger role.

It was very difficult to track down substantial numbers of sets of
twins who really were separated shortly after birth and subsequently
had nothing to do with each other. Moreover, the whole field of twin

studies has suffered for years from severe embarrassment, after the discovery that one of its early practitioners, the late British psychologist Cyril Burt, faked an unknown but probably substantial part of his data. Burt was anxious to prove a large role for genes, and his political agenda seems to have overwhelmed his scruples. Nonetheless, a number of quite good data sets have recently been gathered, particularly by a group of psychologists headed by Tom Bouchard of the University of Minnesota. Their conclusions are remarkably similar to those of Burt, though the effect of genes that they find is not quite so strong.

In these data sets the IQ scores of the twins raised apart are found to be tightly correlated, almost as tightly as the scores of twins that have been raised together. These correlations can be used, after some statistical manipulation, to estimate the influence of the genes, and the results suggest that genes are responsible for at least sixty and perhaps seventy percent of the variation in IQ. This value, called heritability, is a rather complicated one, but taken at face value it suggests that genes play a very large role.

These results were used by the writers of *The Bell Curve* to claim that IQ meritocracies are being established in modern human societies. The children and grandchildren of people with high IQs will tend to have substantially higher-than-average IQs, and the reverse should also be true.

But a powerful argument has recently been made that the contribution of genes to IQ is surprisingly small, certainly smaller than the twin studies would suggest. According to this argument a factor that was downplayed in the earlier studies is much more important than anyone had supposed. That factor is the one part of their environment that even twins who were separated at birth share: the time they spent in the womb.

Bernie Devlin and his colleagues at the University of Pittsburgh School of Medicine and Carnegie Mellon University looked at the data from studies of identical twins raised apart. They also examined data from more than two hundred other studies of IQ correlations between various relatives, some raised together and some raised apart. To begin with, they noticed, as others had done before them, that twin studies by their very nature tend to emphasize the role of genes. The heritability estimates for twins raised apart tend to be much higher than estimates obtained from more distant relatives.

This is because the heritability measurements are different in the two situations. When twins are examined, all the effects of all their genes come into play, including the effects of those genes that depend for their expression on their genetic context—that is, on all the other genes that both twins happen to carry on their sets of chromosomes. However, when the twins marry different people and pass their genes on to the next generation, these genes are combined with the genes from the other parent and thus find themselves in a completely different genetic context. Some of these genes' effects will vanish or will alter in unexpected ways. Heritabilities will be lowered.

Thus, when the correlations between relatives who are more distant than twins are examined, lower heritabilities are found than those seen in twin studies. Only some of the influences of the genes that these relative share—those that can still be expressed in different contexts—can be detected.

This effect undoubtedly decreases the impact of the genes and makes the *Bell Curve* meritocracy idea less tenable. But Devlin and his colleagues discovered an additional reason for the greater heritabilities found in the twin studies.

Dizygotic or fraternal twins are the result of the fertilization of different eggs by different sperm in the same mother, and therefore they are no more genetically alike than siblings born at different times. But despite this resemblance, fraternal twins who have been raised together tend to show a much higher heritability for IQ than nontwin siblings who have been raised together.

This well-known observation had always been attributed to the fact that nontwin siblings, unlike dizygotic twins, are born at different times and as a consequence are raised under slightly different conditions. But Devlin and his group were struck by the fact that the effect was surprisingly substantial, and they wondered whether much of it could be attributed to the shared intrauterine environment of the fraternal twins before they were separated. If it could, then perhaps a lot of the resemblance between identical twins could also be attributed to their shared intrauterine environment.

Statistical analysis confirmed their hunch. When they built genetic models that ignored the effect of the intrauterine environment, and compared these models to all the data from the two hundred studies, the models fit the data very poorly. But models in which the intrauterine environment was taken into account fit the totality of the data very well. These models cut the heritability estimate in half, to a mere thirty-four percent. Thus only a third of the resemblance between IQs seen in relatives can be traced to their genes.

The environment inside the mother's womb is much more important in determining IQ than had previously been supposed. Further, it is astonishingly labile. Like a river that cannot be stepped into twice, a mother's intrauterine environment changes over time. It is this surprising fact that had misled earlier IQ workers—they had not imagined that the environment provided by an individual mother could change so much from pregnancy to pregnancy.

With their finding Devlin and his colleagues have effectively destroyed the IQ meritocracy idea. The genetic effect is so small that while parents and children may resemble each other for genetic reasons, the genetic connection between grandparents and grandchildren is far more tenuous. The genetic meritocracy predicted in *The Bell Curve* could never be established. It would be destroyed by both upward and downward mobility in very short order. There is no way of predicting who will be on the top of the meritocracy in the future.

Delvin and colleagues' findings should come as no surprise to readers of this book, for we have already seen how important the intrauterine environment can be. And that environment can vary even between fetuses carried in a single pregnancy: recall that the schizophrenic member of a discordant twin pair is almost always the lower birth-weight twin.

The fact that twins must share limited resources from their mothers can also work against both of them. Compared with their siblings, twins show a small but significant deficit in IQ, a deficit that various studies have shown to range from four to seven points. The intrauterine environment plays an immense role, and the more we learn about it, the greater we understand that role to be.

Environmentalists, however, need not celebrate a complete triumph. Genes do contribute substantially to IQ, so the evolution of our mental faculties has hardly come to a complete stop. The low heritability means that evolution might be a little slower, but it also means that it is taking place across a broader front.

This is perhaps the most exciting, and encouraging, conclusion that has emerged from the work of Devlin and his colleagues. No brilliant IQ meritocracy has forged ahead in the evolutionary race, leaving the rest of us in the intellectual dust. Instead, we are all playing a role, and a substantial role, in ongoing evolutionary change. The pool of genetic variation in our species is growing in diversity, both from new mutations and from the mixing together of disparate human groups. We are all threads in this grand evolutionary tapestry, and as our genes weave in and out and recombine from one generation to the next, hereditary meritocracies will play no part. We have been shaped, and continue to be shaped, not by a genetic elite but by a vast genetic democracy.

FOURTEEN

✦

Our Evolutionary Future

It would appear, then, that man has, physically
speaking, specialized in unspecialization and by this
means has won himself a new span of evolutionary life
and development. . . . He advances mentally or technically by modifying his environment to suit his needs
instead of, as heretofore, altering his physique and its
needs. . . . So he wins new extension in evolution, but
that change has henceforward to be with increasing
speed and increasingly psychological.

GERALD HEARD, *Pain, Sex and Time* (1939)

Mayflies survive as adults for only a few hours, before dying and falling in uncounted numbers back into the rivers and streams from which
they had so recently emerged. Humans, on the other hand, survive
for the biblical threescore years and ten, and these days most of us
manage to do even better. Still, in an evolutionary sense, both we and
mayflies exist for only a fleeting moment.

Unlike mayflies, humans think about our future, but we tend to
think only a few years ahead. The best prediction we can make about
the more distant future is to say rather feebly that it is likely to be
amazing. Consider our potential physical environment. The inhabitants of the Tibetan plateau have already penetrated the fringes of space,
but some of our descendants will go much farther. In a century or two

259

some of us will probably be living on other planets of our solar system and even on planets circling nearby stars.

Physiologist John West recently reminded our human evolution study group that the astronauts who visited the moon found themselves in a gravitational field one-sixth as strong as that of Earth. They quickly discovered that it was much more convenient for them to hop from place to place than it was to walk. How many generations would it take, on a low-gravity planet, for nature to select for the nimblest hoppers among the colonists? How long before their descendants' legs and pelvis bones changed as a result? Would their skeletons and musculature eventually converge with those of kangaroos? The rate at which such changes happen will depend on the severity of the conditions on the new planet and how necessary it will be to keep hopping in order to stay alive!

The biochemistries of the animals and plants that accompany the colonists will also change with time as they become adapted to new soil and light conditions. Inevitably the biochemistries of the colonists will follow suit. Eventually, perhaps within a few thousand years, the descendants of the colonists will have great difficulty surviving if they return to the earth. Because gene flow will be interrupted, they, like the animals and plants they have taken with them, will be on their way to becoming new species.

The new worlds themselves may be more numerous than we think. Some totally unexpected breakthrough in our understanding of physics may allow us to travel faster than light, or even to discover and colonize alternative universes. Such a breakthrough would greatly accelerate our expansion into many very different environments.

These individual colonists will doubtless be the adventurers among us, who have perhaps always been the ones to evolve most quickly. The adventurous ancestors of the people of the Gran Dolina and the Sima de los Huesos made their way to the farthest reaches of Europe from Africa and the Middle East more than a million years ago. Once isolated, these people went in a different direction from the ancestors of the rest of us and became the Neandertals.

In today's cramped, crowded world, such isolation is not possible, and indeed even the most isolated of living peoples are rapidly blending into the genetic mainstream of our species. But if, in the future, the region of the universe that we manage to inhabit becomes sufficiently large, such gene exchange will be impossible. Our species will once again fragment, as it has done so many times in the past. Science fiction writers have speculated about the possible consequences of such fragmentation. Our children's children are likely to live through the reality.

THE FUTURE OF OUR BRAINS

In the meantime, what about those of us who stay at home? We, too, will not be evolutionary couch potatoes. We will continue to change physically, probably at roughly the same rate that our species has always changed. But our biggest changes will undoubtedly come about because of our growing ability to explore the potentials of that basic human brain that I have been talking about in the last few chapters.

These capabilities of our brains have been building for millions of years. Harnessing them more fully will actually accelerate our evolution, though not necessarily in the directions that we might expect.

To understand this process, we must once again inquire into the various ways in which evolution works. In 1930 one of the great founders of the theory of evolutionary genetics, R. A. Fisher, made a simple but profound observation: The maximum rate of increase in evolutionary adaptation must be proportional to the amount of genetic variation for adaptability present in the population. This observation seems in retrospect obvious, and indeed Fisher was so certain that his insight had all the properties of a mathematical axiom that he named it the Fundamental Theorem of Natural Selection. The more variation there is, the more evolution can take place.

In the decades since, Fisher's theorem has been much argued over. For it to be true, the population's variation must not only be genetic, it must be selectable. Some of the variation detected in twin studies on IQ cannot be selected, as we have seen, because it depends so heavily on the interactions among various genes. As a result, it tends to disappear, or to change its effects dramatically, from one generation to another. This results in a paradox: a population may be full of genetic variation, but if some of this variation is so dependent on genetic and environmental context that it tends not to retain its character from one generation to the next, then natural selection will be unable to act on it.

A second restriction on Fisher's theorem is that a population may have a great deal of selectable genetic variation, but selection simply happens not to act on it in any systematic way. For instance, as we saw, chimpanzee populations are full of genetic variation, but they have changed very little for millions of years. Variation alone, even selectable variation, will not ensure that evolution proceeds in any particular direction. (It must be remembered that Fisher was dealing with the maximum possible rate of increase in fitness, not the rate that might actually be taking place in any given situation.)

A third restriction, and perhaps the most important, is that even if the genetic variation is both present and selectable, it simply may not matter in a particular environmental context. If some of the genetic variation has little connection to the appearance and capabilities of the organisms that carry it, then Mother Nature may as well save her breath to cool her porridge. No amount of selection will get anywhere.

Whether such genetic variation is revealed or not can depend strongly on what happens to the organism as it grows up. This was demonstrated decades ago in a series of remarkable experiments on fruit flies carried out by the embryologist C. H. Waddington.

Waddington began with the observation that fruit flies raised in the laboratory all tend to be very similar to each other; one would be hard put indeed to distinguish among them. But then he took eggs from those same flies and exposed them briefly to a severe environmental insult, a heat shock or brief exposure to ether vapor. The flies that resulted were much more variable in their appearance, and he was easily able to select for strains that had malformed wings or bodies. After only a few generations he could stop using the environmental shock on the eggs—his selected strains bred true even in the absence of the shock.

Waddington had at first supposed that any changes that he saw would be the result of the inheritance of characters acquired as a result of the stress. Such evolution is often called Lamarckian, although this label is not quite correct. Early in the nineteenth century the proto-evolutionist Jean-Baptiste Monet, the chevalier de Lamarck, had famously suggested that the long necks of giraffes have become stretched through their desire to reach tender leaves high in the trees. This lengthened neck was somehow passed down to subsequent generations.

Leaving aside Lamarck's rococo imaginings about desires influencing evolution, he presupposed an evolutionary mechanism in which characters acquired during the lifetime of an organism are somehow passed on to the next generation. This is what is generally meant by Lamarckian evolution. Waddington tried to hybridize Lamarck and genetics by suggesting that the effects of the developmental malformations he induced would somehow be "assimilated" into the gene pool of his flies.

The inheritance of acquired characters is a no-no among evolutionists, who now realize that what matters is changes in the genes, not in the bodies of the organisms that carry them. A sizzling response from Waddington's critics eventually put him right. He had not created the variation by the stresses he had used—the variation was already in the gene pool of the flies he started with. He had simply revealed it by shocking the eggs. The genetic variants were there all the time but did not disturb development except under unusual conditions.

What happens to fruit flies can happen to us. During the 1950s and 1960s, a ghastly and accidental Waddington-like experiment was carried out on our own species. The drug thalidomide, originally developed as a sedative, was also found to control morning sickness. Many pregnant women, particularly in Europe where it had received approval, took it. The dreadful consequences were soon apparent. Ten thousand babies were born who had limb deformities ranging from relatively mild to extremely severe.

Thalidomide has profound effects on fetal development, but not on all fetuses. Hundreds of thousands of pregnant women took the drug, but only a minority of them had deformed babies. Recent experiments on rats have shown that normal rats can be given high doses of thalidomide without detectable harm to their fetuses, while those that carry a mutant gene predisposing them to mild limb deformity prove to be much more sensitive. The human population must also have mutant alleles that, though normally unexpressed or only weakly expressed, will wreak havoc in the presence of thalidomide.*

The evidence is strong that we can reveal the effects of such genetic variation in our own species by drastic manipulations of the

---

* It is ironic that thalidomide is making a comeback—it turns out to be a very useful drug for treating leprosy and a wide variety of autoimmune diseases. But warnings about its dangers are going unheeded. A black market in the drug has sprung up, and thalidomide babies are already appearing again in Brazil, where leprosy is widespread.

environment. The thalidomide story is just one such case. As we saw in Chapter 2 with the Bhopal disaster, massive industrial accidents can reveal the effects of genes that would otherwise have remained concealed. But Bhopal is just the tip of the iceberg. Even milder environmental insults are likely to have profound genetic consequences, both to us and to the animal and plant species on which we depend.

The same process must take place under natural circumstances whenever the environment becomes more stressful. Evolution in any species, not just our own, is able to proceed more quickly under such stressful conditions, since more of the effects of the underlying genetic variation will be revealed.

But this process as it occurs in a state of nature differs profoundly from the experiments of Waddington. In the case of natural selection, the organisms that have the highest fitness in the stressful environment will be selected for, while those that suffer the most from the stress will be selected against. Waddington, who was acting as a clumsy stand-in for Mother Nature, did the opposite. He selected for, rather than against, the physical deformities that he could see under the microscope.

All this means a great deal for our future evolution. Just as physical stresses reveal genetic differences between individuals in their ability to develop normally, psychological stresses can reveal differences in individuals' ability to withstand such pressures, differences that might have had no impact in the absence of such pressures. Hierarchical differences among British civil servants affect their survival and perhaps their reproductive capability as well. Daily news reports tell us of the profound effects of psychological factors on public health around the world.

Such stresses need not be entirely harmful. Even though the environment in which most of us live is stressful psychologically, it undoubtedly is less physically dangerous than that of *Homo erectus* or even those of our parents and grandparents. But it is certainly much more complicated. What is not yet susceptible to precise measurement is how all this increased complexity is affecting our survival and our reproductive capabilities. But there seems no doubt that the environment of the late twentieth century is revealing more than ever of our underlying genetic variation for brain function and perhaps for other characteristics, and is accelerating our evolution as a result.

Our own direct efforts will soon be adding immeasurably to that environmental complexity. Forthcoming "smart" pills will extend the capabilities of our basic brain, enhancing the bells and whistles we all possess in various combinations, but they will only be the beginning. Once we understand basic brain function, nothing will stop us from providing our own bells and whistles. Hands may be wrung about the ethical questions that such biochemical and even genetic manipulations of brain function raise, but parents given an opportunity to produce smarter children will unquestionably take it.

Pills that alter physical characteristics are already widely used. Some parents of children who happen to be a little shorter than the average are demanding that they be dosed with human growth hormone. Physicians are complying in large numbers, and manufacturers are actually promoting this questionable use. Luckily such treatment is a little safer now that recombinant DNA technology is being used to make the hormone. The dozens of children who contracted Creutzfeldt-Jakob disease during the 1970s, from pituitary extracts of growth hormone obtained from cadavers, were early casualties of our apparently overwhelming desire for all our children to become as tall as Wilt the Stilt.

When true "smart" pills are unleashed on our species, it is difficult to imagine the consequences. The reactions to them will vary greatly, resulting in yet another Waddington experiment that reveals all sorts of hidden genetic variation. We cannot predict who will survive this new biochemical onslaught, or what the consequences will be for our gene pool, but the results will certainly be, to put it mildly, interesting.

Future modifications of our mental capacities need not be confined to what goes on inside our own skulls. Soon, probably in a matter of decades, we will be able to tie our brains directly into vast databases such as the Internet without the clumsy intermediary of computers and keyboards. Perhaps little voices will whisper both information and misinformation into our ears, or RoboCop-like screens of data will scroll continuously across our eyeballs.

Already, even without such unnerving aids, some of us can handle the Internet much better than others and sort out the sensible information from the deluge of pseudoinformation and garbage. My rather Luddite guess is that, in the future, the people who are unable to stop up their ears like the crew of Ulysses' ship are the ones who will fall victim to the siren call of the Internet. Most of them will not die, of course, except for those whose paranoia is fed to fatal extremes by unrestricted access to a crazed universe of conspiracy theories. But they will be effectively removed from the gene pool. The people who are able to stop up their ears are the ones who will still be able to interact with other human beings—and get on with the evolutionarily important business of having babies!

STRULDBRUGGS, INC.

The greatest modification to our bodies that is likely to take place over the next few decades will likely be a substantial lengthening of our life spans. As I write this, media excitement is raging about an enzyme called telomerase that modifies the ends of chromosomes known as telomeres. Normal body cells do not make telomerase, which means that with each cell division, their telomeres tend to grow shorter.* Eventually the cells senesce and die. Cancer cells, on the other hand, continue to make telomerase, and many of them can proliferate indefinitely.

When the telomerase gene is turned on in normal human or mouse cells, they go through more than the normal number of cell divisions before they senesce. Newspaper stories greeted this announcement as if the fountain of youth had been discovered, but in fact—as usual—the story is much more complicated. Mice that lack the enzyme altogether have survived for a number of generations in the laboratory, and their rates of aging and cancer seem unchanged. Telomerase, it appears, is only part of the story of aging.

Nonetheless, the mechanisms of aging will certainly be understood in far greater depth very soon. There is no obvious reason why we cannot lengthen our life spans—we already know that they can change and have done so over a relatively brief period of evolutionary time. Our maximum life span of 120 years is twice that of chimpanzees—the oldest chimpanzee for which we have records died at 57. Yet four million years ago the maximum life span of our ancestors must have been similar to that of chimpanzees. If our life spans have doubled over this relatively brief period, they could very well double again.

Just as with "smart" pills, life-span pills will reveal the underlying genetic heterogeneity of our species and will allow selection to

---

* Poet T. S. Eliot anticipated this when he remarked: "I grow old, I grow old—I wear the ends of my chromosomes rolled!"

act more effectively. The effect on our future evolution is likely to be both profound and unexpected.

A pill that could lengthen our life span without converting our bodies into a mass of cancer cells would be a very hot item. At the same time, it is hard to imagine a medication that would have a more destabilizing effect on society. Will the pill have very different effects on different people, lengthening the life span of some and shortening that of others? What will it do to our reproductive capabilities? As with thalidomide, malformed children might result. And the pill is likely to be expensive—will we condemn most of the developing world to short life spans while some of us, like Jonathan Swift's Struldbruggs, grow older and older and hoard the wealth of the planet? Such a situation is unlikely to persist for long: I envision two-hundred-year-old geezers hanging from lampposts while enraged mobs roam the streets.

Life span lengthening is just one of the new regions of phenotypic and genetic modification into which we are moving pell-mell. At a recent symposium on germline gene therapy held at UCLA, biotechnological hubris seems to have reigned supreme. Future technologies in which not just one but dozens or hundreds of genes could be changed simultaneously were touted. James Watson, the co-discoverer of the structure of DNA, insisted that any attempt to regulate such engineering would be "a complete disaster." Leroy Hood, one of the leaders of the Human Genome Project, pronounced: "We are using exactly the same kinds of technologies that evolution does."

Well, not exactly. The law of unintended consequences will ensure that most of our attempts to modify our bodies and our genes in the future will not work out in the way that we expect. Natural selection will render its cool and inexorable judgment on the hubristic excesses of biotechnologists.

In short, as we have done so often in the past, we are sure to go on providing many inadvertent opportunities for natural selection to act on us.

THE FUTURE OF DIVERSITY

The mixing of gene pools, along with continued mutation, means that our descendants will carry more alleles of the many different genes that influence behaviors. In millennia to come, as additional alleles are selected for, our species will continue to become more genetically diverse.

But scientists tend to feel very uncomfortable when confronted with such a genetically diverse population. What, they ask, is going to maintain all this diversity—why should it not be lost by chance or by other factors over time?

This is a real problem. The sickle cell allele, for example, will be maintained in the human population only as long as selective pressure from malaria is present. But sometime in the next century, and probably even sooner, we will conquer malaria. With that selective pressure gone and only its harmful effects left, the sickle cell allele will gradually disappear, as it is already starting to do in the United States. It will disappear even more quickly if we find an accurate and inexpensive way to fix the gene itself. So will all the other genes that depend for their continued existence on this disease—the thalassemias, G6PD deficiency, the elliptocytoses, and the rest of those junky mutations that we met in Chapter 4. The result will be a loss of population diversity, not a gain.

It might be argued that genes for behavioral diversity should be lost as well, since a uniform future awaits us. In science fiction futures, such as that portrayed in Fritz Lang's film *Metropolis*, huge hordes of people behave robotically because of the power exerted by a small elite. In the real world regimentation sometimes works too, at least temporarily. Early in this century Henry Ford regimented the assembly line, which helped to spark Lang's nightmare vision. But as you will remember, Charlie Chaplin managed to escape from that regimentation in *Modern Times*.

In 1968 I was aboard an Air Pakistan plane about to take off from Paris to Rome. At the last minute the plane suddenly filled up with members of a Chinese Communist delegation. They were all dressed in identical blue Mao suits and wore identical red and gold Mao buttons. They had identical stony expressions on their faces as they tried their best to ignore the blandishments of capitalism that were everywhere around them. It was almost impossible to determine the sexes of the delegates.

Such extreme regimentation, of course, has not lasted long, as any visitor to present-day Beijing or Shanghai will attest. And even older and far more rigid societal structures are being destroyed in the current tidal wave of change. The Indian caste system has managed to last for millennia, but it is now breaking down—the president of India is now an untouchable.

A few years ago I suggested a model for the maintenance of behavioral diversity, based on a remarkable phenomenon discovered in the 1960s by the geneticist Claudine Petit of the University of Paris. She found that mutant male fruit flies were more successful at mating with females if they were rare in the population than if they were common.

Female fruit flies are usually courted by a succession of males before they finally mate with one of them. Petit, and many others who later looked at the phenomenon, found that the rare males, regardless of their genotype, were somehow able to short-circuit this process. It would appear that the females become jaded by the repetitious blandishments of one similar male after another, but are more likely to respond to males that are different—regardless of what those differences might be.

This "rare male effect" is just one of many so-called frequency-dependent phenomena that operate in the natural world. Recently, I came across another and much more complex one. Collaborating with tropical ecologists Richard Condit and Stephen Hubbell, I found strong frequency-dependence in a Panamanian forest filled with hundreds of tree species. If a particular species of tree is sparse in one part of a forest and common in another, the sparsely scattered trees tend to reproduce themselves more rapidly than the common ones. We suspect that this may be the result of the action of pathogens that are specific to each tree species—the pathogens spread when the trees become common and cannot spread when they are rare.

Such frequency-dependence will help to maintain the diversity of tree species in the forest, just as the rare male effect should help maintain genetic diversity in a population of fruit flies. As I analyzed the forest data, I began to wonder: Might the phenomenon of frequency-dependence help maintain and even increase human behavioral diversity?

One fascinating and little-noticed aspect of frequency-dependent selection is that it reinforces itself. In a tropical forest, as tree species evolve in various directions and become more and more different from each other over time, they are less likely to be attacked by pathogens.

This is because, since each host species is becoming different from the others, most pathogens will be forced to specialize on particular host species. Given enough time, each tree species becomes so different from the others that there are no similar tree species with which they share pathogens. Thus diversity itself will be selected for. No wonder that tropical ecosystems, with their huge variety of hosts and pathogens, are brimming with diversity.

In our own species we may be seeing something similar, albeit at the behavioral level rather than at the level of host-pathogen interactions. The opportunities for human behavioral diversity are greater than they have ever been. As a result, more and more of our underlying genetic diversity is revealed. It is like a gigantic Waddington experiment, although the way in which this behavioral diversity is currently being revealed is considerably less draconian than the way in which Waddington revealed the underlying variation in his fruit flies.

If this uncovering of underlying genetic diversity for behavior is combined with frequency-dependent selection acting on behavioral characters, then we have a recipe for ever-increasing behavioral diversity.

In our society people with particular skills often do very well if those skills are unusual. There is, it seems, room in our world for only one Rupert Murdoch. If society were filled with many Rupert Murdochs, none of them could possibly be as successful as the original. While the genes that contribute to Rupert Murdoch–ness might be advantageous when they are rare, they would quickly become disadvantageous if they were to be common.

Of course, in the Darwinian sense, Rupert Murdoch can translate his advantage into an evolutionary one only if he or his close relatives have lots of children. There are no genes for acquiring newspapers and satellite broadcast systems, but the alleles that contribute to entrepreneurship in a more general sense, whatever they may be, are likely to be advantageous when they are rare and disadvantageous when they are common. Many other alleles that contribute, in a complex and highly interactive way, to a great variety of skills should show a similar frequency-dependence. Like that tropical forest, our species can be expected to continue to maintain and increase its diversity.

All this, it seems, is happening just at a time when we need it more than ever.

THE BOTTOM LINE

Our species is moving into a period of great but nonetheless manageable danger. Population pressure, while easing slightly, is still a terrible threat to our health and to the well-being of the ecosystems on which we depend. Nuclear, chemical, and biological warfare continue to be perils, particularly in the hands of nationalists and religious extremists trying to preserve intact their antiquated ideas and their already hopelessly contaminated gene pools. But these dangers, as serious as they are, are unlikely to threaten the survival of our entire species. Most of us live much safer lives than our grandparents did.

We have no guarantee that this safety will continue. A real ecological disaster may lie in our future. The ice caps may melt. The Gulf Stream may, as it has done repeatedly in the past, shift its path. A really devastating disease may suddenly appear, or an unexpected environmental contaminant may make us suddenly unable to reproduce. Our planet may at any moment collide with some rock from outer space. All these scenarios would reduce our population drastically. The survivors, like Dr. Strangelove hiding out with his secretarial pool in his cave, would be anything but a random sampling of our species. But the most probable scenario is that for the next few thousand years we will proceed as we have for the last few thousand, lurching from one near-disaster to the next even as most people manage to survive each generation.

Whether our physical environment becomes safer in the future or takes a turn for the worse, our intellectual environment is certainly becoming more challenging. It is in this realm that our future evolution will primarily take place, continuing and enhancing the trend that has continued uninterrupted for the last several million years. The challenges we will face—traveling to other stars, healing our damaged world, learning how to live with our differences—will be met in part because we will be able to draw on that genetic legacy.

We are the children of Promethean ancestors who set us on this remarkable evolutionary course. Fire was only one of the remarkable discoveries that they bequeathed to us. Are we still evolving? Because we must learn to deal with the costs of all those other Promethean discoveries, as well as with their benefits, it is very lucky for us that we are.

NOTES

INTRODUCTION

1 The epigraph quote from Patrick Synge is from Tom Harrisson, *Borneo Jungle: An Account of the Oxford Expedition to Sarawak* (London: Drummond, 1938).

3 Galdikas recounts her adventures in Birute Galdikas, *Reflections of Eden: My Years with the Orangutans of Borneo* (Boston: Little, Brown, 1995). Food use by the orangs is examined in R.A. Hamilton and B.M.F. Galdikas, "A Preliminary Study of Food Selection by the Orangutan in Relation to Plant Quality," *Primates* 35 (1994): 225–63.

5 The Ridley review is Mark Ridley, "Eco Homo: How the Human Being Emerged From the Cataclysmic History of the Earth," *New York Times Book Review* (17 August 1997): 11.

6 The Borneo fire story is told in Seth Mydans, "Southeast Asia Chokes as Indonesian Forests Burn," in *New York Times* (25 September 1997): A1. Orangutans were captured as they fled the fires, as reported by Mydans in, "In Asia's Vast Forest Fires, No Respite for Orangutans," in *New York Times* (16 December 1997): A1.

8 C.P. van Schaik, E.A. Fox, and A.F. Sitompul, "Manufacture and Use of Tools in Wild Sumatran Orangutans," *Naturwissenschaften* 83 (1996): 186–88, recounts a remarkable story of orang tool use. The orangutan project at the National Zoo is described in Bil Gilbert, "New Ideas in the Air at the National Zoo," *Smithsonian*

27 (1996): 32–41. Estimates of orang population size are made in Simon Husson, "On the Move: The Recent Discovery of Orang-utans in the Peat Swamp Forests of Kalimantan Has Highlighted the Urgent Need to Protect This Fragile Ecosystem," *Geographic Magazine* 68 (1995): S2–S3.

9   Differences in Bornean and Sumatran orang chromosomes are reported in O.A. Ryder and L.G. Chemnick, "Chromosomal and Mitochondrial DNA Variation in Orangutans," *Journal of Heredity* 84 (1993): 405–409.

9   Efforts to keep the subspecies distinct are recounted in Lori Perkins, "AZA Species Survival Plan Profile: Orangutans," *Endangered Species Update* 13 (1996): 10–11. John Bonner, "Taiwan's Tragic Orang-utans," *New Scientist* 144 (1994): 10, gives some details of the illegal orang trade.

CHAPTER 1: AUTHORITIES DISAGREE

17–25   The quotations from this chapter come from: Ernst Mayr, *Animal Species and Evolution* (Cambridge, MA: Belknap Press of Harvard University Press, 1963); Jared Diamond, "The Great Leap Forward," *Discover* 10 (1989): 50–60; Richard G. Klein, "Archaeology of Modern Human Origins," *Evolutionary Anthropology* 1 (1992): 5–14; J.S. Jones, "Is Evolution Over? If We Can Be Sure About Anything, It's That Humanity Won't Become Superhuman," *New York Times* (22 September 1991): E17; J.V. Neel, "The Study of Natural Selection in Primitive and Civilized Human Populations," *Human Biology* 30 (1958): 43–72; Edward O. Wilson, *Sociobiology: The New Synthesis* (Cambridge, MA: Belknap Press of Harvard University Press, 1975); Roger Lewin, *Human Evolution: An Illustrated Introduction* (Boston: Blackwell Scientific Publications, 1993); David A. Hamburg, "Ancient Man in the Twentieth Century" in *The Quest for Man*, edited by V. Goodall (New York: Praeger, 1975); and C. Loring Brace, "Structural Reduction in Evolution," *American Naturalist* 97 (1963): 39–49.

19   Some of the papers detailing recent discoveries in pituitary growth hormone production in dwarfism mutants are W. Wu, J.D. Cogan, R.W. Pfaffle, et al., "Mutations in PROP1 Cause Familial Combined Pituitary Hormone Deficiency," *Nature Genetics* 18 (1998): 147–149; G. Baumann, and H. Maheshwari, "The Dwarfs of Sindh: Severe Growth Hormone (GH) Deficiency Caused by a Mutation in the GH-releasing Hormone Receptor Gene," *Acta Paediatrica*,

423 supp. (1997): 33–38; and J.S. Parks, M.E. Adess, and M.R. Brown, "Genes Regulating Hypothalamic and Pituitary Development," *Acta Paediatrica*, 423 supp. (1997): 28–32. An explanation for the short stature of African pygmies is set out in Y. Hattori, J.C. Vera, C.I. Rivas, et al., "Decreased Insulin-like Growth Factor I Receptor Expression and Function in Immortalized African Pygmy T Cells," *Journal of Clinical Endocrinology and Metabolism* 81 (1996): 2257–63.

19   Brace's estimates of the reduction in tooth size are in C. Loring Brace, "Biocultural Interaction and the Mechanism of Mosaic Evolution in the Emergence of 'modern' Morphology," *American Anthropologist* 97 (1995): 711–21.

20   Differences in wild and domestic cats are recounted in E. Fernandez, F. de Lope, and C. de la Cruz, "Cranial Morphology of Wild Cat (*Felis silvestris*) in South of Iberian Peninsula: Importance of Introgression by Domestic Cat (*Felis catus*)" (in French), *Mammalia* 56 (1992): 255–64. Charles Darwin emphasized the role of sexual selection in *The Descent of Man and Selection in Relation to Sex* (Princeton, NJ: Reprinted by Princeton University Press, 1981).

21   The history of eugenics is recounted in Daniel J. Kevles, *In the Name of Eugenics: Genetics and the Uses of Human Heredity* (Cambridge, MA: Harvard University Press, 1995).

26   The cartoon comes from page 106 of C. Loring Brace, *The Stages of Human Evolution: Human and Cultural Origins* (Englewood Cliffs, NJ: Prentice-Hall, 1967).

28   Surveys of rapid cichlid fish evolution are in L.S. Kaufman, L.J. Chapman, and C.A. Chapman, "Evolution in Fast Forward: Haplochromine Fishes of the Lake Victoria Region," *Endeavour* 21 (1997): 23–30. A recent book about the lake, Tijs Goldschmidt, *Darwin's Dreampond: Drama in Lake Victoria*, trans. Sherry Marx-Macdonald (Cambridge, MA: MIT Press, 1996), gives much fascinating information about this remarkable and fragile ecosystem.

34   The book I admired is Theodosius Dobzhansky, *Mankind Evolving: The Evolution of the Human Species* (New Haven, CT: Yale University Press, 1962).

CHAPTER 2: NATURAL SELECTION CAN BE SUBTLE

38   The prevalence of chromosomal abnormalities is documented in P.A. Jacobs, "The Role of Chromosome Abnormalities in Reproductive Failure," *Reproduction, Nutrition, Development*, Supp. 1

(1990): 63s–74s. The Bhopal disaster and new disasters in the making in India's "Golden Corridor" are recounted in Bruno Kenny, "Gujarat's Industrial Sacrifice Zones," *Multinational Monitor* 16 (1995): 18–21, while the effect of the disaster on pregnancies is examined in J.S. Bajaj, A. Misra, M. Rajalakshmi, et al., "Environmental Release of Chemicals and Reproductive Ecology," *Environmental Health Perspectives*, 101 supp. 2 (1993): 125–30.

38 Criticisms of the declining sperm count studies are summarized in Richard J. Sherins, "Are Semen Quality and Male Fertility Changing?" *New England Journal of Medicine* 332 (1995): 327. The Athens study that suggests there may be something to it is D.A. Adamopoulos, A. Pappa, S. Nicopoulou, et al., "Seminal Volume and Total Sperm Number Trends in Men Attending Subfertility Clinics in the Greater Athens Area During the Period 1977–1993," *Human Reproduction* 11 (1996): 1936–41.

40 Anderson's arguments about germ-line therapy are in John C. Fletcher and W. French Anderson, "Germ-line Therapy: A New Stage of Debate," *Law, Medicine and Health Care* 20 (1992): 26–39. R.E. Hammer, R.D. Palmiter, and R.L. Brinster, "Partial Correction of Murine Hereditary Growth Disorder by Germ-Line Incorporation of a New Gene," *Nature* 311 (1984): 65–67, recounts early transgenic mouse experiments with rat growth hormone. The more recent transgenic mouse experiments are found in J.E. Murphy, S. Zhou, K. Giese, et al., "Long-term Correction of Obesity and Diabetes in Genetically Obese Mice by a Single Intramuscular Injection of Recombinant Adeno-Associated Virus Encoding Mouse Leptin," *Proceedings of the National Academy of Sciences (U.S.)* 94 (1997): 13921–26.

41 The astounding results of these athlete surveys are reported in Michael Bamberger and Don Yaeger, "Over the Edge," *Sports Illustrated* 86 (1997): 60–67.

41 The long history of the domestic dog was quantified in C. Vilà, P. Savolainen, J.E. Maldonado, et al., "Multiple and Ancient Origins of the Domestic Dog," *Science* 276 (1997): 1687–89. The literature on the problems engendered by overselection in domesticated animals and plants is immense. One interesting and typical reference recounts experiments that show how outbred cattle outperform Herefords in many different areas: P.F. Arthur, M. Makarechian, R.T Berg, et al., "Longevity and Lifetime Productivity of Cows in a Purebred Hereford and Two Multibreed Synthetic Groups under Range Conditions," *Journal of Animal Science* 71 (1993): 1142–47.

43 The clever detective work that has gone into understanding the first stone tools is recounted in Glyn L. Isaac, *The Archaeology of Human Origins* (Cambridge: Cambridge University Press, 1989). B. Asfaw, Y. Beyene, G. Suwa, et al., "The Earliest Acheulian from Konso-Gardula," *Nature* 360 (1992): 732–35, reports the oldest Ethiopian stone tools, while the old Israeli artifacts and Pakistani tools were described by Avraham Ronen and Robin Dennell at the European Association of Archaeologists meeting, Ravenna, Italy, September 1997.

45 Some of the consequences of cranial synostosis are set out in P. Fehlow, "Craniosynostosis as a Risk Factor," *Child's Nervous System* 9 (1993): 325–27, and A.L. Albright, R.B. Towbin, and B.L. Shultz, "Long-term Outcome After Sagittal Synostosis Operations," *Pediatric Neurosurgery* 25 (1996): 78–82.

46 Details of the Gallic Roman tooth implant are found in E. Crubezy, P. Murail, L. Girard, et al., "False Teeth of the Roman World," *Nature* 391 (1998): 29.

47 The impressive study of the effects of ginkgo extract on Alzheimer's is P.L. LeBars, M.M. Katz, N. Berman, et al., "A Placebo-controlled, Double-Blind, Randomized Trial of an Extract of *Ginkgo Biloba* for Dementia," *Journal of the American Medical Association* 278 (1997): 1327–32.

47 A fruit fly memory-enhancing gene is reported in J.C. Yin, M. del Vecchio, H. Zhou, et al., "CREB as a Memory Modulator: Induced Expression of a dCREB2 Activator Isoform Enhances Long-term Memory in Drosophila," *Cell* 81 (1995): 107–15.

48–53 Ernest Beutler, a pioneer in the study of G6PD deficiency, reviews recent work in E. Beutler, "G6PD: Population Genetics and Clinical Manifestations," *Blood Reviews* 10 (1996): 45–52. The interaction between the genotype and smoking is reviewed in M.D. Evans and W.A. Pryor, "Cigarette Smoking, Emphysema, and Damage to Alpha 1-proteinase Inhibitor," *American Journal of Physiology* 266 (1994): L593–611.

48–49 Some aspects of the remarkably complex paraoxonase story are detailed in H.G. Davies, R.J. Richter, M. Keifer, et al., "The Effect of the Human Serum Paraoxonase Polymorphism is Reversed with Diazoxon, Soman and Sarin," *Nature Genetics* (1996): 334–36; Y. Yamasaki, K. Sakamoto, H. Watada, et al., "The Arg192 Isoform of Paraoxonase with Low Sarin-hydrolyzing Activity is Dominant in the Japanese," *Human Genetics* 101 (1997): 67–68; and M. Odawara,

Y. Tachi, and K. Yamashita, "Paraoxonase Polymorphism (Gln192-Arg) Is Associated with Coronary Heart Disease in Japanese Noninsulin-Dependent Diabetes Mellitus," *Journal of Clinical Endocrinology and Metabolism* 82 (1997): 2257–60.

51 An articulate expression of the concerns raised by environmentalists is found in Richard C. Lewontin, *Not in Our Genes: Biology, Ideology, and Human Nature* (New York: Pantheon Books, 1984). An utterly contrary view can be found in Roger Pearson, ed., *Shockley on Eugenics and Race: The Application of Science to the Solution of Human Problems*. (Washington, DC, Scott-Townsend Publishers, 1992).

51 The book that caused all the fuss is Richard J. Herrnstein and Charles Murray, *The Bell Curve: Intelligence and Class Structure in American Life* (New York: Free Press, 1994).

52 Many of Gardner's seminal ideas about intelligence are found in Howard Gardner, *Frames of Mind: The Theory of Multiple Intelligences* (New York: Basic Books, 1983).

CHAPTER 3: LIVING AT THE EDGE OF SPACE

55 The epigraph is from Tenzing Norgay, *Tiger of the Snows: The Autobiography of Tenzing of Everest* (New York: Putnam, 1955).

56–57 Something about these mysterious remains can be found in V.H. Mair, "Prehistoric Caucasoid Corpses of the Tarim-Basin," *Journal of Indo-European Studies* 23 (1995): 281–307.

58 General surveys of these adaptations are set out in two review papers: L.G. Moore, S. Zamudio, L. Curran-Everett, et al., "Genetic Adaptation to High Altitude" in *Sports and Exercise Medicine*, edited by S.C. Wood and R.C. Roach (New York: Marcel Dekker, 1994); and S. Niermeyer, S. Zamudio, and L.G. Moore, "The People" in *High Altitude Adaptation*, edited by R. Schoene and T. Hornbein (New York: Marcel Dekker, 1998).

58–59 The Caucasus site is reported in L. Gabunia and A. Vekua, "A Plio-Pleistocene Hominid from Dmanisi, East Georgia, Caucasus," *Nature* 373 (1995): 509–12. A readable popular account of the Yangtze find is in R. Larick and R.L. Ciochon, "The African Emergence and Early Asian Dispersals of the Genus *Homo*," *American Scientist* 84 (1996): 538–51.

59 Vadim A. Ranov, Eudald Carbonell, and Jose Pedro Rodriguez, "Kuldara: Earliest Human Occupation in Central Asia in its Afro-Asian Context," *Current Anthropology* 36 (1995): 337–46, and Susan

G. Keates, "On Earliest Human Occupation in Central Asia," *Current Anthropology* 37 (1996): 129–31, give contrasting views about the dating of the Yuanmou finds.

60–61   The complex history of Tibet is recounted in Hugh Edward Richardson, *Tibet and Its History* (London: Oxford University Press, 1962). Descriptions of the Paleolithic finds in the northern plateau can be found in Z. Sensui, "Uncovering Prehistoric Tibet," *China Reconstructs* 1 (1981): 64–65, and A. Zhimin, "Paleoliths and Microliths from Shenja and Shuanghu, Northern Tibet," *Current Anthropology* 23 (1982): 493–99.

62   The function of the Inca roads in the formation and dissolution of their empire is examined in Thomas Carl Patterson, *The Inca Empire: The Formation and Disintegration of a Pre-capitalist State* (New York: St. Martin's Press, 1991).

63   B. Arriaza, M. Allison, and E. Gerszten, "Maternal Mortality in Pre-Columbian Indians of Arica, Chile," *American Journal of Physical Anthropology* 77 (1988): 35–42, details the causes of death among the mummified Andean women. The *manta* cloth and how it works is explained in E.Z. Tronick, R.B. Thomas, and M. Daltabuit, "The Quechua Manta Pouch: A Caretaking Practice for Buffering the Peruvian Infant Against the Multiple Stressors of High Altitude," *Child Development* 65 (1994): 1005–13.

64–65   The remarkable resistance of Tibetan babies to the growth-slowing effects of high altitude is documented in Z.X. Zhoma, S.F. Sun, J.G. Zhang, et al., "Fetal Growth and Maternal Oxygen Supply in Tibetan and Han Residents of Lhasa," *FASEB Journal* 3 (1989): A987.

65–66   The reversal of the fetal reflex in pulmonary arteries of Tibetans was examined in B.M. Groves, T. Droma, J.R. Sutton, et al., "Minimal Hypoxic Pulmonary Hypertension in Normal Tibetans at 3,658 m.," *Journal of Applied Physiology* 74 (1993): 312–18. The striking ability of their babies to take up oxygen after birth is measured in S. Niermeyer, P. Yang, Shanmina, et al., "Arterial Oxygen Saturation in Tibetan and Han Infants Born in Lhasa," *New England Journal of Medicine* 333 (1995): 1248–52. Evidence for the putative gene for oxygen saturation can be found in C.M. Beall, J. Blangero, S. Williams-Blangero, et al., "Major Gene for Percent of Oxygen Saturation of Arterial Hemoglobin in Tibetan Highlanders," *American Journal of Physical Anthropology* 95 (1994): 271–76.

66   The Ladakh studies are in A.S. Wiley, "Neonatal Size and Infant Mortality at High Altitude in the Western Himalaya," *American*

Journal of Physical Anthropology 94 (1994): 289–305, while the comparative Sherpa and Quechua studies are in R.M. Winslow, K.W. Chapman, C.C. Gibson, et al., "Different Hematologic Responses to Hypoxia in Sherpas and Quechua Indians," *Journal of Applied Physiology* 66 (1989): 1561–69.

67   C. Monge, F. Leon-Velarde, and A. Arregui, "Increasing Prevalence of Excessive Erythrocytosis with Age Among Healthy High-Altitude Miners," *New England Journal of Medicine* 321 (1989): 1271, examines the puzzling prevalence of chronic mountain sickness at high altitude in Peru.

68   Breathing rates of Tibetans and Han are compared in J. Zhuang, T. Droma, S. Sun, et al., "Hypoxic Ventilatory Responsiveness in Tibetan Compared with Han Residents of 3,658 m.," *Journal of Applied Physiology* 74 (1993): 303–11. The puzzling lack of a similar high-altitude response among the Quechua of the Andes is recounted in R.B. Schoene, R.C. Roach, S. Lahiri, et al., "Increased Diffusion Capacity Maintains Arterial Saturation During Exercise in the Quechua Indians of Chilean Altiplano," *American Journal of Human Biology* 2 (1990): 663–68. The response of Sherpas and Quechua to artificial manipulations of oxygen level is examined in J.E. Holden, C.K. Stone, C.M. Clark, et al., "Enhanced Cardiac Metabolism of Plasma Glucose in High-Altitude Natives: Adaptation Against Chronic Hypoxia," *Journal of Applied Physiology* 79 (1995): 222–28.

71   Yak pulmonary circulation is compared with that of cattle in A.G. Durmowicz, S. Hofmeister, T.K. Kadyraliev, et al., "Functional and Structural Adaptation of the Yak Pulmonary Circulation to Residence at High Altitude," *Journal of Applied Physiology* 74 (1993): 2276–85. The remarkable history of the yak, and some features of its physiology and genetics, are recounted in Stanley J. Olsen, "Fossil Ancestry of the Yak, Its Cultural Significance and Domistication in Tibet," *Proceedings of the Academy of Natural Sciences of Philadelphia* 142 (1990): 73–100.

71   DNA was used to determine the relationship of yaks to other cattle in G.B. Hartl, R. Goeltenboth, M. Grillitsch, et al., "On the Biochemical Systematics of the Bovini," *Biochemical Systematics and Ecology* 16 (1988): 575–80.

72   The evidence for a possible allele advantageous at high altitudes is presented in H. E. Montgomery, R. Marshall, H. Hemingway, S. Myerson et al., "Human Gene for Physical Performance," *Nature* 393 (1998): 221.

CHAPTER 4: BESIEGED BY INVISIBLE ARMIES

74 The epigraph is from J.B.S. Haldane, "Disease and Evolution," *La Ricercha Scientifica*, 19 supp. (1949): 1–11. The details of how Linus Pauling became interested in sickle cell disease are taken from personal interviews with Harvey Itano and Jon Singer. A straightforward introduction to the history and symptoms of the disease is Miriam Bloom, *Understanding Sickle Cell Disease* (Jackson, MI: University Press of Mississippi, 1995).

77 The paper that resulted from this pathbreaking work is L. Pauling, H.A. Itano, S.J. Singer, et al., "Sickle Cell Anemia," a Molecular Disease," *Science* 110 (1949): 543–48.

77 Ingram's paper on the differences between normal and sickle cell hemoglobin is V.M. Ingram, "A Specific Chemical Difference Between the Globins of Normal Human and Sickle-cell Anemia Hemoglobin," *Nature* 178 (1956): 792–94.

79 The incidence study, one of several encouraging reports, is F.M. Gill, L.A. Sleeper, S.J. Weiner, et al., "Clinical Events in the First Decade in a Cohort of Infants with Sickle Cell Disease," *Blood* 86 (1995): 776–83.

80 Karen K. Kerle, Guy P. Runkle, and Barry J. Maron, "Sickle Cell Trait and Sudden Death in Athletes," *Journal of the American Medical Association* 276 (1996): 1472, reported on the American athlete case. The lack of effect of the allele on athletic career choice is documented in P. Thiriet, M.M. Lobe, I. Gweha, et al., "Prevalence of the Sickle Cell Trait in an Athletic West African Population," *Medicine and Science in Sports and Exercise* 23 (1991): 389–90.

83 Beet's earliest paper on sickle cell and malaria is E.A. Beet, "Sickle Cell Disease in the Balovak District of Northern Rhodesia," *East African Medical Journal* 23 (1946): 75–86.

83–84 The influence of blood cell rosettes on malaria is discussed in J. Carlson, G.B. Nash, V. Gabutti, et al., "Natural Protection Against Severe Plasmodium Falciparum Malaria Due to Impaired Rosette Formation," *Blood* 84 (1994): 3909–14.

85 Agriculture and its impact on the prevalence of disease is detailed in A.F. Fleming, "Agriculture-related Anaemias," *British Journal of Medical Science* 51 (1994): 345–57. The spread of the disease in Madagascar is told in S. Laventure, J. Mouchet, S. Blanchy, et al., "Rice: Source of Life and Death on the Plateaux Of Madagascar" (in French), *Santé* 6 (1996): 79–86.

86 The work of Livingstone and many others on this zoo of mutations is summarized in Frank B. Livingstone, *Frequencies of
Hemoglobin Variants: Thalassemia, The Glucose-6-phosphate Dehydro-
genase Deficiency, G6PD Variants, and Ovalocytosis in Human
Populations* (New York: Oxford University Press, 1985).

89 The recent spread of these genes is examined in J. Flint, R.M.
Harding, A.J. Boyce, et al., "The Population Genetics of the
Haemoglobinopathies," *Baillieres Clinical Haematology* 6 (1993):
215–62. Ewald's views on disease evolution are set forth in Paul
W. Ewald, *Evolution of Infectious Disease* (New York: Oxford Uni-
versity Press, 1994). I take a rather different view in Christopher
Wills, *Yellow Fever, Black Goddess: The Coevolution of People and Plagues*
(Reading, MA: Addison-Wesley, 1996).

90 The long independent evolutionary histories of *Plasmodium
reichenowi* and *falciparum* are documented in A.A. Escalante and
F.J. Ayala, "Phylogeny of the Malarial Genus Plasmodium, Derived
from rRNA Gene Sequences," *Proceedings of the National Academy
of Sciences (U.S.)* 91 (1994): 11373–77. Anopheline mosquitoes have
difficulties in transmitting *falciparum* parasites, which are described
in A.C. Gamage-Mendis, J. Rajakaruna, S. Weerasinghe, et al.,
"Infectivity of *Plasmodium vivax* and *P. falciparum* to *Anopheles
tessellatus;* Relationship Between Oocyst and Sporozoite Development," *Transactions of the Royal Society of Tropical Medicine and Hygiene*
87 (1993): 3–6.

91 Koella's work on the effect of the parasite on mosquito survival is
not yet published but is discussed in V. Morell, "How the Malaria
Parasite Manipulates Its Hosts," *Science* 278 (1997): 223.

92 The impact of human diseases, particularly polio, on Goodall's
chimpanzees is recounted in Jane Goodall, *In the Shadow of Man*
(Boston: Houghton Mifflin, 1971).

92–93 V. Robert, T. Tchuinkam, B. Mulder, et al., "Effect of the Sickle
Cell Trait Status of Gametocyte Carriers of *Plasmodium falciparum*
on Infectivity to Anophelines," *American Journal of Tropical Medicine and Hygiene* 54 (1996): 111–13, measured the effect of being
heterozygous for sickle cell on mosquito transmission. The original Red Queen paper was L. Van Valen, "A New Evolutionary Law,"
Evolutionary Theory 1 (1973): 1–30.

93–94 Cloning of the Duffy-negative gene is recounted in A. Chaudhuri,
J. Polyakova, V. Zbrzezna, et al., "Cloning of Glycoprotein
D cDNA, Which Encodes the Major Subunit of the Duffy Blood

Group System and the Receptor for the *Plasmodium vivax* Malaria Parasite," *Proceedings of the National Academy of Sciences (U.S.)* 90 (1993): 10793–97.

95   The story of how resistance genes to AIDS were discovered is told in Michael Dean and Stephen J. O'Brien, "In Search of AIDS-resistance Genes," *Scientific American* 277 (1997): 44–51.

96   The possible connection between plagues of the past and the HIV resistance allele is explored in J.C. Stephens, et al., "Dating the Origin of the CCR5– 32 AIDS Resistance Allele by the Coalescence of haplotypes," *American Journal of Human Genetics* 62 (1998): 1507–15. Gerard Lucotte, Serge Hazout, and Marc de Braekeller, "Complete Map of Cystic Fibrosis Mutation DF508 Frequencies in Western Europe and Correlation Between Mutation Frequencies and Incidence of Disease," *Human Biology* 67 (1995): 797–804, takes a detailed look at the CF gene in Europe.

97   The workings of cholera on nerve cells can be found in M. Jodal, "Neuronal Influence on Intestinal Transport," *Journal of Internal Medicine*, supp. 732 (1990): 125–32.

97–98 The mouse model for cholera was investigated by S.E. Gabriel, K.N. Brigman, B.H. Koller, et al., "Cystic Fibrosis Heterozygote Resistance to Cholera Toxin in the Cystic Fibrosis Mouse Model," *Science* 266 (1994): 107–109. Quinton's ingenious explanation for how CF mutants might have been selected was put forward in P.M. Quinton, "What Is Good About Cystic Fibrosis?," *Current Biology* 4 (1994): 742–43.

100   The dangers of *H. pylori* infections are recounted in G.C. Cook, "Gastroenterological Emergencies in the Tropics," *Baillieres Clinical Gastroenterology* 5 (1991): 861–86. Some of the many infectious agents associated with heart disease are discussed in R.W. Ellis, "Infection and Coronary Heart Disease," *Journal of Medical Microbiology* 46 (1997): 535–39. A recently discovered connection between a retrovirus and diabetes can be found in B. Conrad, R.N. Weissmahr, J. Boni, et al., "A Human Endogenous Retroviral Superantigen as Candidate Autoimmune Gene in Type I Diabetes," *Cell* 90 (1997): 303–13.

100–01 The confusing story about possible connections between infections and chronic fatigue syndrome is set out in C.J. Dickinson, "Chronic Fatigue Syndrome—Aetiological Aspects," *European Journal of Clinical Investigation* 27 (1997): 257–67.

CHAPTER 5: PERILS OF THE CIVIL SERVICE

104 The basic findings of the two Whitehall studies are set out in M.G. Marmot, M.J. Shipley, and G. Rose, "Inequalities in Death—Specific Explanations of a General Pattern?," *Lancet* 1 (1984): 1003–6, and M.G. Marmot, G.D. Smith, S. Stansfeld, et al., "Health Inequalities Among British Civil Servants: The Whitehall II Study," *Lancet* 337 (1991): 1387–93.

106 Studies that examine the effects of the fear of privatization are in J.E. Ferrie, M.J. Shipley, M.G. Marmot, et al., "Health Effects of Anticipation of Job Change and Non-employment: Longitudinal Data from the Whitehall II Study," *British Medical Journal* 311 (1995): 1264–69. The retirement follow-up is in M.G. Marmot and M.J. Shipley, "Do Socioeconomic Differences in Mortality Persist After Retirement? 25 Year Follow-up of Civil Servants from the First Whitehall Study," *British Medical Journal* 313 (1996): 1177–80.

107 The connection between heart disease and lack of control over one's life was explored in H. Bosma, M.G. Marmot, H. Hemingway, et al., "Low Job Control and Risk of Coronary Heart Disease in Whitehall II (Prospective Cohort) Study," *British Medical Journal* 314 (1997): 558–65.

108 Depression as a large predictive factor in heart diseases is explored in M.M. Dwight and A. Stoudemire, "Effects of Depressive Disorders on Coronary Artery Disease: A Review," *Harvard Review of Psychiatry* 5 (1997): 115–22, and A.H. Glassman and P.A. Shapiro, "Depression and the Course of Coronary Artery Disease," *American Journal of Psychiatry* 155 (1998): 4–11.

109 The connection between socioeconomic status and health in various societies is explored in N.E. Adler, W.T. Boyce, M.A. Chesney, et al., "Socioeconomic Inequalities in Health: No Easy Solution," *Journal of the American Medical Association* 269 (1993): 140–45. The remarkable mortality among unmarried people in Japan is documented in N. Goldman and Y. Hu, "Excess Mortality Among the Unmarried: A Case Study of Japan," *Social Science and Medicine* 36 (1993): 533–46.

111 Baboon hierarchies are explored in C.E. Virgin, Jr. and R.M. Sapolsky, "Styles of Male Social Behavior and Their Endocrine Correlates Among Low-Ranking Baboons," *American Journal of Primatology* 42 (1997): 25–39.

112 J. Altmann, R. Sapolsky, and P. Licht, "Baboon Fertility and Social Status," *Nature* 377 (1995): 688–90, and A. Pusey, J. Williams, and J. Goodall, "The Influence of Dominance Rank on the Reproductive Success of Female Chimpanzees," *Science* 277 (1997): 828–31, present correlations of fitness with hierarchical positions in baboons and chimpanzees.

113 Frans de Waal describes the remarkable behavior of monkey mothers in ensuring that their offspring associate with the "right" friends in Frans de Waal, *Good Natured* (Cambridge, MA: Harvard University Press, 1996).

CHAPTER 6: FAREWELL TO THE MASTER RACE

114–15 Figures on the huge variation in human fertility from tribe to tribe and region to region can be found in Lyliane Rosetta and C.G.N. Mascie-Taylor, ed., *Variability in Human Fertility* (New York: Cambridge University Press, 1996). Neel's computer model of the Yanomama is described in J.W. MacCluer, J.V. Neel, and N.A. Chagnon, "Demographic Structure of a Primitive Population: A Simulation," *American Journal of Physical Anthropology* 35 (1975): 193–207. Chagnon's book about the Yanomama is Napoleon A. Chagnon, *Yanomamo: The Fierce People*, 3rd ed. (New York: Holt, Rinehart and Winston, 1983).

116 Estimates of the age of the tribal branch leading to the Yanomama vary, but some careful calculations are given in R.H. Ward, H. Gershowitz, M. Layrisse, et al., "The Genetic Structure of a Tribal Population, the Yanomama Indians XI. Gene Frequencies for 10 Blood Groups and the ABH-Le Secretor Traits in the Yanomama and Their Neighbors: The Uniqueness of the Tribe," *American Journal of Human Genetics* 27 (1975): 1–30.

118 The huge gap between male and female life expectancies in Russia is stabilizing but not shrinking—and many other societal pressures are operating, as detailed in Bill Powell and Kim Palchikoff, "Sober, Rested and Ready: While Their Men Drown in Vodka and Self-Pity, Russian Women Move On and Up," *Newsweek* (8 December 1997): 50–51.

118 Some of the truly horrifying effects of widespread pollution in Eastern Europe are documented in Irina Norska-Borowka, "Poland: Environmental Pollution and Health in Katowice," *Lancet* 335 (1990): 1392–93.

118–20 Possible futures of the Russian population are charted in Carl Haub, "Population Change in the Former Soviet Republic," *Population Bulletin* 49 (1994): 2–47. Calhoun's famous experiments on the effects of overcrowding on rats are recounted in J.B. Calhoun, "Population Density and Social Pathology," *Scientific American* 206 (1962): 139–48. Freedman's studies on the effects of overcrowding in humans are in Jonathan L. Freedman, *Crowding and Behavior* (New York: Viking Press, 1975). Germany's population future is examined in Gerhard Heilig, Thomas Buttner, and Wolfgang Lutz, "Germany's Population: Turbulent Past, Uncertain Future," *Population Bulletin* 45 (1990): 1–47. The French survey of opinion on future population growth is reported in H. Bastide and A. Girard, "Les tendances démographiques en France et les attitudes de la population," *Population* 21 (1975): 9–50. Population transition in Korea is analyzed in Robert C. Repetto, *Economic Equality and Fertility in Developing Countries* (Baltimore: Johns Hopkins University Press, 1979), and in Taiwan in Ronald Freedman, Ming-Cheng Chang, and Te-Hsiung Sun, "Taiwan's Transition from High Fertility to Below-Replacement Levels," *Studies in Family Planning* 25 (1994): 317–31.

121 Predictions of the world's future population are examined in Barbara Crossette, "How to Fix A Crowded World: Add People," *New York Times* (2 November 1997): WK1. The prescient book written by Darwin's grandson is Charles Galton Darwin, *The Next Million Years* (New York: Doubleday, 1953).

121 Immigration and its impact on Germany are examined in Peter O'Brien, "Migration and Its Risks," *International Migration Review* 30 (1996): 1067–77. R.W. Johnson, "Whites in the New South Africa," *Dissent* 43 (1996): 134–37, is one of many studies on demographic trends in South Africa.

122 Japanese population history and current problems are vividly presented in Suzuki Kazue, "Women Rebuff the Call For More Babies," *Japan Quarterly* 42 (1995): 14–20.

123 An examination of the demographic revolution in California can be found in Dale Maharidge, *The Coming White Minority: California's Eruptions and America's Future* (New York: Times Books, 1996). Recent numbers on interracial marriages can be found in John Leland and Gregory Beals, "In Living Colors: Tiger Woods Is the Exception That Rules. For His Multiracial Generation, Hip Isn't Just Black and White," *Newsweek* (5 May 1997): 58–61.

CHAPTER 7: THE ROAD WE DID NOT TAKE

127   The epigraph is from Jared Diamond, "The Great Leap Forward," *Discover* (May 1989): 50–60.

127   Much information about Atapuerca and its excavations can be found in an issue of the *Journal of Human Evolution* that was devoted entirely to it: vol. 33, nos. 2–3 (1997).

130   The ancient child from the Gran Dolina is described in J.M. Bermudez de Castro, J.L. Arsuaga, E. Carbonell, et al., "A Hominid from the Lower Pleistocene of Atapuerca, Spain: Possible Ancestors to Neandertals and Modern Humans," *Science* 276 (1997): 1392–95.

132   Recent discoveries about *Homo heidelbergensis* are presented in M.B. Roberts, C.B. Stringer, and S.A. Parfitt, "A Hominid Tibia from Middle Pleistocene Sediments at Boxgrove, UK," *Nature* 369 (1994): 311–13.

132   An in-depth review of the meaning of Atapuerca can be found in R. Denell and W. Roebroeks, "The Earliest Colonization of Europe: The Short Chronology Revisited," *Antiquity* 70 (1996): 534–41.

132–33   The discoveries in the Sima de los Huesos are described in J.L. Arsuaga, I. Martinez, A. Garcia, et al., "Three New Human Skulls from the Sima de los Huesos Middle Pleistocene Site in Sierra de Atapuerca Spain," *Nature* 362 (1993): 534–37.

135   Differences in sizes between the sexes in the Sima de los Huesos are examined in J.L. Arsuaga, J.M. Carretero, C. Lorenzo, et al., "Size Variation in Middle Pleistocene Humans," *Science* 277 (1997): 1086–88.

138   The story of the German spears is told in H. Thieme, "Lower Palaeolithic Hunting Spears from Germany," *Nature* 385 (1997): 807–10.

139   Trinkaus's summary of the Neandertals is in Erik Trinkaus and Pat Shipman, *The Neandertals: Changing the Image of Mankind* (New York: Knopf, 1993). The remarkable Neandertal DNA story is described in M. Krings, A. Stone, R.W. Schmitz, et al., "Neandertal DNA Sequences and the Origin of Modern Humans," *Cell* 90 (1997): 19–30.

143   The final days of the Neandertals are described in J.J. Hublin, C.B. Ruiz, P.M. Lara, et al., "The Mousterian Site of Zafarraya (Andalucia, Spain): Dating and Implications on the Paleolithic

Peopling Processes of Western Europe," *Comptes Rendus de l'Academie des Sciences, Serie II: A Sciences de la Terre et des Planetes* 321 (1995): 931–37, and J.J. Hublin, "The First Europeans," *Archaeology* 49 (1996): 36–44. The Neandertal flower burials are described in Erik Trinkaus, *The Shanidar Neandertals* (New York: Academic Press, 1983).

143–44   Modified Neandertal stone tools are described in E. Boëda, J. Connan, D. Dessort, et al., "Bitumen as a Hafting Material on Middle Paleolithic Artefacts," *Nature* 380 (1996): 336–338. The flute is described in I. Turk, J. Dirjec, and B. Kavur, "Was the Oldest Music Instrument of Europe Found in Slovenia?" (in French), *Anthropologie* 101 (1997): 531–40.

144   Advanced artifacts at Arcy-sur-Cure and their association with Neandertals are detailed in J.J. Hublin, F. Spoor, M. Braun, et al., "A Late Neanderthal Associated with Upper Palaeolithic Artefacts," *Nature* 381 (1996): 224–26.

149   The predominance of genetic variation within rather than between human groups is measured by R.C. Lewontin, "The Apportionment of Human Diversity," *Evolutionary Biology* 6 (1972): 381–98. Mitochondrial population genetics of the Ladins is described in M. Stenico, L. Nigro, G. Bertorelle, et al., "High Mitochondrial Sequence Diversity in Linguistic Isolates of the Alps," *American Journal of Human Genetics* 59 (1996): 1363–75.

CHAPTER 8: WHY ARE WE SUCH EVOLUTIONARY SPEED DEMONS?

152   The discovery of our oldest ancestor is recounted in T.D. White, G. Suwa, and B. Asfaw, "*Australopithecus ramidus*, a New Species of Early Hominid from Aramis, Ethiopia," *Nature* 371 (1994): 306–12.

154   The story of the early Australopithecine discoveries is retold in Christopher Wills, *The Runaway Brain: The Evolution of Human Uniqueness* (New York: Basic Books, 1993).

154   M. Brunet, A. Beauvilain, Y. Coppens, et al., "*Australopithecus bahrelghazali*, a New Species of Early Hominid from Koro Toro Region, Chad," *Comptes Rendus de l'Academie des Sciences, Serie II: A Sciences de la Terre et des Planetes* 322 (1996): 907–13, is the source for the recent and surprising hominid discovery in West Africa.

155   The discovery of the First Family and its meaning is told in Donald C. Johanson, *Lucy: The Beginnings of Humankind* (New York:

Warner Books, 1982). The discovery of the fossil footprints can be found in Mary D. Leakey, *Disclosing the Past* (Garden City, NY: Doubleday, 1984).

156   A useful reference for events in human prehistory is S. Jones, R. Martin, and D. Pilbeam, eds. *The Cambridge Encyclopedia of Human Evolution* (New York: Cambridge University Press, 1992). The story of an ancient ape that seems to have had a remarkably upright stance is told in M. Koehler and S. Moya-Sola, "Ape-like or Hominid-like? The Positional Behavior of *Oreopithecus bambolii* Reconsidered," *Proceedings of the National Academy of Sciences (U.S.)* 94 (1997): 11747–51.

157   An excellent popular survey of the contributions of the Leakey clan to anthropology in Africa is Virginia Morell, *Ancestral Passions: The Leakey Family and the Quest for Humankind's Beginnings* (New York: Simon and Schuster, 1995). Susman's examination of the apparent tool-making capabilities of Australopithecine hands is R.L. Susman, "Fossil Evidence for Early Hominid Tool Use," *Science* 265 (1994): 1570–73.

159–60   Some of the confusion surrounding the skulls of *Homo habilis* is set out in G.P. Rightmire, "Variation Among Early *Homo* Crania from Olduvai Gorge and the Koobi Fora Region," *American Journal of Physical Anthropology* 90 (1993): 1–33.

161–62   Examination of Australopithecine brains has been carried out by R.L. Holloway, "Some Additional Morphological and Metrical Observations on *Pan* Brain Casts and Their Relevance to the Taung Endocast," *American Journal of Physical Anthropology* 77 (1988): 27–34, and P.V. Tobias, "The Brain of the First Hominids," in *Origins of the Human Brain*, edited by J.-P. Cagneux and J. Chavaillon (Oxford: Clarendon Press, 1995).

162   G. Philip Rightmire, *The Evolution of Homo erectus: Comparative Anatomical Studies of an Extinct Human Species* (Cambridge: Cambridge University Press, 1990), provides documentation for the increase in brain size of *Homo erectus* during its long tenure. The eventual end of the robust Australopithecines is told in B. Wood, C. Wood, and L. Konigsberg, "*Paranthropus boisei:* An Example of Evolutionary Stasis?," *American Journal of Physical Anthropology* 95 (1994): 117–36.

163–64   The story of the Georgia mandible is told in L. Gabunia and A. Vekua, "A Plio-Pleistocene hominid from Dmanisi, East Georgia,

Caucasus," *Nature* 373 (1995): 509–12. The history of the 'Ubeidiyah excavations is recounted in C. Guerin, O. Bar-Yosef, E. Debard, et al., "Archaeological and Palaeontological Programme on Older Pleistocene of 'Ubeidiya (Israel): Results 1992–1994" (in French), *Comptes Rendus de l'Academie des Sciences, Serie II A: Sciences de la Terre et des Planetes* 322 (1996): 709–12.

164–65 Much controversy surrounds the origin of the heterogeneous collection of archaic *Homo sapiens* remains. A small sampling of the conflicting points of view is set out in R.R. Sokal, N.L. Oden, J. Walker, et al., "Using Distance Matrices to Choose Between Competing Theories and an Application to the Origin of Modern Humans," *Journal of Human Evolution* 32 (1997): 501–22, and S. Sohn and M.H. Wolpoff, "Zuttiyeh Face: A View from the East," *American Journal of Physical Anthropology* 91 (1993): 325–47.

166 The possibility of early human-set fires in Australia is examined in A.P. Kershaw, "Pleistocene Vegetation of the Humid Regions of Northeastern Queensland," *Paleogeography Paleoclimatology Paleoecology* 109 (1994): 399–412.

167 The sad and complex story of Eugène Dubois and his discoveries can be found in Bert Theunissen, *Eugene Dubois and the Ape-man from Java: The History of the First Missing Link and Its Discoverer* (Boston: Kluwer Academic Publishers, 1989).

167–68 Swisher's oldest date for Javan *H. erectus* is in C.C. Swisher 3rd, G.H. Curtis, T. Jacob, et al., "Age of the Earliest Known Hominids in Java, Indonesia," *Science* 263 (1994): 1118–21, and his youngest is in C.C. Swisher 3rd, W.J. Rink, S.C. Anton, et al., "Latest *Homo erectus* of Java: Potential Contemporaneity with *Homo sapiens* in Southeast Asia," *Science* 274 (1996): 1870–1874.

170 Dating of Verhoeven's discovery was performed by M.J. Morwood, P.B. O'Suyllivan, A. Aziz, and A. Raza, "Fission-Track Ages of Stone Tools and Fossils on the East Indonesian Island of Flores," *Nature* 392 (1998): 173–76.

170 Rightmire's *Evolution of Homo erectus* gives the sizes of *H. erectus* brains. A good layperson's account of some of the Australian finds is in Josephine Flood, *Archaeology of the Dreamtime*, 2nd. ed. (Sydney: Collins Publishers Australia, 1989).

173 A brief survey of the recently emerging evidence for ancient cannibalism is Anne Gibbons, "Archaeologists Rediscover Cannibals," *Science* 277 (1997): 635–37.

CHAPTER 9: BOTTLENECKS AND SELECTIVE SWEEPS

174 C. Boesch, P. Marchesi, N. Marchesi, et al., "Is Nut Cracking in Wild Chimpanzees a Cultural Behaviour?," *Journal of Human Evolution* 26 (1994): 325–38, describes this remarkable behavior in detail.

176 Gagneux's discovery of naughty chimpanzee females is detailed in P. Gagneux, D.S. Woodruff, and C. Boesch, "Furtive Mating in Female Chimpanzees," *Nature* 387 (1997): 358–59.

178 Our mitochondrial DNA tree is found in P. Gagneux, C. Wills, U. Gerloff, et al., "Mitochondrial Sequences Show Diverse Histories of African Hominoids," *Proceedings of the National Academy of Sciences (U.S.)* (submitted). If you would like to construct such a tree, you can find out how in M. Saitou and M. Nei, "The Neighbor-joining Method: A New Method for Reconstructing Phylogenetic Trees," *Molecular Biology and Evolution* 4 (1987): 406–25.

181 Frans de Waal, *Bonobo: The Forgotten Ape* (Berkeley: University of California Press, 1997), is a magnficent book about the bonobos.

182 The checkered history of ape nomenclature is recounted in Colin P. Groves, "Systematics of the Great Apes," in *Comparative Primate Biology*, vol. 1: *Systematics and Anatomy*, edited by D. Swindler and J. Erwin (New York: Alan R. Liss, 1986).

183 The similarity between the variation in mitochondrial sequences of humans and of eastern chimpanzees was noted in T.L. Goldberg and M. Ruvolo, "The Geographic Apportionment of Mitochondrial Genetic Diversity in East African Chimpanzees, *Pan troglodytes schweinfurthii*," *Molecular Biology and Evolution* 14 (1997) 976–84.

186 The mitochondrial Eve's name was first suggested by J. Wainscoat, "Human Evolution. Out of the Garden of Eden," *Nature* 325 (1987): 13. He was commenting on a seminal paper by the late Allan Wilson and his colleagues, R.L. Cann, M. Stoneking, and A.C. Wilson, "Mitochondrial DNA and Human Evolution," *Nature* 325 (1987): 31–36.

187 My own calculations about the mitochondrial Eve are in C. Wills, "When Did Eve Live? An Evolutionary Detective Story," *Evolution* 49 (1995): 593–607.

188 The hybridization paper is H. Vervaecke and L. Vanelsacker, "Hybrids between common chimpanzees (*Pan troglodytes*) and pygmy chimpanzees (*Pan paniscus*) in captivity," *Mammalia* 56 (1992): 667–69.

191 Our paper on the transitions and transversions is C. Wills, P. Gagneux, and S. Goldberg, "Selection Against Transversions in the Hominid Lineage," *Journal of Molecular Evolution* (submitted).

193 Doug Wallace's work on Leber's optic neuropathy and other mitochondrial diseases is summarized in D.C. Wallace, "Mitochondrial DNA Mutations in Diseases of Energy Metabolism," *Journal of Bioenergetics and Biomembranes* 26 (1994): 241–50.

194 Hublin's work on the size of human and chimpanzee blood supplies to the brain is not yet published.

CHAPTER 10: STICKING OUT LIKE CYRANO'S NOSE

196–97 The genetic origins of the Ainu are examined in K. Omoto and N. Saitou, "Genetic Origins of the Japanese: A Partial Support for the Dual Structure Hypothesis," *American Journal of Physical Anthropology* 102 (1997): 437–46.

201–02 Loren C. Eiseley, *Darwin's Century: Evolution and the Men Who Discovered It* (New York: Doubleday, 1958), still provides an excellent examination of Wallace's ideas about human evolution. Osborn's ideas about the great antiquity of our species are set out in his introduction to Roy Chapman Andrews, *On the Trail of Ancient Man: A Narrative of the Field Work of the Central Asiatic Expeditions* (Garden City, NY: Garden City Publishing Company, 1926).

202 The molecular time scale for the branch point of the human and chimpanzee lineages is given in V.M. Sarich and A.C. Wilson, "Immunological Time Scale for Hominid Evolution," *Science* 158 (1967): 1200–3.

206 Hundreds of papers have been written about genes found in humans that are also present in chimpanzees, and occasionally vice versa. I know of no functional genes that are unique to one species or the other, although genes of a given type often differ in number.

207 A recent review of the extremely complex regulation of the melanin-producing enzyme tyrosinase can be found in C.A. Ferguson and S.H. Kidson, "The Regulation of Tyrosinase Gene Transcription," *Pigment Cell Research* 10 (1997): 127–38.

CHAPTER 11: GOING TO EXTREMES

211 Glen Dickson and Margot Suydam, "X Games Hit San Diego; ESPN Brings with It Virtual World, POVs and Fake Snow," *Broadcasting*

and Cable 127 (1997): 64–65, provides a brief description of the San Diego games.

214   The book about culturgens is C.J. Lumsden and Edward O. Wilson, *Genes, Minds and Culture* (Cambridge, MA: Harvard University Press, 1981). The rise and more recent decline of industrial melanism in *Biston betularia* is documented in B.S. Grant, D.F. Owen, and C.A. Clarke, "Parallel Rise and Fall of Melanic Peppered Moths in America and Britain," *Journal of Heredity* 87 (1996): 351–57.

216   The fate of normal children of retarded parents is examined in A.M. O'Neill, "Normal and Bright Children of Mentally Retarded Parents: The Huck Finn Syndrome," *Child Psychiatry and Human Development* 15 (1985): 255–68.

216   Pulmonary obstruction in Down's syndrome is detailed in S. Yamaki, H. Yasui, H. Kado, et al., "Pulmonary Vascular Disease and Operative Indications in Complete Atrioventricular Canal Defect in Early Infancy," *Journal of Thoracic and Cardiovascular Surgery* 106 (1993): 398–405. Recent findings about fragile X are summarized in A.T. Hoogeveen and B.A. Oostra, "The Fragile X Syndrome," *Journal of Inherited Metabolic Disease* 20 (1997): 139–51.

217   An excellent recent survey of the effects and genetics of Williams syndrome can be found in H.M. Lenhoff, P.P. Wang, F. Greenberg, et al., "Williams Syndrome and the Brain," *Scientific American* 277 (1997): 68–73.

218   The Williams gene story is in J.M. Frangiskakis, A.K. Ewart, C.A. Morris, et al., "LIM-kinase1 Hemizygosity Implicated in Imparied Visuospatial Constructive Cognition," *Cell* 86 (1996): 59–69.

219   Some of the many genetic deletions that lead to mental retardation are examined in M.L. Budarf and B.S. Emanuel, "Progress in the Autosomal Segmental Aneusomy Syndromes (SASs): Single or Multilocus Disorders?," *Human Molecular Genetics* 6 (1997): 1657–65.

CHAPTER 12: HOW BRAIN FUNCTION EVOLVES

221–22   The connection between genes and narcolepsy is traced in E. Mignot, "Perspectives in Narcolepsy Research and Therapy," *Current Opinion in Pulmonary Medicine* 2 (1996): 482–87.

223   A possible genetic basis for divorce is explored in Helen E. Fisher, *Anatomy of Love: The Natural History of Monogamy, Adultery, and Divorce* (New York: Norton, 1992). Some of the difficulties in

diagnosing ADHD are examined in S.C. Schneider and G. Tan, "Attention-Deficit Hyperactivity Disorder: In Pursuit of Diagnostic Accuracy," *Postgraduate Medicine* 101 (1997): 235–40.

224   P. Hauser, A.J. Zametkin, P. Martinez, et al., "Attention Deficit-Hyperactivity Disorder in People with Generalized Resistance to Thyroid Hormone," *New England Journal of Medicine* 328 (1993): 997–1001, was the first report on the thyroid hormone-ADHD connection.

224   The dopamine-serotonin connection with behavior patterns is explored in S.R. Pliszka, J.T. McCracken, and J.W. Maas, "Catecholamines in Attention-Deficit Hyperactivity Disorder: Current Perspectives," *Journal of the American Academy of Child and Adolescent Psychiatry* 35 (1996): 264–72.

226   Some of the difficulties and advances in tracking down genes for major mental illnesses are discussed in P. Sham, "Genetic Epidemiology," *British Medical Bulletin* 52 (1996): 408–33.

227   The connection between birth weight and schizophrenia, in both twins and in the general population, is explored in J.R. Stabenau and W. Pollin, "Heredity and Environment in Schizophrenia, Revisited," *Journal of Nervous and Mental Disease* 181 (1993): 290–97.

228   The disseminated nature of brain activity is a subject of much research. Some representative publications include H. Damasio, T.J. Grabowski, D. Tranel, et al., "A Neural Basis for Lexical Retrieval," *Nature* 380 (1996): 499–505, and M.I. Sereno, A.M. Dale, J.B. Reppas, et al., "Borders of Multiple Visual Areas in Humans Revealed by Functional Magnetic Resonance Imaging," *Science* 268 (1995): 889–93.

230–32   The Flynn effect, and the author's doubts about its meaning, are set out in detail in James R. Flynn, "Massive IQ Gains in 14 Nations: What IQ Tests Really Measure," *Psychological Bulletin* 101 (1987): 171–91.

231   The example of a Raven matrix is taken from Patricia A. Carpenter, Marcel A. Just, and Peter Shell, "What One Intelligence Test Measures: A Theoretical Account of the Processing in the Raven Progressive Matrices Test," *Psychological Review* 97 (1990): 404–31. The correct answer is 5.

233   Some possible benefits of watching educational television can be found in Aletha C. Huston and John C. Wright, "Educating Children with Television: The Forms of the Medium," in *Media,*

Children, and the Family: Social Scientific, Psychodynamic, and Clinical Perspectives, edited by D.J. Zillmann, J. Bryant, and A.C. Huston (Hillsdale, NJ: Lawrence Erlbaum Associates, 1994).

234 The severe effect of childhood diseases on low birth-weight children is documented in J.S. Read, J.D. Clemens, and M.A. Klebanoff, "Moderate Low Birth Weight and Infectious Disease Mortality During Infancy and Childhood," *American Journal of Epidemiology* 140 (1994): 721–33, and M.C. McCormick, J. Brooks-Gunn, K. Workman-Daniels, et al., "The Health and Developmental Status of Very Low-Birth-Weight Children at School Age," *Journal of the American Medical Association* 267 (1992): 2204–08.

235 Snowdon's study of an aging nun population is detailed in D.A. Snowdon, S.J. Kemper, J.A. Mortimer, et al., "Linguistic Ability in Early Life and Cognitive Function and Alzheimer's Disease in Late Life. Findings from the Nun Study," *Journal of the American Medical Association* 275 (1996): 528–32.

238 The *JAMA* study is "The Infant Health and Development Program. Enhancing the Outcomes of Low-Birth-Weight, Premature Infants. A Multisite, Randomized Trial," *Journal of the American Medical Association* 263 (1990): 3035–42.

238–39 The long-term effects of Head Start are examined by Constance Holden, "Head Start Enters Adulthood," *Science* 247 (1990): 1400–02.

242 Some of these triplet diseases are surveyed in R. Li and R.S. el-Mallakh, "Triplet Repeat Gene Sequences in Neuropsychiatric Diseases," *Harvard Review of Psychiatry* 5 (1997): 66–74. Some possible mechanisms by which glutamine repeats might cause cell death are discussed in C.A. Ross, M.W. Becher, V. Colomer, et al., "Huntington's Disease and Dentatorubral-Pallidoluysian Atrophy: Proteins, Pathogenesis and Pathology," *Brain Pathology* 7 (1997): 1003–16.

243 My book that discusses the evolution of evolvability itself is Christopher Wills, *The Wisdom of the Genes* (NY: Basic Books, 1989).

244 The paper suggesting the term *contingency genes* is E.R. Moxon, P.B. Rainey, M.A. Nowak, et al., "Adaptive Evolution of Highly Mutable Loci in Pathogenic Bacteria," *Current Biology* 4 (1994): 24–33.

245 Richard Moxon and I discuss possible reasons for microsatellite variation in E.R. Moxon and C. Wills, "Microsatellites: The Evolution of Evolvability" (tentative), *Scientific American* (in press).

246 Microsatellite effects are explored in T.G. Tut, F.J. Ghadessy, M.A. Trifiro, L. Pinsky, and E.L. Yong, "Long Polyglutamine Tracts in the Androgen Receptor are Associated with Reduced *Trans*-Activation, Impaired Sperm Production, and Male Infertility," *Journal of Clinical Endocrinology and Metabolism*, 82 (1997): 3777-782. The comparison between chimpanzees and humans is C.A. Wise, M. Sraml, D.C. Rubinsztein, et al., "Comparative Nuclear and Mitochondrial Genome Diversity in Humans and Chimpanzees," *Molecular Biology and Evolution* 14 (1997): 707–16. One of several studies finding differences in microsatellite disease gene repeats is P. Djian, J.M. Hancock, and H.S. Chana, "Codon Repeats in Genes Associated with Human Diseases: Fewer Repeats in the Genes of Nonhuman Primates and Nucleotide Substitutions Concentrated at the Sites of Reiteration," *Proceedings of the National Academy of Sciences* (US) 93 (1996): 417–21.

247 These early experiments measuring the effects of environmental enrichment on rats are detailed in Marian Diamond, *Enriching Heredity: The Impact of the Environment on the Anatomy of the Brain* (New York: Free Press, 1988). The recent experiments of Rusty Gage on mice are in G. Kempermann, H.G. Kuhn, and F.H. Gage, "More Hippocampal Neurons in Adult Mice Living in an Enriched Environment," *Nature* 386 (1997): 493–95.

CHAPTER 13: THE FINAL OBJECTION

250 The book that levels devastating criticism at a role for genes in IQ is Leon J. Kamin, *The Science and Politics of I.Q.* (New York: Halstead Press, 1974).

255 Bouchard's twin studies are in T.J. Bouchard, Jr., D.T. Lykken, M. McGue, et al., "Sources of Human Psychological Differences: The Minnesota Study of Twins Raised Apart," *Science* 250 (1990): 223–28. Devlin's important study is B. Devlin, M. Daniels, and K. Roeder, "The Heritability of IQ," *Nature* 388 (1997): 468–71.

CHAPTER 14: OUR EVOLUTIONARY FUTURE

259 The epigraph is from Gerald Heard, *Pain, Sex and Time: A New Outlook on Evolution and the Future of Man* (New York: Harper and Brothers, 1939).

261–62 Fisher's fundamental theorem is set out in R.A. Fisher, *The Genetical Theory of Natural Selection* (Oxford: Clarendon Press, 1930).

262–63 Waddington's first and mistaken idea appeared in C.H. Waddington, "The Canalization of Development and the Inheritance of Acquired Characters," *Nature* 150 (1942): 563. His important experiments appeared in C.H. Waddington, "Genetic Assimilation of an Acquired Character," *Evolution* 7 (1953): 1–13, and C.H. Waddington, "Genetic Assimilation of the *Bithorax* Phenotype, *Evolution* 10 (1956): 1–13.

265 The ghastly early history of treatment with human growth hormone is set out in S.D. Frasier, "The Not-so-good Old Days: Working with Pituitary Growth Hormone in North America, 1956 to 1985," *Journal of Pediatrics* 131 (Issue 1 pt. 2) (1997): S1–S4. The current ethical dilemmas are examined in Curtis A. Kin, "Coming Soon to the 'Genetic Supermarket' Near You," *Stanford Law Review* 48 (1996): 1573–1604.

265 Some of the media hype surrounding telomerase is described in Robert F. Service, "'Fountain of Youth' Lifts Biotech Stock," *Science* 279 (1998): 472. Experiments with mice lacking telomerase are described in M.A. Blasco, H.W. Lee, M.P. Hande, et al., "Telomere Shortening and Tumor Formation by Mouse Cells Lacking Telomerase RNA," *Cell* 91 (1997): 25–34.

267 A report on the UCLA meeting can be found in M. Wadham, "Germline Gene Therapy Must Be Spared Excessive Regulation," *Nature* 392 (1998): 317.

269 My behavioral diversity model is C. Wills, "The Maintenance of Behavioral Diversity in Human Societies" (comment), *Behavior and Brain Science* 17 (1994): 638–39. Petit's seminal discovery was first described in C. Petit, "Le déterminism génétique et psycho-physiologique de la competition sexuelle chez *Drosophila melangaster*," *Bulletin Biologique* 92 (1958): 248–329. Our rainforest paper is C. Wills, R. Condit, R. Foster, et al., "Strong Density- and Diversity-related Effects Help to Maintain Tree Species Diversity in a Neotropical Forest," *Proceedings of the National Academy of Sciences (U.S.)* 94 (1997): 1252–57.

GLOSSARY

✦

ADHD (attention deficit hyperactivity disorder): This rather vaguely defined behavioral disorder usually involves short attention span, impulsive behavior, and high levels of physical activity. It has been connected with many different aspects of brain metabolism, in particular the effects of thyroid hormone and dopamine.

Ainu: An aboriginal people with rather Caucasian features but firmly in the East Asian gene pool, now confined to the island of Hokkaido in northern Japan.

Allele: An alternative form of a gene. The ABO blood group alleles are good examples, but there are many others. No individual can have more than two alleles of a gene, but a population may have many different alleles. Alleles are shuffled each generation by recombination.

Alzheimer's: A neurodegenerative disease, usually developing in middle or old age but occasionally sooner, in which the brain becomes filled with bits of precipitated protein and other structures forming plaques and tangles. Symptoms include short-term memory deficit and behavioral changes.

Anopheles mosquitos: The chief insect vectors of human malarias.

Aurignacian: Of or relating to late old stone age (Paleolithic) culture. Aurignacian tools, used by the Cro-Magnons and other invaders of Europe, are patiently shaped blades and scrapers, often with a razorlike edge and an elegant appearance. They are far more advanced than the crude stone tools of the Oldowan, but far less advanced than the beautiful stone and bone tools of the Neolithic.

Australopithecines: A heterogeneous group of hominids who lived throughout Africa from more than three to about a million years ago. Some had large teeth and huge jaws, and some were rather less robust. Their brains were roughly the same size as those of chimpanzees, but they may have been able to use tools.

Balanced polymorphism: A situation in which a balance of opposing forces maintains more than one allele in a population. The classic case is sickle cell anemia, in which the heterozygote is fitter than either homozygote. But a balanced polymorphism may arise in other ways. For example, if an allele is advantageous when it is rare but disadvantageous when it is common, this frequency-dependent selection can lead to a balance.

Base: One of the smaller molecular components of DNA. The four types of bases are called, in biological shorthand, A, T, G, and C. Like letters in a word, or words in a sentence, the order in which the bases are arranged along the DNA molecule encodes its genetic information.

Bonobo: A chimpanzeelike primate, found south of the Zaire River in Central Africa. The ancestors of bonobos and chimpanzees probably diverged about three million years ago.

Bottleneck: A dramatic reduction in a population's size, followed by an increase. The result may be a loss of genetic variability.

Chatelperronian: Stone tools and other artifacts used by the Neandertals thirty to forty thousand years ago. They have some characteristics in common with tools used by Cro-Magnons and other later invaders of Europe but are less sophisticated.

Chromosome: A structure in the nucleus of a cell that contains DNA. Mitochondria have chromosomes too, but they are much smaller and simpler than the chromosomes in nuclei and carry far fewer genes.

Codon: A group of three bases making up one unit of the DNA genetic code. Each codon specifies either an amino acid in the protein or a stopping point for the protein.

Contingency gene: A class of microsatellite, found in bacteria, that allows certain important bacterial genes to be switched on or off. These genes, with their high mutation rates, enable bacteria to adapt readily to new environmental contingencies; hence their name.

Cro-Magnon: An Upper Paleolithic group of *Homo sapiens* that migrated into western Europe, starting probably about 35,000 years ago. Cro-Magnons either drove the Neandertals to extinction or genetically assimilated them.

Cystic fibrosis: A genetic disease, common among Europeans, caused by a disturbance in the transport of chloride ions across the membranes that surround cells. This apparently minor defect can have fatal consequences. CF is thought to be maintained in the population by a balanced polymorphism.

Cytoplasm: The region of a living cell that surrounds the nucleus, where protein manufacture and metabolic activity are carried out. Mitochondria are found in the cytoplasm.

Divergence time: The time at which two evolutionary lineages diverged. Neither the fossil record nor the molecular record yet allows us to date such times with precision. Very large errors are associated with the times given in this book and elsewhere.

DNA (deoxyribonucleic acid): The genetic material carried by most living organisms. These long molecules, which form a double helix, comprise genetic information coded in a sequence of bases, like letters in a sentence.

Down's syndrome: A condition marked by mental retardation, heart abnormalities, and many other problems, caused by the presence of three copies of the small chromosome 21 rather than the normal two.

Duffy-negative: An allele that, when homozygous, essentially protects the carrier against malaria caused by *Plasmodium vivax*, because the parasite cannot penetrate their red blood cells. The allele is very common in central Africa.

Elliptocytosis: A condition that changes the shape of red blood cells and protects against malaria. It is particularly common in New Guinea.

Evolutionary tree: A branching diagram that arranges the physical features of organisms, or sequences of genes from organisms, so that the most closely related are nearest to each other and the most distantly related are farther away. Even the best of these trees involve some uncertainty, since the characteristics of organisms that lived in the past must be inferred from those that are living at the present time.

Falciparum malaria: The severest form of human malaria, which can sometimes invade the brain. It is now largely confined to the tropics.

Flynn effect: The inexplicable rise in IQ scores in the developed world over the last few decades.

Fragile X: One of the best-known so-called repeat diseases, in which a small, highly repeated part of the X chromosome (a microsatellite) becomes very long and damages the ability of nearby genes to function.

Fraternal twins: Twins that result from multiple ovulations, in which two eggs are fertilized by different sperm. They are genetically as alike as brothers and sisters born at different times, but their shared intrauterine environment may give them more similar phenotypes. (*See* identical twins.)

G6PD deficiency: A very widespread genetic condition, common in the Mediterranean and Africa, that protects against malaria. Carriers of the gene, however, are very sensitive to oxidizing agents such as those found in broad beans.

Gene flow: An exchange of genes between two or more different populations. If the populations differ in allele frequencies, the result can be rapid genetic change.

Gene pool: The entire collection of genes in a population.

Gene: A stretch of DNA that carries some sort of genetic information. Many genes code for proteins, but some code for information that decides when and how nearby genes are turned on or off, and others code for RNA molecules that aid in the manufacture of proteins or in the movement of proteins from one part of a cell to another. Only about five percent of our DNA seems to have such an identifiable function, which is probably more a reflection of our ignorance than a lack of function for the remaining ninety-five percent.

Genotype: The collection of genes possessed by an organism. These genes interact with each other and the environment to produce the organism's phenotype, its physical appearance and behavior.

Hemoglobin: The pigment-containing protein in red blood corpuscles that carries oxygen to tissues. Hemoglobin genes come in many allelic forms, some of which cause disease.

Heritability: The degree to which a character can be selected for on the basis of its phenotype. If you select for heavier cattle by breeding only the heaviest cattle but get nowhere, then the heritability of that character is zero. But if the offspring are as heavy as the selected parents, then the heritability is a hundred percent. All real heritabilities lie somewhere in between these extremes.

Heterozygote: An organism with two different alleles of a gene, one from each parent.

Homo erectus: A large-brained hominid that first appeared in the fossil record in Africa and East Asia almost two million years ago. (The record of stone tools suggests that it may have made its debut 2.5 million years ago.) In Java there is evidence that it survived almost to the present time.

Homo habilis: The first member of the genus *Homo*, a heterogeneous collection of apparent transitional forms between Australopithecines and *Homo erectus*, with brains roughly intermediate in size. They lived about two million years ago. The fossil record is confusing, however, and much evolutionary ferment was likely going on at that time. Exactly who was related to who is still a matter of conjecture.

Homozygote: A person or other organism with two copies of one allele of a gene. A homozygote for sickle cell anemia, for example, received one copy of the mutant allele from one parent and the second copy from the other parent.

Hypoxia: The effects caused by lack of oxygen. The bodies of Tibetans, well adapted to high altitudes, can respond more readily to hypoxic conditions than those of lowlanders.

Identical twins: Twins that result when, early in development, a single fertilized zygote splits into two. They have identical genotypes, though through environmental vicissitudes they may not have identical phenotypes. (*See* fraternal twins.)

Leber's optic neuropathy: One of a growing collection of diseases caused by defects in mitochondrial genes. In this disease, metabolically highly active nerve cells in the fovea of the eye die prematurely, causing a blind spot in the center of the field of vision.

Melanin: One of several types of pigment, ranging from yellow and red to black. Melanin is also made in the substantia nigra of the brain, though by a different biochemical pathway from that made in the skin.

Microsatellite: A short repeated bit of DNA, often taking a form like . . . CAGCAGCAGCAG. . . . It can readily grow or shrink in length through mutation. Microsatellites are found on all our chromosomes. Little is yet known about their functions, but many of them may be involved in gene regulation.

Mitochondrion: A structure in a cell's cytoplasm that is responsible for much of its energy production. Mitochondria are descendants of bacteria that invaded our cells two billion years ago. They have their own chromosomes, albeit so reduced in size that they no longer carry enough information to reconstruct an entire mitochondrion; genes in the nucleus must also contribute.

Mountain sickness: A disease characterized by subcutaneous and intestinal bleeding, among other symptoms. In people who live at high altitudes, the effects of mountain sickness can be cumulative and debilitating.

Mutation: A change in an organism's DNA that can be passed on to the next generation. A mutation can take many forms, ranging from a substitution of one base for another to the gain or loss of an entire set of chromosomes.

Narcolepsy: A rare condition in which people (or dogs) suddenly and inexplicably fall asleep. Several genes have been implicated in this disease, but its exact nature has yet to be worked out.

Natural selection: A natural process in which individuals best adjusted to their environment survive and pass on their genes. Darwin was the first to realize that this process ensures that organisms best able to leave offspring are the ones that will pass their characteristics on to the next generation. We now know that natural selection depends ultimately on mutations, which provide the genetic variation that selection sorts out. Although mutations appear in all populations all the time, only a tiny minority of them will aid their carrier's survival under any particular circumstances. Genetic recombination shuffles the variation in a population, providing an important source of new genotypes on which natural selection can act.

Neandertals: Heavy-browed, big-boned people who lived throughout Europe and parts of the Middle East from perhaps a hundred thousand to 28,000 years ago. Their brains were at least as large as ours, although their skulls were of a somewhat different shape. Their ancestors, the pre-Neandertals, showed a less extreme morphology; their traces have been found in Europe dating to almost a million years ago.

Nucleus: A globular structure in the center of a living cell that contains the chromosomes. RNA carries genetic information from the DNA of the nucleus to the surrounding cytoplasm, where proteins are made.

Oldowan: A tool-making tradition, named after Olduvai Gorge in Tanzania, in which flakes are struck off stones to produce simple cutting tools. Oldowan-type tools date from as long as 2.5 million years ago, and are found widely through Africa, Europe, and Asia. The process may have been invented more than once.

PCR (polymerase chain reaction): A technique for synthesizing large quantities of a particular DNA segment. Using this technique, one copy of a particular piece of DNA can be multiplied millions of times.

Phenotype: The physical or behavioral characteristics of an organism, produced by a combination of genes and environment.

Plasmodium: A tiny single-celled animal that multiplies in red blood cells and causes malaria. *Plasmodia* are transmitted from one person to another by mosquitoes.

Pleistocene: The geological epoch that encompasses the time from 1.64 million years ago to the beginning of the Holocene, 11,000 years ago. The Pleistocene was characterized by repeated ice ages. Much but not all of our evolution took place during this epoch.

Polymorphism: The ability to take on many forms. Polymorphism may be either visible—as in the great range of hair and eye colors in the human population, or less obvious—as in the phenotypically invisible but very important ABO blood groups.

Population genetics: The study of genetic variation in populations. Population geneticists attempt, by mathematical and experimental means, to disentangle the many processes—selection, chance events, migration, and mutation—that cause allele frequencies to change over time as populations evolve.

Recombination: The shuffling, in each generation, of the genetic variability in a population into new combinations, as chromosomes recombine in the course of the production of sex cells. It is a powerful force in evolution.

Retrovirus: An RNA virus that can insert its genes into the cells of its host. The best-known retrovirus is the AIDS virus, but there are many others.

RNA (ribonucleic acid): A molecule that carries information from the DNA of a cell's nucleus to the rest of the cell. It has a structure very similar to that of DNA but it is usually single-stranded.

Selective sweep: A natural process in which one allele of a gene suddenly becomes extremely advantageous and sweeps through a population, dragging along quite unrelated alleles that simply happen to be nearby on the chromosome. Sweeps have the potential to bring about large, rapid changes in a gene pool, though their role in human evolutionary history is unclear.

Sexual selection: An important subcategory of natural selection, in which individual organisms compete to be chosen as mates. Competition for mates may be so strong, however, that it actually drives organisms in the direction of lower Darwinian fitness—elaborate mating displays and showy colors may increase the likelihood of being eaten, for example. If sexual selection gets out of hand, Darwinian selection is sure to correct the problem.

Sickle cell anemia: A genetic disease, common in Africa and the Middle East, that causes blood cells to assume a sickle shape when the oxygen level drops. The gene for the disease, when heterozygous, confers some limited protection against malaria, but it causes a severe anemia and many other problems when it is homozygous.

Size bottleneck: *See* bottleneck.

Sociobiology: A field of study that assumes that human behaviors, and those of other animals, have a genetic component and therefore an evolutionary history. More recently, this field has given rise to the subdisciplines of evolutionary psychology and Darwinian medicine.

Telomerase: A recently discovered enzyme that maintains the ends of chromosomes. Cells in tissue culture in which the level of telomerase is artificially boosted do not die as quickly as normal cells.

Thalassemia: From the Greek for "the sea"; a genetic disease that can cause severe anemia when alleles for it are homozygous. It is the result of a block in the synthesis of a component of hemoglobin. Heterozygotes have some protection against malaria. Different forms of thalassemia are common in the Mediterranean and Southeast Asia.

Transgenic organism: An organism in which a gene from another organism has been inserted. Molecular biologists now have this capability; already pigs, sheep, and cattle are being used to manufacture clinically useful human proteins. Where all this will eventually lead, of course, knows God, as *Time* magazine would have said.

Transition: A mutation change in which DNA bases of similar sizes are substituted one for the other. A and G are both bases known as purines; a change from an A to a G or vice versa is a transition. Similarly, T and C are pyrimidines; exchanges between these bases are also transitions.

Transversion: A mutation change involving a substitution of a pyrimidine for a purine or vice versa; A to T or vice versa, for example. Transversions are less likely to happen than transitions.

Tyrosinase: An enzyme responsible for making dark pigment in some of our cells.

Williams syndrome: A condition caused by a small chromosomal deletion, often resulting in mental retardation but sparing verbal ability.

INDEX

The Disney Version

The Disney Version

THE LIFE, TIMES, ART AND COMMERCE OF
WALT DISNEY *Revised and Updated*
BY

RICHARD SCHICKEL

PAVILION
MICHAEL JOSEPH

First published in Great Britain 1986 by
Pavilion Books Limited
196 Shaftesbury Avenue, London WC2H 8JL
in association with Michael Joseph Limited
44 Bedford Square, London WC1B 3DU

Schickel, Richard
    The Disney version : the life, times, art and
    commerce of Walt Disney.
    1. Disney, Walt   2. Moving-picture producers
    and directors—United States—Biography
    3. Animators—United States—Biography
    I. Title
    791.43′0232′0924   PN1998.A3D5

    ISBN 1-85145-007-6

Printed and bound in Great Britain by
Billing & Sons Limited, Worcester

Contents

Introduction

IT HAS been a curious experience for me to return to *The Disney Version* in order to prepare this revised and expanded edition. The book caused a certain amount of controversy when it was first published in the United States in 1968, mainly because it was the first extensive work about Walt Disney and his works that held him, and them, up to a critical light. People were not used to that, in part because the Disney organization had created and controlled an exceedingly benign—not to say bland and banal—image of its founder and his accomplishments, in part because the company's products seemed, on superficial examination, so utterly inoffensive. Aside from the occasional cranky complaint from an educator or a child psychologist, or some disenchanted social commentator—usually of either the leftist or the elitist persuasion—there had been no serious examination of what Disney had wrought for many years. Indeed, since the studio's animated films had settled into the highly stylized format of the post-war years, the excitement that had formerly attended their release among the more intelligent film critics—in the intellectual and artistic community in general—had greatly diminished. And the *live action* comedies and adventures that now dominated its production schedule were generally noted only briefly and distractedly by reviewers. Nor was anyone paying much attention to the social impact of the company's other products—its high-rated television shows, for instance, or the books and toys it so assiduously marketed all over the world. And very few writers attempted any very sober analysis of the content and influence of Disney's most important latter-day creation, Disneyland (Disneyworld, in Florida, was still in the planning stages as I wrote). What I was attempting, in *The Disney Version*, was to re-examine this body of work objectively and to penetrate the psyche of the man who had created and ruled this expanding corporate empire, a psyche obviously much more complex than the prevailing imagery. The book surprised and dismayed many readers.

To be sure, there were those who felt I had not gone far enough in condemning the man and his works—especially among those with a Marxist bias. But mostly the complaints came from people like Ray Bradbury, who had graciously helped me with my researches, but had found the end result to be a betrayal of Disney. Bradbury was typical of the large Disney constituency that wished to hear no criticism of a man they regarded as a great visionary and a heroic figure in a national life that was, at the very moment of his death, entering into a period of great moral and intellectual confusion. Finally, perhaps predictably, I learned that the Disney family and his more devoted associates within the company had been outraged by *The Disney Version*, and for some years after the book was published I was banned from screenings of Disney films, a prohibition that was later rescinded, I am glad to say.

I suppose I should not have been surprised by all this, but I was, and rereading my work now, I am, if anything, more astonished than ever by the controversy it caused. It seems

to me, in retrospect, well within the realm of fair comment on
the life and work of a public man whose creations demon-
strably had an influence, for both good and ill, on several
generations of the world's citizens. Moreover, it seems to me
that such comment, as judiciously considered as I could
make it, was long overdue. Indeed, I thought then, and I
think now, that my portrait of Walt Disney flattered him
more than any previous one precisely because it granted to
him a complexity of character and motivation that no one
else had presented. Beyond that, I felt that for all the
criticisms I offered, my portrayal of Disney was shot through
with admiration for him and his achievements. We sprang
from similar middle western American soil and, as a result,
held certain ambitions and ideals in common. I appreciated
his success in capitalizing on them without compromising
them.

If anything, that admiration has increased with the passing years, as I anxiously observed his organization's decline
as a producer of films, and it became clear that what it most
required was a revival of the Young Walt's spirit, which
obsessively, if not always joyfully, embraced risk and experiment and damn (well, darn) the cost. Looking back on my
work from the vantage point of the Eighties, my only regret
about it is that I was perhaps, too grudging in my admiration
for this spirit, not as aware as I now am of how rare it was,
and is, in Hollywood. This, in turn, led me to undervalue,
and perhaps be over-critical of, the studio's work in animation. It may be that it clung too closely to its own conventions
in the post-war years, that it was not as open as it might have
been to alternative styles and techniques developed elsewhere. But there was an elegance to its classic manner that I
did not fully appreciate when I wrote in the late Sixties. I
also think, looking back, that I was too much taken by
various literary condemnations of Disney's work in this
field, insufficiently appreciative of its purely cinematic merits.

INTRODUCTION

It has not been possible, owing to the economic and technical limitations on reprinting older works, to recast the passages where I take up these matters. But as I hope I make clear in the long epilogue I have prepared for this edition, in which I bring the Disney story up to date, I have modified my position on them. By the same token I think this new section makes clear what may not have been clear in the first edition, namely that I regard the Disney organization as a valuable American institution, which, if it can recapture its former status as a purveyor of works that resonate at different pitches, for different segments of the audience, but interest, entertain and stimulate all of them, has yet a valuable contribution to make in our increasingly fragmented and drifting culture.

For the rest, it has been possible to correct errors that crept into the previous edition, to update most of the statistics and eliminate a few predictions that time has proved incorrect. It is also possible to acknowledge here the many kind strangers who have sought me out to tell me that this book was one that they valued and treasured in memory. Nothing else I have ever written has elicited such a response so broadly. Their interest has sustained me through the years and strengthens my belief that there may be others of a newer generation that will find a similar value in this new edition.

Richard Schickel
New York
November, 1985

Foreword

If one thing is more amazing than the warm, wonderful, heart-stopping motion pictures of Walt Disney, it is the man who made them.

What kind of man is this who has won the Medal of Freedom—highest civilian award in the United States—29 motion picture Academy Awards; four TV Emmys; scores of citations from many nations; and some 700 other awards. Who has been decorated by the French Legion of Honor and again by the Art Workers Guild of London; has received honorary degrees from Harvard, Yale and the University of Southern California; wears Mexico's Order of the Aztec Eagle; and counts his citations from patriotic, educational and professional societies and international film festivals by the hundreds?

On the surface, believe it or not, Walt Disney is a very simple man—a quiet, pleasant man that you might not look twice at on the street. But a man—in the deepest sense of the term—with a mission.

The mission is to bring happiness to millions. It first became evident in the twenties, when this lean son of the Mid-West came unheralded to Hollywood [and] began to animate his dreams . . .

—Promotion piece for *The Wonderful Worlds of Walt Disney*, 1966

9

THERE WERE certain words—"warm," "wonderful,"
"amazing," "dream," "magical"—that attached themselves to
Walt Disney's name like parasites in the later years of his life.
They are all debased words, words that have lost most of their
critical usefulness and, indeed, the power to evoke any emotional response beyond a faint queasiness. They are hucksters'
words. This book is an attempt to penetrate somewhat beyond
language of this order and beyond the unthinking but all too
common attitudes it represents. The attempt here is at what
might be called analytic biography. The hope is to create a balanced perspective on the man, his works and the society that
created him and that he, in his turn and in his special way, both
reflected and influenced.

There are problems in any attempt to analyze the creators and
the creations of popular culture. The most serious of these is in
trying to choose which of the ill-shaped and slippery tools of
understanding one wishes to apply to the task. Popular culture
is an impure thing: it is commerce, it is sociology, it is sometimes art. But if the would-be analyst delves too deeply into the
commercial realm, his work ends up reading like the report of a
Wall Street research firm. If he indulges too heavily in the sociological mode, he finds a heavy and dubious mass of statistics
and/or generalities weighing down his work. If he attempts to
use the traditional language and style of literary criticism, he
finds himself trying to apply fundamentally inapplicable standards to his subject, and the discussion soon degenerates into the
easy moralism and the still more convenient subjectivism with
which the literary community customarily discusses the art of
the masses.

In the case of Walt Disney all these problems are magnified.
And there are others, peculiar to his case, that further compli-
cate matters. The most important of these is that the Disney
organization has always had a very ambivalent attitude toward
journalism. Though it encouraged millions of words of the stuff,
it actively discouraged serious objective investigation of the

man and his works. Rarely has so much been written about a public figure; rarely has so little of it been trustworthy. Therefore the sources for this book are almost all somewhat suspect, for the corporate drive has always been toward the preservation of an easily assimilated image, and for the most part, popular journalism has responded to this drive with a limp passivity that is astonishing even to one who is experienced in its ways. The magazines and newspapers, with a few honorable exceptions (see the Bibliographical Note), have preferred to go along with the view of Disney as an avuncular Horatio Alger figure, an ordinary man, perhaps even Everyman, whose career was a living demonstration that the American Dream sometimes works out in a reality stranger than fiction.

It was an attractively reassuring line to take. It made everyone—readers, writers, editors—sleep just a little more soundly to know that Walt was not only on the job but was handling it just the way they would have if they had been in his shoes. Indeed, the reportage implicitly encouraged the notion that they might well have been in his shoes if they had just had a few breaks. He seemed such an ordinary guy—well-meaning, sentimental, a lover of the cute and familiar. No intellectual, perhaps, but no con man, either. And there was just enough truth in the legend that formed over the years to make it seem very persuasive. All you had to do with it, whenever you rewrote it again, was leave out a few questions—and a few answers—about the assumptions, visions and values of the American middle class, which he both represented and served.

As a result, this book may come as a surprise to some people who turn to the lives of figures like Disney as their children turn to familiar fairy stories, in the expectation of once again seeing things come out all right in the end. I have not for one minute conceived of it as an "exposé"—the word is ludicrous in connection with someone like Disney. But it does attempt to see him coolly and objectively and within the context of our developing society. To this end, it partakes of all the disciplines previously alluded to—economics, sociology, cultural and artistic criticism—and a few others as well—psychology, for example, and history. The author does not claim to be an expert in any of these fields and, indeed, cheerfully admits to coursing through the works of many masters in all of them in search of thoughts and material that would help him come sensibly to grips with Walter E. Disney, his life and times. Generally, the material gleaned in this manner is set off from the body of the text as epigraphs heading the chapters. The idea is to indicate that all generalizations are tentative and suggestive, not final. Too many people speak with too much whimsical authority about masscult and midcult (to borrow a couple of words from one of the most whimsical of these spokesmen) for me to want to join their numbers. Here, I have wanted mainly to set forth a large body of previously uncollated information within a context that at least implies an attitude that is more critical of both Disney and his audience than was usually taken while he lived. I hope also to indicate by this examination of the life and work of one purveyor of popular culture that the subject has more—and more interesting—dimensions than many of the blithely critical attitudinizers seem to realize. Most important, I have sought—and believe I have found—in the life and work of Walt Disney a microcosm embodying a good deal of the spirit of our times, including a good many things that disquiet me as a citizen of those times and of the future they portend. In this most childlike of our mass communicators I see what is most childish and therefore most dangerous in all of us who were his fellow Americans.

Many times, as I wrote this book, people asked me why I wanted to devote almost two years of my life to the study of a man whom few writers or critics have taken seriously for more than a quarter of a century. The answer, of course, is that I undertook this work precisely because the period of Disney's greatest economic success, his greatest personal power, coincided with the decline of active interest in him in the intellectual community. As usual, the people who claim to concern themselves the most with popular culture, missed the point. When Disney ceased to make any claims as an artist they dropped him, as if only the artist is capable of influencing the shape and direction of our culture. In America, that seems to me a preposterous proposition. Our environment, our sensibilities, the very quality of both our waking and sleeping hours, are all formed largely by people with no more artistic conscience or intelligence than a cumquat. If the happy few do not study them at least as seriously as they study Andy Warhol, then they will lose their grip on the American reality and, with it, whatever chance they might have of remaking it in a more pleasing style. To me it seems clear that the destruction of our old sense of community, the irrational and unrationalized growth of our "electronic" culture, the familiar modern diseases of fragmentation and alienation, are in large measure the results of the failure of the intellectual community to deal realistically—and on the basis of solid, even practical, knowledge—with the purveyors of popular culture. If, for slightly ridiculous example, some of the easily shocked literary visitors to Disneyland in its early days had really looked at what Mr. Disney was doing there, the work of Marshall McLuhan a decade later would have been infinitely less surprising to them. But enough. The point is made. And I hope to develop it further in the pages that follow.

Though this book has the structure, and the outer appearance, of biography I would be disappointed if readers applied to it the strict, formal standards of that genre. To me, Disney was a type as well as an individual, and part of his fascination for me was that he was a type that I have known and conducted a sort of love-hate relationship with since I was a child—the midwestern go-getter. I believe many of us who were formed by that enigmatic region share certain traits, and it has been particularly interesting to me to find and point out the evidences of those traits in the Disney *oeuvre*. Some of my speculations on these matters may exceed the customs of the biographical form, as may some of my probes into mass culture. Nevertheless, I have felt compelled to proceed with them, for it is in these matters that the value of the book lies—at least for me. I therefore ask the reader to conceive of this as a volume that may be a little less than purely biographical but one that is also, at times, a little more than a biography—a study of an aspect of American culture, perhaps, or a free-form speculation on some qualities of the American mind at work or, less pretentiously, a book that, for good or ill, insists upon setting its own peculiar boundaries.

One final *caveat:* this is a study of a public man. Beyond the courtesies described in the Acknowledgments I had no cooperation from either the Disney family or the Disney organization. Therefore, the reader hoping to discover much about Walt Disney's personal life will be disappointed. In time, there undoubtedly will be an official biography, which will reveal something of his life away from the studio and the limelight. I hope there will be, and I hope the job will be entrusted to someone other than a Hollywood hack—since I firmly believe that people of Disney's power and achievement (however it is valued) are deserving of at least the same standards of scholarship that are automatically applied to the lives of obscure Civil War generals and minor novelists. Unfortunately, I have had to make do with the public record and with such oral reminiscences as I could gather in a limited time, and these, quite naturally, have tended to concentrate on the public man. They have been enough for my immediate purposes, but not enough, I know, to close the record completely.

RICHARD SCHICKEL

New York City
May 28, 1967

PART ONE

A Trial Balance

1   WALT DISNEY, 65, DIES ON COAST:
FOUNDED AN EMPIRE ON A MOUSE
—The New York Times, Dec. 16, 1966

DEATH CAME at 9:35 A.M. on December 15 in St. Joseph's Hospital, directly across the street from the Disney Studio in Burbank. It appears that except for hospital personnel, including the cardiologist who had been summoned to treat the "acute circulatory collapse" that was the immediate cause of death, Disney was alone when he died. The presence of relatives or friends is unmentioned in accounts of his death. Neither the family nor the studio will elaborate on the matter, but if he died alone, it would have been characteristic of the man, for he had always been a loner, especially by Hollywood standards. Equally characteristically, he left very few loose ends untied when he died. The fiscal year of Walt Disney Productions had closed on October 1, and the balance sheet revealed grosses and profits at the highest point in its history. Moreover, the beautifully articulated machine he had constructed over some forty years, many of them frustrating and difficult ones, had, at long last, reached a state so close to perfection that even an inveterate tinkerer like Disney was hard-pressed to find ways to improve it. There were, to be sure, a couple of recreation-*cum*-real estate ventures still in the development stage, and one of them—the Mineral King winter sports center in California—was running into serious opposition from conservationist groups. In the year after he died, there arose a vague uneasiness about the ability of Disney's corporation, now headed in fact as well as in name by his brother, Roy, to continue at quite the high level of financial performance that it had attained under its founder. *Forbes*, a business fortnightly, described the studio, six months after Walt Disney's death, as being "like a fine car without an engine," and added that "the great Disney empire [was] drifting without a leader, as potential successors jockey for power." Others thought, as did Ivan Tors, a producer of TV animal stories for children (*Flipper, Daktari, et al*), that "Without Disney alive, without the personal myth of the artist who created a new form of art in the cartoon, there won't be the same attraction." In short, Disney's death created a vacuum both in the studio and in the hearts of the public that ambitious men were rushing to fill.

And yet it was hard to believe that death would prevent his machine—his beautiful, beautiful machine—from humming steadily along, clicking off profits and banalities at something like the rate it had achieved in the golden years of the early 1960s. The organization has weathered crises at least as serious as the death of its founder. To be sure, Roy Disney was not happy about the "chaos" of 1967. "I know a committee form is a lousy form in this business," he said, "but it's the best we've got until someone in the younger crowd shows he's got the stature to take over the leadership." Roy is, by his own admission, a compromiser, lacking any major creative talent, but he is also a patient and intelligent man, convinced he would do more harm than good if he attempted to run the studio by dictatorship. He also knows the strength of his company perhaps better than outsiders do. In the summer of 1967 he said: "We've never before had this much product on hand.

Walt died at the pinnacle of his producing career in every way. The big thing that's bugging American industry is planning ahead. We've got the most beautiful ten-year plan we could ask for. . . . The financial fellows think we're going to fall on our faces without Walt. Well, we're going to fool them."

He was right—up to a point. By the time Roy Disney made this statement, a production schedule had been hammered out and, to the surprise of veterans, there seemed to be some real interest around the studio in varying what had become an outmoded style. Besides, the Disneys' next of kin could reflect that the quality of the films their company produced, though important, was no longer the key to its success, since films no longer accounted for the major share of corporate income. The situation would have been considerably more perilous if Walt Disney had remained wedded to the notion of "putting over on the audience . . . something from one's own imagination," as Edmund Wilson thought he did in the late 1930s. Instead, of course, Disney placed his not inconsiderable talent in the service of what Mr. Wilson, in the same article, called the search for "an infallible formula to provoke its automatic reactions." By the time he died, he and his associates had found this formula and had managed to adapt it to every medium of communication known to man. They had even invented a new and unique medium of their very own—Disneyland. What was even better about Disney's machine, what made it superior to all its competitors, was that it had the power to *compel* one's attention to a product it particularly treasured. All its parts—movies, television, book and song publishing, merchandising, Disneyland—interlock and are mutually reciprocating. And all of them are aimed at the most vulnerable portion of the adult's psyche—his feelings for his children. If you have a child, you cannot escape a Disney character or story even if you loathe it. And if you happen to like it, you cannot guide or participate in your child's discovery of its charms. The machine's voice is so pervasive and persuasive that it forces first the child, then the parent to pay it heed—and money. In essence, Disney's machine was designed to shatter the two most valuable things about childhood—its secrets and its silences—thus forcing everyone to share the same formative dreams. It has placed a Mickey Mouse hat on every little developing personality in America. As capitalism, it is a work of genius; as culture, it is mostly a horror.

2 "DISNEY'S LAND: DREAM, DIVERSIFY—AND NEVER MISS AN ANGLE"

—The Wall Street Journal, February 4, 1958

SOME FIGURES. In 1966 Walt Disney Productions estimated that around the world 240,000,000 people saw a Disney movie, 100,000,000 watched a Disney television show every week, 800,000,000 read a Disney book or magazine, 50,000,000 listened or danced to Disney music or records, 80,000,000 bought Disney-licensed merchandise, 150,000,000 read a Disney comic strip, 80,000,000 saw Disney educational films at school, in church, on the job, and 6.7 million made the journey to that peculiar Mecca in Anaheim, insistently known as "Walt Disney's Magic Kingdom" in the company's press releases and more commonly referred to as Disneyland. From a state of profitability near zero in 1954 the company has progressed, over the years, to the point where its net income was $12,392,000 on a gross of $116,543,000, which meant that the magic kingdom was very close to joining the magic circle—the 500 largest corporations in the nation—which soon it would do.

All this it had achieved by clinging very closely to the virtues of the Protestant ethic, which is a way of saying that it never went Hollywood, which is a way of saying that it was spared many of the vicissitudes that afflicted that community when, in a single year (1948) it was forced to divest itself of theaters across the nation and saw its old, reliable audience start to replace the movie habit with the television habit. Disney had some luck—the number of children between the ages of five and fourteen doubled in the period between 1940 and 1965, at which point one-third of all U.S. citizens were under fourteen—but he also made some luck. While the rest of the movie business watched in numb horror as television stole its audience,

Disney alone of the moguls—and he was not, at the time, a very big mogul—found a way to use the new medium to his advantage. With two successive shows he hosted himself—as well as the syndicated "Mickey Mouse Club" and the Zorro adventure series—TV became the keystone in a mammoth promotional arch. Each new film, not to mention Disneyland, and, implicitly, the Disney image, received the most delicious sort of publicity— that is, publicity aimed precisely at its proper audience and free of charge. For what was essentially a promotional film about the making of *20,000 Leagues Under the Sea* Disney actually received a television Emmy in 1955. There were other advantages as well. *Davy Crockett*, made as a three-part television drama, proved so successful that it was released as a feature film and generated a merchandising bonanza. Old shorts and cartoons and newer features that failed in the theaters provided acceptable, and mostly amortized, material for the television show, which also functioned much as the B pictures of Hollywood's greener days did—as a testing ground for new talents of all sorts. The show accounted for less than 8 percent of the company's gross, and since it was produced with Disney's lavish attention to detail, it did not do much more than break even. But as a loss leader for the Disney line, it was—and is—a major contributor to the studio's great leap forward. For the best part of this method of handling the studio's product was that it allowed the studio to participate in TV without surrendering control of its precious film library. As Disney's brother, Roy, who was president and chairman of the board, once said: "Since Walt and I entered this business we've never sold a single picture to anybody. We still own them all." In the spring of 1966 that amounted to 21 full-length animated features, 493 short subjects, 47 live-action features, 7 True-Life Adventure features, 330 hours of Mickey Mouse Clubs, 78 half-hour Zorro adventures and 280 filmed TV shows. At a moment when many Hollywood studios were finding that the difference between profit and loss often came from the outright sale of their old films—that is, of its history, its very corporate self—this ability of Disney's to prosper without peddling the inventory placed the studio in an enviable position. Rereleasing two or three oldies every year, the studio has averaged something like four and a half million dollars on them annually, almost all of it clear profit; in effect, the studio has automatically available to it the equivalent of the return on one new, ordinarily successful feature before it begins operation each year—and many of the Disney "classics" are doing better in the new, larger "family market" than they did when they were first released.

So it seems that from the purely commercial point of view, Disney built perhaps better than he knew when he concentrated his animated feature production on timeless tales that are incapable of becoming dated and in placing even his live-action features either in historical settings or in a never-never small town that, because it never existed in reality, can never be seriously outdated by changes in taste in furnishings or dress. His films, for the most part, are endlessly rereleasable.

These films had other uses, too. As early as 1952 Disney's Buena Vista Distribution Co. began renting 16 mm. prints of old Disney theatrical releases to schools and other organizations. That business has grown steadily through the years, and to it a variety of sidelines have been added. The firm has produced 152 35 mm. filmstrips under license by Encyclopaedia Britannica and study prints (13"x18" poster sets on such subjects as safety and transportation for use in the elementary grades) and, lately, has experimented in educational technology with its 8 mm. single-concept film program, employing a lightweight portable projector so simple that a child can operate it. The system makes available to schools some two hundred very short films—one, two, three and four minutes in length, each dealing with a single idea—no more—that a teacher wants students to concentrate on as part of a larger study program. If the school is rich enough, it can buy dozens of the little projectors and send the films home with the child for after-hours study. Most of the tiny films are snipped from old Disney movies, particularly the nature series, while the poster prints use Disney characters and Disneyland attractions as subjects; the Disney name, of course, is prominent on all this material. Thus the studio reaps promotional benefits as well as more conventional profits and the advantages of a foothold in the growth industry of educational technology from a program that may be defined, in a variation of the old saw, as processing everything but The Mouse's squeal.

In the same period the studio slowly developed its capability in industrial film production, in which its capacity for animated and documentary film making was particularly useful. It made one-shot promotional films for American Telephone and Telegraph and the American Iron and Steel Institute, and in 1967 was making a film on family planning for the Population Council and had signed a $1.3 million contract for four 16 mm. educational shorts on health subjects to be underwritten by the Upjohn Pharmaceutical Company—the beginnings of what may be a much more ambitious program in a little remarked but potentially highly profitable field.

Finally, by pursuing an old private dream, dating back to the 1930s—the creation of an amusement park suitable for his own children—he managed to lay the groundwork for the kind of diversification that all the other film companies lust after but rarely achieve. Disneyland was not finished in time for Disney's daughters to enjoy it as children—though undoubtedly they do as mothers—but it pointed Disney toward an ancient entertainment form that had fallen on dreary, even evil, days, a form that was ripe for fresh imaginings and venturings—the amusement park, which in the variation on it Disney worked has been properly renamed "the atmospheric park." The quality of the aesthetic content of Disneyland has been endlessly debated by intellectuals. John Ciardi came away from it crying that he had seen "the shyster in the backroom of illusion, diluting his witch's brew with tap water, while all his gnomes worked frantically to design gaudier and gaudier design for the mess." Novelist Julian Halévy claimed that "the whole world, the universe, and all man's dominion over self and nature has been reduced to a sickening blend of cheap formulas packaged to sell." Aubrey Menen, on the other hand, remarked that "the strongest desire an artist knows is to create a world of his own where everything is just as he imagines it." To Menen, Disneyland was such a creation, a true representation of Disney's truest vision and, beyond that, the kind of pleasure dome that kings and emperors used to create for their private amusement. Only Disney threw his open to the masses.

It is premature to examine these claims and counterclaims here, though it does seem fair to note that Disneyland and the newer projects it inspired claimed the largest share of Disney's psychic energy in the last decade or so of his life. It was here—and not so much in the other areas of his domain—that Disney really lived. The results of this intensive commitment were twofold. The basis of Disney's gift, from the beginning, was not as is commonly supposed a "genius" for artistic expression; if he had any genius at all it was for the exploitation of technological innovation. Thus the man who summarized the real achievement of Disneyland is not a literary intellectual at all but a city planner and developer named James Rouse, a leading figure in one of the latest manifestations of the "New Town" movement —Columbia, Maryland. Rouse put the matter very simply: "The greatest piece of urban design in the U.S. today is Disneyland. Think of its performance in relation to its purpose."

The validity of this judgment was evident to businessmen long before Rouse voiced it, and it was Disney's success in this area that simultaneously led them to him for the creation of what were, by common consent, the best exhibits of the New York World's Fair in 1964–65 and spurred Disney himself to larger related efforts. These, though they retained, as elements of the grand design, large-scale entertainment components, were more accurately seen as attempts to reexamine the customs of urban design (as in the projected "Disneyworld" development in Florida) and of recreational design (as at the projected Mineral King winter recreation area in California's Sierra range). To some degree the beginnings of these concerns were accidentally imposed upon the Disney organization. The creation of something like Disneyland requires the acquisition of the capability of handling large crowds not only efficiently (volume is everything in such an operation) but keeping them in a happy (*i.e.*, spending) frame of mind. This automatically took the company into areas of research and development on which surprisingly little intelligent thought has been expended in this country. From the engineering feats of Disneyland it is not a particularly long step to the kind of engineering required by a Mineral King project or to the creation of a small city as was originally envisioned at the Florida Disneyworld.

But there was another factor operating behind all this. That was Walt Disney's lifelong rage to order, control and keep clean any environment he inhabited. His studio, the last major such facility to be built in Hollywood during its golden age of profitability, was a model of efficient industrial organization and also a very pleasant place to work—at least as a physical environment. In July, 1966, he told a group of journalists that at least part of his recent intensive interest in city planning stemmed from his dismay at the unplanned sprawl of Los Angeles which he observed as he drove about the city. He just couldn't abide a mess, and now, at last, he had the wherewithal to try to do something about the developing national problem of the urban mess—at least as an innovator and perhaps as a creator of standards. He was heard to say that if the nation had put into the development of technologies to deal with urbanization even a fraction of what it had put into aerospace research, the problem might now be solved. As it was, he regarded urban design as the next great frontier of technology, and he wanted to be in on it.

What with the success of Disneyland and the smooth, profitable functioning of the television, music and records, merchandising and publication arms of the company, there was nothing to stop him. Indeed, the trend of his own company as well as the trend of the times urged him on. Revenue from theatrical film rental, which once accounted for 77 percent of the Disney take, had declined by 1966 to 45 percent (though its dollar volume was, of course, higher than ever). One could safely predict that film rental would decline still further as a percentage of income once the huge new element of the great Disney machine—the Florida Disneyworld—were in place and functioning, and one could predict that this state of affairs would be welcome at the company. "Now the Bankers Come to Disney" was the title *Fortune* gave its May, 1966, summary of the company's remarkable economic achievements, and the assumption is reasonable that the bankers would continue to do so. For even with the departure of the master, it seemed that the Disney Studio was well on the way toward solving the ancient Hollywood problem of achieving financial stability when the success of the basic product depends on such unpredictable factors as fads and fashions in stories and stars, the weather in the key cities when a major film is released, the ability to gain major promotional breaks for it, the ever changing whims and fancies and fantasies of the mass audience, even—perhaps in recent years, increasingly so—the attitude of critics.

These, of course, had always been matters of concern and even annoyance in the old days, but before television, when some fifty million Americans had trooped off to the movies every week not much caring what they saw, the problems had hardly been in the life-or-death category. With that kind of an audience you could safely count on making up on the straightaways what you lost on the turns so long as most of your films were kept within reasonable cost, your studio turned them out at a steady, overhead-justifying rate and you had the sure profits of the theater chains to fall back upon at all times. The years that followed the loss of the theaters (under duress of the

25

antitrust laws) and the rise of TV have been spent by Hollywood largely in the search for a new formula to replace the one that had been rendered obsolete. Many were tried. Some studios cut the number of their own productions radically—thereby cutting their overhead—and devoted most of their energies to renting facilities and equipment to independents, financing them and then releasing their products. United Artists sensibly and profitably divested itself of its studio entirely and concentrated solely on the last two functions. A few successful films like *Ben Hur* convinced others that the way and the truth lay in superproductions, in which much was staked on a very small number of films (though a few economically suspenseful ventures like *Cleopatra* and *Mutiny on the Bounty* soon convinced them otherwise). Still others believed in the small, inexpensive film of high quality, of which *Marty* was the classic example (though a few box-office failures like *The Bachelor Party* soon vitiated that faith). Nearly everyone was convinced that the handful of stars who could really draw people to the box office were worth almost anything they asked, and guarantees of one million dollars against a percentage of the gross—usually 10 percent—were by no means rare (though they often went to stars who could not deliver and whose costliness often insured the financial failure of films that, with lower-priced players, might have made money).

Nearly everyone at first reacted to television merely by praying that it would go away. A few saw in TV production a chance to use up overhead through facilities rental, but on the whole the studios were late getting into TV with shows of their own. Many compounded the first error with a second—selling sizable chunks of their film libraries to the competing medium, which improved the cash flow of the studios and kept their balance sheets looking good temporarily but placed the studios in the unenviable position of allowing their old products to compete with their new ones, often to the latter's disadvantage. The studios attempted other forms of diversification in addition to these unplanned forays into TV: they started record companies, they invested in Broadway shows, they did their best to get into merchandising in the Disney manner (though without the stable of wonderfully merchandisable characters he had built up through the years). A few even ventured into real estate, selling off the ranches where they had once shot westerns and, in one case, selling even the backlot of the main studio for real estate development or oil explorations.

But somehow none of these experiments worked really well, except as temporary expedients. Not one of the major studios escaped the last two decades without at least one major crisis; many suffered several, and a few slumped into what appeared to be a permanent state of low vitality. One major company and two minor ones went out of business entirely, and during the 1960s, six of the eight major studios suffered debilitating proxy fights as management confusion and ineptitude were reflected in lower dividend rates and spurred shrewd raiders and manipulators, a breed particularly attracted to entertainment companies, to seek control of the beleaguered companies. Not many of the raiders had any genuine interest in the production of movies; many hoped to make some quick profits—principally through the sale of film libraries to television—and get out, probably leaving the companies sicker than ever, if not dead. A few were at least as interested in the access to starlets studio ownership provided as they were in profitability. Of all the Hollywood studios only one emerged from this lengthy winter of discontent stronger than it had been. Only one advanced in status from a minor corporation to a major one, achieving financial stability at the very moment when the old-line major studios, which had once regarded it as no more than an insignificant industrial curiosity, were losing theirs. That studio was, of course, Disney's.

It was Disney's good fortune to suffer his agony earlier than the other producers. Overcommitted to feature films in the late 1930s and early 1940s, with the foreign market—particularly vital to him—closed by the war and with an expensive new studio sucking up what capital he had, Disney had been forced to make a public stock offering in 1940. The 155,000 six-percent convertible shares had a par value of twenty-five dollars but quickly tumbled to three dollars a share in the market, and only government contracts for training and propaganda films kept the studio functioning, its crisis hidden and the Disney name before the public. The war work also took some of the sting out of an extremely unpleasant strike at the studio, as a result of which The Founder's benevolent image took its first—and possibly only—serious beating. In short, the times were bad for Disney at precisely the moment when they were extraordinarily good for the rest of the industry. The one good thing about the situation—and this became clear only in retrospect—was that the problems that were later to plague the rest of the industry had to be met by Disney at a time when the government could help out and when the general buoyancy of the industry could at least keep him afloat. The result, of course, was a head start in gathering know-how to meet the crisis that was coming—a head start in planning for diversification first of the company's motion picture products, then of its over-all activities. The results were spectacular.

The Disney studio had been, by Hollywood or any other standards, a very small business from the time of its founding in 1928 until 1954, when for the first time it finally grossed more than $10,000,000. After that, the sales chart took a sharply upward curve. There was one minor setback in 1960, when *Sleeping Beauty* awakened only apathy at the box office, but by the end of fiscal 1965 the gross income of Walt Disney Productions, borne aloft by *Mary Poppins'* magic umbrella, had sailed past the $100,000,000 mark (the exact figure was $109,947,000) and by the end of fiscal '66 it was $116,543,000. Profits, which had been no more than a couple of million at the start of the great leap forward and which stayed low due largely to the firm's frugal policy of using its own assets to finance expansion whenever possible, had by the end of 1966 reached a record high of $12,392,000.

As for Disney himself, the man who only a decade before had hocked his own life insurance to finance the early stages of Disneyland, the man who had been, for most of his career, tolerated by *haute* Hollywood as an enigmatic eccentric whose presence was "good for the industry"—this man had gained what all his detractors had lost. When he died, his studio was the only one left that had no truck with the independent producers, that offered no percentage deals to stars, that made no fabulous bids for hot literary properties, that produced, with its own resources and its own full-time staff, all the motion picture products it sent out under its name. A few months before Disney died, Jack L. Warner sold his stock in the studio he had long controlled, and that left Walt Disney as The Last Mogul, the last chief of production who had to answer to no one—not to the bankers, not to the board of directors (it was dominated by his own management men, who were, in turn, dominated by him), not to the stockholders (they were happy—and anyway, one third of them were kids given a share in Mickey Mouse as an introduction to the capitalist way of life). In his domain—in "The Magic Kingdom," as his flacks insistently called it—Walter Elias Disney was the undisputed, and generally benevolent, ruler of all he surveyed.

3 *Walt Disney is one of the best compensated executives in the U.S. these days and, as he says himself, "It's about time."*
—*Fortune*, May, 1966
He was a twentieth century Cellini who supervised the mining of his own gold.
—Joseph Morgenstern, *Newsweek*, Dec. 23, 1966

SOME MORE FIGURES. At the time of his death Walt Disney owned 262,941 shares of stock in Walt Disney Productions. In the open market these were being traded at 69 on the day he died (they subsequently went up 9⅜ points on the expectation that the studio was a prime merger candidate), which means they were worth a little more than eighteen million dollars. In addition to these holdings, which constituted 13.4 percent of the company's outstanding shares, Disney's wife owned another 26,444 shares in the company; Diane Disney Miller, who was elected to fill her father's place on the board after he died, and her husband, Ron, owned another 43,977 shares; Roy Disney and his wife controlled 99,881 shares outright, and a corporation they had set up owned another 50,573 shares. Finally, the Disney Foundation, a charitable institution founded and funded by the Disney brothers, owned another 52,964 shares. Disney's other daughter, Sharon, and her husband owned an undisclosed amount of stock, and various Disney grandchildren held little stakes in the company as well. In all, the Disneys owned approximately 34 percent of Walt Disney Productions, and the Disney Foundation owned another 2.7 percent.

The stock was by no means the sum total of Walt Disney's estate. His basic salary since 1961 had been $182,000 a year, plus a deferred salary accrued at the rate of $2,500 per week that was to be paid to his estate at a weekly rate of $1,666.66 for about seven and a half years. In addition he had, since 1953, the right personally to purchase interest up to 25 percent in any live-action, feature-length film his company produced so long as he exercised this option prior to the start of filming. From 1961 Disney exercised this option—though only up to 10 percent—on almost all the company's films (he also extended a similar right of investment—though up to only 1 percent—to seven key executives). His interest in *Mary Poppins* alone brought him a million dollars in 1965 and, in addition, generated a capital asset estimated at another million. This money, in turn, was assigned to a family-owned company, Retlaw (which it is sometimes amusing to spell backward). The company has reaped dividends from past successes as they have been re-released, and will continue to do so, one imagines, until time itself comes to a stop.

Retlaw also had two other major functions. One was to license Walt Disney's name to Walt Disney Productions for use in merchandising agreements. Disney Productions had the right to use of the name only in its corporate title and on films and television shows in which The Founder personally participated. Retlaw had the right of choice in this matter: it could take 5 percent of any net profits derived from the business or participate in the business through investments in it up to 15 percent, or it could take 10 percent of an amount held to be the reasonable value of Walt Disney's name to Walt Disney Productions in any given situation. In a fairly typical year—1965— the company turned over to Retlaw $292,349 in royalties on deals of this sort. Retlaw's other major function was ownership of the steam railroad and the elevated monorail at Disneyland. This oddity stemmed back to the days when not everyone shared Disney's enthusiasm for the park and he found it necessary, through his own resources, to finance the railroad that circles it—and that was a part of the over-all design he as a railway buff particularly loved. The arrangement worked out so nicely that a similar one was set up when the monorail was added in 1961. The two railways grossed between two and three million annually, in the series and Retlaw paid 20 percent of the gross to Walt Disney Productions as rent for its right-of-ways and keeps the remainder for operating expenses, amortization of its $3.2 million investment in the system, and profits.

Retlaw was, in effect, the successor to the corporation the Disney family previously, held, known as WED Enterprises, Inc. (the initials, of course, are those of Walter Elias Disney). From the time of its organization in 1952 until the 1965 sale for three million dollars, of its name and its most important assets to Walt Disney Productions, it performed all the functions later handled by Retlaw. And there was one more: it served as the design, architectural and engineering arm of Disneyland. In this role it created the exhibits that bore the Disney name at the New York World's Fair of 1964–65, it was the developer of Audio-Animatronics, the system of animating three-dimensional characters that highlight many Disneyland and Disneyworld attractions and have a lovely, licensable, exploitable future, and it has created the WEDway People Mover, which is, in fact, the moving sidewalk so many have predicted and so few have actually experimented with (its first installation was in the new Tomorrowland section of Disneyland). Just how much income Retlaw produces for the Disneys in any given year is impossible to estimate, though John MacDonald of *Fortune* guessed that in 1965 Disney and Retlaw together had an income, apart from his salary and stock dividends, of more than two million dollars. With Disney's death and the ending of Retlaw's ability to participate in new film production deals, that figure, of course, deteriorated. The Disneyland railroads, however, will probably keep chugging out profits for decades to come. And Retlaw's merchandising agreement with Productions runs until 2003.

4 *It is not to Mr. Disney's discredit that, when success and fame rightly came to him, he began to expand as a person and as an ambitious business man. It was natural that he should have flourished under the warm and tinkling rain of public praise, that he should have managed to throw off his shyness, that he should have found it quite pleasant to take bows. . . . He was now moving in the area of the big producer, the Hollywood tycoon, and this was a role he managed with more pretension than with comfort and ease.*
—Bosley Crowther, *The New York Times*, Dec. 16, 1966

AMONG UNSOPHISTICATED people there was a common misapprehension that Disney continued to draw at least the important sequences in his animated films, his comic strips, his illustrated books. Although his studio often stressed in its publicity the numbers of people it employed and the beauties of their teamwork, some *very* unsophisticated people thought he did everything himself—an interesting example of the persistence of a particularly treasured illusion and of the corporation's ability to keep it alive even while denying it.

Disney himself once tried to explain his role in his company by telling a story that may well be apochryphal but is no less significant and no less quoted in company publicity for all that. "You know," he said, "I was stumped one day when a little boy asked, 'Do you draw Mickey Mouse?' I had to admit I did not draw any more. 'Then you think up all the jokes and ideas?' 'No,' I said, 'I don't do that.' Finally he looked at me and said, 'Mr. Disney, just what do you do?'

" 'Well,' I said, 'sometimes I think of myself as a little bee. I go from one area of the studio to another and gather pollen and sort of stimulate everybody.' I guess that's the job I do."

The summary is not a bad one, as far as it goes. But there were a good many more wrinkles in Disney's situation than the smoothness of this explanation would indicate. For one thing, Disney was continually, if mildly, irked because he could not draw Mickey or Donald or Pluto. He never could. Even Mickey Mouse was designed by someone else, namely Ub Iwerks, an old friend from Disney's pre-Hollywood days. Iwerks actually received screen credit for so doing on the first Mouse cartoons. In later years Disney was known to apply to his animators for hints on how to render a quick sketch of Mickey in order to oblige autograph hunters who requested it to accompany his signature. Even more embarrassingly, he could not accurately duplicate the familiar "Walt Disney" signature that appeared as a trademark on all his products. There are people who received authentically autographed Disney books and records but who thought they were fake because his hand did not match that of the trademark—a particular irony in the case of Disney, who had devoted a lifetime to publicizing his name and as we have seen, quite literally capitalizing on it.

Stories like these should not be taken to mean that Disney passed his final years living out the myth that money cannot buy happiness or that when his fortune caught up with his fame he found that all his dreams were a mockery. Far from it. Like many men who have grown rich through their own efforts, he had little use for personal display. His suits were still bought off the peg. His diet still consisted largely of the foods he had acquired a taste for in the hash houses of his youth (hamburgers, steaks and chops were staples; he especially liked chili). He drove himself to work, mostly in standard American cars, though in his last year he took to using the Jaguar that Roddy MacDowall had driven in *That Darned Cat*. He served tomato juice to visitors in his office but allowed beer to be served in the studio commissary—by no means a standard Hollywood practice—and admitted to enjoying a highball or two at the end of a workday that minimally ran twelve hours.

He traveled mainly on business and submitted to vacations with restless grouchiness. His one known extravagance was the scale model train that used to circle his home and that he conceived and then helped to build. In the early Fifties it had been his great pleasure to don a railroad engineer's cap, brandish a slender-spouted railroader's oil can and pilot grandchildren and other visitors around his yard.

The train became, over the years, the subject of a disproportionate share of Disney lore. It was as if he used it to distract attention from other areas of his private life—certainly it made excellent feature copy for popular journalism. Mrs. Disney and his daughters tried to persuade him to build the thing at the studio instead of in his yard, but he went so far as to have his lawyers draft a right-of-way contract for his family to sign giving him permission to build and maintain the train. It was a token of such seriousness of intent that they agreed to sign it, whereupon he declared that in the circumstances their verbal consent was good enough for him. He loved fussing over the train, continually adding to its supply of rolling stock, improving the grades, even digging a tunnel so that it could pass under some of his wife's flower gardens without disturbing them. Mrs. Disney once commented, "It is a wonderful hobby for him . . . it has been a fine diversion and safety valve for his nervous energy. For when he leaves the studio he can't just lock the door and forget it. He is so keyed up he has to keep going on something." She noted that a good deal of their social life in this period revolved around giving people rides on the half-mile line and that a select few were given cards designating them vice presidents of the road. She and her daughters quickly became bored with the train, but Disney did not, and he even suffered some hurt feelings over their indifference. He enjoyed —like a small boy with an electric train—planning wrecks because repairing the damage was so much fun. Once, after he bought two new engines, she heard him enthusing to George Murphy, now the Senator and then one of the toy train line's "vice presidents," "Boy, we're sure to have some wrecks now!"

The train was the only splash of color in the Disneys' quiet home life. Their friends, with the exception of Murphy, Kay Kayser, Irene Dunne and a few others, were not drawn from show business. Indeed, their daughter, Diane, became the first daughter of a motion-picture-industry family to make her debut at the Las Madrinas Ball in Los Angeles (which, in 1967, finally invited its first Jewish girl to participate). To old Angelenos the movies and the Jews were, apparently, virtually synonymous. Of the other aspects of Disney's mature home life little is known. He is recorded as having been a doting father, given to sentimental outbursts on such occasions as weddings and the birth of grandchildren, and the indulgent master of a poodle, for whom he was known to raid the ice box for cold meat. He also forbade the extermination of small pests—rabbits, squirrels and the like—that raided his wife's garden. This was, so far as one can tell, no idle image-protection on his part, but rather an expression of genuine concern for animal life. Outside the home his favorite recreation was wandering around —unrecognized, he hoped—in such places as the Farmer's Market in Los Angeles or on New York's Third Avenue, where he liked to browse the secondhand and antique shops and buy dollhouse-sized furniture. He had, Mrs. Disney reported, "no use for people who throw their weight around as celebrities, or for those who fawn over you just because you are famous."

Among countless perquisites that were available to him as the head of a prospering corporation the only one that seemed to afford him much pleasure was the company's three-plane "airline," the flagship of which was a prop-jet Grumman Gulfstream. He went everywhere on it and took an open—perhaps even childlike—pride in this particular symbolization of his status. In the months before he died, he frequently mentioned the pure jet that Walt Disney Productions had on order and that, typically, was to be a model for the next generation of executive jets—a technological leader.

But the backyard train and the plane were in contrast with the essential Disney personality. The Los Angeles *Times*' obituary editorial speculated that Disney's "real joy must have come from seeing the flash of delight sweep across a child's face and hearing his sudden laughter, at the first sight of Mickey Mouse, or Snow White or Pinocchio." Certainly the sight did not make him unhappy, but the *Times'* own biographical sketch, appearing in the same edition, carried a more nearly true statement from Disney himself on the source of his deepest satisfaction in his later years. A reporter once asked him, according to the piece, to name his most rewarding experience, and Disney's reply was blunt and brief: "The whole damn thing. The fact that I was able to build an organization and hold it . . ."

These are clearly not the words of some kindly old uncle who just loves to come to your child's birthday party and do his magic tricks and tell his jokes and find his kicks in the kiddies' laughter and applause and their parents' gratitude. Neither do they appear to bear much resemblance to anything we might expect from an artist looking back over his career. They represent, instead, the entrepreneurial spirit triumphant. They are the words of a man who has struggled hard to establish himself and his product; who has fought his way in from the fringes of his chosen industry to its center; who has gambled his own money and his own future on his own innovative inspirations and organizational intelligence and more than once has come close to losing his whole bundle. They are, most of all, the words of a man who at last is in possession of the most important piece of information a player in the only really important American game can obtain—the knowledge that he is a sure economic winner, that no matter what happens the chance of his being busted out of the game has been eliminated and that his accumulated winnings will surely survive him.

This knowledge was vouchsafed Disney only late in life, but when it was, he was fond of working variations on the theme that money was merely a fertilizer, useful only to the degree that it could make new crops of ideas and enterprises grow. This is, of course, an image that anyone brought up in the spirit of the Protestant ethic would instantly appreciate and probably applaud; it is also an image that Freud and more particularly his latter-day followers into the dark realms where the relationship between money and feces are explored, would quickly— probably too quickly—understand and explicate at dismal length. To a builder of Disney's character, though, money did in fact perform precisely the function he liked to describe, and its symbolic value was in good part just what it seemed most obviously to be—a measure of distances traveled, a way of keeping score.

Looking back, it is easy to see that Disney was neither a soldout nor a sidetracked artist. He was a man who had obtained what he truly wanted: elevation—at least on the lower levels— to the ranks of the other great inventor-entrepreneurs of our industrial history. He was of the stuff of Ford and Edison a man who could do everything a great entrepreneur is expected to do —dream and create and hold.

5   *When he died . . . Disney was no longer simply the fundamental primitive imagist . . . but a giant corporation whose vast assembly lines produced ever slicker products to dream by. Many of them, mercifully, will be forgotten. . . .*
          —*Time*, Dec. 23, 1966

DISNEY DID NOT CARE. So long as the company prospered, so long as it kept on creating fertilizer for whatever new crops he wanted to plant, he did not care. He was no longer in the art business—if he ever had been—and he was no longer the pure purveyor of a modern mythology, as many intellectuals had once thought he might be. Indeed, the least pleasant aspect of his character in the late, prosperous years was the delight he took in conveying his contempt for art, which he often equated with obscenity.

Speaking to a magazine journalist some ten months before he died, Disney confided: "I've always had a nightmare. I dream that one of my pictures has ended up in an art theater. And I wake up shaking." He buttressed this statement with the rather broad generalization that there was entirely too much depressing and squalid material cluttering up the movie screens of the world. Somewhat defensively he added that he would "stack *Mary Poppins* against any cheap and depraved movie ever made." He proudly told interviewers about all the times he had ordered his projectionist to turn off films when they became too unpleasant for his rather squeamish taste, and admitted that he seldom sat through entire movies but rather ran them "in pieces, just to see a certain actor or actress." In short, it could fairly be said that in the last decade or so of his life he was fundamentally out of touch with the major artistic currents of his times in general and of the motion picture art in particular.

That the period of Disney's greatest economic success coincides with the period of his least interest in contemporary art tells us, perhaps, less about him than it does about American society. Despite extensive and expensive efforts to report the latest fads of the art world—pop, op, camp, psychedelic—and despite the titillated interest of many basically uncultivated people who like to create the appearance of swinging with these presumably sophisticated fads, it would appear that at heart the nation is essentially unchanged in its tastes. It may, somewhat self-consciously, sample the new artistic wares in order at least to be able to condemn them, but generally it turns with relief to the comforting banalities of television (where in truncated and debased form the great constants of our popular art—westerns, detective stories, situation comedies, true romances—abide), and when it ventures forth to the previous home of these forms—the movie house—it gives its most generous patronage to *Mary Poppins* or *The Sound of Music*. Change is surely in the air: audiences are far quicker to laugh at romances and adventures that, a decade or two earlier, they would have taken in deadly earnest. The manufacture of camp films that cue the snicker and the guffaw as carefully as an earlier generation of movie makers cued tears and tension with essentially the same material is now a staple of the industry. Since Walt Disney died, the executives, directors and writers of his studio have been foregathering weekly to see foreign films—ranging from *Georgy Girl* to *Blow-Up*—to study the techniques of the new film makers from abroad just as Disney's animators studied the techniques of the great silent film comedians: to learn what they might borrow and put to their own use. A camera angle here, a shrewd piece of editing there, will perhaps be incorporated into the studio style, which is notably conventional in its filmic techniques. But one does not expect the studio to change its heart even if it does adopt a freer style of expressing itself. Nor is there a need for it to do so. The market for broad, clean humor, for sentimentality and sweetness of outlook, for the hero easily distinguishable from the villain, for adorable animals and children—this will not diminish. Indeed, as the competition drifts away the Disney studio will undoubtedly continue to find—as it has over the last decade or so—that it is almost alone in a very rich market, a market that attracts no critical attention and very little attention from the popular press, which is preoccupied with recounting the latest outrages from the various underground and avant-garde camps. If the studio has any problem to face in maintaining its singular position, it will be one of sincerity. Walt Disney sincerely treasured the values he portrayed in his films and could scarcely credit anyone who saw art or life in more complex, less sunny terms. He was of the old American tradition that believed in keeping its frets and doubts and inner troubles to itself, that blamed its nightmares on something it must have eaten. It is hard to believe that Disney's surviving associates are all of that particular breed. They will most likely continue to give us the outer aspects of the Disney version of life, but will they be able to animate it with the same forceful belief in its honesty and accuracy that the founder did? And lacking that—the equivalent of the much-discussed "X quality" alleged to be present in the work of the great film stars —will they be able to hold their audience's attention as Disney did?

The answer will become clear only with the passage of time. But it must be noted at the outset that it is unjust to criticize Disney for being what he was. Few of Hollywood's industrial pioneers were especially noted for the breadth of their aesthetic or social vision—to put the matter mildly indeed. There is no particular reason why Disney should have been expected to be any different. Except that, as we shall see, there was a rather long moment in time when, if he did not exactly sue the artistic and intellectual communities for favor, he did not reject their heavy-breathing advances either. His personality was at once too simple and too complex for his early acceptance of their favors to be read cynically. He was not using their praise to advance the grand economic design that emerged so forcefully in the last decade of his life: it is doubtful that either he or his brother, Roy, then perceived its full dimensions. Nor should his nastily phrased suspicion toward his one-time allies be read simply as the cries of a hurt soul or of a purist who knows he has somehow gone sour but cannot admit the fact even—or especially—to himself. There may be elements of truth in both these views of the man, but neither is the whole truth. He is, in fact, best seen in a different perspective, one that reaches beyond the confines of the industry *cum* art where he made his career and into the heart of a nation, now changed and still changing, that shaped him and his type.

Walt Disney belonged to a special American breed, middle-class and often midwestern in origin. He was the sort of man who possesses and is possessed by a dream that seems to be particularly and peculiarly of this land of a time only recently past. It is a dream now much satirized but, for all that, no less common among the middle-aging generation. Simply summarized it is that "it only takes one good idea . . ." In the United States one need not even complete the sentence, so clear is the implication: all you need is one winner to get started properly on the road to success.

The dream's natural place of nurture is the workshop in the attic or the basement or the garage in "Gasoline Alley" or "Out Our Way" where a man can be alone with his tools—free of the clock and the cautionary wifely voice and all the other inhibiting forces of family and society. There are several levels at which the dream operates. In some men it calls forth nothing more significant than gadgets that can solve the minor or previously unnoticed problems of organization and efficiency in closets, desks, kitchens and bathrooms and that create the infinite and humorous variety of the Sunset House and Miles Kimball catalogs. And there are the men who, possessed of a slightly higher technical skill (though not necessarily more imagination), create products of wider industrial applicability—a special gasket, say, or a variation on a standard machine tool— and actually move beyond the confines of their own backyards or the rented space in a loft downtown to a little factory of their own.

The modest successes of these people are the unwritten but often discussed footnotes in the folklore of American capitalism. The chapter headings are reserved for those who make it big—the Edisons and the Fords or, in fields closer to Disney's, Joyce Hall, who came down to Kansas City with a little money in a suitcase and the notion that greeting cards should be folded and stuffed in envelopes instead of being simple postcards, or DeWitt and Lila Wallace, who put out the first issues of *The Reader's Digest* all alone in their Greenwich Village apartment. They are the ones with the capacity to build mighty edifices on the dream without losing their proprietary interest in it. They are, most important, the ones who can endlessly replicate its essence—not with its original inspirational force but with sufficient accuracy to insure its continuing acceptability. Indeed, their aim, conscious or unconscious, must be to convert the excitement they initially generated in their audience into something less volatile, since repeated overstimulation is usually fatal to the relationship between the mass communicator and his audience. The trick, of course, is to turn the interest stirred by the original idea into a habit that is beyond the reach of criticism narrowly conceived. It is perhaps for this reason that in the last two decades there has been almost no serious criticism of Disney's work as art; most such comment has been couched almost entirely in the terms of social criticism.

For the purposes of ordinary discussion, these are certainly the most convenient terms, but they do not reveal the essence of the problem posed by mass communication. It is the business of art to expand consciousness, while it is the business of mass communications to reduce it. At best this swiftly consummated reduction is to a series of archetypes; at worst it is to a series of simplistic stereotypes. Since the eyes of the proprietors, especially after the initial act of creative inspiration is finished, are usually focused on something other than this matter of consciousness and sensibility—if indeed, they are aware of it at all —the reductive process usually proceeds erratically. This was so in the case of Disney. Particularly in the beginning, vulgarity and brilliance were inextricably mixed. If, at the end, the balance was tipped rather heavily to the side of the former, the fact remained that there was still enough of the latter present to discourage easy generalizations.

It is fair to say, however, that in the end Disney was a prisoner of his own image. In the days of his first success he was fond of telling interviewers that "we don't bother with a formula . . . I play hunches and leave psychology to others." In his last years he was still leaving psychology to others, but he was also musing out loud to interviewers in terms like these: "Awful good property . . . awful good. But I just don't think a war picture is Disney. It's not what our audience expects from us. Nope, I don't think we can do it." The machine he had created no longer served Disney; he served it, finding his personal satisfactions not in the process of artistic or quasi-artistic creation but in the process of industrial management and development. In so doing, Disney was actually serving a value that ranked, in America in general and in the background that nurtured him in particular, higher by far than an artist's values. Art to Disney's generation was justified only by the uses to which it could be put, and of these uses the social one was tolerable but not nearly so interesting—so magical—as the possibility of building a business on them or their semblance. That Disney did so was, in the popular mind, far more intriguing than any of the products he created. In the last analysis, Walt Disney's greatest creation was Walt Disney. In retrospect it is possible to see that this is precisely what he was working at for some forty years and that the ultimate satisfaction was that he died with the job completed and decorated by as many laurels as his admiring countrymen could bestow upon it. They hardly noticed that the loved object was less a man than an illusion created by a vast machinery. Even when they heard the gears clanking, as they often must have, they didn't seem to care. The trick was everything, and they would have liked to learn its secret if they could.

PART TWO

Touching Earth

---

6   *The country town is one of the great American institutions; perhaps the greatest, in the sense that it has had and continues to have a greater part than any other in shaping public sentiment and giving character to American culture. . . . The road to success has run into and through the country town.*

—Thorstein Veblen

WALTER ELIAS DISNEY was born in Chicago, December 5, 1901. His first name was taken from that of the Congregational minister who baptized him, his second was that of his father. The elder Disney traced his family back to a Burgundian officer named De Disney, who had participated in the Norman invasion of England in 1066, had settled in the conquered country and received a large estate and, as one journalistic historian gently phrased it, "lived and reared his children in a good environment and was classed among the intellectual and well-to-do of his time and age." Be that as it may, the "De" had been chipped off the family name by 1859, when Elias was born in a village in Ontario, the first of eleven children. In the 1870s the family moved to Ellis, Kansas, where they raised wheat and fattened cattle for market. At a comparatively tender age Elias went to work in the railway shops there while also learning the rudiments of carpentry. Within a decade he had made the first of the several unsuccessful moves he was to undertake in search of a competency never to be his. His first venture was in citrus growing in Florida and, apparently, he prospered at first. At any rate, he married an Ohio schoolteacher named Flora Call, who was spending a vacation in Florida, and their first two sons, Herbert and Raymond, were born in Florida. But then the frosts came, killing a crop and forcing Elias to sell out. He drifted back to the Midwest, this time to Chicago, where he worked as a carpenter on the construction of the Columbian Exposition, which opened in 1893.

It is hard to say definitely, but Elias Disney appears to fit the classic description offered by Veblen of the small operator shuttling back and forth between the land to which he was perhaps most strongly attuned emotionally and the city, which seemed to offer more economic opportunities. Essentially these types were, as the great social critic put it, "cultivators of the main chance as well as of the fertile soil," and this "intense and unbroken habituation" imbued them with a "penny-wise spirit of self-help and cupidity" that was their undoing, narrowing their vision to the point where it could encompass only momentary and paltry commercial advantages, preventing them from taking the long view and leaving their work "at the disposal of those massive vested interests that know the uses of collusive mass action. . . ." "Footloose in their attachment to the soil," they were, like the elder Disney, habitually ready "to make the shift out of husbandry into the traffic of towns even at some risk whenever the prospect of some wider margin of net gain . . . opened before their eager eyes."

The margins in Chicago, however, proved to be little wider than they had been in Florida. After the exposition was finished, Elias Disney opened a small contracting firm and ran it hand-to-mouth fashion for something more than a decade. Three more children, Roy, Walt and a younger sister, Ruth, were born in this period. In his spare time Elias served as a deacon of the

Congregational church in which Walt was baptized and oversaw construction of a new church building. Elias was a stern
and frugal man, by all accounts, and never loath to use his family's muscles to further his own ends. Among Walt Disney's
earliest memories was that "my mother used to go out on a construction job and hammer and saw planks with the men." Despite her aid, carpentry proved no more lucrative than orange
growing, and in 1906, before Walt Disney was five, the family
moved on again.

7   *I live in the country. I have no other home. I am impressed
by certain things about farmers. One of them is their destruc-
tiveness. One of them is their total lack of appreciation of the
beautiful—in the main.*
          —Karl Menninger, M.D., *A Psychiatrist's World*

THIS TIME ELIAS DISNEY convinced himself that his
fortune lay in a forty-eight-acre farm near Marceline, Missouri,
in Linn County and on the main line of the Atchison, Topeka
and Santa Fe Railroad, some one hundred miles northeast of
Kansas City. Even in those days a farm of this size could be no
more than a hard-scrabble operation, but Elias apparently rationalized the move morally as much as he did practically. One
friendly biographer of his fourth son states that Elias held
strongly to the belief that "after boys reached a certain age they
are best removed from the corruptive influences of a big city
and subjected to the wholesome atmosphere of the country."

His eldest sons, Herbert and Raymond, objected to the move
and stayed on the farm only a few months before running off
—back to Chicago. According to one version of the official legend the hard calluses of farm labor ill-suited youthful hands

that had recently learned to wield pool cues, but whether it was a taste for this modest depravity or a deeper kind of despair that drove them away, it appears that they were never let back into the family circle or cut in on the profits of the business their younger brothers were to start later. After their runaway, Herbert goes unmentioned in any writing about Disney, and Raymond appears only once, in an unflattering light, although it is known that Herbert became a post office clerk and Raymond, an insurance salesman.

In later life Disney was to remember the farm as a place of enchantment. In the biography of her father which Diane Disney Miller wrote in collaboration with Pete Martin, she reported that Disney could still draw a detailed mental map of the farm and that he had built a replica of the farm's red barn with its lean-tos on either side in the backyard of one of his homes. She also speculates, undoubtedly accurately, that the Main Street, U.S.A., section of Disneyland, the only section of the park through which every visitor must pass, is an expanded and idealized version of Marceline's Main Street. The symbolism is almost too perfect—the strangers forced to recapitulate Disney's formative experience before being allowed to visit his fancies and fantasies in the other areas of the Magic Kingdom.

Life on the farm was certainly more pleasant in retrospect than it had been in actuality. For Elias Disney was a hard man —a believer in physical punishment and harsh economic discipline. The children received no allowances and no playthings either. For Christmas their presents were practical items like shoes and underwear. It was Roy Disney, working at odd jobs, who supplied Walt and his sister with an occasional toy and who, as soon as Walt was big enough to try to handle it, put him on to an occasional good thing. Mrs. Miller tells a story, for example, of Roy's getting a job washing the town hearse and allowing his little brother to participate in the profits of the enterprise despite Walt's having spent most of the time playing dead inside the vehicle. The proceeds were spent at a carnival that passed through town a little later.

Difficult the Marceline years certainly were, but they were formative in several ways. In the best American tradition, they taught young Walter Disney the virtues of hard work or, if not that, at least the idea that one could stand the strain of hard labor without breaking under it. Years later Roy Disney told an interviewer that "as long as I can remember, Walt has been working. . . . He worked in the daytime and he worked at night. Walt didn't play much as a boy. He still can't catch a ball with any certainty."

The small-town life he observed undoubtedly taught Disney the spirit of "self-help and cupidity" that so marked his later business life as well as the need, as Veblen put it, to "be circumspect, acquire merit and avoid offense" that was the basis of his public image. The precariousness of the family's livelihood undoubtedly shaped his own desire not merely to succeed but to do so in a particular way—namely, to avoid surrendering any part of his autonomy to outsiders and to hold his company's stock and its decision-making power as closely as he could. The constant, hopeless indebtedness of the small American farmer is, of course, legendary, and undoubtedly there were in Marceline, besides Elias Disney plenty of cautionary examples of what constant existence in this condition can do to a man's spirit. Certainly the small land-holder's justified prejudice against bankers, the men who kept them endlessly strapped to the wheel of debt, stayed with Disney all of his days (though with his brother's guidance he did learn to live fairly comfortably with the moneymen, who were, in fact, helpful to him—in their special fashion—on several occasions). But a childhood grounding on hard economic bedrock can be invaluable to a man who spends his adult life in a highly speculative enterprise like show business, with its sudden, often shocking, ups and downs. It teaches one how much worse things can be. An often-told Disney story illustrates just this point. It seems Roy Disney one day many years later went to his brother to tell him of his deep concern for the future of their company: they were several million in debt to the Bank of America, it looked as if no more credit would be extended, the cash flow had nearly dried up, and so on and on. When Roy finished his recital, the younger Disney began to laugh—and laugh and laugh till finally his brother was infected with his hilarity. "Just imagine," Walt Disney said in effect when he could finally speak, "a couple of rubes from Kansas City being in a position to owe the Bank of America all that money."

In addition to shaping his economic style, the country town in Kansas shaped Disney's artistic and social sensibility as well. This is quite obvious in *Plane Crazy*, the first Mickey Mouse film he ever attempted. It was set in a farmyard, a setting to which Disney returned time and again in his cartoons, even though Mickey—and the rest of the characters created by Disney's animators—finally took up more or less permanent residence in a small town. A great deal of the humor in all the early Disney cartoons was of the barnyard variety, of which *Steamboat Willie*, the first Mickey Mouse cartoon to be released and the first cartoon ever to have a sound track, provided a perfect example. Its most memorable moment is a one-man band performance by The Mouse, in which he "plays" most of the familiar farm creatures as if they were instruments (the cow's udder, for instance, is turned into a bagpipe). In addition to being low, barnyard humor is often cruel, and the predicaments and punishments to which Disney's animated animals were subjected were often brutal in their initial delineation, if not in their resolutions (in cartoonland, no creature ever dies of the terrible hurts he endures). The country man, dependent on animals for sustenance and accustomed to the bloodier unpleasantries of husbandry (birth, slaughter, gross illnesses) tends naturally to be less sentimental than his urban cousin about them. This was clearly true of Disney, who may be said to have found a way of "using" creatures without killing them. To the degree that he was not, from birth, inured to the more brutal realities of farm life, however, he escaped the totally indifferent attitude toward animal suffering that many farm children seem to acquire from their environment. It could be said, in fact, that Disney blended a citified sentiment about dumb creatures with this practical acceptance of the harshness and the blunt humor of the farm. His most sympathetic treatment of animals was usually reserved for undomesticated wildlife (*i.e.*, *Bambi*, and the subjects of his True-Life Adventure series), while he made the domesticated animals into clowns. He boasted, particularly in his early career, when animals formed the largest part of his subject matter, that he had only once killed a living creature, that the death was accidental and that he learned much about the need for humanitarianism from the experience. The incident occurred on the farm, and the victim was an owl that Disney tried to capture while it slept in the sun of a tree branch. The bird fought back, and the seven-year-old boy, terrified by its beating wings, threw it to the ground, where he stomped it to death. He did not recall the incident more than once in his interviews and, in fact, never acquired any very deep knowledge about animal behavior; rather, he tended, particularly in his animated work, to use them to caricature human types and behavioral patterns, not as fit objects for study or admiration in themselves. The one thing he was convinced of was, as he often said, that "every animal has a separate and distinct personality all its own," an observation that does not apply to the lower phyla and remains open to speculation—some of it semantic, revolving around the definition of the word "personality"—among naturalists, many of whom feel that Disney carried his anthropomorphism much too far in his nature films.

The drive in these films was, of course, toward a sort of multiple reductionism: wild things and wild behavior were often made comprehensible by converting them into cutenesses, mystery was explained with a joke, and terror was resolved by a musical cue or a discreet averting of the camera's eye from the natural processes. It is possible to say that the operative instinct was like the farmer's, which is ever and ever to cut away the underbrush, clear the forest and thus drive out the untamed—and therefore nonutilitarian—creatures. The tradition of the American farmer is to "take up" more land than he can work (hence the endemic condition of being "land poor") and, rather than let it stand—horror of horrors—untouched, to clear it and let it lie fallow until he can get around to it—which may be never. The survival of this instinct may be seen throughout our urban society, which abhors empty land as nature does a vacuum and rushes to fill it with something, anything, even if it is only a parking lot or a hopeful sign promising future development. In the real estate developments that so preoccupied him in his last years, Disney, the one-time farm lad, was every inch the American. Mineral King, where he proposed to build a $35-million winter and summer playground accessible to every moron who could lay hands on an automobile and a picnic lunch (the debris of which he could scatter out the window), is such a place. In its undeveloped state it is "useless"; closed to all but the hardiest most of the year behind a barrier of snow, it has been described as "a spectacular wild mountain valley, an alpine paradise of peaks, ridges and high passes." Dreadful! Bring in the all-weather road (letting the state pay its estimated $20 million cost), bring in the ski lifts and the ski lodges and the restaurants and the ice-skating rink and, for the days when the weather is bad, the bowling alley and the movie theater and, of course, the souvenir shops. Naturally, there is money in it, but there is something deeper than that at work in this national passion to "tame the land," as the cliché goes.

Karl Menninger, the Kansas psychiatrist who, by lucky accident of birth, is one of our best analysts of the midwestern, rural temperament, asks why as a people we have never really learned to love the land as it is, for itself. "What really is the nature of the soil? Is it dirt? Is civilization largely built on overcoming it, or built upon a taboo of dirt, overcoming a natural affection for it?" The question is not idle, and the career of Walt Disney is, as we shall see, much conditioned by the hatred of dirt and of the land that needs cleansing and taming and ordering and even paving over before it can be said to be in genuinely useful working order. It is certainly not unreasonable to suppose that this vision of the land—hardly a unique one and, considering how widely it was held, hardly one for which he can be criticized—began to take shape during the hard years on the farm when, whatever moments of joy Disney found there, the harshness of the life could only have emphasized the filthiness of the land. Somehow it seems appropriate that his first successful creation was a mouse, traditionally viewed as an inhabitant of unclean places and, in his natural state, often an unclean creature himself. In The Mouse, as he was conceived by Disney, all conflict that the animal's real nature might have caused was resolved by an act of creative will: reality was simply ignored. Mickey was a *clean* mouse, right from the start. All inner conflicts about the nature of the land were similarly resolved in Disney's other films: he always, and only, showed us a clean land. Indeed, the whole wide world was scrubbed clean when we saw it through his eyes. "There's enough ugliness and cynicism in the world without me adding to it," he used to say, and one understood his lack of cynicism when he made the statement.

But whatever the precise dimensions of the attitudes toward animals and the land that Disney acquired on the farm, there is no doubt that he did begin to draw in those days. One story has it that his first known work was with tar, a barrel of which, (and a brush for which,) were kept next to the barn for patching roofs and fixing drains. When he was six or seven Disney seized the brush, dipped it into the barrel and proceeded to decorate the white walls of the farm house with large and fanciful drawings of animals. He was punished for the episode, but not long afterward his Aunt Maggie presented Disney with a pad of drawing paper and a box of pencils in which he took an interest that was perhaps extraordinary. Naturally, he received no encouragement from his family, but a nearby doctor often bought his drawings with little presents.

"I recall when I was about seven," Disney said once, "the doctor had a very fine stallion which he asked me to sketch. He held the animal while I worked with my homemade easel and materials. The result was pretty terrible, but both the doctor and his wife praised the drawing highly, to my great delight."

This early encouragement apparently meant a great deal to Disney, for throughout his childhood and youth he continued to enroll in correspondence school cartooning courses and, when he was fourteen, he talked his father into letting him join a Saturday morning art class at the Kansas City Art Institute. Before that happened, however, the Disneys were to enter upon their period of deepest economic decline. In the summer of 1910 Elias Disney was forced to sell his farm, auction off his livestock and, with the proceeds, attempt to build a new life for himself and his family in Kansas City, Missouri.

8
*It takes a particular view of man's place on this earth, and of
the place of childhood within man's total scheme, to invent devices for terrifying children into submission, either by magic or
corporeal terror. . . . Special concepts of property (including
the idea that a man can ruin his own property if he wishes) underlie the idea that it is entirely up to the discretion of an individual father when he should raise the morality of his children
by beating their bodies. It is clear that the concept of children as
property opens the door to those misalliances of impulsivity and
compulsivity, of arbitrariness and moral logic, of brutality and
haughtiness, which make men crueller . . . than creatures not
fired with the divine spark.*
—Erik H. Erikson, *Young Man Luther*

HAVING FAILED IN agriculture at the peasant level,
Elias Disney now sought once again to rejoin the other class for
which he was temperamentally endowed—the *lumpen bourgeoisie*. He bought a Kansas City *Star* newspaper route of three
thousand customers, for which he paid the previous route
owner two dollars apiece. Elias, of course, counted on his two
remaining sons, Roy and Walt, to contribute their labors to the
enterprise, and so, at the age of nine, Walt Disney found himself being routed out of bed at three-thirty each morning in
order to meet the *Star* delivery trucks. The other boys Elias
hired to work with his sons received three dollars a week for
their services. The Disney boys got nothing.

The work was brutally hard, and a couple of months before
he died Disney claimed that he still had bad dreams about it.
Reminiscing to a Los Angeles newsman, he said: " . . . The
papers had to be stuck behind the storm doors. You couldn't
just toss them on the porch. And in the winters there'd be as
much as three feet of snow. I was a little guy and I'd be up to
my nose in snow. I still have nightmares about it.

55

"What I really liked on those cold mornings was getting to the apartment buildings. I'd drop off the papers and then lie down in the warm apartment corridor and snooze a little and try to get warm. I still wake up with that on my mind.

"On nice mornings I used to come to houses with those big old porches and the kids would have left some of their toys out. I would find them and play with them there on the porch at four in the morning when it was just barely getting light. Then I'd have to tear back to the route again." Disney claimed, on another occasion, that the newsboy nightmares of his adulthood frequently centered on forgetting to leave a paper for one of his customers and having to get back and rectify the error before his father found out.

There were other rigors besides the physical and the psychological. Eager to make some money that he could keep for himself, Disney at one point talked his father into ordering fifty extra papers for him to sell on street corners. When he finished his route at six-thirty he would run home, bolt some breakfast and start hawking his own papers. The only trouble with the arrangement was that Elias insisted on taking the boy's money "for safe-keeping," and Walt never saw it again. The boy finally began ordering papers for himself, without his father's knowledge, in order to retain such small earnings as he could make. He also got a job in a candy store during the noon recess from school—and apparently kept it, too, secret from his father.

In none of the accounts of Disney's childhood is much mention made of either his mother or his younger sister. The emotional poles of his life were his father and his older brother, Roy, "one of the kindest fellows I've ever known," in Walt Disney's latter-day estimation. In this period of Walt's transition from what amounted to one American stereotype of boyhood—the barefoot farm lad—to another—the slightly ragged newsboy—Roy Disney, eight years his senior, was his confidant and mentor. "When we were kids," Disney told one of his associates many years later, "Roy and I slept in the same bed. I used to wet the bed and I've been pissing up Roy's leg ever since." In effect, Roy was the intelligent and worldly comptroller, not only of the company they formed and nurtured in later years but of Disney's personal growth. At times he was the conservative, the one who urged him to go slow, to build with care. Yet he was also, like any good surrogate father, the one who would find, when necessary, the ways and means of clearing the way for the younger man to express himself. The record is full of tensions between the brothers, but it was, for the most part, a healthy tension, of the sort that nurtures intelligent creative effort. Displayed on Roy's office wall, in later years, was the peace pipe Walt gave him to patch up one of their fights.

As for the childhood years, it is indubitable that Roy Disney offered his younger brother the clearest available vision of the possibilities of decent and humane—if rough-hewn—manhood. When Roy graduated from high school in 1911 at eighteen, he almost immediately decided to run away from home as his elder brothers had. Before he left, however, he did Walt the singular favor of telling him he did not have to stand for any more beatings from his father.

The old man's habit had been, upon the occasion of discovering a real or imagined failing, to order the boy to the basement for a strapping. According to his daughter's account, Walt had submitted to these beatings "to humor him and keep him happy"—a rather strange way of putting the matter and perhaps yet another example of the Disney inclination to reduce life's unpleasant side to casually manageable proportions. In any case, Roy advised his brother that the punishment was unjust and that he no longer had to submit to physical discipline. The boy preceded his father to the basement and waited. When Elias began to work him over with his leather strap, his son seized his hands in a grip the father could not break. They struggled briefly, and then the old man began to cry. He never again raised his hand to his son.

Two other incidents stand out in the Kansas City period. The first is shrouded in ambiguity and is not spoken about in any of the written accounts of Disney's childhood. It came to light in the late forties when he fell into a political conversation with one of his screenwriters, who was strongly oriented toward the left. Disney inquired if the man had suffered some childhood experience that conditioned his strong beliefs and was told that there had been none—that he had, in fact, arrived at his political stance in his maturity and under the impress of the depression. Disney then countered with the information that his own political conservatism was the result of a street fight. As he told it, his family was strongly Republican in its leaning, while most of the neighborhood was persuaded toward the Democrats. One day on the way home from school, he was ganged up on, beaten and forced to submit to what can be described only as a quasi-sexual assault by some Democratic kids. He told his listener that from that day on, he had never been able even to rationally consider the possibility of voting for anyone but a Republican.

What is odd about Disney's recital of the incident is that he had told his daughter that his father had been a socialist who voted consistently for Eugene Debs and had subscribed to *The Appeal to Reason*, a well-known leftist organ. He even recalled that among his first cartooning attempts had been a representation of a conventionalized capitalist bloated and wearing a vest decorated with dollar signs standing in opposition to a laboring man wearing the traditional hand-fashioned square paper hat such as newspaper printers still make and wear. What is the truth of the assault incident? Is it possible that Disney imagined it merely to win a debating point with his employee? It could be, for there are variant versions of many of the important turning points in his life in the written records of his career. Contradicting this, however, is the fact that the drift of his reminiscences was always the same, leading to the establishment of a mythic, if not literal, truth about his background. Is it possible that he was, in fact, beaten for upholding his father's socialist views? That seems a little more likely, since a Republican was hardly an exotic figure in Kansas City, while socialism might possibly have represented an intolerable eccentricity to a group of boys. In opposition to that interpretation one must recall that only a few years later Disney was perfectly willing to let his daughter discuss her grandfather's political views in a book that was serialized in *The Saturday Evening Post* and widely read. Was he, perhaps, loath to admit, in argument, that he had turned his back on a leftist heritage or that his politics were yet another repudiation of his father—an act he could never commit, possibly because of the impositions of a career devoted to upholding the virtues of family life. Even if his story were pure fantasy, it is an interesting one, expressing a need to create an intense and deep-rooted emotional justification for his political views.

Whatever the truth of this matter, there can be little doubt that among Disney's happiest moments in Kansas City were his introduction to formal art instruction and his first theatrical experiences. The former took place on Saturday mornings when he escaped his family and journeyed to the art institute for young people's classes in painting and drawing. On this one matter his father proved to be indulgent, for, as Disney later said, "he would go for anything that was educational. He was determined to improve his sons, whether they liked it or not." Art instruction, Disney liked; there seems to be no evidence that he particularly cared for anything else in the educational line. As for theater, this was a matter of amateur nights in the neighborhood movies, where he teamed with a young man named Walt Pfeiffer in an act called "The Two Walts." They won a few prizes—the largest seems to have been two dollars—and Disney won some similarly modest fees for his impersonations of Charles Chaplin when Chaplin was at the height of his first great success in the period of 1914 to 1917. Disney later tried an act in which he attempted a comic Dutch accent, flopped and retired from the spotlight until he began hosting his own television shows (although he always spoke for Mickey Mouse on his soundtracks).

The young Disney also carried away from Kansas City the memory of a showing at a special newsboys' matinee of the Marguerite Clark version of *Snow White and the Seven Dwarfs*, which was released in 1917. He traced his selection of *Snow White* as his first animated feature directly to the strong impression the silent film had made on him. Typically, it was an accident of technology more than the story line that caused the picture to catch in his mind. The film was shown in a large auditorium with a four-sided screen set up in the center and the audience grouped in a circle around it. Disney was seated, by chance, at a point where he could see two of the screens. The projectors were not perfectly sychronized, and so he had the odd experience of seeing the film twice, but with the time lapse between screenings reduced to a matter of seconds—instant *déjà vu*. He could not forget the story because he could not forget the oddity of projection.

At about this time, Disney's father again decided that opportunity was beckoning from afar. This time it was a jelly factory back in Chicago that caught his eye. He sold his newspaper route, invested the proceeds in the factory and became its chief of construction and maintenance. The move gave his son his first real taste of freedom, for he stayed behind in Kansas City to finish his school term and help the new owner of the paper route learn the business. That summer he got a job as a candy butcher on the Santa Fe Railroad, traveling mostly between Kansas City and Chicago but occasionally venturing as far afield as Pueblo, Colorado. His suppliers frequently cheated him, and he was sometimes the butt of practical jokes, as when an older candy butcher gave him a card of introduction to a hotel in Pueblo, which turned out to be a whorehouse. ("I was just a kid," Disney said later, "but I caught on. When I got through the door I broke into a run!") He loved the trains, and it was a love that stayed with him throughout his life, as his model rail-road proved. A screenwriter who wrote a movie scene set in a Pullman car of roughly this vintage recalls that discussion of the scene sent Disney off into a long reminiscence about the elegance of this plush and velvet world he glimpsed for the first time in the summer of 1917 and the lasting impression it had made upon him. Disney used to delight in telling reporters of eating up all his profits by snacking on his candy and of bribing the engineers with plugs of tobacco so that he could ride in the coal car behind the locomotive and watch the countryside rushing up in front and then streaming away behind him. For a boy who had been forced to live as meanly as he had, it was obviously a golden summer—one that glowed even more brightly in memory.

When it ended he joined his family in Chicago. There he attended McKinley High School, where he did both drawing and photographs for the school paper. He also, of course, worked for his father at the jelly factory, running the bottle washer and the capper and mashing apples to make pectin. Somehow he managed to find time for further art instruction, first under a man named Carl Werntzl, then with a newspaper cartoonist named Leroy Gossett. The following summer Disney got a job in the post office, sorting and delivering mail from seven-thirty in the morning until midafternoon, then trying for extra work on the mail pick-up routes. If there was none, he would grab an elevated line gateman's cap out of his locker, ride out to a terminal of the Wilson Avenue line and stand in the shape-up for rush-hour jobs that paid forty cents an hour. "I thought it was a gold rush," he later said.

As a result of his labors Disney was able to buy a seventy-dollar camera on the installment plan, though there is no evidence that he did more than fool around with it in a boyishly inconclusive way. The tone of such reminiscences as there are of the brief second period in Chicago is predominantly one of aloneness. He seems to have had few friends—his schedule probably precluded any close attachments—and he seems to have occupied himself very much in the manner of a man waiting for something more important and interesting to happen. Girls, he once recalled, were a nuisance. "I was normal," he said, "but girls bored me. They still do. Their interests are just different."

It is possible to speculate that his relationship with his father and mother prevented the formation of the habit of trust in his fellow man. There is nowhere in his history any record of any long-standing or deep intimacy with anyone, with the exception of his brother Roy and his own wife and children. And even his immediate family seem to have been kept rigidly compartmentalized from his work, into which so much of his psychic energy was poured. He appears to have confided his troubles and his inspirations to them only rarely. Though he often exhibited a streak of gruffly sentimental generosity to old associates, there is not one to whom he ever offered more than a brief, tantalizing glimpse of the forces that formed him or of his innermost thoughts. The same is true of his relationship to the hordes of interviewers who trouped through his office in the course of an extraordinarily lengthy public career. He told dozens of anecdotes about himself—and he told them well—but they had a way of getting flattened into archetypal experiences, promoting an image of a standardly mischievous American boy of the sort who peopled his live action movies by the score. They were a screen, behind which hid a man who fundamentally—and with good psychological reason—mistrusted the human animal, rejected intimacy and discovered early that he could rely completely only upon himself. "I count my blessings," he said on numberless occasions, and surely one could say that among them, paradoxically, was his habit of distrust and his basic estrangement from people: they were qualities fundamental to his business style, enabling him to retain control of an enterprise that might easily have slipped away from a man of another sort.

During the summer of 1918 when Disney was working for

the post office, his brother Roy reappeared briefly in his life, as a recruit headed for the Great Lakes Naval Training Station outside Chicago. Walt met him at the station and was very nearly taken out to Great Lakes with him when a petty officer mistook him for one of his brother's boot camp group. Since getting away from home at about this age was very much in the family tradition, the mistake gave the youngest Disney boy an idea: why not join the Navy? He was told by recruiters that he was too young, but a friend told him he needed to be only seventeen to join the Red Cross Ambulance Corps. He was a few months shy of that age, and in any case, his father refused to sign his application for a passport. But his mother intervened, saying that she preferred knowing where her son was going to having him run off blindly as his elder brothers had. She signed the application and then turned her back as Disney jiggered the true birth date she had felt compelled to write in.

Disney passed the rest of the war at a staging area in Sound Beach, Connecticut, where, as he recalled it, the service was not much fun. Though the drivers were not officially soldiers, they did have military discipline, complete with a guardhouse which, he claimed, he frequently occupied as a result of "clowning." The group was disgusted when the armistice was signed before any of them got a chance to get overseas. There was, however, still a need for truck drivers in Europe in the period immediately after the guns ceased firing, and Disney was one of fifty drivers in his group who were chosen for this duty—he claimed his was the last name on the list read out in camp—and sent to France after all.

It was, apparently, good duty. He drove all sorts of vehicles, including five-ton trucks, and he spent a good deal of time on the road, free of close supervision, delivering relief supplies all over France. He had to face a board of inquiry one time when a truck he was driving broke down, his assistant got drunk instead of reporting the incident to headquarters as he was supposed to, and the exhausted Disney fell asleep in a nearby shack at precisely the time help finally arrived and towed the vehicle away without him. Disney later claimed it would have been a dreadful disgrace to be kicked out of the Red Cross. "If you couldn't make it in that easy-going outfit," he said, "you were considered hopeless."

In due course all was forgiven and he was assigned as a driver to a canteen at Neufchateau. It was there, with time on his hands, that he made his first real money as an artist. As he recalled it he painted a cowboy on the canteen truck, producing a mild sensation and causing his local bosses to put him to work drawing signs for the canteen. Then he painted a replica of the Croix de Guerre on his leather jacket, soon he was duplicating the decoration for everyone in his outfit at ten francs per job. All this activity brought him to the attention of a con man known in the Disney annals only as The Georgia Cracker. He was doing a nice little business selling German "snipers' helmets" to young American replacements then streaming through Neufchateau to relieve combat troops from their occupation and garrison duties in Germany. The Cracker had acquired a supply of new German helmets into which he carefully shot a bullet apiece, adding just the right touch of authenticity to the item, which was a great favorite with the recruits. He now asked Disney to paint them with phony camouflage colors to add to their realism, after which he banged them up with rocks and rubbed them in the dirt to complete the illusion of battle wear. Again, Disney received ten francs apiece for his work, and it was welcome, particularly since he was getting nothing but rejection slips from such back-home humor magazines as *Life* and *Judge*, to whom he started sending cartoons at this time.

What with his art earnings and his savings from his pay envelope and the proceeds from one particularly good night at the poker table—he won something like three hundred dollars—Disney was able to return home in 1919 with a small grubstake. His father had a job waiting for him in the jelly factory—a steady twenty-five dollars a week—but Disney had made up his mind to have a go at commercial art. His father was against the idea, and presumably to escape his ongoing opposition, perhaps to avoid the competition he might encounter in Chicago, Walt decided to leave most of his capital in his mother's safekeeping and head back to the familiar ground of Kansas City. He had nothing more definite in mind than attempting to get a job as an apprentice artist on what was then regarded as the greatest of the midwestern newspapers, the Kansas City *Star*.

PART THREE

K.C. to L.A.

9

The atmosphere of . . . the prosperous Midwestern city, as a whole, after all, was primarily one of literate optimism.
— Charles Fenton, *The Apprenticeship of Ernest Hemingway*

The personalities involved were not always mean, hypocritical, evasive, fearful individuals; frequently they were afflicted with a kind of frustrated pride, more often by an excessive narrowness of cultural experience, a lack of the opportunity to educate themselves.
— Frederick J. Hoffman, *The Twenties*

THE POPULATION OF Kansas City almost doubled in the first two decades of the twentieth century, rising from 163,-752 in 1900 to 324,410 in 1920. Situated at the junction of the Kansas and the Missouri rivers, it had been a trading and distribution center since the middle of the nineteenth century, when it had also served as the starting point for wagons setting out on both the Santa Fe and Oregon trails. By the beginning of this century it was a place of booming enterprise, with its stockyard, its grain market, its heady interest in real estate. As much as any American city of the time, it was a place where the go-getter, the hustler, the man who was on the rise, could find a congenial atmosphere in which to test his salesmanship and the quality of his commercial vision; in Veblen's terms it was the country town grown up, the principal trading center for a vast and vastly prosperous agricultural region. Yet the city had about it an open and spacious quality that, perhaps, reflected a style and mood generated by the surrounding prairies. In any case, there was—and is—a generosity to the design and layout of Kansas City that contrasts vividly with that of the more constricted-seeming midwestern metropolises farther east. It was one of the first American cities, for example, to incorporate a great deal of open land within its boundaries for development as parks, and it therefore remains one of the few cities to meet what experts regard as the minimum standard for recreational space—one acre for every three hundred inhabitants. It is not and never has been a city particularly noted for its interest in the high culture—among its major contributions to the modern American way of life is the shopping center, the first of which was located there in the 1920s—and its leading citizens have traditionally been modest to a fault when it comes to civic do-gooding, but it is, in general, a handsome and pleasant place. "Who in Europe, or in America for that matter, knows that Kansas City is one of the loveliest cities on Earth," André Maurois once asked, probably thinking particularly of the Country Club and Mission Hills residential districts that were built around the shopping center in the 1920s and where, as one journalist recently put it, "among the imitation Tudor houses there are running brooks, broad boulevards embroidered with statuary and country clubs as pretty as a Hollywood set."

Behind these façades, things were not always so pleasant. For instance, Roy Wilkins, present head of the NAACP, came to Kansas City in the same period as Disney and, working on a Negro newspaper there, discovered, as he recalls, the first deep racial prejudice he had ever encountered (he had lived previously in Minnesota). Downtown, where the Prendergast machine held sway, building toward that free-swinging period

of its total dominance in the late 1920s and early 1930s, ve-
nality and corruption were becoming a standard part of the
business style. In a few years former convicts would be able
to find a place on the police force, the uniforms of the wait-
resses in one of the more popular clubs consisted of nothing
more than a pair of shoes, and the city, in general, had a regional reputation second only to Chicago's for sinfulness. The
present Mayor, Ilus Davis, recalled recently that the downtown
atmosphere was "corrosive." "We had 104 unsolved murders in
a twelve-month period. It got so they didn't even report it in
the paper. There were slot machines in every drugstore. It was
a near complete breakdown of society." The nice people turned
their back on this corruption, doing business with the machine
when they had to and never seriously opposing it. About all that
came out of that era was a school of jazz, to which the city lent
its name and of which Bix Biederbecke was the most famous
practitioner. It is interesting that this, one of the few significant
footnotes Kansas City has contributed to our cultural history,
was nurtured by the city's criminal element.

In addition to jazz, Kansas City had another claim to glory—
the Kansas City *Star*, which promptly turned down Disney's
application for work as an artist and even as an office boy. The
paper was indeed a paragon among the provincial journals of
the time. Less than two years before Disney's return to Kansas
City, its city room had been serving as one of the principal
training grounds for the young Ernest Hemingway (who curi-
ously enough shared with Disney, though at a much more intense level, Chicago, Kansas City and the Ambulance Corps as
formative backgrounds). Through it had passed, and continued to pass, some of the great reportorial and executive names
of the new style of journalism then developing. The *Star*, in
fact, contributed institutionally to that development—the lean,
terse prose its legendary style book demanded of reporters becoming something of a model for other newspapers striving to
shuck off the leisurely, excessively ornamented and overly liter-

ary style that had afflicted American journalism during the latter portion of the nineteenth century and the early years of the twentieth. The *Star* did not have a terribly wide-ranging world view, and its owners and managers were constitutionally incapable of seeing life in a particularly sardonic or satirical way. To temper the style he absorbed in its service Hemingway "required," according to Fenton, "a sustained encounter with provocatively deceitful situations" not obtainable in Kansas City.

What was true of him was undoubtedly even more true of the town's less-gifted citizens. Kansas City appears to have been a good place to come from. Its spirit was more open, perhaps even a shade more sophisticated, than that of the other regional metropolises of the Midwest. A young man could find friendly encouragement for his ambitions there, even, as Hemingway did, some solid hints about how to proceed toward their practical realization. What was missing from its atmosphere is what is often missing from cities of its type even today—a sense of life's more dangerous and tragic undertones, its more fantastic overtones and, most of all, an opportunity to sense the patterns, philosophic and poetic, that an artist can sometimes apprehend in the seemingly random events of the passing days and weeks.

The American Midwest is a highly practical place. Its habit is to ask how much, how big, how far, and, sometimes, simply how. It rarely asks why. The bluff and hearty manner it affects is not a conscious pose: it believes in its own friendliness and good spirits—and it is as surprised as anyone when clues to its hidden, darker side escape and intrude upon the carefully slow and easy surface it genuinely prefers for its habitat. It may or may not be afraid of the depths, particularly the personal ones, that intellectuals and easterners and Europeans (terms which are, among the folk of the region, virtually synonymous), but it certainly dislikes them, finding them depressing to contemplate and the work of explicating them of dubious practical value.

For a young man of developing artistic potential, then, Kansas City offered two possibilities. If he knew where to look for them and had some interest in exploiting them, he could find a short stay there quite educational: it was not so closed and so narrow an environment as it may look to some who have never experienced such a place. If, however, the young man had neither the background nor the temperament to seek in its environs the intellectually or artistically shaping experience, then the town had a natural tendency to send him in quite a different direction. It seems fair to say that Disney, unlike the conveniently comparable Hemingway, lacked the necessary preparation to make good use of such opportunities as Kansas City could indeed offer the young artist. Hemingway, for example, was the scion of a comfortable and literate middle-class professional family; Disney was the son of a man whose principal preoccupation was a search for the peripheral chance, and there is no evidence that there was anyone in his family to supply either a softening or a broadening of that mean and narrow sensibility. Success, economically defined, was the one universally agreed-upon goal of his young manhood, and the lack of it clearly soured his father and turned him into a potent negative example.

As a practical matter, the difference between Disney and Hemingway can be simply put: the writer's father had enough influence to obtain, through friends, a place for his son on the *Star;* Elias Disney did not.

Although neither Hemingway nor Disney pursued formal education beyond high school, there were differences in their opportunities here, too. The former at least received his diploma, and he had the good fortune to obtain it from a suburban public school system that prided itself on its quality. Disney, on the other hand, received a catch-as-catch-can schooling in several undistinguished places and did so while devoting his spare time to neither reading nor sports nor imagination-stretching idleness. He had to work after school and on vacations in order to contribute to the family economy. The upshot is clear: one man's imagination was free to grow as it might;

71

the other's was early forced into the most practical and least-elevated channels. One man was able to use the not inconsequential resources of Kansas City as a place to begin the exploration of a much wider world; the other, although he moved through it almost as quickly, was never able to grow intellectually or emotionally beyond it. Hemingway was influenced by Kansas City, as he was by many youthful experiences. Disney was easily absorbed by it and its values. The quality of his sensibility, of his fancies and fantasies, even the means of his later success and the measures he applied to it, remained the most practical and the most obvious ones he observed in Kansas City in particular and the Midwest in general.

One thinks of Disney's Americanism, of the kind of clean, moral, simple and innocent stories he most often chose to present on the screen, of the right-wing politics of his later years, of his broad, gag-oriented sense of humor, containing no elements of social or self-satire, as entirely typical of the tastes of the region that formed him. The geographic center of the nation is also, broadly speaking, the most passionately American of the American regions. Life on the prairies seems to nurture an intense, if innocent, ethnocentricity. The Cheyenne Indians, who previously wandered the plains states and claimed the territory for their own, were in the habit of referring to themselves as "the human beings" and denying similar recognition to strangers.

As a result, within a few years the Midwest would become, as Frederick J. Hoffman put it, "a metaphor of abuse" and, for a time, it would indeed seem that "the progress of the American had . . . been reversed" with the bright young men and women—in particular the would-be artists—fleeing from the banality, the decadent puritanism, the suppression of life, the fear of death and the hypocrisy that seemed to infect both the small towns and the larger cities of the midlands, and heading east, if not all the way to Europe, in search of freedom, sophistication, style, culture and moral maturity. This view of the life of the region and its meaning is certainly no longer news, and though what was once a living metaphor to an entire literary generation has now become an unthinking stereotype, it still contains a modicum of truth. The young man who, like Disney, chose to stay in the region and who, when he finally decided to move, went still farther west instead of back east, was making an unconscious choice of sides in the decade's great confrontation between philistinism and art. There is a clue in this choice to the aesthetic he would eventually embrace—a clue available even before his career presented him the opportunity to make that choice in any significant way.

In short, Disney was not an exceptional young man, and his discontent was not with the system but with the failure of his family to rise within it. There is no record of what he thought of the literary expatriates, but one imagines his being puzzled by them. He was, without knowing it, on his way to proving anew the truth of a statement by Lewis Mumford, that prescient man, made in the very year Disney finally abandoned Kansas City and headed to the new outpost of midwesternism, Los Angeles. "The truth of the matter is that almost all our literature and art is produced by the public, by people, that is to say, whose education, whose mental bias, whose intellectual discipline, does not differ by so much as the contents of a spelling book from the great body of readers who enjoy their work." This was to be the source of Disney's strength and of his weakness, and his unthinking ability to make the right choice for himself at this time is not to be lightly dismissed. All those who found real power in the new world of mass communications showed a similar sense of identification with the new masses. It would have required a young man tuned and tempered in a very special way—a way quite different from Disney's—to seek out and absorb the lesson, say, of cultural relativity in Kansas City or Chicago. An exceptionally exceptional young man.

Such criticism of the cultural climate of the region as the foregoing may imply should be tempered, however, with an appreciation of the virtues Disney could find there in his apprenticeship. Against the satiric vision of that time and place we have inherited from the Sinclair Lewises and H. L. Menckens, we should balance—more often than we usually do—the remarks of a figure like Eric Hoffer. "Imagine an American writing about America and not mentioning kindness," he said recently, "not mentioning the boundless capacity for working together, not mentioning the unprecedented diffusion of social, political as well as technological skills . . . not mentioning the breathtaking potentialities which lurk in the commonest American." All of these considerable truths about himself and his kind were made manifest to Disney in Kansas City. There, for the first time he glimpsed his own potentials and the potentials of a new and by no means highly regarded medium of communications, and in that unlikely place, he discovered people who could help him begin to realize both. That he glimpsed a commercial gift in the first instance and a commercial potential in the second is not completely surprising. He had, after all, never aspired to be anything loftier than a cartoonist, and such training as he had was as a *commercial* artist. In addition, his father was opposing his plan to try art as a career, and Disney probably felt that he had to demonstrate to the old man that he could make a living as an artist, perhaps even become what he had always wanted to be—an independent entrepreneur, beholden to no one.

In short, the history of Disney's three-year sojourn in Kansas City—not a struggle for artistic expression but rather a struggle for commercial stability—suggests that it was fatuous ever to see him as a man who prostituted an artist's potential on the altar of commerce (one has only to try to imagine him in a Left Bank garret). The miracle is rather the opposite—that there was enough force in his pen, and in those that he hired, to be mistaken for so long as an honest primitive.

The joke, really, is on the intellectuals, who for so many

years mistook him for their own. He was an energetic man, and it is possible that, swayed by their excellent opinion, he may have, for a time, tried to live up to their image of him, but it is almost certainly more accurate to surmise that Walter Elias Disney finally became what he apparently wanted to be from the start—a wildly successful entrepreneur of art or, more properly, its simulacrum. Indeed, it begins to seem, given his family and his geographic background, that he had very little choice in the matter. Certainly in 1919, he had none. At that point, and throughout most of the 1920s, Walt Disney was running so hard just to stay alive and ahead of the bill collectors that he had no time to reflect on the niceties of art versus commerce. His major problems were (a) to eat and (b) to stay free, free as his father never had been.

10 Gotta hustle
—George F. Babbitt

DISNEY'S FIRST JOB in Kansas City paid only fifty dollars a month and it lasted for only about six weeks, but it was very educational. He was an apprentice in a commercial art studio doing work mostly for one of the town's advertising agencies. He got the job by showing, as samples, the cartoons he had done in France, and then only after he worked for the studio without salary for a week. There was nothing exalted about the job—he did rough pencil layouts of ads dealing with farm supplies, for the most part—but he liked the atmosphere, and he learned some valuable tricks of the trade.

"When you go to art school you work for perfection," he was to say later. "But in a commercial art shop you cut things out, and paste things over, and scratch around with a razor blade. I'd never done any of those things in art school. Those are time-saving tricks."

The other valuable thing that happened to Disney in this brief period was his meeting with Ub Iwerks, who worked in the studio specializing in lettering and in air brush work. He was undoubtedly Disney's superior in the techniques of their common craft, and more important in the long run, he was a man endlessly fascinated by the technical problems of motion picture production and played a key role in developing many of the devices and techniques that later gave the Disney product the technical edge it consistently maintained over its competition.

Iwerks told an associate later that the first time he ever saw Disney, Walt was seated at his drawing board, practicing variations on his signature, an activity often associated with willful attempts to resolve the identity crises of adolescence (and Disney, not quite eighteen, was still an adolescent, by any means of reckoning). Surely this first experience of turning what had been a knack into a profession was a significant moment for Disney and one in which the creation of a suitably mature and impressive signature was a symbolically correct, if faintly comic, gesture.

Disney's job in the art shop did not last. The Christmas rush for advertising material petered out toward the end of November, and he passed most of December once again in the employ of the post office department. At night he worked up a somewhat more sophisticated set of art samples employing the slickeries and trickeries he had acquired in his first job. Just after Christmas, Disney and Iwerks went into business for themselves.

They acquired free desk space in the office of a publication called *The Restaurant News*, the owner of which also paid them intermittent fees of around ten dollars for doing illustrative line cuts and occasionally talked his advertisers into buying their services. Disney appears to have provided such capital as the partnership needed—for drawing boards, desks, an air brush and a tank of air—out of the five hundred dollars his mother was holding for him. In the great family tradition she naturally inquired precisely what he intended to do with the money before she released it to him, and he was forced to write her, with some asperity, that he was about to follow another great family tradition and set up in business for himself and that, in any case, the money *was* his.

At any rate, the partnership was a modest success from the start. In their first month the young men took in $135, slightly more than their combined previous salaries. A few weeks later, however, they saw in a classified ad that an artist was needed at a concern called the Kansas City Film Ad Company. Disney who was then, as always, a good salesman despite his occasional moodiness, answered the ad, evidently hoping to get the firm to take on both Iwerks and himself. There was room for only one artist, however, and since the salary was $40 a week, there was no question that it would represent a substantial rise for who-ever seized the opportunity. Iwerks deferred to Disney in the matter—taking their business for himself and letting his part-ner take the higher-salaried job. In theory it was an equitable arrangement since the business gross, undivided between two men, came close to equaling the salary Disney would earn. Un-fortunately, Iwerks had none of Disney's sales talent, and over the next couple of months the business dried up.

Meantime, however, Disney had found his métier. Kansas City Film Ad was, in its little way, in the animated cartoon business, turning out crude one-minute advertisements—silent and in black and white, of course—for showing on local theater screens. It used one of the oldest and crudest techniques of ani-mation—in effect, paper dolls, cut-out figures whose arms and legs worked on dowels and could thus be moved infinitesimally as each frame of film was shot, giving the illusion of movement. The technique, of course, has none of the flexibility or subtlety of drawn animation, but the state of the art in those days was quite primitive, even beyond the confines of Kansas City. Indeed, it is fair to say, as one veteran animator, recently did in recalling the era, that "nobody was doing good work in those days—everybody was doing the same kind of thing."

Nevertheless, it was a naturally attractive business to many young commercial artists of the time, for the field seemed to be—though actually it was not—wide open for innovation, far more so than many of the other areas for which their talents were suited. Disney was no exception—he was genuinely fascinated by the work and its problems—and, within a couple of months, Iwerks joined him at Kansas City Film Ad. Together they devised sundry improvements on the cut-out animation technique, hiding the joints of their figures to give them more fluid movement and making sure that they were placed accurately so that such bendings and twistings as the figures were subjected to were closer to nature.

There was no time for experiment on the job, however, and Disney soon talked his boss into letting him borrow a camera to play with in his garage in his spare time. He worked up a sample reel of local jokes and announcements, apparently something like a satirical newsreel, which he sold to the Newman Theater in Kansas City. It was called The Newman Laugh-O-Gram, and its successors ran on a regular basis at the theater, with a by-line credit for Disney on it. He also did special work for the theater—one short celebrated the theater's anniversary with movie stars popping out of a cake, another featured a little professor who announced he had a sure cure for people who read titles out loud—a Rube Goldberg invention that caused the offender's seat to disappear and sent him down a chute and out into the street. It was not very high-level stuff, obviously, and Disney, when asked to name a price for his Laugh-O-Grams, had forgotten to calculate a profit margin into his figure. As a result, he was doing his basic moonlighting at cost.

However, in the pricing of his other jobs he was able to compensate somewhat for this oversight, and he was soon doing well enough to return his borrowed camera to his daytime employers and buy one of his own. Shortly thereafter he felt he was ready to leave Kansas City Film Ad and launch a full-time production company of his own. He retained Laugh-O-Gram as his corporate name, somehow rounded up working capital in the surprisingly affluent neighborhood of fifteen thousand dollars, moved out of the garage where he had been working and engaged a group of helpers. These he paid nothing, promising them only the opportunity to learn and, perhaps, a share of future profits. This was the beginning of the basic relationship with employees that he was to develop more formally in Hollywood later on, where an elaborate apprentice system supplied his studio with young talent at low prices. It was not a wholly exploitative system—the opportunity to learn was genuine—but the system was ultimately a factor in creating the extraordinary bitterness that attended the great strike at the Disney studios in 1940. His first assistants were soon making salaries because Laugh-O-Gram quickly sold a series of seven animated fairy tales, each seven minutes in length; among the subjects were *Puss 'n' Boots* and *Little Red Riding Hood.*

Motion picture distribution in those days was a free-booting enterprise (it still is to a surprising degree, especially among the small independents), and though Disney's representative in New York was getting the prints around, somehow his share of the grosses was not getting back to Disney in Kansas City. Rather quickly he was forced to let his staff go, and shortly thereafter he was forced to abandon his apartment and live in the Laugh-O-Gram offices, sleeping on a pile of pillows. The experience was only the first of several bad ones with the people and firms who distributed his films through the years, and each of them reinforced his essential suspicion of outsiders—an attitude that the native midwesterner has to guard against anyway —and it surely reinforced the passion with which he invested

79

all his efforts, in later years, to control his own economic destiny.

In the years prior to 1923, when he left Kansas City for good, Disney was forced to a wide variety of desperate expedients to keep alive in his profession. Whenever he needed a haircut, he traded cartoons to a barber, who displayed them in his shop window. He used his camera, his only asset and one to which he managed to cling right up to the end, to take little movies of babies for their parents to project in their living room. And he served as stringer for such newsreel companies as Selznick, Pathe and Universal. Whenever an assignment came his way, he would hire a flivver, paste a press sticker on the windshield and head for the event—often some natural disaster—he was supposed to cover. The usual request was for one hundred feet of film, for which he was paid a dollar a foot if it was acceptable. If it was not, the newsreel firms sent him an amount of unexposed film equal to what he had shot.

One of his principal sources of support in those days was his brother Roy, who for a time was at a veterans' hospital in Tucson, then at one in Los Angeles, recovering from tuberculosis. He periodically dispatched a blank check to his younger brother with instructions to fill it out in any amount up to thirty dollars. Disney customarily went the limit. His other major source of aid was two brothers who ran a Greek restaurant that occupied the ground floor of the building in which Disney had his studio. He had credit there up to a limit of sixty dollars, and as he recalled, he was usually close to it, with the use of his brother's checks going mainly to keep the restaurant owners appeased. At one point, at least, they cut off his credit, and Disney was forced onto a diet of dry bread and beans he scrounged from an abandoned photographic studio next door. A couple of days later one of the Greeks happened to wander in as Disney was downing one of these meals, took pity on him and invited him downstairs for something more substantial. Disney's daughter Diane, once asked him if he didn't regard this as the low point

80

of his life, only to have him deny that it was any such thing. "No, it wasn't bad," he is reported to have said. "I love beans" —a statement his later dietary habits tended to prove.

In the lean years after the failure of Laugh-O-Gram, Disney actually managed to produce two animated films. One was an educational effort, designed to teach children the advantages of dental hygiene. It was called *Tommy Tucker's Tooth* and was underwritten, in the amount of five hundred dollars, by a Doctor Thomas McCrum, who later sold it to the Deener Dental Institute, a local clinic. When McCrum called Disney to tell him that the money was available and to invite him over to work out details, Disney was forced to tell him that he couldn't leave his studio to meet him. His only pair of shoes had fallen apart, were in a shoe store in his building (he had padded back to his studio in his stocking feet after leaving them for repair) and that he needed $1.50 to retrieve them. The dentist came to him.

Far more significant to Disney's future was another film he had made as a last effort to salvage Laugh-O-Gram. It was called *Alice in Cartoonland*, and it was a twist on what was then a popular animated cartoon series, *Out of the Inkwell*, created by Max Fleischer, who was to gain somewhat more fame later with his *Betty Boop* series. Fleischer's characters, in the earlier effort, literally popped out of an inkwell or up off a drawing board to perform their little bits of crude comedy against real or seemingly real backgrounds. Disney's notion was to place a live actor—in the first instance a child who did some modeling at the Kansas City Film Ad Company—into a cartoon. She was never seen except against drawn backgrounds, and everyone she associated with was a drawn figure. (The trick was simply accomplished: she was photographed only against white backgrounds, and the drawings were made on white backgrounds; the two films were then integrated in the printing process.)

Disney claimed that he secured an agreement from his creditors, when he went bankrupt, allowing him to keep a print of the film as a sample of his work, but whether anything as formal as a legal agreement actually existed, or was necessary, is unclear. The fact is that even the proceeds of *Tommy Tucker's Tooth* and of his other free-lance activities did not come close to covering Disney's debts when it became clear to him that his future in Kansas City was limited and that he would have to repair to the West Coast if he wanted to make good in the movies. He finally had to sell his camera—at a profit—to finance his trip west, then went around to his creditors, asking them if they would accept a partial settlement in order for him to make a fresh start. Most of them refused his offer, telling him that he would need the money for a grubstake and to repay them later, when he could afford it. "They were wonderful people," he said, recalling the period.

In later years Disney invited the barber with whom he had bartered drawings for haircuts to come to the Coast for a visit at his expense. He also lent quite a bit of money, after his great success with Mickey Mouse, to one of the restaurateurs who had carried him in the lean days. Indeed, he drew a little lesson in folk economics from the experience. The man wrote to him, sometime in the 1930s, asking for a thousand dollars to help re-establish the restaurant business after he had spent some time as an automobile dealer. Disney advanced him the money. A little while later he wrote again, this time for five hundred dollars to buy an air conditioner for the place. Again the money was advanced. The next time the man applied for funds—to buy out a partner—Disney turned him down. The point was, as Disney phrased it, "that the $60 worth of food [he] let me have on credit cost me $1500." It should be added, however, that Disney, for all his latter-day moralism about lending, had been a habitual borrower and raised a critical $250 he needed in his early days by writing for a loan from the organist of a Kansas City theater for whom he had done animated sequences of music and lyrics to flash on the screen during the community singing that used to be such a familiar part of moviegoing. He also borrowed money from the Kansas City girl whom Roy Disney ultimately married and who had also helped out with free meals during the days of his bankruptcy, and at one point he got $60 from his brother Raymond, then working as a bank teller in Kansas City—a loan that was to have some consequences for both of them.

All of this is significant only to demonstrate how much Disney was a product, however unwillingly, of the climate of small enterprise well beneath the attention of banks and bankers. It is, of course, a world of suspiciousness relieved by sudden, often inexplicable, generosity, a world where the basic salesmanship must be done on the friends one turns to for backing, where the rules insist that debts incurred in this way are debts of honor, but where one must be ever wary of the con man willing to violate them. It is a world of people lured on by tales of others like themselves who became millionaires on the basis of a small investment in some inventor's dream and of people made cautious by the equally prevalent tales of promoters who skipped town with everybody's hard-earned savings. Nowhere is the fulcrum on which we balance our dream of wealth against our nightmare of insecurity more precariously—or more poignantly—balanced, for this is a world of borrowers who can't afford to lose yet can scarcely bear to go on as they are. One does not pass through this world, especially after a childhood like Disney's, unscarred. His future economic style—frugality in day-to-day matters, willingness to plunge on his own ideas, distrust of the outsider who might somehow take it all away from him—was surely an acting out, on a large scale, of the truths he had learned in Kansas City.

11   *Arthur Koestler suggests that there is in the revolutionary*
      "some defective quality" which keeps him from growing up. The
      indications are, however, that the present trend toward juvenile
      behavior has been gathering force for over a century and has
      affected people who cannot be classed as revolutionaries. . . .
      the American go-getter, though he has no quarrel with the
      status quo, is as much a perpetual juvenile as any revolutionary.
      —Eric Hoffer, *The Temper of Our Times*

HOFFER'S THEORY OF the juvenile nature, which he sees
as a mass phenomenon and as a principal factor in the creation
of the temper of our times, suggests that we are not at the beginning of an era of radical change but well along in a transformation that began with the industrial revolution at the end of
the eighteenth century. The urbanization of the millions who
were scooped up off the land and set to work tending the new
machines has been, in his view, the central fact not only of our
own historical moment but that of our fathers and grandfathers
as well. It may be said to account for the gropings of Elias Disney as much as it did his son's. It is not that the machines are, in
themselves, dehumanizing but that the changes they bring to
every area of human existence have a vastly upsetting effect,
forcing uprooted millions to seek new identities in a world that
appears no longer to offer them traditionally reassuring values.
The result is a human type who, as Hoffer says, is "juvenile,
primitive and plastic," for the juvenile is "the archetypal man in
transition" and whenever we are subjected to drastic change,
we recapitulate to some degree the adolescent's passage from
childhood to manhood, a process that "results in some degree of
primitivization." *En masse* this means we attempt to seek identity by resort to mass movements, nationalism, charismatic
leaders, medicine men and, of course, the kind of nostalgia for a

84

carefully falsified past which Walt Disney, among others, was later to trade in so successfully.

One warm July day young Disney, twenty-one years old, already an experienced, if failed, go-getter, waited on the platform at the Kansas City station to board the Santa Fe's California Limited. If ever there has been an archetypal man in transition, it was he. He wore a checked coat and a well-worn pair of pants that did not match the coat. In his imitation-leather suitcase there were a couple of pairs of socks, a couple of pairs of underwear, an extra shirt and some drawing material. In his pocket he had, so goes the official legend, forty dollars, all that was left from the sale of his camera. "But his head was packed with ideas," his daughter writes. "Someway, somehow, when he got to Hollywood, he'd get to the top of the movie heap."

He had no realistic reason to believe he would be any more successful on the Coast than he had been in the Midwest. Indeed, he had no precise notion of what he wanted to do once he got there, beyond that standard dream of glory, "breaking into the movies." By any reasonable method of judgment he was no more than a journeyman at any of the motion picture crafts in which he had dabbled, and he had none of the interest in self-expression or cultural and social observation that usually drives and sustains any artist worthy of the name. He had, to be sure, accidentally found a congenial medium in which to exercise whatever talents he might have, but at this stage the motion picture was to him what real estate was to George Babbitt—a means to some ill-defined, but economically rewarding end. It was by no stretch of the imagination an end in itself. He saw the movies only as a living, not as a life. There is no way of knowing how or what he thought of himself or, indeed, if he made any attempt to find in himself a unique quality. In retrospect the three years in Kansas City seem to have been a last chance for him as an artist. Had they been less harassed and desperate, it is possible—just barely possible—that he might have been able to integrate into a unique artistic personality the ego so badly shredded in his loveless, rootless childhood. Instead, he adopted, with very little need of modification, a prevalent mass style—that of the go-getter—and then he got out to get.

That Walt Disney remained perpetually the juvenile that he was chronologically when he came to Kansas City for the second time and still was psychologically when he left it for good, turned out to be no bar to his economic success. Quite the contrary. It enabled him instinctively to purvey to his audiences precisely what they wanted—perhaps needed—in a way that few others so consistently managed. That he seemed to remain psychologically preoccupied with his own isolation and therefore be cut off from most forms of social intimacy was no problem, either. At first he played the part of the primitive artist allowing "the professors" to explain his own work to him and then scoffing boyishly at the explanations. It was a most satisfactory public relations device, tiding him over until he could fashion the more avuncular but still homespun public style of his later years. There were plenty of public relations men to help him with it, and, of course, the mass media were only too happy to conspire in the maintenance of this image. Their readers and viewers liked it, for it was untroubling; the proprietors of the media liked it because Disney was such a reassuringly clean and decent mogul—and Protestant, as few other moguls were. He could be used to blunt a lot of generalized moral and religiously prejudiced criticism about mass communications.

But all this was in the future. The young Walt Disney might have been a rube, a cornball, a hick, on anybody's scale of sophistication, but he was something else as well. "History is made," says Eric Hoffer, "by men who have the restlessness, impressionability, credulity, capacity for make-believe, ruthlessness and self-righteousness of children. It is made by men who set their hearts on toys."

PART FOUR

Back to the Drawing Board

---

12 *Know ye that at the right hand of the Indies there is an island named california, very close to . . . the Terrestrial Paradise. . . . The island everywhere abounds with gold and precious stones. . . .*

—Ordóñez de Montalvo, *Las Sergas de Esplandian* (1510)

All too many of us Americans who took part in the conquest of the empty continent, when we finally got to Los Angeles, found we had put the emptiness inside.

—George W. Pierson, *The Humanities and Moral Men*

INSIDE. Where it was safely hidden, even from oneself, particularly if oneself was very, very busy. You might not notice it for years and others might never notice it at all, especially in Los Angeles, where there were so many important things going on.

Out of the chaos of many small competitors, the large studio system was beginning to evolve, a business was becoming an industry—"*The* Industry," as it is still reverently referred to on ceremonious occasions. In 1923, a few months before Disney arrived, another young man named Irving Thalberg, less than three years older than Disney, left Universal to join a one-time junk dealer named Louis B. Mayer as his chief of production at the small Metro Studios. Within a year they would merge with the Goldwyn Studio, under the aegis of Loew's, Inc., to form the firm that would quickly become the biggest of the sausage factories, Metro-Goldwyn-Mayer. Famous Players-Lasky had sometime before absorbed Triangle, which had itself been a merger of three of Hollywood's leading creative forces—D. W. Griffith, William Ince and Mack Sennett—to form Paramount, which numbered among its assets the strongest theater chain of the time. Griffith had meantime—and unhappily—joined with Douglas Fairbanks, Mary Pickford and Charles Chaplin to form United Artists in 1919. In the years that followed, other combinations of producers, and of producers and theater chains, were effected. The result was a recapitulation of a process that had been going on elsewhere in American industry for some time—a reduction in the number of independent production units but with each of the remaining units becoming larger and at least potentially more efficient and more stable than they had been separately. Disney, when he finally arrived in Hollywood, felt that he was almost a decade too late to do what a slightly older group of pioneers had done—set up shop completely on their own. It would be something like a quarter of a century later, for example, before he had his own distribution corporation and was free of the percentages of the gross which the larger companies exacted from him—as well as from other independents—for getting their work into theaters.

The changes being wrought in the industry were not solely economic. However quaintly a silent film from the middle of the 1920s may flicker before modern eyes, the fact remains that in the years between 1919 and 1923, they had begun to lose their innocence. Erich von Stroheim had, for example, begun to make the series of films (*Blind Husbands, The Devil's Passkey, Foolish Wives*) that Lewis Jacobs was to characterize as "melodramas of lust" in all its forms. "Executed with a hard unrelenting honesty, they were by turns sordid, scathing, mocking, ironic," and they "brought to the screen an individuality, a maturity, and a meaning not to be found in the pictures of the DeMilles, the Inces. . . ." In the very summer that Walt Disney arrived in Hollywood, von Stroheim, his cast and crew, were toiling in the incredible heat of Death Valley to complete the shooting of his blighted masterpiece, *Greed*, in which for the first time he was to direct his mordant gaze upon an America that, as J. K. Galbraith has put it, was "displaying an inordinate desire to get rich quickly with a minimum of physical effort."

DeMille himself had turned to sex comedies that were very racy by pre-war standards, and his famous bathtub scenes were already becoming a legend. The very names of the stars who were born in the post-World War I era—Valentino, Gloria Swanson, Clara Bow—evoke for us today a faintly comic image of an industry both reflecting and helping to create a new American atmosphere. The year 1923 was fairly typical of the period. Pola Negri (whose last screen appearance was in Disney's 1964 film, *The Moonspinners*) was driving men to ruin, insanity or suicide in *Mad Love*, released in March of the year, then suffering most deliciously, a few months later, for her transgressions in *The Cheat*, about which *Time* magazine, founded that same year, commented: "The spectacle of a hot iron sinking into the white contours of [her] shoulder should be 50¢ worth to anybody." At that, it was no more sadistic than Lon Chaney's writhing under the lash in *The Hunchback of Notre Dame*, which appeared the same month.

In short, film content was changing, trying to grow more sophisticated. Many of the great names of the movies' early days —men who had been at the pinnacle of the industry only five or six years before—were suddenly racing to catch up with the new mood. D. W. Griffith, to whom, as Charles Chaplin said, "the whole industry owes its experience," simply could not make the transition. His 1923 release, *One Exciting Night*, was deemed jumbled and pointless and failed; it was neither the first nor the last of his increasingly desperate attempts to respond to the new spirit of the times. William S. Hart, "the greatest cowboy that ever faced a camera," had been forced into retirement two years earlier and now emerged to find that, although the critics still liked him in *Wild Bill Hickok*, the public was indifferent to his realistic vision of the West. They wanted a more romantic, less gritty version of their pioneer heritage. Even Buster Keaton, who was perhaps the greatest of the silent comedians, ran into trouble with his first long film, *The Three Ages*, in which he had difficulty in sustaining his comic line for an hour and a half.

Of the great early established names, only a few emerged from the year unscathed. Douglas Fairbanks had abandoned the attempt to deal with the contemporary scene several years earlier and in 1923 released one of his most successful historical romances, the marvelously athletic *Robin Hood*. His new bride, Mary Pickford, emerged from a brief retirement with a new image: she stopped trying to be America's Sweetheart Forever in this transitional year and, her famous golden curls tucked on top of her head, appeared in *Rosita* as a Spanish street singer who rises to the nobility. Even Chaplin changed. His first release of 1923 was *The Pilgrim*, and in it his "Little Fellow," the beloved tramp, appeared as an escaping convict turned minister ("custard piety," one critic called it). Chaplin did not even appear in his second 1923 effort, *A Woman of Paris*, which starred the leading lady of his shorter, earlier comedies, Edna Purviance, and dealt with a Parisian mistress whose tinseled world comes tumbling down when the youth she used to love suddenly reenters her life. For her director it represented a considerable shift from his usual mode.

No one knew quite what to make of it all. Intelligent commentators declared that there was still something awfully naïve about the new cinematic sophistication, that by and large the spirit of *True Romance* had won out over the spirit of romantic truth. "Breathes there a man with a brain so dead that he has not repudiated those curly co-eds, those red-blooded 'Society folk' with midnight bathing parties, those flat-footed vampires?" *Time* inquired with the iconoclastic exuberance of its youth. Even so, the cry of the censor was heard in the land, and Will Hays, the Indiana politician who was hired in 1921 to cleanse the blot placed on the industry's banner by the Fatty Arbuckle scandal, and who, with his high collars and *pince nez*, was the personification of midwestern rectitude, was hard put to keep up with the demands for moral purity that the older America was making on the movies.

In this period he got very little help from his new constituents. After two hung juries, Arbuckle had just suffered through his third trial for the manslaughter death of a bit player named Virginia Rappe, who had been his companion during a wild San Francisco party in 1921; director William Desmond Taylor was murdered in 1923, and two well-known actresses, June Taylor and Mary Miles Minter, were unpleasantly linked with his free-wheeling life, though not with his death. And clean-cut Wallace Reid, an important leading man who had been revealed to be a drug addict, died in 1923 in the hospital where he was attempting to find a cure. Hays and the industry leaders drew up a blacklist of personalities whose morals might cause them future difficulties, DeMille wrapped a soothing Biblical cloak around the industry by releasing *The Ten Commandments* (which nevertheless contained a very satisfactory orgy sequence), and there was a general effort to tone down Hollywood's excesses both on screen and off. Adolph Zukor gave out an interview in which he said, "Hollywood is a very quiet place. A *very* quiet place. No drinking—very little smoking. And as for the evenings . . . they're practically inaudible. No sound at all but the popping of the California poppies."

Others were less bland about the situation in Hollywood. In the wake of the Reid tragedy, Miss Elinor Glyn, the English novelist who coined the word "It" as an expression of the ultimate in libidinal energy and who was almost single-handedly responsible for the cycle of Ruritanian romances so much a feature of this period, was asked what would happen next in the sorely tried film colony, and she sagely replied: "Whatever will bring in the most money will happen." As for the intellectuals, some of whom were beginning to sound as if they would like to like the movies, they were disappointed. It might seem to one journal of opinion that "almost every hamlet has a good film every week," but Joseph Wood Krutch spoke for the majority when he declared that "the inanities blessed by Mr. Hays are more genuinely corrupting than any pornography."

It is impossible to say what, if anything, Walt Disney thought of all this when he arrived in Los Angeles. Since, like many Americans who were essentially technological innovators, his social imagination was rather underdeveloped and since he faced such a difficult battle for survival, it is likely that it all seemed rather remote to him. Emotionally he was as far away from the heights where these alarums and excursions were taking place as he had been back in Kansas City. Indeed, the small circle into which he moved was one of emigres from the midlands—rootless, envious, nostalgic, intellectually and emotionally juvenile—like those who, since the 1920s, have been the source of Southern California's reputation for crankish, cultish excess. It is likely that if he thought about the morality of the movies and the industry that made them, he was vaguely disapproving. All his life he remained essentially alienated from the industry's power structure and unwilling to join it either socially or in its endless, frantic searches for the formulae of box-office success. To borrow Isaiah Berlin's famous metaphor, the industry's leaders were the foxes who knew many things and he was a hedgehog who knew one important thing. They were quick men with the peddler's instinct to vary his stock and his spiel the minute he senses a slight restlessness in the crowd and the smallest clue to their next whim. They were, many of them, children of the ghetto—colorful, volatile, emotional and, if not what one would call cultivated, then in love with the game and cheerfully cynical about its rules and its outcome. If they had a flaw it lay, not in the ignorance and the occasional brutality and trickery of their business style, which provided so much good copy for journalists and other malcontents through the years. In retrospect, there was something refreshing about all that chicanery in the context of a nation that has always been rather too pious about the role (and rule) of business.

No, their problem was that they were still immigrants—even those who had been born here. They did not have, bred in their bones, the culture of the American majority; they always had to go out and look for it—a quest that frequently left them a half-step behind its shifts in mood, and rarely more than a half-step ahead of it. They nearly always came out ahead, because the art of survival *was* bred in their bones, but they were nervous men, and that is what made them adopt the more obvious manners of the autocrat—their style in dealing with underlings, their enormous homes, cars and offices, the retinue of ubiquitous sycophants.

Disney, the hedgehog, had no such insecurities. He knew what he knew about Protestant, middle-class, midwestern America, and he knew that what he liked, it liked. He needed no public opinion polls (though he used them, in later years, because they were so reassuringly scientific), few sycophants (he had them, but that was more their problem than his) and, above all, no atmosphere of excessive materialism to reassure himself about the value of his own opinions or work. The determined informality of his manner was no affectation, at least at first, though at the end, when he was a multimillionaire, it was a trifle disconcerting, in the same way it is to see pictures of Nelson Rockefeller chomping on a hot dog. It was this in-bred sureness of his audience (which was merely himself multiplied) that led to his ultimate success and led people to misunderstand him as one kind of genius when he was actually a genius of quite another sort.

What is most significant about the atmosphere of Hollywood in the 1920s as it relates specifically to Disney's history is that in this period the town acquired its reputation for wickedness among Disney's class and kind, a reputation that has never been completely expunged. We have observed that the industry was at expensive pains to cleanse its reputation, and it is precisely in this area that Disney was most useful to the lords of celluloid creation. Whenever the voices of the moralists grew too shrill, Disney could always be pointed out as a purveyor of clean, uplifting family entertainment and he could be counted upon, too, for statesmanlike comments on the need to maintain a climate of moral excellence and the pleasure he took in contributing to it. These, too, were apparently quite sincere in their motivation.

As a result of this natural cleanliness of outlook, he was responsible in some measure for the demise of the one great contribution Hollywood made, without knowing it, to our cultural heritage. This was, of course, the short comedy. The great comedians—Chaplin, Keaton, Harry Langdon, Harold Lloyd, Laurel and Hardy—were beloved by both the plain people and by those intellectuals who were paying attention. And with good reason. They had developed to a precise and subtle degree the art of physical comedy, often known, imprecisely, as "slapstick." But despite the high regard in which many held their work, there was a countercurrent. James Agee, in his autobiographical novel, *A Death in the Family*, has the mother of a small boy object to Chaplin as "that horrid little man. . . . He's so *nasty* . . . so *vulgar!* With his nasty little cane; hooking up skirts and things, and that nasty little walk!" In 1923, a magazine noted that "our good old uncles and funny old aunts" demurred from the view that Chaplin and the rest represented art. "They said that when one comedian dropped a lighted cigar down another comedian's trousers it was not art. And for their part they couldn't see anything funny in one man hitting another in the seat of what they termed 'pants.' In their day [it] was a disciplinary objective; they refused to admit the right of Charles Chaplin to make it simply the butt of a jest." Disney, when he borrowed the balletic movements and timing of the silent comics but changed the actors from human to animal form in his cartoons and removed the sexual references from the great routines, effectively disposed of many of their objections. What had been nasty as a form of human behavior became acceptably adorable as obviously fictive animal behavior.

Disney was even useful, as a silent example, to the maniac fringe who insisted that control of the movies was vested in the international Jewish conspiracy. One had merely to wave in his direction to indicate that the *goyim* were not barred from success in this field. There was a certain irony in this, since Disney appears to have shared, in mild form, some of the anti-Semitism that was common to his generation and place of origin. His studio was notably lacking in Jewish employees, and at least once he presented a fairly vicious caricature of the Jew on screen.

In 1923, whatever his fantasies of the future, Disney's practical, immediate ambitions appeared modest enough. He made no attempt, on first arriving in California, to set up in business for himself. He did not even turn first to animation in his search for work. He reasoned that the studios, now beginning generally to work on the rationalized, production-line basis that Thomas Ince had pioneered a few years earlier, would need plenty of directors and that there might be a considerable turnover in the field. He therefore presented himself to the man at Universal who had been handling his newsreel submissions and got from him a visitor's pass to the studio. He then proceeded to hang around the lot, trying to pick up tips on directing and also trying to discover where, if anywhere, opportunities might lie. After a couple of weeks he scaled his ambitions down somewhat, but even at the lower levels of Universal, there were no jobs for him.

He was at least halfway convinced that he was too late, by perhaps six years, to break into animation, but since that was the only area in which he had any prior experience, he rented a battered camera, made an animation stand out of scrap lumber and set up a studio in the garage of his Uncle Robert Disney's home, where he was staying (room and board was five dollars a week). He was still getting occasional small loans from brother Roy, who was by now in a veterans' hospital in Sawtelle, California, where he was drawing eighty-five dollars a month in disability pay from the government. The only hope Walt Disney had—and it seemed a faint one—was his old *Alice in Cartoonland* sample, which the Lloyds Film Storage Company was attempting to sell to distributors in New York. He certainly could not afford to wait any longer for action on it. Nor did it appear that a lengthy siege of the employment offices of the other studios would be a much better gamble than trying to set up his own business again. There are, he later philosophized, two kinds of people. "The first kind are licked if they can't get a job. The second kind feel they can always do *something*, even if jobs are scarce." He was definitely one of the latter types.

Even so, he cannot have felt very hopeful as he set up shop again. He was acutely aware of the successes Paul Terry had achieved with his two series, *Aesop's Fables* and *Farmer Al Falfa*, as well as Pat Sullivan's crude but effective *Felix the Cat*, not to mention the work of such prominent pioneers as Raoul Barre and Fleischer. Then, too, a number of the popular comic strips of the day, among them *Mutt and Jeff* and even George Herriman's *Krazy Kat* (which proved to be as literally inimitable in animation as it was figuratively so on the page), were getting considerable attention from the producers, who then, as now, were mightily inclined toward products on which the audience was possibly presold. Disney later commented that he felt that the field, neither large nor profitable, was overcrowded, even though its growth, in the decade since its birth, had been far slower than that of the rest of the movie industry.

13   *"Energy is eternal delight."*
      —William Blake

IT IS CUSTOMARY to begin most histories of animation—indeed, of the movies in general—with a reference to the Lascaux cave paintings or, at the very least and latest, to the pictographs of Egypt or the decorations on some Greek vases which, if spun in the hand, imparted the illusion that a figure, drawn in the various attitudes of movement, was actually moving. All of these art objects, as well as many others, like the Bayeux tapestry, are claimed by historians seeking to attribute to the movies, a suitably long cultural pedigree, though these objects may be more accurately said to have represented the wish for the ability to make pictures move rather than the actual accomplishment of the deed.

In fact, animation may reasonably claim a longer history than the movies themselves, which had to wait upon the invention of photography. Man has been drawing since prehistoric times, and the principle of persistence of vision, on which all illusions of motion are based, was known to Ptolemy. Thus animation of a sort was used commercially for the first time as early as the sixteenth and seventeenth centuries, when publishers brought out books—often pornographic in nature—containing little figures that seemed to move when the pages were rapidly riffled. It was not until 1825, however, that the first toy based on the principle of persistence of vision was developed. In that year Dr. John Ayrton Paris demonstrated a simple little gadget known as the Thaumatrope, which was nothing more than a cardboard disc with a string on either side, so that it might be twirled between thumb and finger. There was an image on each side of the disc, and when the disc was rotated, these drawings would seem to combine; thus a child might be able to observe a bird that seemed to pop into a cage or a rider jumping onto his horse. In 1832 Professor J. A. F. Plateau of Brussels came forth with something he called the Phenakistoscope, which may be regarded as the first mechanical animation device. It was no more than a notched wheel mounted on a handle. If you held it up to a mirror and peered through the notches the figures painted on the reverse side seemed to move as you observed them in the mirror. Two years later virtually the same device was independently conceived by Simon Ritter von Stampfer of Vienna, who called his version the Stroboscope. Actually, both men had been preceded by W. G. Horner of Bristol, England, who perfected, as early as 1834, a somewhat more sophisticated toy he called the Zoetrope. It was a slot-pierced drum, revolving on a pivot. You placed a flat disc in the bottom of the drum and a paper band around the sides. On each disc were painted figures in various stages of a movement. By peering into the slot and rotating the drum, a very satisfactory illusion of motion was obtained. Horner's gadget did not reach the market until 1867, and in a decade it was improved upon by Professor Emile Reynaud of Paris, who did away with the slots and notches all the preceding devices had utilized. His Praxinoscope had a series of mirrors in the center of the drum, and you kept your eye on them while rotating the outer disc, where the band of carefully drawn figures was placed. He even devised, a few years later, a method of projecting his moving figures onto an enchanting little table-top stage.

Neither he nor his various competitors, who called their gadgets by such names as "the wheel of life" and the "wheel of the devil," would ever master large-scale theatrical projection. The line of inventive work that led up to the virtually simultaneous invention of the movies by the Lumière brothers in

France and by William Kennedy Dickson, working for Edison in the United States, stemmed directly from still photography and proceeded through George Eastman's vital creation of flexible celluloid film and thence to the Kinetoscope, which was a true movie in its creation but not in its presentation, because it could be viewed, like a peep show, only by a single person. At last came the Vitascope, which Edison marketed in 1893, and the Cinematograph of the Lumières. But all these men were wedded to the photographic reproduction of movement and had little interest in animation, which no one seems to have thought of as a possibility for the new medium.

It was not until 1905 that J. Stuart Blackton, a one-time artist for the New York *World* and later a producer for Edison, attempted an animated film. It was called *Humorous Phases of Funny Faces* and consisted of, among other actions, a man rolling his eyes and blowing smoke rings. In the flood of short films emanating from the studios in those days it was largely ignored as interest focused on the first story films following in the wake of the hugely successful *The Great Train Robbery*, produced in 1903. A number of other artists did attempt animated films, but none was particularly successful until Winsor McCay, creator of the exquisitely drawn Little Nemo comic strip, animated *Nemo*, did a little film about a mosquito, and then made his great breakthrough with *Gertie the Dinosaur. Gertie* was conceived as part of a Vaudeville act. Her image was projected on a screen, and she seemed to respond to commands McCay issued from the stage. The big climax of the routine was her catching and eating an apple that McCay appeared to toss to her (but actually palmed).

After *Gertie* (whose other wondrous trick was to seem to drink a whole lake dry and then spit it all out—a simple matter of reversing the film) the animation market began to open somewhat. John R. Bray had a success with his *Col. Heeza Liar in Africa*, a 1913 satire on Teddy Roosevelt's big-game hunting expedition, and developed it into a series. At about the same time, Earl Hurd, who worked for Bray, provided the infant industry with a technological breakthrough to match the one McCay had had achieved with audiences. Until Hurd appeared, it had been necessary to draw, *in toto*, each frame of the film—a tedious and expensive business that effectively prohibited films of any length and, indeed, any sophistication of animation. Hurd developed the idea of doing his animations on celluloid and photographing them against a background that needed to be changed only when the scene changed. This eliminated a great deal of redrawing and concentrated the artist's attention on the principal problem of the cartoon, which is movement. The cells, as they quickly came to be called, presented a fairly serious fire hazard at first, when nitrate celluloid was used as the basic material, but with the introduction of acetate safety film, this problem was soon eliminated. Even though the early animators frequently saw their work literally go up in a puff of smoke, the development of the cell technique was essential to getting the business moving in even the low gear that pertained throughout the second and third decades of this century.

There was a great deal wrong with the animated film business at the time Disney decided to go back into it. Some of the trouble was artistic, some was financial. To be sure, a couple of the basic conventions of the animated cartoon had already been adopted. The repeal of the law of gravity had occurred in Raoul Barre's atelier as early as 1917. He was then producing the *Mutt and Jeff* series and one of his teams was doing an Alpine sequence in which Mutt was to be shown leaning against a guard rail with a steep drop below. A careless cameraman forgot to place the cell containing the rail before his camera, and so the character appeared, in the rushes, to be leaning on thin air. Barre, who was French and formal, was not amused, but his American gagmen laughed to see such a sight. Before long cartoon characters were walking on air, on the ceilings, anywhere they might get a laugh, and, of course, the basic device is still much used today. Similarly, the emphasis on anthropomorphizing animals was discovered early, especially in the work of Paul Terry. Because they are rounder- and softer-looking than clothed humans, they were thought to be easier to animate and caricature; they were not bound by realism the way their competitors—live human actors—were, and audiences could easily feel simultaneously affectionate and superior to them. Finally, the uses of cruelty without consequences were discovered. A cartoon character could take a great fall, break into a dozen pieces and be put together again, no harm done, with a few strokes of an animator's pen, thus conventionalizing and therefore deemotionalizing sadism.

Still, despite this establishment of the basic aesthetic of the industry, the animated cartoons, five minutes or so in length, constituted no more than a curious appendage to a curious industry. Drawing remained crude and hasty, the technique of animation reflected no more than a suggestion of the skills that were to come, while the plot lines were generally no more than a string of gags. There was no attempt to develop character in the course of the little films, and there was a fairly firm belief that only animals were suitable as characters, since their spontaneity and physical characteristics made them almost automatic caricatures when they were placed in human postures and predicaments. Finally, the lack of color and sound was never the advantage in animation that it occasionally was in some silent films, and both ingredients were nearly always more sorely missed in the cartoons than in feature films. Lacking color, the animators were working without one of the graphic artist's basic resources. Without sound, they lacked the dramatist's basic tool. And lacking the artistry of some of the actors and directors who worked in features, as well as the wealth of the major studios, cartoons had nothing with which to compensate for these disadvantages. In particular, the need for titles to tell jokes and explain the action was a serious disadvantage and slowed down the comic movement.

But it was the economic situation of the animators that was

perhaps the most distressing aspect of their fledgling art. The days were past when Paul Terry received just $405 outright, with no possibility of additional royalties, for his first cartoon, *Little Herman*, which he had done entirely by himself over a period of several months. But they were not much past. The going rate for an ordinary cartoon was around $1,500, and though a shop with a series of proved popularity might get somewhat more, there was still no participations in the grosses of the films. What was worse, the theater chains preferred to buy their short subjects in blocks (a situation that continues today) and so had no interest in the odd film of more than routine interest. What they wanted were series about which they could say that if you had seen one you had seen them all, thus obviating the need for screening the product carefully or even thinking much about it. As for the distributors, they were interested only in keeping costs and troubles at a minimum. Quality, experiment, any innovation that might run up the cost per foot of the shorts, these were in no way pleasing to them.

That any young man was willing to attempt such a business must remain as a permanent tribute to his stubbornness. That Disney, alone of all the men who went into animation was able to emerge from it a full-fledged tycoon (several made a bit of money out of it eventually) must stand as a tribute to organizational abilities of a very high order.

14
*The young man walks fast by himself through the crowd
. . . blood tingles with wants; mind is a beehive of hopes buzz-
ing and stinging. . . .
The young man walks by himself, fast but not fast enough,
far but not far enough. . . .
No job, no woman, no house, no city.*
—John Dos Passos, *U.S.A.*

DISNEY'S FIRST MOVE, once he had made up his mind to
go into business for himself again, was to renovate his old
Laugh-O-Gram idea—a series of topical joke anthologies—and
try to peddle it to the local theaters. He began hanging around
the offices of Alexander Pantages, who controlled a chain of
West Coast movie houses, and finally got a junior executive to
listen to his idea. The man informed Disney that Mr. Pantages
would not be interested in the series. But unknown to both, the
boss overheard the conversation from his office and called out,
"How do you know I wouldn't?" He then appeared, listened to
Disney's pitch and told him to make up a sample reel.

Disney got to work. Since he had no choice but to do all the
work himself, he kept it very simple—using white stick figures
on a black background and no more than a suggestion of backgrounds—a tree, a sliver of a moon in the sky, a simple horizon
line. Pantages seemed to like the sample—which utilized over-
printed titles to tell its jokes—and it is said that he was on the
point of closing a deal when word arrived from the East that a
distributor named Charles Mintz was willing to pay Disney
fifteen hundred dollars apiece for a series of *Alice in Cartoonland* subjects.

The offer placed Disney in a predicament. He had what
looked to be a sure thing with Pantages, and his brother who
had helped to finance the sample reel out of his modest govern-

103

ment stipend was reluctant to pull out of it for what seemed a more chancy venture. In addition, it was clear to Roy that they would need capital from an outside source if they were to undertake a series of the longer (nine hundred feet) and more technically difficult *Alice* reels. Their discussion on this occasion set the pattern for many discussions that were to follow as the Disney enterprise grew, with Walt pressing for the expansive project, Roy at first resisting him and then giving in and somehow finding the ways and means of financing it.

Their needs for *Alice* were small—five hundred dollars in order to make their first reel—but their resources were equally modest. Indeed, about the only person to whom they could turn was their Uncle Robert, who was essentially as unyielding a type as their father. Disney had had an argument with him on his first day in Los Angeles when the older man had asked him if he had come through Topeka on his way west and had been told that the train bypassed the city. Uncle Robert insisted that *all* Santa Fe trains ran through Topeka, and the argument was not resolved until his wife checked with the railroad and found out that indeed Disney's train had entirely missed Topeka. The uncle thereupon nursed a grudge for several weeks.

Now, when the brothers applied to him for a short-term loan to tide them over until the distributor paid for their first reel, he refused, claiming that "Walter doesn't pay his debts." Quizzed on the point, he replied that he knew Disney had never repaid his sixty dollar debt to his brother Raymond in Kansas City. It seems that after that loan had been made, Roy Disney had the idea of getting his brothers to chip in and buy their mother a vacuum cleaner and wrote to Walt asking for a contribution and also asking him to solicit one from Raymond. The latter replied that Walt could make a contribution to the Christmas fund in his name, deducting it from the amount Walt already owed to him. Characteristically, Walt did no such thing. He simply wrote Ray out of the whole transaction and split the cost of the machine with Roy. There is no record that he ever paid Raymond back, either; the incident was apparently the final split between them. Now it very nearly cost him his big chance. But Roy smoothed things out; they got their money from Uncle Robert, and Disney rented a tiny office nearby, enabling him to move out of his garage workshop. Roy had to dig up another $250 to finish the first *Alice*, but somehow it was done, and in short order they received their first fee from New York—representing a 100 percent profit on the $750 invested. Uncle Robert was paid back immediately, and he, too, now disappears from the Disney story.

The first *Alice* films were only about one-third pure animation. The rest of the footage was made up of Alice playing with her pals, who were neighborhood kids Disney recruited for fees of about fifty cents apiece. Uncle Robert's police dog was also featured in their activities. Only Alice was actually integrated with the cartoons, usually in dream sequences or in stories that she seemed to make up and tell her playmates. About the only thing the pictures had going for them was their inexpensiveness to the distributor and the gimmick, quite crudely done, of mixing a live actor in the same frames with cartoon characters.

Still, no one was getting rich on them. Disney's drive for technical perfection, one of the most important elements in his career, manifested itself early. He did all the animation himself, of course, functioned as director and cameraman on the live sequences and even constructed the sets single-handed. Even so, costs mounted as he put more money into the settings and paid a little more to his child actors (at their demand). Finally, he was forced to hire a cameraman, because Roy was never able to master the steady rhythm required to crank the old-fashioned, hand-powered silent movie camera. He tended either to overcrank, which slowed down motion as it appeared later on the screen, or to undercrank, which caused motion to appear jumpy and unnaturally speedy. The best executive decision he made in this period, however, was to send to Kansas City for Ub Iwerks, who could draw.

105

By the time Iwerks reached the West Coast the Disney brothers had managed to turn out a half-dozen *Alice* subjects, but they were barely making enough to live on comfortably, since Walt insisted on instantly plowing each month's modest profits back into their next release. They borrowed money— from the organist back in Kansas City for whom Disney had done his animated song sheets and from Edna Francis, the girl whom Roy would eventually marry and from whom Walt had got an occasional free meal in Kansas City. (Walt, incidentally, sent the letter asking for her loan without his brother's knowledge and then had to talk him into accepting it when it arrived.) The brothers lived with great frugality, sharing a small apartment, where Roy did the cooking and beans again were the mainstay of their diet. When they went out to eat, they tended to buy a single tray of food in a cafeteria and then split its contents.

But neither personal discipline nor an earnest attempt to improve their product could change the fact that the *Alice* series was not doing well in the theaters. Indeed, after the sixth one had been released, Charles Mintz appeared on the West Coast and attempted to obtain a release from the contract. The Disneys were close to seeing his point and agreeing to the cancellation. The one thing that caused them to resist his blandishments was the fact that *Alice* Number Seven was already in the can, and they wanted at least to get a return on that investment. By this time Iwerks was on the job, and his skill at the animation board and the know-how gathered through trial and error over Disney's first half-year of work combined to produce their first solid commercial success. Mintz kept the *Alice* contract in force and then renewed it twice more, with slight raises in the price paid to the Disneys.

It is difficult to imagine just what anyone saw in the *Alice* films. Even compared to the low grade maintained by their competitors, the Disney pictures were feeble. Alice was nice and she was conventionally pretty, but she had no real character beyond a certain willingness to try anything and therefore get into "situations." The animals that accompanied her were, at best, sketchily drawn and lacked the strong element of caricature, in both features and movement, that is the mark of good animation; indeed, most of the time their definition was so weak that nothing could be seen except the broadest outline of their shapes and movements. The Disneys, even quite well on in their career with *Alice*, had still not learned how to find ways for the animation camera to approximate the flexibility of the live-action camera. There were no close-ups, no odd angles, few variations in the camera's basic point of view, which was well back from its subjects and observed them head-on. Since backgrounds were still being kept to a minimum, the over-all impression one gains on seeing an *Alice* film today is of tiny figures cavorting in a white void. Typically, its jokes look as if they had already been stale at the time Disney and his helpers originally made use of them, and so the films have the rude, crude, flat air of the old comic strips about them. Finally, there is none of the fast-paced rhythm, none of the careful construction of a gag line, building, building, building to an inevitable, and inevitably hilarious, climax as the silent comedians were so brilliantly doing at this time and that Disney's organization was itself to master within a very few years. In short, *Alice in Cartoonland* was, by and large, a limp, dull and cliché-ridden enterprise. All you could really say for it was that it was a fairly ordinary comic strip set in motion and somewhat enlivened by a photographic trick. It seems worthwhile to note that Disney owed his survival, at least in part, to this trick of blending live and animated action, just as he owed each leap forward during the rest of his career to his ability to seize upon similar technological advances and exploit them while his rivals were still fearfully contemplating their possible drawbacks.

Disney later claimed that when he arrived in Los Angeles he had worked out a program for himself, calling for him not to get married until he was at least twenty-five years old and had ten thousand dollars saved—certainly a very sensible program for a go-getter and one often advised and even practiced by the breed. However, the natural expansion of his little studio, once the *Alice* series was established, turned out to be the instrument that spoiled his plan. He asked a young woman in the office if she knew of anyone who might be added to the staff, and she suggested a girl named Lillian Bounds, who lived near enough to the studio so that she could walk to work and thus avoid paying carfare. It was a worthwhile saving, since Disney could pay her only fifteen dollars per week, a fact she failed to establish when she was hired. She was newly arrived in town from Lewiston, Idaho, where she had attended business college, and was living with her sister and brother-in-law while getting established. Legend has it that the girl who told her about the job did so only on condition that she promised not to marry the boss, upon whom she had fixed her own eye. Another legend, often told by Disney himself, was that he fell so far behind in salary payments that he had no option but to marry Miss Bounds.

In any case, it is clear that they had one of those poverty-stricken, faintly comic romances that were later to become something of a staple in Disney's live-action films. The Disney brothers at this point were paying themselves salaries of just thirty-five dollars a week and putting the rest back into the studio. Walt Disney was putting a fair share of his wages into the maintenance of a Ford roadster, since a car was already a necessity in Los Angeles. It was also useful in advancing romance. Because they all frequently worked nights, Disney had fallen into the habit of taking Miss Bounds and the friend who had got her the job at the studio home in his car and "when he started dropping the other girl off first so he could talk to me," the future Mrs. Disney knew he was interested, she later recalled to a reporter. One night when they were working alone at the office, Disney suddenly interrupted dictation to plant a firm but slightly surprising kiss on her lips. The only thing holding their relationship back was that Disney still didn't have a suit.

"We would sit outside the house in the jalopy because Walt had nothing but his old sweater and trousers and he wouldn't go in the house," Mrs. Disney said. "Finally one evening he gulped: 'If I get a new suit—will you let me come in and call on your family?' " She said she thought that would be fine, and he forthwith acquired the clothes by appealing to his brother, who declared that both of them needed suits and even indulged his sibling to the extent of letting him buy one for forty dollars, five dollars more than the top limit they had previously agreed upon. Disney, so that story goes, was so pleased by the purchase that very nearly the first thing he said upon finally being presented to his future in-laws was "How do you like my new suit?"

One of Mrs. Disney's memories of this period is of a night when she and her sister were visiting friends, and her fiancé, with a rare night free from both work and romance, decided to drop in at a movie. "A cartoon short by a competitor was advertised outside, but suddenly, as he sat in the darkened theater, his own picture came on. Walt was so excited he rushed down to the manager's office. The manager, misunderstanding, began to apologize for not showing the advertised film. Walt hurried over to my sister's house to break the exciting news, but we weren't home yet. Then he tried to find Roy, but he was out too. Finally he went home alone." The reason for Disney's excitement was that he had never had a chance to see one of his pictures actually playing in a theater. Mrs. Disney wrote, many years later, "Every time we pass a theater where one of his films is advertised on the marquee I can't help but think of that night."

On the one occasion when Disney discussed his marriage publicly he was determinedly unromantic and unsentimental about it. He thought his future bride was a good listener and that she would provide him with "companionship," he said. He claimed that what really decided him was that his brother had asked Edna Francis to come out from Kansas City and marry him and that it was obvious that he, Walt, would "need a new roommate." His manner of proposing was as utilitarian as his attitude toward the enterprise. He and Lillian had been planning to purchase a new car together, but one night he simply said to her, "Which do you think we should pay for first, the car or the ring?" She replied that the ring was uppermost in her mind and that settled the matter. With their customary frugality the Disney brothers located a cut-rate jewelry merchant in an upstairs backroom somewhere, and Disney passed over some thirty-five dollar rings and, after some haggling with both his brother and the merchant, settled on a seventy dollar diamond set with sapphires. Roy and Edna Disney were married in the spring of 1925 in Los Angeles, Walt and Lillian Disney in July of the same year in Lewiston, in the parlor of her brother, who was the town's fire chief. Their wedding night was spent on a train heading toward Seattle, where they planned to board a steamer for a leisurely trip back to Los Angeles. Disney, alas, developed a toothache and spent the night sitting up with a Pullman porter, helping him polish shoes.

Back in Los Angeles the young couple settled first in a tiny apartment, then purchased a prefabricated house in the tudor style. It cost seven thousand dollars and was located in the Silver Lake district, which lies between Hollywood and Los Angeles and was then a fashionable middle-class area. The house was convenient to Hyperion Avenue, where Disney had acquired a store and converted it into the nucleus of his first large-scale studio. Before it would begin to grow into the almost legendary establishment it became for an entire generation of young animators, however, Disney was to endure—and learn from—a failure of striking bitterness.

By 1927 it was becoming clear to Mintz that the *Alice* series had run its course, and he dispatched his brother-in-law, George Winkler, to Los Angeles to encourage Disney to create something to replace it. He and his little team came up with an idea for a series featuring a rabbit, and his name was selected by throwing all the suggestions into a hat and inviting the distributor to pick one blind. The name he drew was "Oswald," and thus was *Oswald the Rabbit* born.

He was an advance over *Alice*, but he was not exactly *Mickey Mouse* or even a *Bugs Bunny*. He was round and rather cuddly though reasonably naturalistic in design. He owed as much to the tradition of the *Peter Rabbit* series as anything. That is to say he was innocently mischievous and rather more childlike in manner than the furry and feathery anthropomorphs that were to come after him, both from the Disney studio and its competitors. By this time the price on the Disney product had reached $2,250 a reel. *Oswald*, much more sophisticated in technique than *Alice*, was instantly more successful than the earlier series had ever been.

The only thing the Disneys could not understand was the almost monthly visit they received from George Winkler. He customarily brought them their check and received from them their latest film and the art for the lobby poster, which the studio also did. The visit represented a job of no importance, and even in the nepotistic movie business, messenger work seemed beneath the dignity of the most incompetent in-law. The Disneys soon discovered the meaning of Winkler's visits.

Full of confidence, taking his wife along for the ride, Disney entrained for New York in 1927 to renegotiate his contract with Mintz (who, by this time, was releasing the films through the facilities of Universal Pictures). There was still six months to run on the original *Oswald* contract, and Disney planned to ask only for a modest raise in fees—to twenty-five hundred dollars a film. He was instead asked to take a cut back to eighteen hundred. He refused—and then found out the reason for Winkler's trips. He had been instructed to ingratiate himself with as many of Disney's small staff as he could, hinting to them that there might be raises if they were willing to leave Disney at the right moment. He told Disney that he must sign or risk losing some of his key people—and *Oswald*, too, since he was copyrighted in the Mintz firm's name (a common practice in the animation

111

industry then, one probably borrowed from newspaper cartooning, where the practice insures the life of a popular character after his creator's death or desertion). Disney fought back. He allegedly told Mintz that if the four men Winkler had been able to recruit would leave him they would undoubtedly leave Mintz just as readily. He also told him that they would be no real loss to him—that he would be able to replace them easily. In this, he was right. Animation was still a relatively unsophisticated art, and a reasonably competent and well-trained commercial artist could be trained to the work fairly easily. It was not until later that the techniques of animation advanced to the point where a long apprenticeship—as much as fifteen years, according to some artists—was necessary to create a full-fledged animator, and it was the Disney studio that evolved both the techniques and the training program for the young men who needed to master them.

Even so, Disney later recalled that he actually felt none of the confidence about his own future that he expressed to Mintz, and he was apparently deeply hurt, as he always was, by his employees' willingness to leave him. Even in later years, when his employees left for better cause than this, he still read ingratitude and personal disloyalty in their decamping. In point of fact, his first deserters got their come-uppance. Within a year the veteran animator Walter Lantz had replaced them on the *Oswald* series.

This incident further deepened Disney's natural distrust of the people with whom he was forced to do business. In the future he was always exceedingly careful to control all rights in his creations. Indeed, even when he began buying literary property from writers outside his organization, as feature-length animation and live-action feature projects virtually forced him to do, he did everything in his power to make the original story and characters his own. Part of this was accidental: his well-geared merchandising organization, with its many "versions" of original stories like *Bambi* and *Mary Poppins* (designed to catch all age groups) and with its intensive program of licensing dolls, toys and games based on the subjects, had a natural tendency to blanket the original. But part of it was anything but accidental. Disney naturally got credit above the title and in far larger type than the original author, who could not get a contract from the organization without agreeing to submit his creations to the merchandising process. Even worse, they, their spouses and even their children traditionally had to sign a release preventing them from ever making any future claims of any sort against the Disney organization for the material sold. When writer Cynthia Lindsay referred to Disney as "the well-known author of *Alice in Wonderland*, the *Complete Works* of William Shakespeare, and the *Encyclopaedia Britannica*," she was not unfairly summarizing popular opinion of the authorship of the works his studio had taken over.

In short, Disney learned an important lesson in New York in 1927, though he was certainly not aware of how right he was when, before boarding the train for California, he sent a wire to his brother that read: "Everything O.K. Coming home." He thought he was merely trying to prevent Roy from worrying during the several days they would be out of touch while Walt was traveling west, but he was in fact predicting their future— and rather modestly at that.

PART FIVE

Bringing Forth
The Mouse

---

15   *The age demanded an image*
Of its accelerated grimace,
Something for the modern stage,
Not, at any rate, an Attic grace;

Not, not certainly, the obscure reveries
Of the inward gaze;
Better mendacities
Than the classics in paraphrase!
                    —Ezra Pound, *Hugh Selwyn Mauberley*

THERE ARE UNCOUNTED versions of the birth of Mickey Mouse, for Disney, his flacks and hacks, could never resist the temptation to improve upon the basic yarn. This much seems to be true: that the idea to use a rodent as the principal character for a cartoon series came to Disney on the train carrying him back from his discouraging meeting with Mintz in New York. The most flavorsome telling of the tale appeared under Disney's own byline in an English publication called *The Windsor Magazine* in 1934 (though it is doubtful that he did more than glance at the draft his publicity department prepared for him).

The key section begins with his boarding the train, with no new contract and no discernible future.

"But was I downhearted?" he inquires. "Not a bit! I was happy at heart. For out of the trouble and confusion stood a mocking, merry little figure. Vague and indefinite at first. But it grew and grew and grew. And finally arrived—a mouse. A romping, rollicking little mouse.

"The idea completely engulfed me. The wheels turned to the tune of it. 'Chug, chug, mouse, chug, chug, mouse' the train seemed to say. The whistle screeched it. 'A m-m-mowa-ouse,' it wailed.

"By the time my train had reached the Middle West I had dressed my dream mouse in a pair of red velvet pants with two huge pearl buttons, had composed the first scenario and was all set."

It is well known that the name Disney first gave his creation was Mortimer Mouse—borrowed, it is said, from that of a pet mouse he kept in his Kansas City studio. Disney himself never claimed this, though he frequently confessed "a special feeling" for mice and readily admitted that he had kept a fairly large family of field mice in his Kansas City offices. Originally he had heard their rustlings in his waste paper basket, but characteristically, he refused to let them have the run of the place. Instead, he built cages for them, captured them and allowed one of them, who seemed especially bright, the occasional freedom of his drawing board. He even undertook a modest training course for the little fellow, drawing a circle on a large piece of drawing paper and then tapping him lightly on the nose with a pencil each time he attempted to scamper over the line. Before long he had trained him to stay within the circle, though he would venture right up to the line and at mischievous speed, at that. When it came time to leave Kansas City, Disney set all of his mice free "in the best neighborhood I could find," as he later put it. Of his parting with his special pet, Disney said that he "walked away feeling like a cur. When I looked back he was still sitting there, watching me with a sad, disappointed look in his eyes."

There are two versions of the renaming of the cartoon mouse. In the more common one Mrs. Disney is reported to have found Mortimer too pretentious and insisted on a less formal-sounding title for the little chap; some say she suggested the name Mickey, others that Disney named his new character and she approved it during the course of their long train ride back to California. Yet another story is far more prosaic; it is simply that one of the first distributors Disney approached with his new idea liked it but not the name and that it was his objection that caused Disney to rename his creation.

In any case, it is certain that immediately after he returned from New York, Disney set his little studio to work on a cartoon that had a mouse as its principal figure even as the studio was fulfilling the last demands of his contract with Mintz. "He had to be simple," Disney later said, in discussing the details of The Mouse's creation. "We had to push out 700 feet of film every two weeks [actually, the production schedule was slightly less frantic than this], so we couldn't have a character who was tough to draw.

"His head was a circle with an oblong circle for a snout. The ears were also circles so they could be drawn the same, no matter how he turned his head.

"His body was like a pear and he had a long tail. His legs were pipestems and we stuck them in big shoes [also circular in appearance] to give him the look of a kid wearing his father's shoes.

"We didn't want him to have mouse hands, because he was supposed to be more human. So we gave him gloves. Five fingers looked like too much on such a little figure, so we took one away. That was just one less finger to animate.

"To provide a little detail, we gave him the two-button pants. There was no mouse hair or any other frills that would slow down animation."

117

In short, The Mouse was very much a product of the then-current conventions of animation, which held that angular figures were well nigh impossible to animate successfully and that clearly articulated joints were also too difficult to manage, at least at the speed of drawing demanded by the economics of the industry. Hence the odd appearance of so many old cartoons when we glimpse them today on the programs our children watch on television—the thick round bodies contrasting so oddly with the rubbery limbs of the characters. In a few years, thanks largely to the leadership of the Disney studio, these conventions were abandoned, but whatever advantages their adoption provided economically was offset, as Disney later said, because they "made it tougher for the cartoonists to give him character."

Indeed, it is possible that The Mouse would have had a life no longer than many of his competitors if a technological revolution had not intervened and presented Disney with an opportunity that was particularly suited to his gifts and interests and that he seized with an alacrity shared by few in Hollywood. *Plane Crazy*, in which both Mickey and Minnie Mouse made their first appearance, was on the drawing boards—in a silent version, of course—on October 6, 1927, when Warner Brothers premiered *The Jazz Singer*, the movie that broke the industry's resistance to sound. A practical sound system had, in fact, been available as early as 1923, when inventor Lee DeForest toured the lecture circuit showing sound movies of prominent vaudeville acts; since he was working outside the customary commercial channels, the industry could safely ignore him. It also managed to ignore a series of short films, featuring the voices of famous opera singers, the movietone newsreels of William Fox, who achieved the first sound-film box office success with his coverage of Lindbergh's transatlantic solo flight, and even Warner Brothers' *Don Juan*, which had a musical score and sound effects and even a little introductory speech by Will Hays hailing the new technological miracle. Warners' at that time was a studio desperately trying to stave off bankruptcy, and the success of *Don Juan* led them to extend their experiments by filming some of the sequences in *The Jazz Singer* with sound film, making it the first film to integrate sound with the story. Only a modicum of the dialogue was recorded, but all the musical sequences were, and the result was a sensation. Warners' immediately added similar sequences to three other movies the studio then had in production and laid plans for an *all*-talking picture. The rest of the industry formally agreed to fight the "Warner Vitaphone Peril," but by the spring of 1928, opposition had collapsed and all the studios were rushing into production with sound pictures. There was nothing dignified or even very intelligent about the transition period. It was, indeed, a hysterical scramble in which Walt Disney and his problems, never in the forefront of anyone's mind, were almost entirely lost from view.

At this point the previous frugality of the Disney Brothers paid off. Between them they had somewhere between $25,000 and $30,000 in personal assets, and their studio, though not wildly successful, could afford to go ahead with its first three Mickey Mouse films even though they had no contract for them. or any real idea of who might be persuaded to distribute them. In addition, they were in somewhat the same position as the nearly bankrupt Warner Brothers had been: they had nothing to lose by experimenting with sound; their investment in unreleased product was negligible; they had no investment in actors whose vocal qualities might not be suitable to the microphone, and in the animated cartoon they had a medium ideally adapted to sound. The early sound camera was virtually immobilized in the soundproof "blimp" necessary to prevent its whirrings from being picked up by the microphone. Disney, of course, had no such problem with the camera. The animation was shot silently as always, and sound was added later. This meant that his little films retained their ability to move while all about them were losing theirs. Just as important was the control he could exercise over the relationship between pictures and sound. They could be perfectly integrated simply by matching the musical rhythm to the rhythm of the drawn characters' movements. (Later on, the planning of music and movement would be carefully integrated from the earliest moment of production as the influence of an idea in one area flowed to the other. Far from being a simple accompaniment, the music became in the Disney studio product an almost equal partner with the visual imagery, the sound effects and the dialogue in creating the total effect of the film.)

True to his taste and talent for technological innovation, Disney hesitated only briefly before plunging into sound-film production. After *Plane Crazy*, he shot a second Mickey Mouse, *Gallopin' Gaucho*, silently, but in his third Mickey Mouse, *Steamboat Willie*, he began to experiment with sound.

His chief assistants were Ub Iwerks, in these early days getting a "drawn by" credit on the title card of each film, and a young man named Wilfred Jackson, a newly employed animator at the studio who liked to play the harmonica in his spare time. The picture was plotted to the tick of a metronome, which set rhythms for both Jackson, who played standard, public-domain tunes on his mouth organ, and Iwerks, who got from the metronome a sense of the rhythms he would have to use in his animations. The calculation of the ratio of drawings to bars of music was thus arrived at far more simply than it would be only a few years later. To color the sound track further, Disney rounded up tin pans, slide whistles, ocarinas, cowbells, nightclub noisemakers and a washboard. When the animation was finished, the co-workers invited their wives to come over to see something new one evening, then ducked behind the screen and played their "score" live, through a sound system Iwerks concocted out of an old microphone and the loudspeaker of a home radio. In later years Disney recalled that the ladies had been vaguely complimentary about their efforts but had not allowed their husbands' novelty to distract them from what they considered the main business of the evening—girl talk about babies, menus, hairdressers and so on. The story, like so many of Disney's reminiscences, seems a little too patly similar to situation comedy, but, on the other hand, the performance was probably not terribly impressive at this early stage.

By early September, 1928, however, he felt he was ready to head for New York again in search, first of all, of someone to put his perfected score on a sound track and, second of all, of a distributor. By this time Disney and his co-workers had on paper their complete score, down to the last rattle of a cowbell, and had devised a system by which a conductor could keep his beat precisely on the tempo of the film. Sound speed was standardized at ninety feet of film per minute, or twenty-four frames per second. The musical tempo was two beats per second, or one every twelve frames. On the work print Disney took with him, a slash of India ink had been drawn at every twelfth frame, causing a white flash to appear on screen every half-second. All a conductor had to do was key his beats to the flash, and in theory, if he never missed a flash, he would reach the end of his strange-looking score at precisely the moment the film ended.

But first Disney had to find someone willing to record his sound track, and then he had to find a conductor willing to sacrifice the most important variable at his command—tempo—to a gang of musically illiterate cartoonists. Neither was easy to find. The best sound system was controlled by RCA, and though the company was perfectly willing to take Disney's work, it was unwilling to follow his score. Its technicians had already added tracks to some old silent cartoons and were convinced, apparently, that close synchronization between music and sound effects and the action on the screen was, if not impossible, then certainly not worth the effort. Disney knew otherwise and refused to relinquish his precious piece of film to them. They, on the other hand, were not about to let a twenty-seven-year-old stranger tell them their business.

Disney then started on the rounds of those entrepreneurs who

121

owned outlaw sound equipment—that is, recording devices either not covered by the patents controlled by RCA or its chief rival, Western Electric, or which those giants had not moved against in the courts. Here, again, history was repeating itself, for the first silent cameras and projectors had been controlled by Edison and licensed only to producers and theaters that belonged to his so-called "trust." His invention had been so attractive that long before he gave up the fight to enforce his rights, the trust had been effectively broken by a number of competitors too vast to bring them all to book. The situation never grew quite so unreasonable in the sound business, for both the quality of the patented systems and the power of the firms that controlled them were such that it was to everyone's advantage to rationalize the market. But it was impossible in those early days to control everyone, and Disney had no real difficulty in locating someone to take on his recording chores. The trouble was that the man he found was a semilegendary figure named Pat Powers, who had learned the movie business—and his code of ethics—in the freebooting days before World War I. It is said, perhaps apocryphally, that Powers, who at one time distributed Carl Laemmle's Universal Pictures, actually resorted to throwing his books out the window (he was on the twelfth floor) rather than let Laemmle take a look at them at a moment when the producer got to wondering where his profits had gone. It is also said that Powers had the foresight to have a man waiting in the street below to retrieve the books and make a getaway with them. Disney was to be victimized by the casual unseemliness of this style during his association with Powers, but he appears to have known what he was getting into, and Powers did have a sound system he was willing to put in Disney's service—the "Powers Cinephone."

Powers was amenable to allowing Disney to supervise the recording session and engaged a conductor, Carl Edwardi, who led the band at New York's Capitol Theater, rounded up the musicians and handled all the details—which included a price of $210 per hour for the thirty members of the orchestra, plus fees for the four men who were to handle sound effects and for the technicians. The first session went badly. The bass player's low notes kept blowing tubes, and Disney himself, doing, as he was always to do, the Mickey Mouse voice, blew a tube when he coughed into an open mike. Worst of all, Edwardi flatly refused to key his beat to the flashes on the screen, no doubt reasoning that a man of his reputation ought to be able to conduct, unaided by mechanics, a score consisting of "Yankee Doodle," "Dixie," "The Girl I Left Behind Me," "Annie Laurie," "Auld Lang Syne" and so on. After three hours and some $1200 in costs they still had nothing usable on the track, and Powers, who had promised to pay all excess costs should Edwardi not live up to his reputation, declared that his offer had not included the musicians' salaries—it was to cover only the cost of the sound equipment and the film. Disney had no option but to wire his brother for more money, and to raise it, Roy Disney sold, among other things, Walt's Moon Cabrolet, an automobile with red and green running lights of which Disney was particularly fond. At the next session Edwardi agreed to try following the flashes on the screen, Disney cut the orchestra almost in half (dispensing with the bothersome string bass among other instruments), dismissed two of the sound effects men and took over some of their functions himself. This time, everything worked out as planned, and within days Disney was making the rounds of the New York distributors, trying to get someone to handle *Steamboat Willie* for him.

It was a disappointing business. In the crisis atmosphere of that year a man with a cartoon short subject, even one with an artful sound track on it, did not have high priority. The tendency was to stall Disney and then to look at his product, if at all, only after screening more commercially promising projects. Disney recalled peeping out through projection booth windows, trying to gauge reaction in the screening rooms, on those rare occasions when his work was given a showing. He also recalled getting his best laughs from the projectionists next to him.

But at one of these screenings a figure, if anything, more legendary than Powers happened to be present, and it was he who sensed the possible exploitation value of Disney's film. His name was Harry Reichenbach, and as a press agent he had previously promoted an utterly ordinary piece of *kitsch* called *September Morn* into a scandal of international proportions. At this time he was managing the Colony Theater in New York, and he asked permission to book *Steamboat Willie* as an added attraction. Disney was hesitant, fearing that with the New York cream skimmed off, he would have an even more difficult time in persuading a national distributor to take the film. Reichenbach argued, however, that this was a special film demanding special treatment. Unless extraordinary attention such as he was prepared to engineer was focused on it, the little picture would simply be lost in the flood of new sound films. He offered Disney a two-week run beginning September 19, 1928, and Disney talked him into paying five hundred dollars a week for the privilege.

The strategy worked. Reichenbach got the press to attend and write stories about his added attraction, and Disney spend a good deal of time at the Colony listening to the laughter of audiences responding to the first genuinely artful use of the new technology of sound film. After the run at the Colony, *Willie* moved to the S. L. Rothafel's two-year-old Roxy Theater. Now the distributors started coming to Disney, asking him what he wanted to do and what they could do to help him. They got only part of the answer they were hoping for: Yes, he did want to go on making Mickey Mouse cartoons; no, he did not want to sell them the films outright. The loss of *Oswald the Rabbit* was still fresh in his mind. This time, he insisted that he retain complete control of his product. Again, it appeared that he had reached an impasse.

124

It was Pat Powers who rescued him. He was looking for markets and promotion for his Cinephone device, and he sensed in the Mouse exactly the sort of gimmick he needed. In return for 10 percent of the gross Powers offered to distribute the Mickey Mouse films through the system of independent or "state's-rights" bookers. It was one of the simplest, and oldest, methods of film distribution, and indeed, some of the major figures in the movie industry got their start in the state's-rights business, which still exists in truncated form. What happened was that an independent distributor bought the right to distribute an independent film in a given territory and then, taking a percentage of the gross for his trouble, peddled the films thus acquired to theaters in the district, which generally comprised several states. The advantage was obvious: a salesman for an operator like Powers could easily canvass the relatively few state's-rights bookers in several territories efficiently and frequently and sell more film more quickly than he could if he had to approach each theater owner separately. Of the several disadvantages, the most obvious was that the system took three sizable slices off the top of the gross—one each for the theater owner, the state's-rights man and the national distributor—before any profit dribbled back to the original producer. Worse, the state's-rights distributors usually did not have access to the very best theatrical outlets, many of which were owned outright by the large production and distribution companies, many more of which were locked into more or less exclusive agreements with the major firms under the block booking system. This system demanded that the theaters take large packages of indifferent films in order to obtain the opportunity to show the ones that contained major stars and, of course, offered the greatest potential profit. These people simply did not have the time or the inclination to handle independent films. Finally, with film and money changing hands so often, there were many opportunities all along the line to hide profits without much fear of subsequent audits. (This problem has risen again to plague a new generation of independent producers who must release their pictures through established distribution firms and must depend on their word about the grosses against which their percentage is figured.)

Despite all these disadvantages, Disney had no real choice if he wished to retain control of his product: he had to release on a state's-rights basis or not at all. He was asking no more than customary terms—a five-thousand-dollar advance for each negative print furnished a would-be distributor and a 60–40 split of profits, the larger figure going to the distributor once the advance was paid back. But still there were no other takers, and one gathers that the Powers offer was not only financially attractive but gratefully received by a man tired of worrying over money and deals at a moment when he sensed that he had, for the first time, a real hit, one on which he might build a future of incalculable brightness. The main thing, at this point, was to exploit his sudden advantage, to get more films into the works and before the public. Disney behaved in this period much like a man anxious to leave the details until later, a man who knew that large immediate returns were less important to him than immediate production on as large a scale as he could manage. He seemed to sense that it was this that would establish the value of his name and thus improve his bargaining position in later rounds.

So he closed with Powers and got to work, while still in New York, on the creation of scores for his two unreleased silent Mickey Mouses, *Plane Crazy* and *Gallopin' Gaucho*, as well as for a brand-new vehicle for the Mouse, *The Opry House*. Again he turned to Kansas City for talent he could rely on: the faithful theater organist who had previously lent him money was brought east to score the films. By the time he left the city early in 1929, Disney had a package of four films ready for release. Their cumulative impact when they went into national distribution was simply tremendous, so much so that Disney was emboldened to attempt an animated short without Mickey or Minnie but one that, however primitive it may seem to a modern audience, nevertheless represents an extremely important advance for the animated cartoon.

That picture, of course, was *The Skeleton Dance*, the first in the famous series of *Silly Symphonies*. In *The Opry House* Disney had already moved beyond "Turkey in the Straw"; the musical feature of the film was a Rachmaninoff Prelude. Now he decided to do a sort of "Danse Macabre," which would ultimately include some quotations from Grieg's "March of the Dwarfs" (from the *Peer Gynt* Suite) among other bits and pieces of serious music. Obviously this was not suitable material for Mickey and his gang, whose activities were firmly located against an American small town and country background. Disney would have to break with the most firmly held of all the conventions of animation—the use of animals as the principal characters.

It is difficult to say just what was driving Disney culturally at this time. Although he had studied the violin briefly as a boy, he had no real musical knowledge or, indeed, standards, as his cheerful chopping and bowdlerizing of music testifies, not only in *The Skeleton Dance* but in all his later work, not excluding his mightiest effort at uplift, *Fantasia*. Indeed, when he was working on *Fantasia* he so felt his lack of musical training that he subscribed to a box at the Hollywood Bowl concerts, where, he recalled to a co-worker, he invariably fell asleep, lulled by the music and the warmth of the polo coat he liked to wrap around himself in those days. One suspects that his desire to bring serious music into an area where it had not previously penetrated was based on several considerations. First there was a commercially intelligent desire to differentiate his work from that of competitors who showed neither the ability nor the ambition to move beyond their low origins. Then there was the technical challenge that complex music presented to the animator. As early as 1925 he had made a little short in which a cartoon character seemed to conduct a theater's live orchestra from the screen. He also had toyed with the notion of supplying musical cue sheets to theater musicians so his silent shorts could be properly accompanied. There was also, perhaps, that vaguely defined, yet keenly felt, lust for "the finer things" that was so common among people of his *lumpen bourgeoisie* background. His father, after all, "would go for anything that was educational," and Disney himself proved throughout his career that he was chipped off much the same block (though his ideas of what was educational and what was not were often eccentric). Finally, and perhaps most important, there was his own shrewd sense of what, precisely, was the basis of his new success.

To understand this, one must take a closer look at Mickey Mouse. There is no more useful tool for this purpose than the print of the silent version of *Plane Crazy* owned by the Museum of Modern Art in New York. The film is not a bad piece of animation when judged against the standards of its time. Particularly in the anthropomorphizing of inanimate objects, like the airplane in which Mickey takes Minnie up for a ride, it is quite good. The machine, like others that followed it, actually registers emotions: it strains forward in anticipation of action, cowers when it sees obstacles looming ahead, and so on. The construction of the story, too, is superior to that found in most competing cartoons. It is, like most silent comedies, live or animated, nothing more or less than a string of physically perilous disasters, each more absurd and more dangerous than the last but all stemming from a perfectly reasonable premise—the natural desire of a young swain to prove his masculine mastery by taking his girl friend for a ride in an airplane he has designed and constructed. Some of the gags along the way are crude: a dachshund, for instance, is wound up tight, like the rubber bands boys use to drive their model planes, and serves as the power plant for Mickey's plane. And none of the nuances of gesture and expression that the great silent comics had developed in the course of shaping their art were within the range of Disney's animators. But the naturalness and inevitability with which the temporary and ill-conceived solution to one problem

128

of flight leads to another are extremely clever. When, finally, the tension generated by the possibility of the ultimate disaster is resolved innocuously (Minnie parachutes to safety with a great display of patched bloomers, Mickey successfully makes a forced landing), it is clear that one has been in the presence of a work in which talent and intelligence, however crude, have been manifested.

One also observes that Disney and his people made shrewd use of satire by giving Mickey's hair a Lindberghian ruffle when he stands in his plane taking bows for his achievements. And one is impressed, too, by the fact that, faced with the need to create a truly original cartoon—under particularly trying circumstances at that—Disney had the wit to draw upon his own background for subject matter. There was, of course, his remembrance of the mice in his Kansas City workshop, but more important were his drawing upon the world of the barnyard for background and the spirit of the independent tinkerer-inventor for the film's psychological motivation.

But having observed all this, one is still bound to believe that The Mouse, as originally conceived and silently presented, was a very shaky basis for the empire that was to come. As *Time* observed in its acute biographical sketch of Mickey (December 27, 1954), "he was a skinny little squeaker with matchstick legs, shoe button eyes and a long, pointy nose. His teeth were sharp and fierce when he laughed, more like a real mouse's than they are today. . . ." Mickey's disposition matched his appearance. He was quick and cocky and cruel, at best a fresh and bratty kid, at worst a diminutive and sadistic monster, like most of the other inhabitants of that primitive theater of cruelty that was the animated cartoon. Five minutes with The Mouse, however diverting, were quite enough. His amazing—indeed, impossible—voyagings, in various guises were often amusing enough in their broad conception, and they supported some excellent comic situations, but the truth is that at the outset it was the witty, constantly surprising use of sound to punctuate the stories and the technical genius of the animation—and its visual orchestration—from which the best laughs derived.

This is nowhere more obvious than in *Steamboat Willie*. In it The Mouse has, if anything, regressed as a character; he represents nothing more than the spirit of pure, amoral and very boyish mischief. Employed as an assistant to the villainous Peg-Leg Pete, he takes it upon himself to rescue Minnie, who appears to be the river boat's sole passenger, from Pete's lustful advances. In the course of the ensuing chase the mice find themselves in the boat's hold, where domestic animals are being kept. A goat nibbles some sheet music, Minnie cranks his tail as if it were the handle of a street organ and out of his mouth comes "Turkey in the Straw." There then ensues a formidable concert—in which a cow's teeth are played as if they were a xylophone, a nursing sow is converted into a bagpipe and so on, in a marvelously inventive and rapidly paced medley. The gags are not very subtle ones, and indeed, there is something a little shocking about the ferocity with which Mickey squeezes, bangs, twists and tweaks the anatomy of the assembled creatures in his mania for music. But even today the cumulative effect of the sequence is catchy; in 1928 it must have been truly stunning.

What happened in *Steamboat Willie* was both a rediscovery and a discovery. The rediscovery lay in the effects of animation on an audience. As Professor Erwin Panofsky says in his justly famous essay, "Style and Medium in the Motion Pictures," "the primordial basis of the enjoyment of moving pictures was not an objective interest in specific subject matter, much less an aesthetic interest in the formal presentation of subject matter, but the sheer delight in the fact that things seemed to move, no matter what things they were." Throughout the 1920s animators were extending the range of movement the camera could record; they made inanimate objects move, they made animals move like humans (and, sometimes, humans move like animals), they gave visual expression to emotion through contortionistic (and therefore caricaturistic) movement of every sort of creature. Now Disney was forced by the rhythmic patterns of music to orchestrate this movement more carefully. Instead of being a series of random effects, the cartoon achieved through music more solid structure than it had been possible to acquire from an unadorned story line. The audience was therefore able to rediscover some of the joy in observing pure movement that had first attracted it to film—a sense of delight that, with the exception of the work of the silent comedians, it had been increasingly denied as movies grew more literary in orientation throughout the twenties.

As for Disney's discovery, it is very simply expressed: he saw that sound was not merely an addition to the movies but a force that would fundamentally transform them. The silent film, as Panofsky points out, was never truly silent: the humblest neighborhood theater employed a pianist to underscore its shifting moods. It was then rather more like a ballet than any of the other traditional art forms. Most people supposed that with the addition of sound the film would come to be, perhaps even should be, more like a stage play. What Disney sensed was that it was about to become a totally new form, one in which, as Panofsky put it, "what we hear remains, for good or worse, inextricably fused with that which we see; the sound, articulate or not, cannot express any more than is expressed, at the same time, by visible movement. . . ." In short, and in less elevated terms, Disney was the first movie maker to resolve the aesthetically disruptive fight between sight and sound through the simple method of fusion, making them absolutely "coexpressible," with neither one dominant nor carrying more than its fair share of the film's weight. To put it even more crudely—and rather as one imagines many of his first audiences might have put it— "he made things come out right." The animal concert in *Steamboat Willie* presages those wonderfully intricate animation sequences that were to be the high points of his studio's work in years to come—a rendition of "The William Tell Overture" continuing aloft in a twister in the Mickey Mouse short *The Band Concert* (1935); the house-cleaning sequence in *Snow White;* the ballet of the hippopotamuses (to "The Dance of the Hours") in *Fantasia,* for example. All are infinitely more sophisticated than Mickey's musical interlude in *Willie.* But their fundamental appeal is the same: "they come out right." And it is perhaps not inappropriate to note that their fascination, both terrifying and thrilling, is not unlike the kind one derives from observing a brilliantly engineered assembly line functioning at its best.

It would be preposterous to assume that Disney knew, philosophically, where he was going in *Steamboat Willie.* He, too, simply liked things to come out right. But he knew better than his distributors where the basic appeal of his films lay. They were convinced that Mickey Mouse was, like the characters who had preceded him as the leading figures in cartoon series, an exploitable personality, a "star" of a sort. Disney went along with them—though for reasons of his own, to be explored later —but grasped that it was technique, not personality, that was bringing the people in (at least in the first year or two). We shall return later to his attempts to overcome the problem of The Mouse's fundamentally cruel character (by turning him first into a straight man, then a supporting player and finally by simply utilizing him as a corporate symbol). Even though within months Mickey Mouse cartoons were getting billing on marquees, an almost unprecedented commercial tribute for a short subject, Disney was determined not to be Mouse-trapped. Whether he was consciously aware of The Mouse's basic weakness for the long economic pull, whether he was impelled by a genuine desire to broaden his experiments or to upgrade his medium and his culture, or whether it was a combination of these and other factors that drew him on, it is difficult to say. What is known for certain is that in May of 1929—with The Mouse series barely under way—he had *The Skeleton Dance* ready for release. He sent the print off to Powers, who rejected it, telling

Disney to stick to mice. Disney then took it to a friend who ran one of the theaters in the United Artists Los Angeles chain and got him to show *Dance* to a morning audience. One of the man's assistants sat with Disney during the screening and when it was over, told him he could not possibly recommend it for regular showings, even though most of the customers seemed to like it. Disney's companion claimed it was simply too gruesome.

Still Disney pressed on. He tracked a film salesman all the way to a pool hall because the man was alleged to be influential with Fred Miller, owner-manager of the Carthay Circle Theater, one of the most prestigious houses in Los Angeles. The man actually managed to get *The Skeleton Dance* to Miller, who took a fancy to it, showed it and collected an excellent set of notices. Back the film went to Powers, this time accompanied by its West Coast reviews and with the advice to get it seen by Roxy personally. The famous showman also liked it, put it into the Roxy Theater and found himself with such a successful short on his hands that he held it over through several changes of features.

The Skeleton Dance has no story and no characters at all. It is set in a graveyard in the smallest hours of the night, when the skeletons emerge from their graves and vaults, dance together for a few minutes and then, with the coming of dawn, climb back into their resting places. It is such an innocently conceived movie that it is hard to imagine its frightening any child, and the connections of Gothic and grotesque art, to which Disney was frequently attracted, must have seemed quite advanced to an audience that rarely saw such iconography at the movies.

Its dividends for Disney were not to be counted merely in box-office terms. Indeed, after a year of work and twenty-one completed films, Disney economically had very little to show for his efforts. What it did was add another string to his bow, forming the basis for a series of films in which, free of the artistic conventions and story lines imposed on him by Mickey and his "gang," he could experiment with techniques, and whether he planned on this or not, he could begin building his reputation with the intellectual community. He even had one or two very large financial successes with The Silly Symphonies, though at first his distributors insisted on a very strange billing: "Mickey Mouse presents a Walt Disney Silly Symphony."

Meantime, it was becoming clear to Disney that his deal with Powers was not working out at all as he had hoped it would. Throughout this year of extraordinarily hard labor, he and his brother had been unable to get a full financial report from their man in New York. Instead, Powers sent them checks for three or four thousand dollars, enough to keep going, but nothing like what the Disneys felt they should be getting if the Mickey series was the success everyone was convinced it was. Even more disturbing was the rumor that Powers was about to try to make off with Ub Iwerks. Powers had been led to believe, not totally erroneously, that Iwerks was the real talent at the Disney studio and that if he could be lured away and set him up in his own shop, he might be in much the same position Disney himself was now in—that is, as a producer with a possibly unlimited future. In any case he hoped to use the threat of hiring Iwerks as a way of bringing Disney to satisfactory terms on a new contract.

Disney and his wife again entrained for New York, this time taking along Gunther Lessing, an attorney who was to become a powerful force within the Disney organization (a persistent story identifies Lessing as a one-time aide of the Mexican bandit Pancho Villa, a tale that gained both wide circulation and credence among the Disney strikers in 1940 when Lessing undertook the role of management's most unyielding negotiator and spokesman). In Powers, they found a man who had outsmarted himself. He had not expected The Mouse to be the great success that he was. Planning to use Disney's creation as no more than a loss leader in the promotion of his sound system, Powers had neglected to make a tight contract with Disney. It ran for only one year, with no renewal options. From the moment he realized that Mickey Mouse was far more valuable than his Cinephone, he had been trying to squeeze Disney. It was this attempt to bind Walt to him with extralegal ties that had led to his pursuit of Iwerks and to his withholding, throughout the year, of the full amounts due Disney on his contract.

Now Powers' strategy became clear. He greeted Disney by showing him a telegram from his Los Angeles representative indicating that Iwerks had now signed a contract with Powers to produce a cartoon series of his own. He also informed Disney that he could not examine his books to determine what money was due him unless he agreed to a new contract, although, of course, he was free to undertake a costly, time-consuming court fight for a fair accounting. On the other hand, Powers did not wish to appear unreasonable. His true aim, after all, was not to drive Disney to the wall but to continue to participate in an extremely profitable arrangement with the young producer. If Disney agreed to sign the new contract, he would not only receive the sums due him for his last year's labor, but Powers would tear up his contract with Iwerks and agree to pay Disney $2,500 a week in the future.

Disney asked for a little time to think things over, and he later recalled that it was not until he was walking back to his hotel that the full magnitude of the $2,500-a-week offer struck him. It was more money than anyone had ever mentioned to him before, and he remembered muttering the figure aloud, to the wonderment of various passersby. The impressiveness of the sum was undoubtedly a mistake on the part of Powers, as indeed was his whole strategy, since it was based on a fundamental misreading of his man. The $2,500 had the effect of confirming Disney's own estimation of the value of what he was doing. It also indicated to him that he was not without bargaining strength himself.

Before he had left home his brother had told him that he must obtain an immediate cash advance from Powers in order to keep the studio operating, and so, when he returned to the distributor for further discussions, he casually mentioned the need for cash. Anxious to indicate his good will, Powers drew a check for $5,000, and Disney stalled him until it cleared. Then he broke off negotiations with Powers.

He made no attempt to retain Iwerks, who, with Powers' backing, set up his own shop and, for a while, produced a series called *Flip the Frog*. It did not catch on, perhaps because he lacked the one talent everyone who ever worked for him agrees that Disney possessed in abundance—that of a story editor. He certainly lacked the Disney brothers' ability to control costs on the one hand and to master the fast-talking financial world beyond the studio door. Within a few years Iwerks was back at work for Disney, where he supervised special effects on many of the studio's major productions and was instrumental in developing improved animation technique as well as the multiplane camera, on which he had actually begun experimentation during his brief independent fling. He won two Academy Awards for his technical achievements. He remained at the studio until his death in 1971, and in his late years, besides working with Hitchcock on *The Birds* figured prominently in the development of an improved matte process that made possible not only the technically perfect blending of animation and live action in *Mary Poppins* but also its brilliant illusions of flight and the sequence that was the film's highpoint, the dance over the rooftops of London. Personally, Disney's relationship with Iwerks always remained on a business-only basis. Witnesses report Disney carefully looked the other way when passing Iwerks on the lot or, at best, spoke to him in monosyllables. Iwerk's technical genius was valuable enough to Disney for him to tolerate his presence, but he never could forgive his long-ago moment of disloyalty.

PART SIX

Everyone Grows Up

---

16 *And every week, every day, more workers joined the procession of despair. The shadows deepened in the dark cold rooms, with the father angry and helpless and ashamed, the distraught children too often hungry or sick, and the mother, so resolute by day, so often, when the room was finally still, lying awake in bed at night, softly crying.*
> —Arthur M. Schlesinger, Jr. *The Crisis of the Old Order*

I encountered nothing in 15,000 miles of travel that disgusted and appalled me so much as this American addiction to make-believe. Apparently, not even empty bellies can cure it. Of all the facts I dug up, none seemed so significant or so dangerous as the overwhelming fact of our lazy, irresponsible, adolescent inability to face the truth or tell it."
> —James Rorty, *Where Life is Better* (1936)

OBVIOUSLY THE BUSINESS to be in was the illusion business. At least at first. Buoyed by the novelty of sound films and by the need of its audience for inexpensive escape, the American movie industry maintained its general prosperity until 1933, slumped briefly, recovered (eighty million admissions were pur-

137

chased in 1936), joined the rest of the economy in the 1938 recession and thereafter matched the rest of the nation's business cycle until the postwar period, when, as noted, it failed to participate fully in the general boom. It cannot be said that the Disney studio performed exceptionally well from a business analyst's point of view during the industry's period of retained prosperity, 1930 to 1933. By their very nature it is difficult to make a great deal of money on short subjects, especially animated cartoons. The shorts share an underdeveloped land on the margin of the industry, and the only hope of continuing success with them is to achieve efficient, large-volume production at low unit cost and with rapid turnover. Animated films have the additional problem of being more costly to produce than live-action films of the same length. Disney's strategy, given this situation, was to sacrifice any hope of large immediate profits and strive only to keep operating while building up diversified sources of income and a reputation for quality production. This, he hoped, might sustain a break-out from shorts into feature production, where real profits and genuine stability lay, if they lay anywhere in so chancy a business as the movies. In pursuit of this goal it was necessary for the studio to attack simultaneously on several fronts—the economic, the technological and the aesthetic—and it had to be prepared for a long struggle.

Economically, Disney's first need, in 1930, was for a new distributor. He went first to Metro-Goldwyn-Mayer, which had no major cartoons series of its own and was interested. M-G-M decided, however, to turn Disney down for fear of deepening its already substantial reputation for mercilessness, the inevitable concomitant of its rapid growth to first place in size among the studios. It simply did not want to be in a position where it might look like a bully stealing a property from a small operator like Powers.

Harry Cohn, the president of Columbia Pictures, had no such qualms. He was a little guy himself, but even if he had not been, he was not a man to hesitate when Frank Capra, the decade's leading comedy director, praised Disney's work and indicated that his studio's product was available for distribution. This time Disney had no trouble striking a bargain that allowed him to retain ownership of his films and the independence of his studio. The trouble was that his drive for an improved product continued to push costs up. At the time he signed with Columbia a Disney short cost $5,400 to produce. By late 1931 he was spending $13,500 on each little film and doing no better than breaking even on them.

He was also beginning to break down, emotionally. Each film he turned out barely paid for the next one, and the problems presented by his leading character were also extremely frustrating. The Mouse more or less fitted the description of him, written three years later by no less a literary figure than E. M. Forster, who found Mickey "energetic without being elevated" and added, "No one has ever been softened after seeing Mickey or has wanted to give away an extra glass of water to the poor. He is never sentimental, indeed there is a scandalous element in him which I find most restful." The trouble was that some people were scandalized without refreshment by Mickey's style. It was also true that although people wanted to identify with him—and many succeeded—it was not possible to make his alter ego into the universal hero Disney seems to have wanted him to become. His popularity, as Gilbert Seldes said in 1932, had "some of the elements of a fad, where it joins the kewpie and the Teddy Bear" and that, obviously, was not good enough for Disney. Mickey had to become both more verbal and somewhat softer in manner if he was to satisfy Disney—and if he was to be transformed from his fad status into the kind of symbol that can ride above any shift in the wind of fashion. By the time Forster wrote, this process was already well under way.

It seemed to disturb Disney, almost as if it were his own personality that was being tampered with. It also presented him with purely aesthetic decisions that were painful to him. Likable as Mickey was becoming, his new sweet self was as difficult to use as the core of a film as his old sharp self had been. He was still not funny in and of himself: he was merely unfunny in a different, though perhaps more generally appealing, way. This humorlessness as well as his naïveté and his enthusiasm for projects were perhaps the first traits he inherited from Disney, who insisted that he had a sense of humor, put down those who lacked it, but was never the author of a genuinely funny remark that anyone ever recorded.

Thus Disney apparently saw as early as 1931 that, despite strenuous efforts to avoid it, he was, as he later said, "trapped with the mouse . . . stuck with the character." Mickey "couldn't do certain things—they would be out of character. And Mickey was on a pedestal—I would get letters if he did something wrong. I got worried about relying on a character like the mouse—you wear it out, you run dry." The answer for him was diversification. "We got Pluto and the duck. The duck could blow his top. Then I tied Pluto and Donald together. The stupid things Pluto would do, along with the duck, gave us an outlet for our gags."

The process of developing new characters, however, was slow. As of 1931, Disney had in addition to Mickey and Minnie only the villainous Peg-Leg, a couple of rubber-limb and circle combinations known as Clarabelle Cow and Horace Horsecollar, who were serviceable but not memorable and who were, in any case, the products of a style of animation Disney wanted to move beyond. It is true that a prototypical Pluto had appeared as one of a pair of bloodhounds chasing Mickey in *The Chain Gang* in 1930 and that he had appeared again alone with The Mouse in *The Picnic*. He was known only as "Rover" in that picture, however, and his useful presence did not make itself felt until he acquired his one, true name and the beginnings of his marvelously eager, innocent and therefore troublemaking identity late in 1931. Goofy did not make his first appearance—as an extra in a crowd scene in *Mickey's Revue*—until 1932, and he, too, had another name at first—Dippy Dawg. The Duck did not arrive until 1934, when he had one line of dialogue ("Who—me? Oh, no! I got a bellyache!") in *The Wise Little Hen*. His genesis is, however, interesting. It seems Disney heard Clarence Nash, who has always done Donald's voice, on a local radio show, reciting "Mary Had a Little Lamb" as a girl duck. He discovered that Nash was employed by a local dairy to put on shows for schools, hired him away and then tried to create a character to match his duck voice. It was not until some forgotten genius changed the duck's sex and temperament that Nash found his métier. The other members of the stable (or "gang," as Disney's people preferred to call it)— Chip and Dale, the team of comic chipmunks, Daisy Duck, Donald's inamorata, and his three nephews Huey, Louie and Dewey —all came along several years later than the duck.

Which is to say diversification of characters—the creation of new stars, if you will—required most of a decade and therefore was not the immediate answer that Disney required to his problems. The same may be said of his attempts to diversify stylistically. He clung steadfastly to the Silly Symphonies idea, about half of which were original material, the other half based on traditional folk material (with a heavy reliance on Aesop). But until *The Three Little Pigs* became his biggest short-subject success in 1933–34, the series did not make a large, direct contribution to his financial progress. If anything, it did the reverse.

The temporary solution to the problem of keeping Mickey fresh and amusing was to move him out of the sticks and into cosmopolitan environments and roles. The locales of his adventures throughout the 1930s ranged from the South Seas to the Alps to the deserts of Africa. He was, at various times, a gaucho, teamster, explorer, swimmer, cowboy, fireman, convict, pioneer, taxi driver, castaway, fisherman, cyclist, Arab, football player, inventor, jockey, storekeeper, camper, sailor, Gulliver, boxer, exterminator, skater, polo player, circus performer,

141

plumber, chemist, magician, hunter, detective, clock cleaner, Hawaiian, carpenter, driver, trapper, whaler, tailor and Sorcerer's Apprentice. In short, he was Everyman and the Renaissance Man combined, a mouse who not only behaved like a man but dreamed dreams of mastery like all men. His distinction was that he got the chance to act them out. As Forster said, "Mickey's great moments are moments of heroism, and when he carries Minnie out of the harem as a pot-plant or rescues her as she falls in foam . . . he reaches heights impossible for the *entrepreneur*."

Unfortunately, it was as an entrepreneur that he finally found his more or less permanent identity. By the middle of the 1930s he more and more frequently appeared as the manager and organizer of various events—often variety shows and concerts (where at least he got the chance to conduct)—and one cannot help but suspect that this was a reflection of Disney's strong identification with The Mouse. By then Disney himself was occupied principally as the manager, organizer and coordinator of an organization that had grown from a handful of people to a complex of some 750 employees when *Snow White* was in full production in 1937. Here there was still opportunity for mastery but hardly for the heroism implicit in the life of the pioneer Disney had been. The Mouse was by this time rounder, sleeker (and far more humanoid in design). And like his creator he was also more sober, sensible and suburban in outlook—better organized. His "gang" was carrying the whole comic load.

But in 1931 all this was still in the future, and though Disney undoubtedly perceived something of the direction in which he wanted to go, he certainly did not see it in any detail. All he knew was that he wanted more and he wanted it better, and he was driving himself in the time-honored tradition of the American success ethic. He recalled later that he often took his wife out for dinner, then suggested dropping in to the studio to do a few minutes' work. She would stretch out on a couch in his

As an admittedly self-serving analysis it was not a bad one, and if he had stopped there the piece would not have much claim on our attention. But the young man, "bewildered but wholly unspoiled by his sudden rise to fame," as a contemporary chronicler put it, could not resist mentioning the important people who were known to adore The Mouse. "Mr. Mussolini takes his family to see every Mickey picture. Mr. King George and Mrs. Queen Mary give him a right royal welcome; while Mr. President F. Roosevelt and family have lots of Mickey in them, too. Doug Fairbanks took Mickey with him to savage South Sea Islands and won the natives over to his project. Mickey is one matter upon which the Chinese and Japanese agree."

Alas, however, there were still people who had somehow resisted The Mouse's charm, which grieved Disney. "Mr. A. Hitler, the Nazi old thing, says Mickey's silly. Imagine that! Well, Mickey is going to save Mr. A. Hitler from drowning or something some day. Just wait and see if he doesn't. Then won't Mr. A. Hitler be ashamed!"

So much for geopolitics as viewed from Disney's land. (What Hitler's propagandists had actually said was that The Mouse was "the most miserable ideal ever revealed . . . mice are dirty.") The coy style, the determined reduction of the world view to a level even lower than the very simple one Disney privately tended toward, the downright other-worldly quality of the piece—all were fairly typical of the sort of thing Disney's ghost writers and public relations people put out under his name all his life. Judged even by Hollywood standards, the Disney studio's off-screen pronouncements reached a level of ickiness unparalleled anywhere. No matter how self-serving a great man's speech or a great corporation's press release, one can usually find at least the shadow of the man or the institution behind the words. That was generally not the case with Disney. No studio put out more words of public-relations copy, yet none was less informative than this highly secretive organization. Indeed, the function of all those words went beyond that of a smoke screen—the usual *desideratum* for PR—and became a full-scale diversionary action. The words were designed to portray the organization as an open, happy, sunny institution, presided over first by a bashful boy artist, then (as he aged) an avuncular genius of the masses. Neither image could have been further from the truth about this complex man or his remarkable corporation. But despite the style in which his discussion of "Mr. A. Hitler" was couched, the piece is truthful in at least one respect: it revealed a man almost totally disengaged from the realities of the larger world even as that world was reaching out toward him, fairly begging him to let it bestow its favors on him.

19 *I see no reason why there should not be a theatrical season providing my proposed plan for using pressed figs and dates as money goes into effect fairly soon.*
—Robert Benchley, 1933

A THEATRICAL SEASON was begun in the fall of 1933, and there was no need to resort to Mr. Benchley's plan for a substitute coinage or to any of the more seriously suggested proposals that were offered, mostly during the brief bank holiday of March, to tide the country over until more conventional currencies were available. There was, however, a good deal of barter going on in the U.S.: a newspaper offered to exchange subscriptions for bushels of wheat, a hotel accepted "anything we can use in the coffee shop" as room rent, and two tubes of toothpaste were good for a trolley ride in Salt Lake City. Credit, too, was extended in often ludicrous fashion: at one Philadelphia department store you could charge streetcar tokens to your account, and in New York, although toothpaste wouldn't buy you a ride on the subway, it was possible to buy a three-months' supply of Pebeco with a check for a dollar postdated by three months.

The movies, as previously noted, continued to do well through 1934. With admissions at only twenty-five or thirty cents in the neighborhood houses, they provided, next to the thirty thousand new miniature golf courses, where the fee for nine holes was only fifteen cents, the cheapest possible sort of escape. They also provided other come-ons, among them Bank Night (usually Tuesday), when one was rewarded only if the number that one was assigned on the previous Bank Night was drawn and only if one had bought a new admission and could claim the reward personally. Administered from a central office in Denver, the scheme offered prizes sometimes as high as a thousand dollars and had the effect of encouraging habitual attendance on a traditionally slow night at the Bijous and Orpheums. There were also Dish Nights and Bingo games at some theaters, and then that peculiar mingling of pain and delight—the double feature—was introduced in these days as an additional method of luring the penny-wise, hour-foolish out of their homes (and away from that popular new appliance, the radio). In short, though the movie business could be good, it was necessary for one and all to scramble very hard for their shares of those minuscule admission charges.

None had to work harder for his share than Disney, despite the acclaim he was winning. Luckily, he had found, in various sidelines, the equivalent of Benchley's pressed figs and dates, which if he did not use directly as money, he could easily convert into cash. On several occasions he publicly complained that the short subject was far too often short-changed in the marts of his trade. To be sure, Radio City Music Hall, with which Disney quickly established a mutually admiring relationship based on their parallel pursuits of the family trade, paid as much as $1,500 a week for one of his shorts, but the more usual rental fee in the first-run houses was in the neighborhood of $150 a week. This compared very unfavorably to the $3,000 or so which a feature could command in the same house. And when impartial observers in the trade considered that most of his playing dates were not in the first-run houses but in the neighborhoods and the small towns (where for an old short the rental might be as small as three dollars and where the double-feature movement was beginning to cut out shorts altogether), they could detect a certain justice in Disney's complaints. Nor were these the end of them. He admitted that his rentals were higher than those of his competitors, but his costs were more than proportionately higher than theirs. He even groused that *The Three Little Pigs*, his biggest hit till that time, was not really doing very well. It was renting at black-and-white rates, about one-third those charged for color films then, even though Technicolor prints were much more expensive to make. And because of its sudden hit status, Disney had to pay for more prints to satisfy the demand, prints he felt would not be useful once that demand slackened.

He covered his pain on this subject by talking like an artist. "We will continue to follow our rule to put every cent of profit back into the business, for we believe in the future and what it will earn for us," he told an interviewer in 1934. "I don't much favor commercialization. Most producers think it is better to get while the getting is good. We have not operated that way. . . ." And, in that same year, *Fortune* reported that Disney felt that the only way to make short subjects pay was to make his product so good that audiences would demand them in preference to his cut-price competitors.

He did have a few other assets, though, not the least of which was its merchandising program. He had been in it from the moment Mickey Mouse achieved his initial success. In New York on business in 1930, he was waylaid in his hotel lobby by a nameless fellow who wanted to put The Mouse's likeness on school tablets, and offered Disney $300 for the privilege. Needing money, as usual, Disney signed the contract that was thrust at him and was launched on a sideline that was, at times, more profitable than the picture business.

The same year King Features came forward with an offer for a syndicated cartoon strip, and by 1934, Disney had fifteen people employed in his New York office to oversee licensing of products that were either reproductions of his characters or bore their likeness in some form or other. At this point, some eighty U.S. concerns, including such blue chips as General Foods, RCA Victor, National Dairy and International Silver, were moving some seven million dollars' worth of stuff carrying what amounted to the Disney imprimatur. For this they paid royalties in the range of 2½ to 5 percent—occasionally as much as ten percent—and Disney was grossing $300,000 on the operation and clearing half that amount, which made up almost one-third of his net profit. By the 1960s the percentage contributed to the company's profits by its merchandising and publications program has dropped to something less than 10 percent of its gross income, and it has stayed near that level in the years since. It remains, however, a tidy little profit center.

In the early days, Mickey Mouse naturally was the leading figure in these sidelines and performed at least two certifiable economic miracles in the depression. The Lionel Corporation, manufacturers of toy trains, was rescued from receivership by the sale of 230,000 handcars that featured the figures of Mickey and Minnie pumping away at the handles. The Ingersoll Watch Company was saved from deep financial trouble when it put out Mickey Mouse watches, of which something like ten million have been sold to date. (The originals, the ones on which Mickey's pointing fingers told the time, are now collector's items for trivia specialists, bringing as much as fifty dollars in the market. Despite their proliferation the watches were never meant to live past the recipient's puberty, and so they were thrown away, as they were bought, by the millions.) From the start, merchandising was carefully handled by the Disney

163

organization, which exercised more rigid control than was usual in the field over the ethics and the advertising standards of licensees and guarded against price cutting and corner cutting in manufacture. This careful guardianship of the Disney name occasionally reached ludicrous proportions, as when the rights to the "Who's Afraid of the Big Bad Wolf?" tune were denied the producers of Edward Albee's *Who's Afraid of Virginia Woolf?*, presumably on the ground that the song would be injured if it were exposed to the insalubrious moral climate of George and Martha's house. Instead the words were sung in the play to the tune of "Here We Go 'Round the Mulberry Bush," which is in the public domain. But as a general rule, and as anyone who has raised a child during the last thirty-five years can testify, Disney-licensed products have tended to be somewhat more durable, attractive and well-designed than most of the junk with which they compete in the drug and dime stores. But, of course, they remain junk by any intelligent standard of toy manufacture.

In the beginning, The Mouse and his friends did not confine their activities in aid of commerce only to the wares of childhood. Someone put out a Mickey Mouse radiator cap, and for a brief moment in history it was possible to purchase a Mickey Mouse diamond bracelet for twelve hundred dollars. And before Mary Poppins thought of the idea, Mickey was, in effect, the spoonful of sugar helping the medicine to go down; that is, you could glimpse his cheery countenance on the label as you downed a certain brand of milk of magnesia. Later on, no Disney character was allowed to appear on products with an unpleasant connotation for children, and that ruled out most medicines. Beyond setting general policy on merchandising, Disney paid little attention to this endless flow of goods and confessed, after it had reached the flood stage during the prosperous Fifties (when The Mouse alone had appeared on five thousand different items, which had contributed a quarter of a billion dollars to the gross national product), that he simply couldn't keep up with it all. When his own children, Diane and Sharon, were small, he occasionally presented them with a selection of the items at Christmas time, and he often had it delivered in carload lots to the John Tracy Clinic for deaf children in Los Angeles—a favorite charity, founded by Spencer Tracy, a fellow polo enthusiast, and named after the actor's deaf son, who was for a time employed by Disney in a minor capacity. But essentially the toys, so long as they were safe and well made, were of little concern to Disney; they helped him to enlarge his profit margin painlessly and to publicize the work he most cared about—his movies.

By the middle thirties merchandising was extremely useful in broadening the wedge into the foreign markets that Disney's films had begun to open. The cartoons were, of course, universal in their appeal and very easy to understand. By 1934 Disney had opened small offices in London and Paris to aid in the distribution of merchandise as well as film, but as late as 1937 his films were being dubbed into only three foreign languages, French, Spanish and German. Elsewhere, Disney cartoons had to appear with subtitles, at first no great problem since there was so little dialogue, but the growth of verbal humor in the pictures proved an increasing handicap in the quest for easy universal understanding. It was easier to build an audience with the printed word, and so Disney comic strips, comic books and story books appeared in no less then twenty-seven of the world's tongues. It was as a printed comic and as merchandising figures that The Mouse and his cohorts won many of their first audiences in out-of-the-way corners of the world (thus reversing the order of recognition that had pertained at home), and in the process increased Disney's revenue in a way that no other movie maker could so easily and consistently exploit.

By 1937, indeed, The Mouse had become something of a political figure. Hitler, whose wrath had been kindled because in one of Mickey's films his animal friends appeared wearing the uniforms of German cavalrymen, thus dishonoring the nation's military tradition, had been forced, by popular demand, to allow Michael Maus, as he was known in Germany, back into the country. This triumph, however, was balanced by The Mouse's expulsion from Yugoslavia, where he was suspected of representing a too attractive revolutionary figure (the correspondent who reported this extraordinary chapter in the history of bureaucracy's march was also thrown out by the nervous Slavs). The Mouse was never banned in Russia because he was never much of a threat to the new Socialist order there, trade relations between the U.S. and the U.S.S.R. being almost nonexistent in the Thirties, but at one point the Russians denounced him as a typical example of the meekness of the proletariat under capitalist rule, then shifted the line on him and decided his films were "social satire" depicting "the capitalist world under the masks of pigs and mice," then shifted again to call The Mouse "a war monger." Finally they created a competitor known as Yozh, or porcupine, who had a suitably prickly— and correct—Marxist-Leninist attitude toward world conditions.

If nothing else, the political passions occasionally stirred by The Mouse during this decade indicate the folly of overinterpreting essentially innocent popular culture material in the light of any ideology—political, psychological, religious or even literary. Such work has a way of rebounding against the critic in an awkward, even comic, fashion—a curious process that has the unhealthy byproduct of rendering the purveyor of popular culture immune to serious attempts at understanding the effect of his work. In any case, by the middle of the decade, Disney was reaping benefits he had in no way counted on, thanks to the international favor his creations had won. England's King George V refused to go to the movies unless a Mickey Mouse film was shown, and Queen Mary was known to have appeared late for tea rather than miss the end of a charity showing of *Mickey's Nightmare*. At his screening of movies in the upstairs hallway of the White House, Franklin D. Roosevelt usually tried to show a Disney cartoon; other statesmen who were on record as favoring Disney in general and his Mouse in particular included Canada's Mackenzie King, Jan Christian Smuts of South Africa, the aforementioned Mussolini and the Nizam of Hyderabad. *The Encyclopaedia Britannica* devoted a separate article to The Mouse, Madame Tussaud included a replica of him among the figures of the mighty in her wax museum, and Disney himself had been admitted to the Art Workers' Guild of London—the first movie maker ever to be so honored. All over the world there were Mickey Mouse clubs (not to be confused with the later television version of the promotional gimmick), whose members carried a Mickey Mouse emblem, took a Mickey Mouse oath, sang a Mickey Mouse song and used a Mickey Mouse handshake. There were fifteen hundred of these clubs in the U.S. alone. In Africa it was discovered that some native tribes would not accept gifts of soap unless the bars were stamped with The Mouse's outline, while many other primitives carried Mickey Mouse charms to ward off evil spirits. A traveler in China reported seeing his likeness peering from a window in Manchuli, the transfer point between the Chinese Eastern Railway and the Transsiberian Line. And it was reported that in Japan he was the most popular figure next to the Emperor. As a child, the present Duchess of Alba, a direct descendant of Goya's mistress, had her beloved Mickey Mouse doll included—along with her cat, dog and horse—in her first formal portrait. Thus, Mickey now hangs in her Madrid palace, along with works by Goya, Rembrandt, Titian and other masters in the family collection. In all, Mickey Mouse was available in one form or another in thirty-eight of the world's nations by 1937, and the newspapers took considerable pleasure in printing lists of his foreign aliases. In France he was Michel Souris and, popularly, Mickey Sans Culotte, because of the shortness of his pants and because that was the vernacular term for a revolutionary. In Japan he was Miki Kuchi; in Spain, Miguel Ratoncito; in Italy, Topolino; in Sweden, where one theater ran a program made up entirely of Disney cartoons for seventeen weeks, he was Musse Pigg; in Greece, Mikel Mus; in Brazil, Camondongo Mickey; in Argentina, El Raton Mickey; in Central America, El Raton Miguelito. "I guess," said Walt Disney, "the cartoon is something everyone knows and likes."

That simple sentence speaks volumes. There was no character in the movies more American than Mickey Mouse, no environment more American than the small-town locale he inhabited with his gang. And yet the appeal of the Disney cartoons was universal. Their international success must be attributed to the surmise that the message of their medium—the animated cartoon—was so easily understood by everyone. Dialogue was minimal; action, drawn in high definition and never needing any verbal explanation, was as easily understood by an Italian peasant as it was by a New York intellectual—more easily perhaps, given the propensity of the latter to look for complications where none exist. To some degree the international popularity of American action films of all sorts—a popularity which has persisted throughout the postwar dolors that have afflicted those films at home—is based in large measure on simple abhorrence of ambiguity, on preference for straightforward action rather than needless, possibly misleading talk. The bold colors of the Disney cartoons, their use of animals, which have no nationality, as the major characters, their broadness and brevity, the commonality of the problems they encountered and the solutions they invented, and above all, their cheerfulness—all contributed to the success of the films in the international market. What is true of high art is also true of mass art to the extent that universality of appeal is based on success with particularities. In their uncomplicated way, the Disney craftsmen caught something of the truth of the American scene and situation at the time and, lo, found that they had touched a chord that set up responsive vibrations all over the world.

By 1935 The Mouse had become, in the words of a League of Nations scroll that Disney traveled to Paris to accept (along with a gold medal), "an international symbol of good will." Disney was accompanied by Roy on the European trip, which produced another of his "aw-shucks" anecdotes. The Disney brothers were told that the award would be presented at a formal garden party at which the accepted dress would be morning coats. The Disneys protested but to no seeming avail and so stopped off in London to have the required formal wear tailored for them. They showed up at the party in their new finery only to discover that their hosts had embraced American ways to the extent of abandoning their cutaways and striped pants in favor of business suits.

But the trip produced more than an anecdote and some free publicity. Abroad, Disney noted that his cartoons often received billing above the features with which they played, and in Paris he and his brother attended a showing of a half-dozen of his cartoons playing alone, without support of a feature of any kind. This confirmed his growing belief that the surest way out of the mouse race was to get into feature production himself.

The story behind the show on which the two Disneys dropped in actually begins almost a year earlier in Stockholm, where a young construction engineer named Willard G. Triest was trying his hand at film promotion after several years of working on extensions of the New York subway system. Disney had been unhappy about his Scandinavian grosses and pressured United Artists for some special action there. Local custom decreed but two shows a day and these, of course, were dominated by whatever feature the theater was showing. Disney shorts thus received only a small number of screenings and only a tiny share of the gross receipts. Triest decided to put together a 55-minute program of Disney cartoons, in both color and black and white, and to play them on a continuous-run basis from one in the afternoon until 11 at night. The innovation was so radical that only a second-class theater could be engaged to try it, and Triest tried some American-style promotion to lure the people. Music shops were talked into playing

"Who's Afraid of the Big Bad Wolf?" on loudspeakers directed at passers-by, little Mickey Mouse and Pig cutouts—a million and a half of them—were distributed to school children. Opening night was a Kliegslighted, red-carpeted gala, with the diplomatic corps in attendance, white tie obligatory and a champagne reception afterward. The result was a 14-week run, a trebling of the theater's house record, and even, one night, attendance by a goodly share of the Swedish royal family. Triest even helped locate two children who had been missing from the Japanese embassy for some eight hours—he discovered them, at nine in the evening, sitting through his program for the seventh time. He was called upon to repeat his success in Paris, and it was his *L'heure Joyeuse de Mickey avec Les Trois Petits Cochons* that captured Disney's eye and imagination late in its run.

He had been toying with the idea for a feature for some time. There had been rumors in Hollywood of an attempt by Disney to make a version of *Alice in Wonderland* combining live action and animation—and starring Mary Pickford. From 1932 onward Disney had increased his staff to a level higher than he strictly needed to maintain his regular schedule, and he was also putting quite a bit of his earnings into an effort to educate his people in the fine arts—again something that was not absolutely required by the sort of film he was producing. Obviously, something was stirring in his brain, and by 1934 he was quietly discussing a proposed feature-length production of *Snow White and the Seven Dwarfs* with visiting reporters. He thought it might cost as much as $250,000 and take eighteen months to make. "We will use a full symphony orchestra and fine singers," he promised and added: "We've got to be sure of it before we start, because if it isn't good we will destroy it. If it is good, we shall make at least a million."

In every respect, Disney's estimates were far too modest.

PART SEVEN

Disney's Folly

---

20 *Of all the things I've done, the most vital is co-ordinating the talents of those who work for us and pointing them at a certain goal.*

—Walt Disney, 1954

FROM THE TIME—roughly—of *Snow White*'s conception to the time that it was finished, a period of approximately five years, the staff at the Disney studio grew from about 150 workers to some 750. And most of the growth was from the bottom. Unlike most growing industrial enterprises, the largest number of new employees were not in executive or administrative work but were, instead, animators, assistant animators, breakdown men and in-betweeners (the last being the traditional starting point for a young man learning the art, and whose work, logically enough, consisted of doing the drawings that come "in between" the drawings of the main action, executed by the men above him on the ladder). There is no doubt that the animators were the glamorous figures of the infant industry, the men who drew the best salaries ($150 to $200 a week) and among whom were a few who came to work wearing polo clothes, ready to ride with the boss at the end of the day. But there were plenty of other new faces around the Hyperion

171

Avenue studio, on which Disney spent $250,000 before he moved to his new studio in 1940. There were layout and background men, art directors and story men, directors and musicians in profusion. And always there were apprentices— about 20 of them at any given moment—studying in a wide variety of art classes right in the shop, doing routine work on the films and getting paid as little as $18 a week for their efforts.

It was a young shop—Disney himself was only thirty-six when *Snow White* was finished—and full of exuberance. The employees were confident they were working in the best studio of its kind in the world, and full of belief they were in on the ground floor of an art that was going to be one of the important forms of expression in this century. Older men by the dozens who cared about animation deserted New York, the former capital of the art, to lay siege to Disney's door. They resorted to every sort of trick to gain admittance and usually didn't get it until they indicated a willingness to accept a cut in salary and status and perhaps even to join the apprentices in their art classes. Younger men, just out of art school, responded by the hundred to the ads the studio placed in newspapers around the world, seeking recruits. Don Graham, an instructor at the Chouinard Art School in Los Angeles (and latterly its dean), was the man in charge of the apprentice program and also the man who brought Disney into as much contact as he could stand with the finer realms of graphic expression. Chouinard once received, according to an old Disney hand, thirteen hundred pieces of mail—containing sample art—in response to a series of recruiting ads placed in the foreign press: twenty of the senders were invited to join the program and two finally attained the status of animator. The attrition rate was similarly high among domestic recruits, whom Graham found in the course of cross-country recruiting drives.

The studio to which these men reported looked like "a jerry-built bee-hive," as one of them later put it. To accommodate his burgeoning staff Disney was constantly forced to remodel, cramming more and more people into tiny offices, acquiring space in adjoining buildings and then burrowing through the walls in order to connect the outlying rooms with the main building. Even so, the place had amenities that other animation studios did not offer. There were no time clocks because, Disney said, he had hated them when he was a young man. All the equipment the men used was of the highest quality; there were plenty of fifteen-hundred-dollar movieolas (one Disney graduate was shocked when he went to work for Max Fleischer on his Gulliver feature and found there was only one such machine— absolutely vital to editing film—for the entire production), and every artist had plenty of light, a commodity in short supply at some competing studios. Nor was there any minimum footage requirement for the animators. They were, instead, encouraged to throw out a whole day's work if they did not like it. The desire for quality was further stressed by bonuses ranging from four dollars to twelve dollars per foot for especially good bits of animation.

Finally, there was at first very little resentment of the employer, despite the long hours and low pay he enforced. In later years Disney's informality may have seemed a trifle studied, but in those days it appeared to his multitude of bright young men as completely genuine. The Disney brothers, to be sure, had paneled officers fronting on Hyperion Avenue, but they were modest in scale and were reachable through a tiny reception area guarded by just one receptionist-switchboard operator. In any case, Walt Disney spent only a minimum amount of time behind the drawing table he used as a desk. He insisted on being first-named by everyone, and he constantly dropped in on all and sundry to see how they were coming along. His story conferences were models of democratic give-and-take, and everyone who ever sat in on one seems to agree that it was as an editor and critic of stories that he had his finest creative hours. He had a fine sense of pacing, a gift for stretching and embroidering a basic gag or situation that some have compared to that of the great silent comedians, and above all, in infectious enthusiasm for ideas, even bad ones, that kept the ideas bouncing until, somehow, the plot or situation or character was sharpened to a satisfactory but not necessarily preordained point.

For an interviewer of the time, he summed up what he was looking for in these sessions. "We have but one thought, and that is for good entertainment. We like to have a point to our stories, not an obvious moral but a worthwhile theme. Our most important aim is to develop definite personalities in our cartoon characters. We don't want them to be just shadows, for merely as moving figures they would provoke no emotional response from the public. Nor do we want them to parallel or assume the aspects of human beings or human actions. We invest them with life by endowing them with human weaknesses which we exaggerate in a humorous way. Rather than a caricature of individuals, our work is a caricature of life."

Disney was perfectly capable of overseeing the story elements of his operation himself. He had an unfortunate predilection for bathroom jokes, which were usually edited out, and for jokes involving either an assault upon or the adoration of the posterior. (The Rube Goldbergish paddling machine by which the wolf is punished in *The Three Little Pigs* is a classic example of Disney's *derrière*-assault propensity, and the little boy unable to button the drop seat of his pajamas in one of Disney's early Christmas specials is a representative instance of the *derrière*-adoration school of animation. The lad's trouble in maintaining his modesty is the latter film's most heavily emphasized running gag, and the climax of the picture is the present that Santa Claus leaves for him—a tiny chamber pot. Even when he was not being as explicit as this, Disney's interest in the posterior was a constant in his films. Rarely were we spared views of sweet little animal backsides twitching provocatively as their owners bent to some task; often were the buttocks of some child or sprite, like Tinker Bell, emphasized in the character's basic design.) But within his own terms, Disney performed his editorial tasks with undiminished enthusiasm until the end of his life. He was not a terribly articulate man and often resorted to pantomime, mimickry and the sketching of vague shapes with his hands, which were often described by colleagues as "restless" or "constantly in motion." Subtle changes in the rhythm of a tapping pencil were studied by aides as an indication of shifts in his mood. In later years certain of his associates seemed to do little more than read and interpret statements of opinion and desire that were often Delphic in their ambiguity. ("I want black on black," he told the people animating the fight between the stags in *Bambi*—which is, of course, impossible. But the artists were able to give a black-on-black impression by keeping colors dark and illuminating the scene with lightning flashes.)

At all times Disney remained remarkably unwilling to release scripts or storyboards to the production people. He liked to huddle over them as long as possible, hoping they might be improved in some way that was not entirely clear, even to him. In the days when Technicolor facilities were limited and one had to sign up months in advance for their use, these shooting dates served as arbitrary deadlines, forcing him to give projects up to the fates even though they did not fully satisfy the ideal version that existed only in his mind and that he could never express. Later, the demands of his carefully worked-out releasing schedule had somewhat the same effect on him, though here he could sometimes juggle things around to gain more time. Somehow, however, he made what can only be termed the literary elements of his films work—and he did it largely through his own efforts. One animator later recalled a session at which Disney acted out the entire story of *Snow White*, playing all the parts in a performance that required several hours to complete and was to serve as the basic guide to the studio. It was referred to throughout the film's production whenever anyone had a creative question.

Strangely, for a man who was generally thought of by the

public as an artist, it was with the graphic elements that he depended on others to do the hardest labor, both practical and theoretical. He, of course, sat in on everything, most particularly on the "sweat box" sessions (so-called because the projection room at the Hyperion Avenue studio was tiny and lacked air conditioning), where all the animators gather to criticize one another's rushes, often no more than pencil sketches of what was to come. There he was very much like the amateur gallery-goer who knows little about art but knows what he likes. What Disney wanted was more and more imitative realism in the movements of his characters, more and more detail (and lushness) in the backgrounds, greater and greater fidelity to nature in special effects ranging from lightning to the fall of raindrops. It is worthwhile remembering that Disney's first experience in show business was doing "impressions" of Charlie Chaplin at amateur nights. One animator has said that Disney's taste in art was for work at approximately the level of a competently illustrated nineteenth-century children's storybook—perhaps too narrow a judgment. Disney achieved something of this quality in *Snow White* and *Pinocchio*, and it had a distinct charm on the screen. The trouble was that he kept trying to improve upon it without ever breaking cleanly away from it. Some peculiar mishmashes resulted from this development. For instance, *Bambi*'s forest had some of the green-and-gold lushness of the antique style, but the deer themselves were rather too carefully naturalistic in appearance while many of their smaller woodland friends turned out to be conventionalized, commercialized images of cuteness—fluffy and cuddly and saucy. Indeed, it is fair to note that with the smaller creatures Disney and his animators never had much success. In the early days of animation it was believed that circular forms were the only ones that could be successfully animated. The Disney studio did much to disprove this theory but somehow clung to it when drawing the smaller animals—chipmunks, squirrels, rabbits and so on. If anything, they grew rounder and softer and therefore cuter as the years wore on. At first glance they were natural representations, but over the course of a long film it became clear that the true intention of the artists was easy adorability. Their use in the typical Disney film was always to reduce dramatic tension at any point where it excessively threatened audience sensibility. In *Snow White*, for instance, the gathering of the little creatures around the exhausted figure of the girl, after her frightened dash through the woods, assuages not only her alarm but that of the audience as well. Later on, the animals join the dwarfs in their attempt to rescue Snow White from her wicked stepmother, and their gamboling presence on the mission serves to reassure, even to provide a few small laughs, despite the desperate urgency of the situation.

To some degree the style of the earlier animators, who combined rubbery stick limbs with the round heads and bodies of their animal creations, served to check this tendency toward the cuddlesome; those creatures were much more clearly—if crudely—cartoon conventions, bearing only superficial resemblance to real animals. You knew where you stood with them; they did not encourage you to want to gather them to your bosom or to think of them in human terms. In fact, it would have been impossible to do so: they were too spiky, too hard in outline, too weird. The animals of the Disney features films, as they grew more real, paradoxically grew more subtly false —Thumper, the rabbit, and Flower, the skunk, for instance, in *Bambi*.

This tendency to adorableness was the most consistent problem of Disney's later work, and one suspects that the style in which the animals were drawn dictated the dialogue they spoke and the situations in which they found themselves. Sweetness called forth sweetness. There were, however, other problems, too. For a while, in the late Thirties and early Forties, he and his artists attempted to push on to styles new to them. There was, for instance, the massively "modernistic" manner of parts of *Fantasia* and the wartime *Victory Through Air Power*, with which they never achieved more than mechanical competence.

177

Dumbo, in the same period, was essentially an attempt to use the bright, uncomplicated style of the Mickey Mouse cartoons in an almost feature-length production and was quite successful in its unpretentious way, though apparently not exciting enough creatively for Disney and his people to stay with exclusively in their future films. Sequences were done in this manner in *Peter Pan* and *Alice in Wonderland*, stories to which it was not nearly so well suited as it had been in *Dumbo*, but the basic drive was toward an extensive elaboration of the old storybook style. In films like *Cinderella* and *The Sleeping Beauty*, detail was piled on detail, technical effect on technical effect, until the story was virtually buried under their weight. It was an art of limited—some would say nonexistent—sensibility, a style that labored to re-create the trifles of realistic movement, that fussed over decorative elements, that refused to consider the possibilities inherent in the dictum that less is more. The wonderful simplicity that Disney's graphic art naturally possessed in the beginning and that he may have distrusted as betraying its humble and primitive origins, disappeared. In the late films complexity of draftsmanship was used to demonstrate virtuosity and often became an end in itself, a way of demonstrating what was a kind of growth in technical resourcefulness but not, unfortunately, in artfulness.

But however one disagrees with his goals, Disney's ability to analyze what he wanted—however unclearly he sometimes expressed his analysis—and his ability to find the people who could give them to him must be admired. He was probably quite right in believing that an audience's attention could not be held over the span of a long film done in the crude—if forceful —style of the early Mickey films and that he would have to add something to the art itself if he was to grow beyond the short subject.

The first thing he understood—and it was a basic insight— was that he would have to train his men himself, as none of his competitors was willing or able to do. "I decided," he said, "to step out of their class by setting up my own training school. We had enough revenue coming in so we could plan ahead, and I laid out a schedule of what I want to accomplish over the years." "It was costly," he added on another occasion, "but I had to have the men ready for things we would eventually do." Some have placed the cost of his art classes at $100,000 a year.

At first he sent his artists to Chouinard itself for night classes, occasionally, in the early days, driving them to the school himself in his car and then returning to pick them up afterward. Then, with the advent of prosperity and the discovery of Graham, the classes were held within the studio. They learned, in some cases relearned, the basics of art—the essentials of capturing the illusion of three-dimensionality in a two-dimensional medium, those tricks of exaggeration and perspective that are necessary to capture the illusion of reality whenever one seeks to re-create it in an artistic medium.

Disney had, in addition, a special problem, one that was never faced by any other serious graphic artist in quite the way he had to face it. That was the problem of movement. The techniques of suggesting it in the static art forms had, of course, been worked on for centuries; the problem of simply capturing it on movie film had been refined in a matter of decades. But motion in animation was a new challenge. And it was not a simple one at that. Photographic realism, oddly enough, looked unreal in animated films. It was a subtle matter to obtain a realistic effect that lapsed into caricature only intentionally. "Look, pull your hand across your face and you'll see what I mean," Disney told one interviewer. "You don't see a single hand; it's sort of stretched and blurred. We had to learn the way a graceful girl walks, how her dress moves, what happens when a mouse stops or starts running."

To begin with, Disney's animators studied motion by attempting to have models break down a simple movement into a series of poses representing each of its components. But this resulted in a style of animation that lacked a smooth flow. Graham's answer to that was what he called "action analysis," and the basic technique of teaching it—to apprentices and senior staff members alike—was to have a model execute a full motion and then have the animators attempt to sketch it from memory, giving the impression of the movement rather than an overly detailed rendering of it. The same technique was applied to the problem of drawing the movement of the mouth as it speaks. The trick here was to avoid overanalysis of its workings, which led to a constant twittering of lips as the artists attempted to draw the shapes they assumed in pronouncing each word. Instead, they were encouraged to draw an impression of a whole phrase. Finally, there was work to be done on the whole repertory of animal movement, which is the basic vocabulary of animation. For a while Disney kept a small zoo in the studio so the artists could draw from nature. Then it was noted that a zoo is not exactly a natural habitat and that the captive animals responded differently to stimuli then did their wild brethren. So he employed nature photographers to bring back footage of the animals in their natural surroundings. This was, incidentally, the genesis of his highly regarded series of nature films, which he began releasing in 1949.

Disney was especially eloquent on reasons why animals dominated the animation field and why animators, at their very best, were never very successful with people. It was because, he said, the reactions of animals to stimuli were always expressed physically. "Often the entire body comes into play. Take a joyful dog. His tail wags, his torso wiggles, his ears flop. He may greet you by jumping on your lap or making a circuit of the room, not missing a chair or a divan. He keeps barking, and that's a form of physical expression, too; it stretches his big mouth.

"But how does a human being react to stimulus? He's lost the sense of play he once had and he inhibits physical expression. He is the victim of a civilization whose ideal is the unbotherable, poker-faced man and the attractive, unruffled woman. Even the gestures get to be calculated. They call it poise. The

180

spontaneity of animals—you find it in small children, but it's gradually trained out of them.

"Then there's the matter of plastic masses, as our animators put it—mass of face, of torso, and so on. Animation needs these masses. They're things that can be exaggerated a little and whirled about in such a way as to contribute the illusion of movement, you see, like a bloodhound's droopy ears and floppy gums or the puffy little cheeks and fat little torsos of chipmunks and squirrels. Look at Donald Duck. He's got a big mouth, big belligerent eyes, a twistable neck and a substantial backside that's highly flexible. The duck comes near being the animator's ideal subject. He's got plasticity plus.

"For contrast, think of the human being as the animator sees him. It takes the devotion of a whole boyhood to learn to wiggle an ear as much as three-sixteenths of an inch, which isn't much. The typical man of today has a slim face, torso and legs. No scope for animation, too stiff, too limited. Middle-agers tend to develop body masses—jowls, bay windows, double chins. But you can't very well caricature a fat man. Nature has beaten the animator to the punch."

This uncharacteristically long speech, delivered in the 1950s, represents a theoretical summary of some two decades of hard, practical work at the Disney studio to learn the techniques of an art and then to refine them into a style that existed at first only in the mind of Walt Disney. The unfortunate thing was that when his people finally achieved what he wanted, it was not, by any objective standard, as excellent as he thought it was, though he continued to cherish it. What was missed by him— though not by some of his associates—were a good many promising stylistic possibilities that developed out of the natural limitations of their previously unexplored art. These, unfortunately, were consumed by the drive for that realistic surface, that mechanical slickness that Disney desired and that evidently pleased his audiences. Even so, the foundation was laid in those days for an art the possibilities of which have still to be fully explored; both a technology and techniques applicable to almost any problem one can imagine arising in animation were formulated.

In the 1930s Disney was willing to investigate almost any possibility that promised to bring him closer to his goal. His mind was still open and educable. He hired Rico LeBrun, the well-known muralist and an expert on animal anatomy, to instruct in his school. He brought in guest lecturers of all sorts to speak on all manner of subjects. There was, for example, an academic with a heavy German accent, hired to deliver a talk on the theory of humor. He began his lecture with the declaration, "Ve vill now explain vot iss a gak"—and quite unintentionally brought down the house of assembled gagmen, rendering them incapable of attending the finer points of his analysis.

Alexander Woollcott fared somewhat better before this tough audience, and so did Frank Lloyd Wright, though the latter somewhat nonplused Disney. The architect had in his possession a Russian animated cartoon that Disney was anxious to see, and since this was late in the decade, it is possible that Wright was interested in getting the commission to design the new studio Disney wanted to build. In any case, a meeting was arranged, and Wright was full of enthusiasm for the work he observed going forward as he toured the studio, though Disney was reportedly puzzled by the architect's suggestion that he ought to distribute his animators' black and white roughs that were projected for his edification and not bother with polishing them to the customary high gloss—an excellent idea, by the way, since the roughs are full of a vitality and an individuality that the overrefined finished products lack once they have passed through the factory.

Then Wright showed the Russian film, announcing, by the way, that its score was by Shostakovich, which caused Disney to inquire, "Who the hell is he?" When it was over, the studio's artists were abuzz with excitement. As one of them later recalled it, the shapes in the film were deliberately flattened, forcing them out of normal perspective and imparting an expressionistic quality to the work. Wright thereupon stood up and cried, in his most imperious manner, "Walt Disney, you too can be a prophet!" To which the honest rustic replied, in genuine perplexity, "Jesus Christ, you want me to make pictures like that?"

Stories of Disney's naïveté in matters of high and traditional art are legion. His daughter reports his leafing through a book of paintings, stopping to study a page and remarking, "I like this guy. Who is he?" Informed that the guy was Goya, Disney registered recognition and added comfortably: "Good man, Goya." One of his animators kept in his office a Cezanne reproduction, which Disney spotted one day. He asked the name of the artist and commented that "the fellow can't draw—that vase is all crooked." Some time later, however, the same animator was having trouble getting a scene just the way Disney wanted it. Finally, in desperation, Disney told him to "get some of that 'Seezannie' quality" into the work, which, it developed, was exactly what was required to create the desired mood. The mispronunciation, the alertness to the useful possibilities in artistic traditions alien to him and the ability to store intellectual and aesthetic oddments he might someday find useful were all characteristic of Disney's role as coordinator and ultimate arbiter of studio activity. He was trying—and he was learning. The trouble was that he borrowed from the great traditions of art only what was immediately useful to him, the superficials of manner and style. Nothing that he saw in it broadened or deepened his sensibility.

This limitation in him did not seem to affect the spirit of the young men who learned animation in his studio in the short golden period that was probably over by 1940, when Disney opened his new plant in Burbank, and was certainly finished by the strike of 1941. In the memories of those who worked there in the great days the experience is regarded as something far more than a merely educational one; it created a sense of community among them that is felt only by those who have shared the excitement of exploration and experiment, those who have known what it is like to be young in a young field. It sets these men apart, "as if we were all members of the same class at West Point," one of them recently said.

There was, as a result, a passionate commitment to the new art on the part of many of the employees that far exceeded the kind of loyalty most companies can command. Particularly when *Snow White* began to overtax the studio's facilities, Disney was able to obtain overtime in huge amounts with no more than a vague promise of bonuses, in unnamed amounts, if the film succeeded. The impression is that he could have got the extra time even without dangling this dubious carrot before his workers. The desire to do the work well was—at least to hear them tell it—goad enough. Visitors to the studio were always amused by the sight of animators endlessly trying out facial expressions and movements before their mirrors and then trying to get down on paper what they had just done. One tourist recalled recently: "You'd see a guy sitting quietly at his animation board [different from an ordinary drawing board in that it could be illuminated from below and in that the glass circle on which the paper was placed could be revolved a full 360° so the artist could gain easier purchase on difficult lines] and when your guide asked him to do the thing he had just invented for Donald or The Goof he'd suddenly jump up and start bouncing around the room, his face working like a maniac and emitting these weird sounds. When he was finished he'd sit down quietly and pick up his pencil as if nothing had happened." They were, as the cliché went, "actors with a pencil," and they would risk any loss of dignity to study their lines, as it were. One of them was once arrested by the police for lying down on a suburban sidewalk in the midst of a rainstorm; he claimed he was studying the lightning and he escaped punishment.

For many this period ended in the bitterness of layoffs and strikes; for others, even—perhaps especially—those who stayed on with Disney, there is a sense of alienation that has grown as the corporation has grown and animation has passed from the center of its life to the periphery, where its expense has been tolerated (at least while Walt Disney lived) out of sentiment and a desire to keep the organization alive in the field where it began and with which it is still closely associated in the public mind. For a very few there is an anger—about the way it all ended, the way animation as an art has slipped once again to the edge of everyone's consciousness, the way they feel Disney treated them—that the passage of time has only slightly dulled. They speak in the passionate tones of unrequited love. But one senses, talking to all of them, that however the experience ended, nothing can really touch or spoil the intrinsic quality of those days. Mack Sennett, whose work, incidentally, Disney unwittingly helped drive from the screen, once said of his apprenticeship with D. W. Griffith: "He was my day school, my adult education program, my university." Many of his animators had the same feeling for Disney. Just before he died, one of these men wrote to Disney, for whom he had not worked and whom he had not seen in some thirty years. He said, in effect, that he had lived an extraordinarily happy life as an animator and that he felt he owed both his happiness and his success in the field to the training he had received at the Disney Studio and that he simply wanted Disney to know what he had done for him.

Which is not to say that all was seriousness or that emotional and intellectual intensity was the prevalent personal style at the Hyperion Avenue studio. It was hard work, especially for the younger men, who had their regular duties to attend in the office plus a full educational schedule (more than one of them remembers that "you had to fight to hang on in the classes" even as you were "fighting off the resentment of the older men who had never had any formal training and thought it was pretty effete"). But the place had, in compensation, "something like the mood of a college fraternity house." Disney himself had a reputation as a practical joker—though no one seems to remem-

ber any he played—and the pail poised to fall from the top of a
half-open door when it was swung open appears to have been a
basic tool of interpersonal relations around Hyperion Avenue,
while the art of throwing push pins at targets attached to the
cork boards with which the place abounded attained one of its
highest flowerings there: several of the artists actually learned
to fling four push pins at once and make them stick. Given the
small size of the pins and the awkwardness of their balance,
one's admiration of their dexterity is boundless.

As for more elaborate japes, there were too many of them to
relate here. But a couple may give the flavor of the institution.
There was, for example, the sad case of the artist who had an
obsession with clean water and insisted on having a water
cooler of his own, which no one else was allowed to drink from.
One day he came to work to discover a school of goldfish using
it as a swimming pool. Then there was the naïve and nervous
gagman, not long out of high school and so tense at having to
present his first storyboard at a conference with Disney that he
fell victim to what is now known as "distress of the lower tract."
His colleagues solemnly told him that the only sure cure for his
condition was to drink a can of sauerkraut juice immediately
before entering the conference. It had a predictable effect on
him, and he had to break off in midsentence to dash for the
nearest bathroom—which had been locked by the conspirators.
He has his revenge, however. He acquired a large washtub,
wedged it into the back seat of the car belonging to his principal
tormentor, ran a hose in through the window, filled the tub to
the brim with water and sat back to enjoy the man's efforts to
extricate the tub without damaging the car (an enormous si-
phoning operation was finally undertaken).

Obviously, controlling an organization abounding in such
youthful high spirits was no easy task, and through the years
memos designed to establish clear lines of authority and efficient production methods fluttered down from the front office in
an unending snowstorm. In his early study of Disney, Robert

D. Feild quotes two such documents, which give some sense of the seriousness of Disney's effort to get things organized:

> In order that the Supervising Animators may be relieved of as much management responsibility as possible, a secretary-manager of this function is maintained who works in close collaboration with the Production Operations Manager, who (under the direction of the Supervising Animators on the productions with which they are concerned) has the responsibility for maintaining a supply of work for all animators, which involves the making of the necessary contacts to see that men are properly and promptly cast, and the maintenance of statistics relative to productivity, the issuance of deadlines, the operation of adjusted compensation systems, and the formation of such routines, procedures, etc., as are necessary for the uniform movement of production.

All clear? If not, consider this definition of the "Advisory Function to Production" of the Director of Technical Research:

> He is authorized to hold direct contact with any Production Department Head in whose department he sees the opportunity for instigating technical progress. In order that a single control of each department may be maintained, he works through the department head involved, advising the department head of ways and means of achieving desired technical results. The department head is responsible for making an analysis of the suggestions offered and for submitting same to the Production Supervisor. In the event that a routine necessary for the accomplishment of a technical advance affects more than one department, the establishment of that routine is to be discussed at the earliest Production Board meeting.

The trouble with all this was that it was a studio that constantly discussed problems around the water cooler or in the courtyard or over lunch, which meant that everyone was always stepping out of the narrow channels Disney was trying to dig for them. He was himself one of the great offenders in this matter, being unable to resist peering over people's shoulders or just dropping in to see how things were going in this office or that. Thus one artist, unhappy at his assignment as a storyboard sketcher, looked up one day to see Disney in his door and, replying to the inevitable question about how things were going, replied that they were just terrible and that he hated his work. Asked what he thought could do, he replied, simply, "Pluto." He was given a one-month trial with the team that had the dog in hand (or on a leash), succeeded, and thereafter was one of the studio's three Pluto experts.

Gradually, although he was trying to stay on top of an organization that could not help outrunning its organization charts, Disney, with the indubitable but immeasurable help of Roy Disney, began to bring everything under control. The rapid influx of new employees, though they presented a housing problem, actually helped in this process. With so many people to oversee there was an obvious rationale for bureaucratizing the chain of command without hurting the feelings of the old hands, who were used to a more informal method of doing business. The art instruction, far from being the boondoggle many thought it was, contributed directly and soon to the efficiency of the new order. Besides providing cheap labor for the low-level jobs, its students achieved a technical mastery that eliminated a great deal of waste effort. One of its graduates said recently, "Before I started there I animated well and badly both. But I couldn't reproduce the former at will and I couldn't fix up the latter. What the classes did for me was to give me consistency." His sentiments are often echoed by his fellow students. Very simply, Graham's efforts produced, more quickly than anyone predicted, a studio style of steadily high technical quality, and those who could not meet the new standards were either weeded out—though Disney always hated firing people—or, if they were lucky, a small niche in one of the new specialty occupations was found for them.

Specialization was the new watchword on Hyperion Avenue. Without, apparently, giving a moment's conscious thought to the matter, Disney in the late Thirties was recapitulating the history of the entire industrial revolution behind his Spanish stucco walls, breaking down film production into smaller and smaller components, just as all other forms had been broken down since Eli Whitney began experimenting with the production line. Some of this was inevitable; obviously a sound effects man could not perform the tasks of a musician or the actors who did voices. Neither was it entirely reasonable to suppose that an animator could necessarily swing around and do backgrounds or special effects or camera work. It would have been equally wasteful to put high-priced creative artists to work on such routine chores as inking and painting the cels, which are the finished products that the animation camera photographs, and which, in any case, were best completed by feminine hands. It is also clear that the natural gifts of the artists within all departments would vary, making one man more successful with Donald Duck than with Mickey Mouse (known to the artist who usually drew him as "The Varmint"), another better with chipmunks than with turtles.

On the other hand, something distinctly valuable is lost in overspecialization—namely, the artist's personality. Ruskin, writing as the flood of industrialism reached its first great crest, said that when evidence of the single, shaping hand disappeared from an object intended as art, its aesthetic value also vanished. It is a tenet of criticism now so deeply ingrained that most people are scarcely aware of its comparatively recent origins or that Disney's production method had ample precedent in the ateliers of the Renaissance masters and in those remarkable products of community artistic endeavor, the cathedrals of Medieval Europe. If he had been interested in such abstract matters, Disney could have said that The Mouse Factory (as people took to calling it) was compromised from the beginning, as all film studios are, because the art begins in machinery (the camera) and because it depends on instant public acceptance for survival. Looked at in this light, his efforts become only the next logical extension of the trends within his art *cum* business and, indeed, of the trend toward mass production that has been the central fact of existence in Western culture since the end of the eighteenth century.

But of course Disney was both too preoccupied by practical matters and too uninterested in theory to join any such high-flown discussion. David Low, the great English political cartoonist, in effect spoke for him when he said, in a much quoted statement: "I do not know whether he draws a line himself. I hear that at his studio he employs hundreds of artists to do the work. But I assume that his is the direction, the constant aiming after improvement in the new expression, the tackling of its problems in an ascending scale and seemingly with aspirations over and above mere commercial success. It is the direction of a real artist. It makes Disney, not as a draftsman but as an artist who uses his brains, the most significant figure in graphic art since Leonardo."

All of which has a certain validity. But it is not the whole truth, for as the decade wore on and Disney's artists and artisans began to see the shape of things to come, resentments grew. There were whimsical variations in the pay of people who were performing essentially similar tasks, and these appeared to be the result of favoritism on the part of the modern Leonardo. Then there was the matter of credit; excepting only Ub Iwerks' brief appearance on the credit card, no name but Disney's ever appeared on his films, a rule that perhaps was good showmanship but not one calculated to win the hearts of the employees. With *Snow White* he finally yielded on this matter. But on his first feature he carefully gave so many credits—and his name, in contrast, appeared in such very large type—that the effect was the same. Lost in the crowd, the individual craftsman merely suffered a new form of anonymity. Finally, and perhaps most dangerous of all for the future, the system had the effect of crushing individual initiative. The best artists chafed under the restraints of their specialities and, undoubtedly, began to withhold something of themselves from the group creative process (some took to moonlighting for other producers or working on children's books of their own). Others grew more and more vocal in their resentments of Disney. The seeds of what now seems an inevitable revolt were planted and growing.

Disney appears to have been unaware of the subtle change in the atmosphere. He continued to see that his men had the best available equipment and appointments and occasionally dreamed of a new studio that would be more than just a place to work. He seemed to think of it as a community of artists, a variation on the old American dream of a Utopia, where work and leisure—perhaps even family life—would be totally integrated to the benefit of all. He seemed not to notice that some of his men were no longer offering him ideas and executions of ideas representing their best judgment on the correct solutions to artistic problems. They were instead beginning that familiar, dispiriting process endemic to all mass media—attempting to read the boss's mind, anticipate what he wanted and give it to him without his knowing it. The little stratagems for sneaking things past him were also beginning to be devised. For example, if you wanted to distract Disney from some element on the storyboard that seemed weak you did the sequences around it in bright colors, hoping his eye would miss the trouble spot. There was a variant on this technique: to create a problem, indicated by a blank sheet in the board, which at Disney's insistence (he could be childishly stubborn on the point) always had its lower right hand corner turned back and pinned up. He would then focus on filling in this blank and—it was hoped—skip lightly over the areas about which the artists were insecure.

Such foolery was dangerous, for Disney had a fantastic

memory for the details of a story, and woe indeed to anyone who tried to skip over some story point or bit of characterization Disney remembered having agreed to. One screenwriter, sitting at rushes of a day's shooting, recalls Disney's noting that a small piece of business prescribed in his script was missing from the scene as shot. The writer had not noticed the absence himself, but Disney insisted on the point, the script was checked and he was proved right. He ordered the scene reshot. Indeed, he would order almost anything reshot, at no matter what cost, in order to get a film exactly the way he wanted it to be. He may have paid low salaries, but when it came to the product to be sold under his name and, psychologically, made his own through his incessant interventions, no corners could be cut.

Of course, for a man as intense as Disney in his desire to control his environment, the childlike desire to make a little world all his own, animation was the perfect medium psychologically. You can redraw a character, or even a line in his face, until it is perfect; you need never settle as the director of the ordinary film must, for the best an actor—imperfect human that he is—can give you. If the best an animator gives you does not meet your standards, you can get a new animator and set him to the problematic task. And another. And another. They are infinitely interchangeable, and operating as Disney was, with everyone cloaked in anonymity and working within the confines of a house style, no one except those directly concerned knew where one pencil began and another left off. As a writer who worked for Disney in the great days of animation recently put it: "Animation is where screen direction gets down to matters of detail unheard of in live action. In animation you deal with the glint in the eye, the twitch of an eyebrow, the tic of a muscle. You're dealing with microscopic fractions—if you want to." Disney wanted to, perhaps even needed to. For animation, to borrow from the unfortunate jargon of psychology, is a compulsive's delight.

21 *During the nineteenth century artists proceeded in all too impure a fashion. They reduced the strictly aesthetic elements to a minimum and let the work consist almost entirely in a fiction of human realities. In this sense all normal art of the last century must be called realistic. . . .*

Works of this kind are only partially works of art, or artistic objects. Their enjoyment does not depend upon our power to focus on transparencies and images, a power of the artistic sensibility; all they require is human sensibility and willingness to sympathize with our neighbor's joys and worries. No wonder that nineteenth century art has been so popular; it is made for the masses inasmuch as it is not art but an extract from life. Let us remember that in epochs with two different types of art, one for minorities and one for majority, the latter has always been realistic.

—José Ortega y Gasset, *The Dehumanization of Art*

The film is, moreover, an art evolved from the spiritual foundations of technics. . . . The machine is its origin, its medium and its most suitable subject. . . . Films . . . remain tied to an apparatus, to a machine in a narrower sense than the products of the other arts. . . . The film is above all a "photograph" and is already as such a technical art, with mechanical origins and aiming at mechanical repetition, in other words, thanks to the cheapness of its reproduction, a popular and "democratic" art.

—Arnold Hauser, *The Social History of Art*

THE ARISTOCRATIC PHILOSOPHER and the Marxist historian, so opposed in their opinions of the validity of realism as a mode of artistic expression, at least agree that popular art tends to be realistic and that the artist who wishes to be popular must be realistic. From this, much unhappiness has followed, especially among artists and aesthetes. It must never be thought, however, that Walter E. Disney shared this unhappiness. "All we are trying to do is give the public good entertainment," he

193

said late in the 1930s. "That is all they want." And that was all he wanted, for he *was* his public, and when he knew his own mind—which he usually did—he knew theirs. It generally pleased them to think of his works as magical creations rather than technological ones, and that did not bother him in the slightest. Indeed, he took a great deal of pleasure in allowing publicists and journalists to reveal the technical secrets of his magic making. He was proud of it, and he knew that all of us are rather like children, so intrigued by the magic of clocks that we must take them apart in an attempt to divine the secret of their trick, yet still able to retain our awe at the total effect even when the pieces are strewn all over the floor around us.

Superficially, this may seem something of a paradox, since the content of Disney's work was usually the highly unrealistic fairy tale or children's story often folkish in origin—that is, its authorship was probably communal in the first place and whatever the circumstance of its creation, it was, by the time Disney got to it, community property. Writing in what some impressionable commentators have been pleased to call the Age of McLuhan, this is a less troublesome point than it might have been in the dark ages prior to 1965, when the Canadian professor came into vogue and taught us that the content of a new medium (as the movies still were in the 1930s) is nearly always the product of an older culture. We may add to this the basic insight that the fantastic is always more acceptable to plain people—and sometimes to sophisticates—when it is rendered in the most realistic possible style. So, when offering time-tested mythic material, Disney was careful to present it in every day, down-to-earth artistic terms that offered no difficulties of understanding to the large audience—that in fact gentled them with the familiar instead of shocking them with the aesthetically daring. It was the way he personally preferred to arrive at the state in which disbelief is willingly suspended.

And so the enormous effort at art education, the enormous expenditure on improved sound and color film processes, the care lavished on the details of production, must be seen always as a drive toward realism, no matter what subject matter Disney chose to feed into the elaborate and expensive machine he was constructing. As he went to work on *Snow White* he began developing another technological breakthrough, a new camera that he believed would mightily enhance the believability of *Snow White* and, indeed, of all his future work. It was called the multiplane camera, and though it was not an achievement of the magnitude of sound or color film, it was a most useful device for Disney. So it will be for the student of his work, for by analyzing the effect of the camera on the Disney Studio's product as well as the symbolic meaning of the effort that went into the development of this, the only important technical device to be created solely at his studio, we gain an important insight into Disney's sensibility.

Until the perfection of the multiplane camera, the conventional animation camera, even in its most advanced state, had an important defect. We can begin to understand this defect by drawing on a contemporary account of the customary animation camera in operation:

"When all of the celluloid drawings for the entire footage have been finished they are sent to the camera room. Each camera is mounted above a table lighted by mercury bulbs. A frame the size of the cels is ready. First the background is laid down. Then a cel with Mickey, another with Minnie, a third with the villain and a fourth that is blank are slipped over the pegs, which hold them in perfect register (and which were, and are, the same everywhere in the studio, from the animator's board through the ink and paint department to this point). To the eye and the camera the picture appears to be on one sheet.

"Compressed air clamps a glass pane over the drawings to remove wrinkles, the operator's hand touches the control button of the camera, a click is heard as the lens shutter blinks, a tiny bell rings, the air lifts the glass and the photographer removes the cels and replaces them with the next set. . . ."

The trouble with this method of photography—if, indeed, it was a trouble—was that the camera could not truck into the frame without spoiling such small illusion of depth as it presented. In particular, there was a problem with objects at infinity. Foreground objects could simply be removed cel by cel in order to give the illusion that the camera was moving past them. But what about objects that were supposed to be behind the object or character toward which the camera was supposed to be moving closer and closer? Obviously they would grow in size, as this movement continued, at exactly the same rate as the foreground object.

The answer to the problem posed by the disconcerting visual effect of the early technique was obvious, theoretically at least. Separate the cels, putting each object on a separate plane and arrange these planes in a scale corresponding to the natural order of things. Thus, a cel representing a foreground tree might be only an inch or two in front of the camera's lens, while the cel on which, say, a house was drawn would be six inches from the lens. The horizon line, in turn, might be a foot or more behind the house. With a setup of this kind the camera could obviously prowl at will through a scene just as it could through a full-scale, three-dimensional setting or a natural location, or, indeed, just as the human eye observes the world.

But the multiplane camera, as it was called, was more easily described than built. Ub Iwerks had made a primitive model of the machine during his time of independence from the Disney organization (using, it has been said, railroad tracks with a jalopy mounted on them respectively to guide and to power the camera). When he came back to Disney he led the team that perfected first the vertical multiplane camera, which peered down through an iron framework fourteen feet tall, through layers of cels set in grooved shelves "like a baker's pie-wagon," according to early description. Later Iwerks participated in the development of the horizontal multiplane, which allowed the camera to pan simultaneously across the frame as well as into it, and which also permitted one very large background to be used for many scenes.

It was the work of many years to develop this camera fully for practical use, but Disney won a 1937 Oscar for the first film shot totally by the multiplane camera, a short cartoon called *The Old Mill*. It had no real plot, and it consisted, not unnaturally, of a succession of trucking shots that showed the activities of the animal inhabitants of a deserted windmill from sunset to sunrise on a night enlivened by a storm. That year the studio also won a special Academy Award "for scientific achievement" for developing the multiplane. Disney had hoped that the camera would be ready for use throughout *Snow White*, but most of its photography had to be done on the old camera. The multiplane was simply not ready as the finished cels began to fall off the assembly line, and Disney, heavily in debt on the film, was under unremitting pressure to get it into release and begin recouping as soon as possible.

Nevertheless, the camera was used for special effects—and to good effect—at several points in the picture—for example, in the sequence where Snow White sings (in the disconcerting coloratura that meant class to Disney) "Some Day My Prince Will Come," and her tears are seen falling down a wishing well and splashing in the water at the bottom. For Disney himself the camera represented a triumphant status symbol. "It was always my ambition to own a swell camera," he said, "and now godammit, I got one. I get a kick just watching the boys operate it, and remembering how I used to have to make 'em out of baling wire."

There is no doubt that the camera was "a scientific achievement" of sorts (more properly it was an engineering achievement). The question is whether it was a genuinely useful artistic tool. There is a school of cinematic theory that holds that the attempt to create a greater and greater illusion of three-dimensionality in the movies is a feckless one. Rudolf Arnheim, who has been the most determined proponent of this notion argued as early as 1933 that "the lack of depth brings a very welcome element of unreality into the film picture. Formal qualities, such as the compositional and evocative significance of particular superimpositions, acquire the power to force themselves on the attention of the 'spectator' when there is no 'leeway' between objects and they appear to be on the same plane." In other words, Arnheim is applying to film a theory almost as basic to art criticism as Ruskin's theory of the shaping hand. Simply stated, the notion is that art is defined by the natural limits of the medium in which the artist is working. These limits serve to separate art from reality perceived in the raw, as it were, and they form the basis for the formal rules of its critical evaluation. As Arnheim sees film history, each step taken by the film industry toward a closer approximation of reality (sound, color, new techniques of photography and direction) paradoxically brought the movies further from the state of artistic purity and closer to a realism that could never transcend its subject matter as high art must. For him, the silent, black-and-white screen had a potential for achievement through the austerity of technological self-denial that someone like Disney, with his primitive aesthetic notions and his ruling passion, equating progress with invention, could never comprehend. Arnheim understood this mentality and observed that it fitted well with the predisposition of the audience. "Engineers are not artists," he wrote. "They therefore do not direct their efforts toward providing the artist with a more effective medium, but toward increasing the naturalness of film pictures. It vexes the engineer that film is so lacking in stereoscopic quality. His ideal is exactly to imitate real life. . . . The general, artistically untrained public feels much the same. An audience demands the greatest possible likeness to reality in the movies, and it therefore prefers three-dimensional film to flat, colored to black-and-white, talkie to silent. Every step that brings film closer to real life creates a sensation. Each new sensation means full houses. Hence the avid interest of the film industry in these technological developments."

It cannot be said, of course, that the multiplane camera was a sensation. Indeed, most people were probably unaware of its use in Disney's later films. Nor is it possible to apply Arnheim's arguments *in toto* either to the movies in general or to the work of Disney in particular. He wrote in a historical moment when the art of film photography was in a regressive state, with the camera imprisoned in an unwieldy booth where its whirrings could not be picked up by the microphone and when the director's authority was temporarily superseded by that of the sound technician. This imparted to the first talkies a static quality, visually, that was extremely disconcerting to some aestheticians of the film, and for a time, the theoreticians were inordinately fearful of the new technology. This basic fear informs Arnheim's work to a degree greater, perhaps, than even he was aware.

Then, too, the addition of sound to film brought the film closer than it had been to the traditional literary arts, a disappointing development to critics who had been schooled in the visual arts and were predisposed to interpret film largely in terms borrowed from them while ignoring the fact that even silent film contained heavy elements of plot, characterization and drama, which, however unsuitable to painting, do have a right to live in a mixed medium like film. From this, however, it should not be imagined that the sound film was generally welcomed by the literary intellectuals, either. Some of them did noble service in its cause, but many were as hostile to the talkies as their ancestors had been to the printing press, which converted literacy from a mark of class distinction to a universal right. Since, as Eric Hoffer says, it is "natural for the scribe to limit proof of individual worth to fields inaccessible to the mass," intellectuals who had come to regard the novel and the drama as private preserves tended to look with dismayed distrust at a medium that made something like the pleasures of these arts available to a wider audience.

More directly to the point, Arnheim's theory—and it must be

emphasized that his work was quite typical of the period—did not take into account the rise of a new grammar of film, a process that has greatly accelerated since 1945 and has now quite transcended his original objections. The addition of sound and color were not, in practice over the years, nearly so restrictive as he predicted they would be. Even as they came into wider and wider use, film makers, largely through the development of bold new editing techniques, discovered ways of telling a story that were uniquely cinematic. And these had the effect of marking off the boundary between film art and life's reality far more effectively—and interestingly—than did his rather narrow proscriptions. Anyone seeing the highly personal statements of a Resnais, an Antonioni, a Bergman, a Fellini, a Truffaut, even an Orson Welles, can have no doubt that what he is seeing is the statement of an artist in film, not the simple reproduction of real life. Nor can he doubt that these men have constructed a new and more flexible aesthetic of the film than Arnheim ever dreamed of.

Still, with all these *caveats* entered, Arnheim's work has a peculiar applicability to Disney, who once said "When we do fantasy, we must not lose sight of reality." So, like his audience's, his principal definition of art as an imitation of life and the drive toward greater and greater realism, carried out at such vast effort and expense throughout the Thirties, begins to seem needlessly, heedlessly narrow. To be sure, the desire to reproduce nature more accurately led, at first, to a higher quality of work at every stage of the film-making process, and the Disney Studio must receive credit for seizing upon technical advances of all kinds, developing many of its own and then synthesizing all the discoveries in the field into a method of production that was commercially viable and wildly attractive to the public. Indeed, Disney's position became so dominant so quickly that it is fair to say that without his efforts it would not now be possible to call animation either an art or a profession; it would still be merely "the cartoon business." Moreover, as we have seen, color added an element to the cartoon that no artist, even the most abstract abstractionist, can do without, while the synchronization of visual imagery with musical sound definitively separated the film cartoon from its heritage, the newspaper cartoon, and gave it an expressive potential far above that of its ancestor.

But once launched on the realistic course, Disney was unable commercially and unwilling artistically to deviate from it, or to press beyond its limits as he had earlier, when he was trying to move beyond the limits of the short, silent cartoon. After *Snow White* there were no breakthroughs comparable to those that preceded it; there were only fussy improvements on the basic structure Disney and his people had already built. The multiplane camera thus becomes a symbolic act of completion for Disney. With it, he broke the last major barrier between his art and realism of the photographic kind. He could at last give his audience the illusion of three-dimensionality, and he was exceedingly pleased.

In animation—at least at the Disney Studio—the rest was clean-up work. A man who worked on *Bambi*, which went into production shortly after *Snow White*'s release, even though it was not finished and publicly shown until 1943, said recently: "He might as well have gone out and taken pictures of real deer, that was the quality he was driving for in the animation." During the war years animator Ward Kimball, who had been called by one of his colleagues "Disney's artistic conscience," began experimenting with limited animation, which is deliberately two-dimensional in style and makes no attempt to copy movement from life (and is therefore quite inexpensive compared to full animation). He used it successfully in the "Baby Weems" sequence of *The Reluctant Dragon* and in parts of *Victory Through Air Power*. But the style did not appeal to Disney.

If the freshness, the sense of excitement that had once attended Disney's efforts in animation diminished, so did the quality of the humor contained in his short films. With many of his best men gone and those remaining concentrating on fea-

201

tures, the little cartoons responsible for his first fame suffered seriously. The Warner Brothers cartoon department, which had among its directors the gifted Friz Freling and a genius named Chuck Jones, moved into what has been termed "La Grande Epoque." Its star, Bugs Bunny, was as urbane as The Mouse was rural, and infinitely more versatile than any of the Disney gang, each of whom had limits. The Duck, for example, was barred from verbal humor because of his deliberately garbled voice; The Goof had, built in, a stupidity problem; Pluto, of course, had neither voice nor brains; Mickey, finally, was now too nice to be a comedian. Bugs, on the other hand, could do everything. The inimitable voice, supplied by Mel Blanc, could push home a punch line in the same deft manner that the radio comedies of the time had taught us to appreciate; the body was built for speed, and the chases that were the high points of his films were masterfully constructed. Finally, and most important, The Bunny's personality perfectly suited his times. He was a con man in the classic American mold, adept in the techniques and ethics of survival, equally at home in the jungle of the city and in Elmer Fudd's carrot patch. In the war years, when he flourished most gloriously, Bugs Bunny embodied the cocky humor of a nation that had survived its economic crisis with fewer psychological scars than anyone had thought possible and was facing a terrible war with grace, gallantry, humor and solidarity that was equally surprising.

Around Bugs the Warner craftsmen created a memorable stable of similarly breezy spirits—most notably the maddest fantasist—victim in all cartoon land, Daffy Duck, and later, the greater cartoon comedy team, Wile E. Coyote and The Road Runner. The aforementioned Fudd, with his angry lisp and his ever-present shotgun, was the perfect foil for Bugs and Daffy, and for the generation that passed its prepuberty in the Forties, they were the super stars of the cartoon screen. The Disney stable—slower, less wildly inventive, possibly more subtle—were distinctly of the second rank. As for the rest of the competition—the mindlessly sadistic Tom and Jerry created by the Hanna-Barbera team (in those pre-TV days working out of M-G-M), Walter Lantz's hopeless Woody the Woodpecker—they were out of sight.

Thus, in the short-cartoon field, the popularity contest was lost by Disney to his competition by 1950. Shortly thereafter, the failure to develop stylistically cost him his artistic leadership as well. In the early Fifties it passed from him to UPA, a small firm founded by former Disney employees during the war. Using limited animation against backgrounds both impressionistic and expressionistic, the firm produced cartoons splashed with bold colors and enlivened with caricatures of motion as delightful as those Disney had offered in his early works. Though Disney paid grudging compliments to his new competitors and his brother, feeling that well-produced competing films made by commercially solid concerns were good for everyone, was free with advice and assistance on financial matters, many felt that Disney was quite hurt by his loss of artistic leadership and critical acclaim.

He fought back in a sulky, unprogrammatic and desultory fashion. He allowed Kimball's unit to produce a film in the new style, *Toot, Whistle, Plunk and Boom*, and it won an Academy Award as the best short cartoon of 1952, but he allegedly told Kimball no film like it would ever again be produced at his studio. He was not quite so unrelenting as that, for a number of limited animation shorts have since been produced at the Disney Studio as educational films and as features of various attractions at Disneyland. Limited animation is also frequently used to tie elements of the Disney television show together. But Disney never let Kimball set up a purely experimental unit, as many animators, both within and without the studio, hoped he would. Instead, he grew more and more preoccupied with Disneyland, where, ironically, Kimball found a part-time outlet for another of his talents—as director of the "Firehouse Five," a Dixieland band originally formed as lunchtime recreation at the studio but soon professionalized and turned into an "attraction" of some power at the amusement park.

By the end of the decade he was out of the short-cartoon field entirely. It had become too chancy and—as those who survived him for a while discovered—its profits were too small to bother about. By that time no major studio had an animation unit working on films for theatrical release. Friz Freling has gone into partnership with David De Patie and they spun a reasonably successful theatrical series off the title sequence they did for *The Pink Panther*. Released through United Artists, the Panther was an excellent creation. He did not speak and all his humor was, therefore, visual. Backgrounds were simple, bright and imaginatively used; indeed, a splash of color—a pure, formless abstraction—was frequently used almost as a character, as a foil or a partner in the Panther's activities. He was, in short, a character who lived only in and on film, never in a simulacrum of reality as the Disney gang did. De Patie-Freling became the only firm—as a small one—then doing interesting work in the theatrical-cartoon market on a consistent basis. A former Disney employee, the late John Hubley, in partnership with his wife, Faith, did an occasional exquisite short film that got a commercial release, but did the bulk of his business as a producer of TV commercials. Occasionally one of Norman McLaren's fine abstract shorts got imported from Canada, and, even less frequently, American audiences got a chance to see in their theaters some of the excellent animation being done in France and in the Iron Curtain countries. But, since the best American animators refused to be locked into a series with a running character, and since the distributors only wanted packages with such characters, and since profit margins remained as thin as ever they were in the post-war period, no one offered the animators much commercial encouragement. The gifted Ernest Pintoff, having made two splendid shorts, *The Violinist* and *The Interview*, for example, left the theatrical short-subject field and tried without great financial success, to make the transition to features.

Lacking the government subsidies that kept the art healthy, or at least alive, in other countries, most American animators could be found hiding out in television in the 1960s. As Neil Compton has pointed out, the best of their shows—the non-horrific ones aimed at the younger children—were regarded by the networks as Saturday and Sunday morning fillers and often escape the attention of their vice-presidents in charge of blandness. As a result a certain amount of satire went on down there in the bargain basement. It was not great stuff, but it was often more lively and pointed than what went on in prime-time situation comedies. Bullwinkle, Dudley Do-Right of the Mounties, and Roger Ramjet (whose very short films were sold in syndication and dropped into cartoon shows at the whim of local stations) were healthy spoofs of traditional mass-media heroic attitudes and one was grateful for them as small favors, slight correctives to the steady drip-drip-drip of banality in most children's TV programming. One also wished that budgets had been a little more generous so that the creators of these shows could offer a new generation something akin to the elaborate visual humor that earlier generations took for granted at the Saturday matinees.

As for Disney, his television program contained only a modicum of animated work—much of it quite weak—and in the Fifties the number of animated features he produced for the theaters dropped to one every four or five years, and they were of indifferent quality. Indeed, so locked into the patterns of the past were these features that the animal characters of the latest film looked almost exactly like the animals of the first full-length efforts to come out of the studio. There was, very simply, a Disney bear, a Disney deer, a Disney chipmunk and so on, predictable and unvaried from film to film. By this time their relationship to their natural models was vague indeed; their relationship to past Disney products, all-important.

There was also a structural rigidity about the Disney animated features that grew increasingly obvious as the years passed. The editing principles applied to *Snow White* were those of the conventionally well-made commercial film of the time. There was nothing particularly daring about the way it was put together: its merit was based on other skills. In general, a scene would open with an establishing or master shot, then proceed to an intermediate shot, then to close-ups of the various participants, with conventional cut-aways to various details of scenery or decor as needed. Confusing flashbacks or dream sequences were avoided, and special effects were introduced in such a way that every child was aware that something out of the ordinary was about to happen. It was a good enough way to shoot a film, but, especially in recent years, a far more flexible editing principle has come into play, particularly in the foreign films. It has been discouraging to see the Disney Studio cling to the old conventions of telling a story on film, particularly when the animated film and the stories it usually adapts seem particularly suited to the deftly allusive new style, with its bold leaps through time and space, its sudden juxtapositions of seemingly unrelated material, its quickness of mind and spirit, its sheer pleasure in the film as film. The fact that children are constantly amused by this style as used on television commercials, and seem completely able to understand it, would seem to undercut most arguments in favor of the clarity of traditional editorial techniques.

Again, Disney knew what he liked and refused to change with the fashion. About the only development one sees in his films is the increasingly heavy use of what has come to be known in the trade as "Disneydust," those sparkly highlights that burst from any object touched by any magic wand in any

Disney animated film. Of late years the dust seems to have settled on almost every flat surface in sight, and it is, of course, a very close cousin to the stardust that flakes off any Disney rendering of a heavenly phenomenon. Walt was known to have liked the effect.

22

Clearly, mythology is no toy for children.
—Joseph Campbell, *The Masks of God: Primitive Mythology*

IN RECENT TIMES most people have preferred to think of mythology in general and the folktale in particular—the latter is admittedly not quite such a serious matter—as children's playthings. In a much quoted simile, J. R. R. Tolkien has pointed out that the fairy tale has been retired to the nursery as old furniture is retired to that place, not because the children like it so much, but because their elders no longer care for it. They have thus achieved, in our time, a kind of cultural neutrality, particularly in the minds of the unbookish, mass audience. They are considered safe, moral, entirely suitable for innocent eyes. It is a very naïve view, as Disney was to discover once he began trying "to lick" the story (as they say in Hollywood) and as at least some members of his audience were to discover when their children had to be carried from the theater screaming at even the fairly mild, often gently charming version of *Snow White* that Disney finally made.

It is necessary to conceive of fairy tales in a more complex way if one is to understand the way they work on us. As C. S. Lewis observed, "Many children don't like them and many adults do." Which means, of course, that their appeal is more universal in one way than many people suppose (it is not confined to a single age group) and less universal in another way (there are some people who are constitutionally unable to respond to them no matter what their age, education or station in life). It is easy to see what Disney liked about the form. It imposes on the teller certain restrictions cited by Lewis—"brevity, its severe restraint on description, its flexible traditionalism, its inflexible hostility to all analysis, digression, reflections and 'gas.'" These were criteria Disney insisted upon in all the works that emanated from his studio, no matter what their subject matter. Liking the fact that folk stories are the most nonliterary of all literary forms, he also appreciated the informality of their composition, for as Joseph Campbell says, the folktale is "told and retold, losing here a detail, gaining there a new hero, disintegrating gradually in outline, but re-created occasionally by some narrator" in an entirely democratic fashion—"an art on which the whole community of mankind has worked."

Disney, in the ineffable style of his early days as a theoretician of aesthetics, seemed to agree with this definition when he offered his own definition of culture to a wandering scholar. The very word, he said, "seems to have an un-American look to me—sort of snobbish and affected—as if it thought it were better than the next fellow. Actually, as I understand it, culture isn't that kind of snooty word at all . . . a fellow becomes cultured, I believe, by selecting that which is fine and beautiful in life, and throwing aside that which is mediocre or phoney. Sort of a series of free, very personal choices . . .

"Well, how are we to recognize the good and beautiful? I believe that man recognizes it instinctively. . . ." How could a man holding such views avoid the fairy tale? Even more deeply ingrained in him, however, was nostalgia for the vanished past, and this, too, surely played a part in his choice of material. Study of the origins of myth and of folklore was begun intensively almost precisely at a time when the world that supported the creation of legends and fairy tales ended—that is when the rise of rationalism, then of science, then of mass industrialism and urban life called into question the primitive systems of belief that had sufficed to explain the natural world and its workings to earlier generations. Then, says Ralph Harper, "it was appropriate that, when all old values and beliefs were being discredited by revolution and by the new confident bourgeois civilization, some men should go back, surreptitiously, to the past, for help in surviving in a time when everything spiritual had disappeared but self-confidence." The brothers Grimm worked at their self-appointed task of systematically collecting European folktales even in the midst of the Napoleonic wars, and throughout the nineteenth century, with its wars and revolutions, mass migrations and intellectual upheavals, others continued in the same vein. The disciplines of philology, anthropology and archaeology made vast strides in less than a hundred years, in the process cutting a trail to man's communal past that we continue to widen.

Paradoxically, this effort has all along been aided by the very forces that had exploded the old world in the first place. Improved methods of transportation opened up previously obscure corners of the world, and improved medicine made it safer for explorers and scientists to visit them, while the need for raw materials of every kind drove the Europeans into the most all-encompassing effort at colonialism the world has ever witnessed. Meantime, scientific advances of all kinds gave the scholars and scientists who followed in the wake of this effort—and whose work was often used to justify it—a set of tools with which to examine the phenomena of the past with a new sophistication. Finally, the self-confidence of the times, of which Harper speaks, filled all who participated in this last great age of exploration with a missionary zeal, an eagerness to convert distant brethren not just to Christianity but to an entire way of life. The other side of this coin was an unwavering belief in the superiority of Western ways, which led to an almost casual expropriation of tons of primitive art and artifacts and its convenient placement in institutions like the British Museum, which to modern eyes is a veritable monument to the spirit of cultural rape that characterized so much of colonialism.

Out of this vastly important intellectual movement there slowly evolved a popular nostalgia for the life of an imagined past—a nostalgia that had its basis in a superficial reading of the results of scholarly and literary endeavor and in an innocently enthusiastic approval of the spirit in which the work was carried out. This nostalgia informed much of the new popular culture—also a product of industrialism—and it continues to inform present-day popular culture as well as the critical effort to understand it. (This applies both to defenders of it, like McLuhan, and to opponents of it, like Marya Mannes and Dwight Macdonald; the chief difference between them is the past cultures they select for invidious comparison to our own.) In effect, this nostalgia for communities past is an antidote to the pace and disruptive quality of modern life as well as the pressure it places on the traditional, individualistic values of the nineteenth century—factors that have spawned alienation, the disease of our time, which is often expressed, as Harper puts it, "in personal homesickness and longing for lands never seen." What we are dealing with here is a short chain of paradoxes, summarized by Harper in this way: "If fulfillment must somehow precede longing, it is nevertheless fitting that an understanding of homelessness must precede an understanding of longing and fulfillment." In other words, one of the great functions of culture, both high and low, in both the nineteenth century and the twentieth century, has been to make the links in this chain easily visible. It has, in effect, given us a detailed and not entirely erroneous vision of a past, a previous fulfillment, the loss of which we can sorrow for, the recapture of which we can work toward.

Very little of this is consciously expressed by any of us, and certainly it was never articulated by Walt Disney. Yet, viewing his work and his life, it is impossible to doubt that it was operating in him, and on two levels at that. He was not, as we have seen, a very well-read man or a very subtle observer of the society around him. Yet almost from the beginning of his career he presented us with images of longing. There was the barnyard and small-town environment of Mickey Mouse and his companions, and then there were all the adventurous uprootings of The Mouse that caused him so much trouble and that he overcame principally by asserting the so-called old-fashioned virtues. There was, all along, the emphasis on the innocence and playfulness of life in the forests and fields where every prospect pleased and only man was vile. There was the determinedly uncomplicated statement of aesthetic aims and beliefs of Disney himself, harking back inevitably to simpler times. Finally, there was the seemingly deliberate isolation of the man and his studio from the major currents of his art and his industry and his times, for it must not be forgotten that in the very period that he was creating his first major statement of the nostalgic theme in *Snow White*, the rest of the film industry was turning toward a kind of documentary realism about comtemporary life that was unprecedented in its short history. Arthur Knight puts it this way: "As the thirties wore on . . . growing tensions produced a notable series of films that rode the mounting wave of liberalism without recourse to either the 'fantasy of good will' or 'screwball' subterfuge. Labor unrest, slum housing, unemployment and dislocation aggravated by the dust storms of the mid-thirties—all of these were put on the screen with a directness that stressed the social and economic sources of such hardships. There was sympathy for the common man and new hope for a better tomorrow. In place of the contrived and improbably 'happy endings' of the depression musicals and 'back to the soil' films, there was now a forthright expression of belief in the inherent strength of democracy. . . ." The fierceness of the embrace in which both the public and the intellectuals wrapped Disney in this period resulted, not from any conscious effort by him to state social themes, but because his fantasies were capable of interpretation as dream works symbolizing and allegorizing the feelings that other film makers, other artists of all kinds, were then expressing in more direct terms—and, of course, because his visions appealed to the longings for the vanished past that were universally present in the national psyche.

The second level of Disney's appeal was stylistic rather than ideational, and it was also present in his work from the beginning of his fame. We have already spoken of his concern for realism and of a conception of beauty very close to the style favored by illustrators of nineteenth-century children's books. These were part of a larger kind of cultural conservatism to be found typically in small-town, middle-class Americans. It is necessary to note only in passing the predilection of this class for building banks and public buildings in the Greek and Romanesque styles, their strange admiration for home architecture in styles even less congenial to the American landscape—the gothic and medieval structures affected by the well-to-do, and the Tudor and Colonial homes their less affluent neighbors began building early in this century. In this tradition also fits the suburban "ranch house" of today, momentarily re-creating, for their more imaginative owners, the vision of a more spacious style of life, now vanished. Nor is there need to dwell on the other cultural ideals of this middle class—the genteel tradition of its literature, the Romanticism of the music it preferred to hear at its concerts and musicales, the preference for sentimentalized portraits and nature painting in its art. All of these were "culture" to the striving middle class of America. All partook of an element of nostalgia and all remained in lively vogue, at least away from the urban centers, until very recently.

All of them represented "culture" to Disney too, even if he did find the word snooty and un-American—which, in fact, it sometimes was as some people used the term. All these examples of culture found expression in his work—in the sentimental cuteness of his characters, in the studio's corporate artistic style, in his unremitting desire to juxtapose his art with "classical" music—remember the "Danse Macabre" in *The Skeleton Dance* as early as 1929—and, most important, the choice of *Snow White* as the subject for his first feature. Its use of the sleeping beauty theme automatically makes it a work that caters to nostalgia, for all of us would like to go to sleep, as Snow White does, and then awaken to find the world unchanged, indeed improved, since it is her lost lover's kiss that awakens her. Moreover, as a nursery story, first heard in childhood, it is bound to stir the sweetest memories when it is revisited in adulthood, especially with one's own offspring in tow.

But there is something more to this tale and to all folk literature than the matters we have so far discussed—a much deeper current that Disney himself could never catch and that he would never allow his employees to toy with either. It was the inability—or was it unwillingness?—to see folk literature as "the primer picture-language of the soul," to use Campbell's phrase, which, of course, guaranteed the success of Disney's film. But it was also the factor that kept it—and him—from greatness. The limitations of his background provided an excuse for this ultimate failure. And so did the conditions under which he created his first long film.

23 *We've bought the whole damned sweepstakes.*
— Roy O. Disney, 1937

IF HE HAD NOT been in the midst of *Snow White*, 1936 might have been the best year of Disney's life, financially speaking. In bargaining to renew their distribution contract with United Artists that year, the Disney brothers insisted that they must control the future television rights to their products—a remarkable piece of technological foresight, since most Hollywood independents were signing those rights away even after the new medium was well established. Unable to secure them from UA, they turned to another studio, RKO Radio, and made the best deal they ever had. The distributor advanced the full cost of each short production to them and, in return recouped its investment and turned its profit out of a kitty limited to 30 percent of the gross. Without the killing overhead *Snow White* added to their operations, the Disneys would have been—temporarily at least—as secure as it is possible to be in the motion-picture business. On the other hand, they would have achieved stability at the price of future growth, and they knew—better than many of their critical competitors—that they dared not stop growing. At first they had financed *Snow White* out of the accumulated earnings of the studio, but Disney's first cost estimate of $250,000 was soon surpassed, as was the $500,000 figure next projected. By the time the shooting was done, the negative cost of the film was to be in the neighborhood of $1.7 million and the total gross needed before it would become a Disney asset was placed at $2.5 million.

Racing to get the picture into the theaters in time for the 1937 Christmas season—traditionally the best movie-releasing time—Disney finally had to go to his principal backer, The Bank of America, and request an additional $250,000. In an incident he loved to retell, Disney was forced to show an unfinished print of the film to one of the bank's vice presidents, Joseph Rosenberg, in order to secure the fresh backing. The print available was not color-corrected, and many of the most important sections were not finished. So that Rosenberg could get some idea of the film's continuity, pencil roughs of the unfinished sequences were inserted throughout. Disney recalled the Saturday he showed *Snow White* to Rosenberg as a nightmare. For once, it was cold in the sweat box—or at least he *felt* cold —and he kept up a desperate line of chatter to explain and perhaps distract the banker from the film's uncompleted moments. The session must have been agony for a man as compulsive as

Disney was in his love for finished products of perfect smoothness and technical brilliance. Rosenberg appeared unmoved, responding to Disney's line with grunts and monosyllables. When the ordeal was finally finished Disney took Rosenberg on a tour of the plant, on which the banker volunteered not a single comment about the movie he had just seen. Finally, as he was getting into his car, he turned to Disney and said, "Good-bye. That thing is going to make a hatful of money."

Rosenberg later confessed to Disney that he was new to the film business and afraid to trust his instinctive response to *Snow White*. He was also discouraged by several very dubious opinions among rival producers he polled about the project. Among them, the only one he later recalled as optimistic about the commercial possibilities of a full-length animated cartoon was Walter Wanger. Also on Disney's side was the manager of Radio City Music Hall, who booked the film sight unseen, and a United Artists promotion man named Hal Horne, who had stayed friendly with Disney even after he had taken his business elsewhere. He told Disney that the speculative gossip about *Snow White*—some of it openly derisive—was in fact good publicity, bound to heighten curiosity over it and to pay off ultimately at the box office. He also advised Disney to follow his own instinctive desire and sell the film as a fairy tale, not as a prince-and-princess romance, which many promotion people, with the love of the cliché sales device endemic to their craft, were urging him to do. It was, after all, a fairy tale, and that form, however new to movies, had had a certain popular appeal through the centuries.

This handful, however, were the only people outside the studio whom Disney could later recall as standing by him in the loneliest of his many lonely periods. Even his own wife was dubious about the project. When he told her he was going to do a story involving dwarfs, her immediate reaction was one of repulsion. "There's something so nasty about them," he later recalled her saying. But that was the least of his worries. He admitted, after the fact, that "when we did *Snow White*, we weren't really ready." To begin with, the story as the Grimms set it down was quite a violent one, difficult to tell in visual terms acceptable to the commercial audience. There was the bloody business of the substitution by the huntsman of the stag's heart for Snow White's to prove to the wicked queen, her stepmother, that he had carried out her instructions to kill the girl. Then there were no less than three attempts on Snow White's life by the stepmother herself—disguised as a peddler crone—after the girl found refuge with the dwarfs. Each of these was bound to make for unpleasant viewing—choking her with a corset, giving her a poisoned comb and, finally, the successful gift of the poisoned apple. Then, too, there was the Grimm ending, calling for the stepmother to attend Snow White's wedding to the Prince and then undergo a fatal torture: "But iron slippers had already been put upon the fire, and they were brought in with tongs, and set before her. Then she was forced to put on the red-hot shoes, and dance until she dropped down dead."

The solution to this multitude of grisly problems was to emphasize the roles of the dwarfs, who were not even named in the Grimms' tale, and to play down the Queen's part. There is infinitely more footage devoted to the funny little men than there is to the stepmother, whose attempts on Snow White's life are reduced, in the movie, from three to one. Nor did Disney linger long over the scene with the huntsman. The grieving of the dwarfs over the death of Snow White is also sharply cut. And, of course, the original ending was totally eliminated (Snow White and her Prince simply ride off singing).

Added to the film were some brilliant animated sequences—in particular, Snow White's flight through the forest after her escape from the huntsman, during which the trees seem to reach out to hold her back. Delightful, too, were her discovery of the animals who guide her to the dwarfs' home and their ingenious cleaning of the place: in which a squirrel's tail becomes a bottle brush, a deer's becomes a duster, the birds trim pie crust with their beaks or use them to reel up a cobweb. Finally, there were the famous characterizations of the dwarfs—Doc, Happy, Grumpy, Sneezy, Bashful, Sleepy and Dopey—which became almost a comedy of humors.

The original balance of the tale, however, was destroyed by these inventions, as well as by the songs for which Frank Churchill, who did "Who's Afraid of the Big Bad Wolf," contributed the music (lyrics were by Larry Morey). Three of them were hits—"Whistle While You Work," "Heigh-Ho," and "Some Day My Prince Will Come"—and most of them integrated fairly well into the Disney version of the tale, although like the tamperings with the original story, they made the film fundamentally different in mood from the tale the Grimms told.

Nor were the story and its mood the only problem the studio faced. It was feared, for example, that audiences could not tolerate, for the length of a feature, the hard brightness of color that was customary in the short cartoons, and so the entire tonal range had to be adjusted. After much experiment a muted palette, with browns and greens predominating, was arrived at— and it remains one of the most pleasant aspects of the picture. Equally difficult problems were presented by animation. The house-cleaning sequence, very long in relation to its importance to the plot, was probably the finest sustained choreography of movement achieved up to that time by Disney, and it remains to this day a classic of its time and its field. Watching it, one senses not the industrial pressures one feels in some Disney studio work but a kind of inspiration that is as rare in this field as it is in any other. Infinite, expensive care was lavished on it as well as on some other, shorter pieces of animation that approximate its quality—in particular, the scenes of the dwarfs washing up for dinner, the dance in which they celebrate Snow White's arrival and acceptance in their house, and their chase through the forest in a vain attempt to rescue her from the wicked queen—and it was by no means wasted. This material was drawn and redrawn, and Disney confessed that "I had to use different artists on various scenes involving the same dwarfs, so it was hard to prevent subtle variations in each dwarf's personality." Some of the other graphic problems were never solved. For reasons suggested earlier, it was impossible to do Snow White and her Prince Charming in acceptable fashion. They could not be caricatured because they had to meet conventional standards of beauty and handsomeness, and except for Snow White's race through the woods and her dance with the dwarfs, the plot did not provide them with actions and movements of the sort that animation is notably successful in rendering. As a result, all the prince's action and most of Snow White's were Rotoscoped—a process that has remained one of the deeper secrets of Disney's land, never admitted to in public. Very simply, it is a process of photographing live action against a blank background and then integrating these shots into animated sequence by tracing them frame by frame onto the film and inking and painting the photographed figures so they will look as if they were animated. No matter how subtly done, these figures never look quite right in the context of an animated film. Their movements are jerky and hesitant, lacking the smoothness, force and purposefulness of figures that are wholly the creation of an animator's pencil. In later years reliance on the Rotoscope was sharply reduced, but whenever the animators have had to deal with conventional human beauty, they have resorted to the device. Disney's publicists have always claimed that the photographing of live models—word of which got around Hollywood, particularly among the aspiring actors who accepted the work—was only for "study," never for actual use in a film. Apparently it was felt that knowledge of this process would spoil the purity of the studio's image.

Indeed, in the context of the total film, few of its momentary imperfections mattered. Otis Ferguson of *The New Republic*, who was perhaps the most sensitive film critic of the time, noted, for instance, that when the queen transformed herself into a crone, she bore a rather uncomfortable resemblance to Lionel Barrymore (shortly thereafter Disney went through an embarrassing phase of caricaturing well-known movie stars in his short subjects). Ferguson also stated his disappointment in seeing the comedy occasionally falter. "Such things as running into doors and trees on the dignified exit, the jumbled consonant (bood goy, I mean goob doy, I mean . . .), headers into various liquids, etc., are short of good Disney." But Ferguson found, in contrast to these lapses, many examples of finely observed comedy, based on the detailed study of both animal and human behavior: "Take the young deer in the little scene where the forest life first gathers around Snow White: shy but sniffing forward, then as she starts to pat it, the head going down, ears back, the body shrinking and tense, ready to bound clear; then reassurance, body and head coming forward to push against the hand—half a dozen movements shrewdly carried over from the common cat. Or take the way (later) the same deer moves awkwardly and unsteadily on its long pins in the crush of animals milling about, as it should, but presently is graceful in flight, out in front like a flash."

In the last analysis the distinction of the film—and for all its imperfections, it is a major cinematic achievement—rests on hundreds of details of this sort, small touches that one scarcely notices on a first viewing but that must finally be seen as the movie's true subject matter. In the end, they disarm all criticism in much the same casual way that Disney brushed aside the dark mists that enveloped the basic legend. It does not matter that there is an occasional lapse into overly broad humor or sentiment. There is a genuine sunniness of outlook here, a sense that one sometimes gathers from works in the higher arts, of the artist breaking through to a new level of vision and technical proficiency and running joyously, freely before this wind of change, the agony of creation for a moment at least in a state of remission. Ferguson found the film to be "among the genuine artistic achievements of the country" and, in general, that opinion was echoed by other critics. Audiences responded—for the most part—with uncomplicated delight.

To be sure, there was one subartistic current of criticism that has plagued all of Disney's animated films. That was the claim that the film was too violent, too frightening, for children. Journalist Bill Davidson, for example, claims that "the witch in Snow White probably has caused more children's nightmares than Frankenstein's Monster and 'Godzilla' combined." Dr. Benjamin Spock, discussing the alleged terrors of the film, went so far as to tell an apocryphal story against it: "Nelson Rockefeller told my wife a long time ago that they had to reupholster the seats in Radio City Music Hall because they were wet so often by frightened children." All over the country there were earnest debates about the appropriateness of *Snow White* for children, and everyone seemed to have a story about someone's child who had hysterics in the theater or bad dreams for weeks after seeing the film. Disney received hundreds of letters from protesting parents. Typically, his reply was that "I showed *Snow White* to my own two daughters when they were small, and when they came to me later and said they wanted to play witch I figured it was all right to let the other kids see the witch." (Actually Diane Disney remembered *Snow White* a little differently: "When the wicked old witch flashed on the screen I was so terrified that I hid my face in my hands.") Indeed, of the many criticisms leveled at Disney throughout his long career this one about the scares he gave children seems the least valid, though the most widespread. (In the same way, the most common criticism of movies in general—which has to do with their alleged sexiness—also seems the least intelligent, for the moral transgressions of movies are much too complicated and interesting to be discussed in such simple-minded terms.)

In any case, the offending moments of *Snow White* occurred when the queen-stepmother transformed herself into a crone

(not really a witch, as most people seem to remember h
when the huntsman's knife seemed to flash down on the
it was unclear, for an instant, whether or not he had kille her.
Most children recall these scenes with a kind of delicious shud-
der—the sort of thing that, before psychology and do-gooding
liberalism combined in attempt to smooth all the interesting
edges off childhood, used to be the source of our best remem-
brances. A good many youngsters—those who knew their own
limitations—were wise enough to do exactly what Disney's
daughter did: avert their eyes and request a companion to tell
them when the scary part was over (children are remarkably
more sensitive to their own fears than most people credit them
with being). Only a few found themselves so startled by fear
that traumatization followed, for as Donald Barr has observed,
"Disney terror is mere scare. Unless it happens to touch on a
particular child's particular phobias, it simply does not disturb
children, for it has no reference to any of the unease and tension
that arise from a child's private trials."

Disney's critics seemed to forget several things about *Snow
White* and about the other animated features to which similar
objections were raised (notably *Pinocchio*, with its tiny hero
being swallowed by a quite enormous whale, and *Dumbo*, with
its frightening storm and ensuing separation of the mother elephant from her baby). The first and most obvious of these
things is that the majority of children should not be denied the
very real pleasure of exposure to the quickly and pleasantly resolved excitements of these films because a minority cannot deal
with them. Then, too, there is the simple fact that we need to
learn to deal with things that frighten us, even as children. The
ability to separate that which is, after all, "only a story" or "only
a picture" from that which is reality is not a bad one to cultivate
early. Nor is it wrong to introduce children to symbolic representations of evil, in restricted doses and in the right context of
course, in the hope that it will provide them with the imaginative tools to deal with it intelligently when they encounter it in

its more chilling manifestations in life. As for the bad dreams *Snow White* was supposed to have caused, it is a simple fact that it is humanly impossible not to dream, that the materials entering into them are many and varied, and that no child and no adult can escape this particular working of his unconscious. If Disney had not been putting things into the child's stream of consciousness the day he saw *Snow White*, someone else would have been, perhaps to better effect, perhaps to worse, but certainly to *some* effect. One cannot protect a child from experience, and it seems preposterous to protect him from one as generally innocuous, as often rewarding, as *Snow White*.

Indeed, looking back on *Snow White* from the perspective of thirty years, one is inclined to believe that its defects are unimportant in terms of the film itself; it easily triumphed over them. Their real effect was on the future when, uncorrected and allowed to deepen, they vitiated much of the potentially promising work that emanated from the Disney studio. Large numbers of people who knew very little about the movies were encouraged to make sweeping statements about Disney's genius. Mark Van Doren for instance, wrote that "his technique, about which I know little, must of course be wonderful, but the main thing is that he lives somewhere near the human center and knows innumerable truths that cannot be taught. That is why his ideas look like inspirations and why he can be good hearted without being sentimental, can be ridiculous without being fatuous. With him, as with any first rate artist, we feel that we are in good hands; we can trust him with our hearts and wits." The professionals of film criticism sensibly confined themselves to the work at hand, which Frank Nugent of the New York *Times* summarized as "delightful, gay and captivating." If it was not quite fit to rank "with the greatest motion pictures of all time," as Howard Barnes of the New York *Herald Tribune* thought it did, it certainly did "more than match . . . expectations," as Nugent put it. But it also raised expectations for the future, and amidst the chorus of unmeasured praise only one voice was raised to point out the already obvious—but still correctable—defects in the Disney technique.

That voice belonged to Gilbert Seldes, who was one of the best critical friends Disney had. Seldes deserves recognition not only for acumen but for prescience, for he observed and isolated the difficulties Disney had with folk material *before* the release of *Snow White* and based his strictures on two short cartoon renderings of traditional tales, *The Golden Touch* (in which King Midas begged for a hamburger) and *The Country Cousin*, an adaptation of the country mouse-city mouse tale. His immediate concern was a statement by another dabbler in the popular arts—Mortimer Adler. In his book, *Art and Prudence*, the Chicago philosopher declared that Disney's work "reaches greatness, a degree of perfection in its field which surpasses our best critical capacity to analyze. . . ."

Seldes begged to differ. You could at least "murmur a word of warning," he declared, and he did so through a quite precise analysis of the folkish films in comparison with three recent Disney originals involving The Mouse, the duck and the incomparable Pluto. The Mickey Mouse film was *The Band Concert*, which Seldes called "Disney's greatest single work," a judgment that stands the test of time. "It has," wrote Seldes, "comedy of detail—such as the sleeve of Mickey's oversized uniform continually slipping down to conceal his baton; it has comedy of structure based on the duck's persistent attempts to break up the concert by playing a competing tune on the flute—a tune to which the band, against its will, has to turn; it has comedy of character in the stern artistic devotion of Mickey contrasted with the unmotivated villainy of Donald; it has comedy of action when the tornado twists the entire concert into the air and then reverses itself and brings the players back to the grandstand. In no other picture have I observed so many flashing details—during the tornado itself there is the general line of comedy that in the whirlwind or caught on treetops, all the players continue to follow their music, but there are moments in this scene when the screen seems to be animated by dozens of separate episodes. They are miraculous if you catch them and even if you do not, the total effect is miraculous still." He goes on to pay deserved tribute to *Mother Pluto* and *Alpine Climbers*. In the former the good-natured, slightly moronic dog conceives "the fantastic idea that he might by accident hatch a brood of chicks and far from being aghast comes to be more protective than the chicks' legitimate parents." In the latter the ineffable duck and that magnificently shaggy, unnamed St. Bernard who frequently appeared with him encounter a series of angry mountain fauna in the course of an expedition that is a great comedy of errors and terrors.

To these films Seldes contrasts the folk adaptations—"a little pretentious and a little dull," with their heavy backgrounds and "absence of hilarious fun." He correctly sensed what was troubling Disney and his artisans. The bright, simple style of the Mickey Mouse cartoons, despite the great strides that had been made in them, was not suited to traditional folk material; you could not suggest with it the depths the folk material often contained. Neither was the answer to be found in standard storybook illustration, which is where Disney now seemed to be turning. He needed, Seldes implied, more eclectic taste in his study of models. A wide-ranging, informed animator might profitably study Tenniel's great drawings for *Alice in Wonderland* or even those of Peter Arno or James Thurber. Then, too, there were Goya and Hogarth and Adolf Dehn and even George Grosz—if not precisely to copy, then to indicate the range of possibilities available. Disney had found, said Seldes, a gold mine in legend and myth, but he must not confine his operations to rapacious strip mining. He must be prepared to dig deeper and farther. And for that he needed graphic tools of greater range and greater subtlety than he then possessed.

It was sound advice. Perhaps *Snow White* presented too many difficulties, financial and technical, to apply it immediately. And immediately thereafter, Disney found himself on a sea of economic troubles that were not conducive to thoughtful experimentation. But Disney was a man capable of infinite effort, under all sorts of pressures, when he wanted to do something new. The inescapable conclusion is that he was perfectly content with his artistic style and saw no reason to expand its capabilities or its flexibility or even its terms of historical reference.

What one might have dared to hope for—what his critics implicitly did hope for—was not a slavish imitation of the great masters of drawing but rather an expanded capability of fitting a variety of styles to a variety of stories in many moods. Sometimes these styles might have been borrowed, sometimes they might have been invented on the spot. The point is that such eclecticism might have served as a brake on "Disneyfication," that shameless process by which everything the studio later touched, no matter how unique the vision of the original from which the studio worked, was reduced to the limited terms Disney and his people could understand. Magic, mystery, individuality—most of all, individuality—were consistently destroyed when a literary work was passed through this machine that had been taught there was only one correct way to draw.

What was consistently wanting in all the Disney versions of folktales and classic children's stories was something of the spirit with which the brilliant young English actor and director, Jonathan Miller, recently approached *Alice in Wonderland*. Sensibly declaring that there is no good reason to be loyal to the text with which one begins since "the book remains after all," he points out that "the scriptures . . . have survived the creative depradations of the Renaissance painters, and the Greek myths are reborn each century through the imagination of a Seneca, a Racine or a Joyce." The best children's books are like these sources in that "they have a simple, imaginative hospitality which somehow invites the mind to make them over again." The point is, however, that these timeless images must "excite some sort of answer" on the part of their re-creator that is similar to the original work in imaginative intensity and depth of feeling. This is precisely what *Snow White* failed to draw from Disney and his craftsmen, though in the general excitement over what they were attempting, this was overlooked. Disney, the man who could never bear to look upon animals in zoos or prisoners in jail or other "unpleasant things," was truly incapable of seeing his material in anything but reductive terms.

Joseph Campbell has offered a brilliant description of folk stories which can and should be extended to include other material Disney chose to animate later, works like *Pinocchio*, *Bambi*, the Uncle Remus tales (used in *The Song of the South*), *The Sword in the Stone*, *The Jungle Book* and *Winnie-the-Pooh*. Says Campbell: "The tale survives . . . not simply as a quaint relic of days childlike in belief. Its world of magic is symptomatic of fevers deeply burning in the psyche: permanent presences, desires, fears, ideals, potentialities, that have glowed in the nerves, hummed in the blood, baffled the senses, since the beginning. . . . Playful and unpretentious as the archetypes of fairy tale may appear to be, they are the heroes and villains who have built the world for us . . . all are working in order that the ungainsayable specifications of effective fantasy, the permanent patterns of the tale of wonder, shall be clothed in flesh and known as life."

This knowledge need not be intellectualized. Indeed, it better not be, since in the overthought adaptation "the spiritual sap of the work dries up, and the whole thing falls apart like so much dead wood," as Jonathan Miller says. But it must be present, it simply must be present, in some form or another. Perhaps if Disney's psyche had been a little less the product of *petit bourgeois* striving, perhaps if he had had a little less education of the wrong kind—had been more truly a primitive—he might have found the inner spirit to comprehend the power of the material he was going to depend upon and then have found the graphic style, or styles, suitable to it. Perhaps if he had had a little less technological imagination—and a great deal less business acumen—to distract him, he might have found a way to be faithful not to the letter of his material, which is a feckless demand for any movie maker engaged in the process of translation from one medium to another, but faithful instead to its true spirit, its *animating* spirit. Unfortunately, he lacked the tools, intellectual and artistic, he needed for this task. He could make something his own, all right, but that process nearly always robbed the work at hand of its uniqueness, of its soul, if you will. In its place he put jokes and songs and fright effects, but he always seemed to diminish what he touched. He came always as a conqueror, never as a servant. It is a trait, as many have observed, that many Americans share when they venture into foreign lands hoping to do good but equipped only with know-how instead of sympathy and respect for alien traditions.

PART EIGHT

Troubled Times

24 *When banks came into pictures, trouble came with them. When we operated on picture money, there was joy in the industry; when we operated on Wall Street money, there was grief in the industry.*

— Cecil B. DeMille

WALT DISNEY AGREED. Not long before he died he fell to discussing bankers with a reporter and declared, "They are fellows who don't understand your business—my problem all along. I don't mean to be heavy on bankers, but they just don't understand." He was fond of declaring his sympathy for his brother, who had to spend so much time with the money men and once told him, "You've got to get away from those fellows. They'll get you down. Stay as far away from them as you can."

Had they chosen to, the Disneys could have been in a position to avoid "those fellows" for years, perhaps for the rest of their lives after the success of *Snow White*. For the people set aside whatever fears for their children's mental health the film stirred and seemed to have none of the other reservations since expressed about it. Six months after its release, *Snow White* had returned a gross of some $2 million to the Disney studio and was on its way to grossing the $8.5 million it produced on its first run. (As of 1985, the picture had returned $41. $4 million to the studio's coffers.) It would have been possible to finance a modest, steady expansion of the studio out of its profits alone. But Walt Disney saw a chance for a next leap forward. "Success is hard to take," said Roy Disney. "We thought we would do two animation features a year." Before *Snow White* went out into the world, Disney was already beginning work on two features—*Pinocchio* and *Bambi*—and, shortly after they were placed in full production another project began to grow—and grow—in his mind, a project that would wed animation to serious music on a scale never before attempted, in an effort to popularize such music for a very wide public. The result was, of course, *Fantasia*.

All this naturally required huge outlays of money, and as if the films were not enough, the Disneys decided to go ahead with the construction of Walt's $3.8 million dream studio. As a result, by 1940 not only had Disney run through most of the *Snow White* profits, but his organization found itself in debt to the banks by roughly $4.5 million. Worse, the prospects of paying off such an amount grew dimmer by the hour as the spread of World War II closed off the foreign markets where Disney counted on gaining about 45 percent of his return. And this was not the end of his troubles: he had labor unrest to contend with; *Bambi*, in which he wanted animation of almost photographic naturalism, was seriously delayed by technical problems, and *Fantasia* turned out to be at least a temporary financial calamity.

What appears to have happened is that in the first flush of big success, Disney set aside his habitual, populist distrust of bankers and plunged ahead with a recklessness unprecedented even for him. In 1938, for the first time, the banks were genuinely receptive to him, even encouraging. And, after all, their aid had been instrumental in the completion of *Snow White*. Why not use them again to press forward in this most propitious of all moments? Times were, if not good, certainly better than they had been in a long while. And the worth of Disney's labors had been blessed not only by an honorary degree from the University of Southern California but by similar honors from those citadels of the eastern intellectual establishment, Harvard and Yale. At the latter, no less a personage than William Lyon Phelps declared, in writing the citation, that "one touch of nature makes the whole world kin, and Walter Disney has charmed millions of people in every part of the earth. . . ." Why, he was even a Chevalier of the French Legion of Honor. Best of all, it was beginning to seem that he was gaining acceptance from Hollywood's power elite. When *Snow White* premièred at Hollywood's Carthay Circle Theater in December, 1937, he at last had a full-length feature that could be given a klieg-lighted send-off of the sort the older moguls habitually gave their pictures. "A long, long time before," his daughter wrote, "he had promised himself that someday a cartoon of his would have such an opening. . . . So . . . with Hollywood's brass turning out [it] was a triumph for Dad." The same may be said about the special Academy Award that was presented to the studio in 1939 for *Snow White*. It came in the form of one big Oscar and seven little ones—for each of the seven dwarfs—and it was presented by none other than Shirley Temple. Disney's head was turned: to a reporter who referred to *Snow White* as a cartoon he allegedly snapped, "It's no more a cartoon than a painting by Whistler is a cartoon." It was a very blunt statement, somewhat at odds with his modest public character of the time. Still, Disney was entitled to his euphoria, and to relish his new power, even though the mood was to prove costly.

Of all the projects in hand, only *Pinocchio* presented no insuperable problems. This film was shot entirely on the multiplane camera, and indeed, its opening sequence, with the camera panning in at night across the rooftops of the village where the toymaker Geppetto lived and then down on a dusty, wayfaring cricket named Jiminy, who was looking for a place to sleep,

231

was regarded as something of a marvel. No less than twelve planes were employed. The shot cost about $25,000, and the accountants talked Disney into using less elaborate setups in the rest of the film. The problems Carlo Collodi's original story presented were not so easily resolved. The Italian writer had written his book directly for magazine serialization, and since he had been paid, as was the nineteenth-century custom, on word rates, he had allowed himself to ramble considerably. As a result his tale of a puppet, given life by a Blue Fairy but denied real boyhood until he proved himself "brave, truthful and unselfish," needed considerable pointing, though even in its sensibly trimmed version, there was plenty of action to fill the scenes. Before the film was finished the puppet fell in with such comically evil companions as J. Worthington Foulfellow, a con-man fox; his henchman, Gideon, who was an avaricious cat, and a juvenile delinquent named Lampwick. He gave himself up to the debaucheries of Pleasure Island and was very nearly turned into a donkey before he escaped. Finally, he was transported on the wings of a great dove to the middle of the ocean, where he allowed himself to be swallowed by the great whale, Monstro, in the hope of rescuing his foster-father, Geppetto, who had suffered the same fate in the course of a voyage to save Pinocchio from the evil temptations placed before him.

One must suspect that Disney found in this story elements of autobiography, since he had himself been a child denied the normal prerogatives of boyhood. It is certainly possible that at least some portion of his drive for success was a compensation for his failure to find the father who had, in the psychological sense, been lost to him since childhood. This may also explain why he so relished the paternal role he was now beginning to play with such earnestness—and such ineptitude—with his employees. In any case, such an interpretation suggests why *Pinocchio* is the darkest in hue of all Disney's pictures and the one which, despite its humor, is the most consistently terrifying. The menacing whips that crack over the heads of the boys who are turned into donkeys after their taste of the sybaritic life may have had their origin in his recurrent nightmare of punishment for failure to deliver his newspapers. And, of course, one suspects the dream of winning the neglectful father's approval by a heroic act—such as rescuing him from a living death in the whale's maw—must have occurred to Disney at some point in his unhappy youth.

Whatever problems the range and sweep of the story presented, it had a rhythm and a drive far stronger than *Snow White* and more like that of the short cartoons. There was more adventure in it, more spectacle and, perhaps, a better balance among all its elements. The most sympathetic and animatable characters were also the main characters in *Pinocchio*, as contrasted to *Snow White*, in which the heroine's role had to be cut back in favor of the dwarfs' because of the difficulties she presented to the artists. In addition, the art, in all its aspects, was smoother and more elaborate, though not yet overslick— perhaps the high point of Disney's chosen mode and, this time, well adapted to all the film's moods instead of just a few of them, as had been the case in *Snow White*.

The picture was also notable for its introduction of a device that was to be employed frequently in future Disney animated features in order to make them more easily palatable to the family audience. In the original Collodi story, there was a cricket who offered to serve as the puppet's conscience but who was indifferently squashed by Pinocchio after very little discussion. It was not exactly an endearing gesture, nor did Pinocchio's character prove entirely winning as the story proceeded: he was, from Disney's point of view, too much the out-and-out delinquent. He was consequently softened into a good-natured lad, easily led astray but trying in his inept way to discover the principles of correct behavior. More important, the cricket was not squashed but developed. Appointed the puppet's conscience by The Blue Fairy, he became Pinocchio's worldly wise mentor, ally and rescuer—serving him, in fact, in rather the manner that Roy Disney served his younger brother. He also served as a kind of chorus commenting on the action and as an admirable foil to the hero. Most important, he was an adorably bright little creature, cute as a bug as it were, and probably more popular with audiences than Pinocchio, who was rather square and simple in manner (the cricket also had the film's most memorable song, "Give a Little Whistle").

The success of this wee creature showed Disney and his people a convenient way to brighten and lighten any story they feared might grow too serious or unpleasant for audiences. Jiminy Cricket's direct spiritual descendants were thereafter rarely absent from Disney's animated films. In *Bambi* there were Thumper the Rabbit and Flower the Skunk, who introduced the young deer to his woodland environment and instructed him in matters of safety and convenience; in *Dumbo* there was Timothy the Mouse, who was both conscience and theatrical manager; in *Cinderella* there were the mice who were the heroine's only friends; in *Peter Pan* Tinker Bell, who had always been a tiny light on the stage, was characterized as a midget (but physically well endowed) nymphet; while a talkative owl was luckily present in the original of T. H. White's *The Sword and the Stone*, the least well known of Disney's animated films. It is perhaps significant that Disney's greatest animated failure, *Sleeping Beauty*, contained no tiny creature for audiences to love or to tell them how to respond to what they were seeing. In that film they were left floundering among the ambiguities like so many intellectuals. In the first instance, however, Jiminy Cricket was an effective little character and an effective little storytelling device as well.

The reviews of *Pinocchio* were almost as ecstatic as those accorded *Snow White*. Indeed, Frank Nugent thought *Pinocchio* "superior to *Snow White* in every respect but one: its score. . . . [It is the] best thing Mr. Disney has ever done and therefore the best cartoon ever made." His colleague on the *Herald Tribune*, Howard Barnes, thought it "a compound of imagination and craftsmanship, of beauty and eloquence, which is to be found only in great works of art." But one senses in all the notices a slightly reserved note, almost as if the critics were prisoners not only of their previous enthusiasm but of the general unwillingness to criticize the works of such a universally esteemed citizen. Barnes noted, for instance, that the new film "lacks the element of surprise and something of the emotional depth" of its predecessor. Franz Hollering in *The Nation* said that "one misses the lyrical parts of *Snow White* . . . and also its stronger central idea." It seemed to him, therefore, "smaller than its parts," which was the exact opposite of the common response to *Snow White*, where technical difficulties were ignored because the total effect was so enchanting.

Such minor reservations were not the cause, however, of the film's failure to return more than its $2.6 million cost on its first release. The trouble was that by the time it was released, Germany had marched into Poland, beginning World War II. *Pinocchio* therefore became the first Disney film to feel the effect of the curtailment of the overseas, though ultimately it would gross a handsome $30.4 million. For the moment, though, it represented the beginning of a squeeze that was to grow tighter and tighter in the next few years.

The inability of *Pinocchio* to perform at the box office did not prevent Disney from moving into his new studio, but it certainly darkened what was a major event in his corporation's life. Here, at last, was the rationally planned factory he had wanted for so long, here at last was a structure symbolizing a stability and an importance never achieved before by an animator. To be sure, he had been forced to make some compromises. Delighted as they had been by *Snow White*'s acceptance, Disney's bankers were still not absolutely certain he was here to stay. They would advance him money for a new studio only on condition that its principal structure (still called the animation building, but nowadays housing mostly executives) could easily be adopted to some other use. This particular section of Burbank needed a hospital, and so the bankers insisted that the building be constructed for easy conversion into such a place. As a result, it is one of the oddest edifices to be found on any movie lot. The central corridor on each of its three floors is about twice the normal width, the better to trundle beds and wheelchairs about. Scattered along it are niches that nowadays hold Coke and candy machines but that could easily serve as nursing stations. Entrances to eight short wings, four on either side of the hallway, are placed at regular intervals, and one of them, on the top floor, was given over completely to Walt Disney's suite of offices. It, like all the other wings, could be turned into a ward with the greatest of ease. In fact, this peculiar design entails no great inconvenience for anyone, although the ease with which movement along the corridor could be controlled—so important to hospitals—was to become a sore point among the employees in very short order. It should also perhaps be noted that the bankers were right about the need for a hospital in the area, for St. Joseph's, where Disney died, was built directly across the street from the new studio a few years later.

There were two other major compromises with Disney's initial vision. The first of these was immediate: the idea of an amusement park was abandoned. Resident archaeologists have found rough sketches for such a park dating from the early 1930s in the files. It was to have been two acres in size, containing train and pony rides, and there was talk around the studio, according to veterans, of singing fountains and perhaps statues of Mickey Mouse, Donald Duck and the rest of Disney's stars. This conceit, at least, was carried out to a modest degree by naming the studio's principal streets Mickey Mouse Boulevard, Minnie Mouse Boulevard, Donald Duck Drive, Snow White Boulevard and Dopey Drive. The main building stands at the intersection of Mickey and Dopey, where the wrought-iron signboard has a tendency to tilt in an untidy fashion. The second compromise with the dream came later on, when the building designated on the original ground plan as a school became, ignominiously enough, the headquarters for the publicity department.

Still and all, it was—and is—a neatly functioned place, "a mass of clear-cut buildings, with a horizontal emphasis of design, not self-consciously asserting what we are told at school is the 'modern functional style' but merely workmanlike," to quote an early description. "Under the California sun they show up clean, in tones predominantly pink, cream-colored and gray, with the sun glinting on the dull metal used in conjunction with brick." Such lack of true architectural distinction as the place may have was compensated for by pleasantries of landscaping. Plots of well-tended, very green grass separate the buildings, cheerful flower beds abound, and there are plenty of areas useful for various informal games. Disney was wise enough to purchase enough land—forty-four acres—for uncrowded expansion, and over the years three sound stages have been added to the original single stage as the bulk of production has switched from animation to live action. Reflecting the same trend, a scene dock has been added, along with an expanded carpentry shop; a warehouse has become a casting office, and the short-subject building has become the makeup and costume room serving the influx of actors. Even with this change in emphasis, however, the place still fits the criteria that were, according to Robert D. Feild's constantly burbling study of the man and his works "inseparable in Walt's mind"—"formal planning, which would allow the picture to be manufactured as efficiently as possible, and an arrangement whereby those employed could retain the greatest possible amount of creative independence." Feild's line was that "it may be a factory, but in this twentieth century, by so becoming, it has dignified the artist's calling." As he saw it, the studio was a step toward breaking down "the tradition that art is only messing about with paint; that the artist is normally to be found in an 'attic' or a 'barn'; and that studio or art school must center around a model, in the nude perhaps, regardless of whether its activities are 'pure' or 'applied' or 'commercial' or

237

'industrial.' " He wrote this in 1942, and the involuntary wince with which one reads these words today is a measure of how the hoped for marriage of art and technics has failed under the perverting pressure of worldly concerns.

Disney's pride in the new plant was immeasurable. But, ironically, the assaults against that pride were incessant. His father, for instance, was unimpressed. "When the new studio was finished," a contemporary chronicle reports, "Walt took his father around and proudly pointed out all the gadgets. The little boy who had carried papers was grown up now and this was his dream come true. 'It's all air conditioned,' he boasted. 'You can get any kind of weather, any time you want.' The old man thought it over. He seemed unimpressed. Taking in the whole architectural triumph with a wave of his hand, he inquired: 'What else is it good for?' " Disney is elsewhere reported to have replied, with some weariness, "A hospital." Of this same visit, Disney later said, "Pa was of the generation that didn't believe in borrowing money, and he offered to help me out with $200 when I was eight million dollars in the hole trying to build my studio." The last figure is an exaggeration, but the tone of the reminiscence rings true; Disney appears to have seen as little of his parents in these years as was decently possible.

25 *Oh*, Fantasia! *Well, we made it and I don't regret it. But if we had it to do all over again, I don't think we'd do it.*
—Walt Disney, 1961

Fantasia BEGAN TO preoccupy Disney before either *Pinocchio* or the new studio were finished. It has been said that the film grew out of a meeting between Leopold Stokowski and

Disney at a Hollywood party, where the former repeated an oft-made wish to work with Disney on something and requested a tour of the Disney studio. At the time, Disney was in the midst of one of his periodic, sentimental attempts to restore Mickey Mouse to something like his early popularity. The instrument of this rehabilitation was to be a short subject based on Paul Dukas' scherzo for orchestra, "The Sorcerer's Apprentice." Stokowski was entranced, and agreed to conduct the score. He persistently wondered, moreover, why Disney should stop there. Why didn't he make a full-length film, using several other musical works to suggest "the mood, the coloring, the design, the speed, the character of motion of what is seen on the screen," as the conductor later expressed it. In short, a fantasia, in the precise sense of the musical term, which means a free development of a given theme. (In the beginning the word was merely a working title for the film, but no one could improve on it, so it stuck.)

Among the first ideas the oddly matched collaborators had was to dispense with written credits. Their wedding of art and music was to be uncluttered by any dependence on the literary arts. So Deems Taylor, sometime composer and musicologist then at the height of his fame as an explainer of music for the masses on the Metropolitan Opera broadcasts, was hired to give a spoken introduction to the film. Soon he was sitting with Stokowski and Disney in a three-week conference attended by the heads of various Disney departments, at which hundreds of recordings of hundreds of pieces of music were played in a search for the film's program. By 1939 Stokowski was recording the selections they had made, working in Hollywood and in Philadelphia, where he was nearing the end of his long tenure as conductor of The Philadelphia Orchestra.

Not unnaturally, it was decided to open the film with the maestro's own transcription of Bach's "Toccata and Fugue in D Minor." Stokowski had begun his career as a church organist.

he had built and conducted his orchestra so that critics often spoke of its organ-like sonorities, and his transcription of Bach's great work for the organ not only had been a labor of first love but was well received by audiences right from the start, despite some purist quibbling by critics. Indeed, it is not too much to say that the elevation of Bach to a proper place of esteem among American music lovers in this century began with the popularity of Stokowski's lush and viscerally exciting work, so different in its effect from pure Bach, whose appeal lies in austere intellectuality.

In any case, the "Toccata and Fugue" was followed in *Fantasia* by excerpts from "The Nutcracker Suite," "The Sorcerer's Apprentice," "The Rite of Spring," "The Pastoral Symphony," "The Dance of the Hours," and a sequence that combined "A Night on Bald Mountain" with the "Ave Maria." Musically speaking, it was a very mixed bag, and it was even more so as a piece of animation.

As Stokowski said, "*Fantasia* was created and drawn by artists most of whom have no knowledge of, or training in, music." He thought that as "enthusiastic listeners" they were able to penetrate the inner character of the music, there to discover "depths of expression that sometimes have been missed by musicologists." Even he had to concede, however, that the visions contained in *Fantasia* were just one way "out of many possible ways" of visualizing the music. All too often, these were the crudest ways. In the opening sequence, for example, Mickey Mouse appears, shakes hands with Stokowski (in silhouette) and welcomes him to Hollywood. The orchestra, also in silhouette, begins to play the "Toccata and Fugue," and images suggestive of musical instruments being played by hundreds of hands begin to appear on the screen. These shapes grow more and more abstract, then clouds begin to form into strange shapes, comets flash, a rippling mass fills the screen and finally shapes itself into a pipe organ—just in time for the climax of the piece—when it disappears, as Deems Taylor later put it, in "a blaze of light, against which stands the tense, vibrant silhouette of the conductor. A last great chord. Bach has spoken."

Someone—or something—surely has. But it is a travesty to imply that it is Bach, since the music heard is a transcription of his work and the animation is a transcription of the transcription, into another medium entirely. Disney enjoyed working on the sequence, perhaps, because its basic concept was his: "I said, 'All I can see is violin tips and bow tips—like when you're half asleep at a concert." He thought they were abstractions, but they were not, of course. They were merely a form of literalism different from any he had attempted before.

But if the opening sequence was unsuccessful, it was at least not offensive in *Fantasia*. There is some debate about which one really represented the nadir of its taste. Some hold for the animated vision of the creation of the world with its garish volcanic eruptions, its cheaply scarifying dinosaurs, its attempt to set general-science level education about geology and evolution to Stravinsky's "Rite of Spring." Others are most distressed by the revel on Mount Olympus, which formed the "story" for a truncated version of the Beethoven *Sixth Symphony*, a cheerful nightmare compounded equally of the oppressively cute and the depressingly heavy-handed. To make sure everyone got the idea that this was art, the girl centaurs were originally drawn bare-breasted, but the Hays office insisted on discreet garlands being hung around their necks, a decision that satisfied puritans but did nothing to hide the grotesquery of their conception. The torsos and heads that topped the horse bodies of these creatures belonged to adolescent girls styled to resemble the teen-ager down the street—surely the weirdest blend of classicism and Americanism ever observed in a nation too long devoted to the feckless enterprise of reconciling these irreconcilable impulses. The sequence ends with the most explicit statement of anality ever made by the studio, which found in the human backside not only the height of humor but the height of sexuality as well. Two of the little cupids who scamper incessantly through the sequence finally—and blessedly—draw a curtain over the scene. When they come together their shiny little behinds form, for an instant, a heart.

Probably the worst thing in the film is its conclusion, in which the terrors of "A Night on Bald Mountain"—bats and gargoyles and devils and the other creatures of the gothic demonology engage in a black mass—are dispelled by the coming of dawn and a procession of the churchly singing the "Ave Maria"; it is essentially the structure of "The Skeleton Dance" enormously elaborated. Taylor thought that "incongruous as this juxtaposition may sound at first blush, it is astonishingly successful; for the two works are so antipodal in mood that they complement each other perfectly . . . an engrossing tone picture of the struggle between good and evil, and the ultimate triumph of the good," which proves that nothing is sacred, not even the sacred, when an eager popularizer gets busy. In itself, the idea of this juxtaposition is not offensive, but as the climax of the film it seems insincere—a conventionalized invocation of religiosity, an arbitrary resort to a surefire sentiment, rather like the placement of a brassy patriotic number at the end of a musical review. The execution of the sequence is arty, the musical arrangement throbs with false emotion achieved through an excess of stringed instruments and a lush choral setting of Schubert's song, which is, of course, intended for a solo voice.

Having observed all this, it is only fair to point out that there is much in *Fantasia* representing a distinct advance in the studio's work and some small things that succeed on almost anyone's cinematic terms. There are a couple of pleasant things in "The Nutcracker" animation, especially the dance of the Chinese mushrooms; "The Sorcerer's Apprentice" sequence, featuring Mickey, with which the whole project began, is a very good one, full of wit and invention, and "The Dance of the Hours," with its ostrich and hippopotamus ballerinas remains a lovely piece of low comic animation, a joke on a piece of music that had become something of a joke itself as well as a broad satirical comment on the absurdities of high culture. Indeed, in the context of the film this short section can be seen as a mockery of the very material that surrounds it. Finally, even in the most pretentious of *Fantasia*'s sequences there is an eclecticism, a reaching out beyond the studio's standard style that is most welcome. The massive renderings of the earth in upheaval in "The Rite of Spring" and the battle of the dinosaurs that follows do have a certain rude power one might not have thought the animated film capable of attempting, however flawed the results. And up until the tasteless modulation into the "Ave Maria," Disney's venture into the tradition of the grotesque in "A Night on Bald Mountain" is a rewarding one. Wolfgang Kayser points out that there is a relationship—and not one of opposites, either—between Disney's chosen métier, the fairy tale, and the grotesque. Indeed, the former appears at first glance to be very like the latter, a world where we are exposed to the "alienation of familiar forms." On closer examination we see that those elements of the fairy tale that are already familiar to us are usually not transformed; they are comfortingly familiar as they are in our own experience. Only the truly imaginary creations of the tale are strange to us—imaginary toads in a real garden, as it were. The genuinely grotesque takes us only one significant step farther, transforming everything, the familiar and the unfamiliar alike, into terrifying symbols that have as their goal "a secret liberation." "The darkness has been sighted, the ominous powers discovered, the incomprehensible forces challenged," leading to a definition of the grotesque as "an attempt to invoke and subdue the demonic aspects of the world," as Kayser puts it. Until the willful intrusion of the Pilgrim chorus, this is precisely what Disney's artists attempted in the last moments of *Fantasia*, and it is because they succeeded so well—though not perfectly—that the "Ave Maria" intrudes so disturbingly on their work. It is as if Disney could not bear the implications of what they did—or was afraid that his audience could not. To intrude religious sentimentality on such a potent vision of blackness is to spoil it, to negate all the good work that went into it. Had Disney not pulled back, it is possible that, at least with the sophisticated audiences, he might have swept aside, in one last burst of power, all the objections that the earlier sequences of the film raised. Instead, he merely reinforced them.

It is possible, of course, to say that if the studio had not been ready for *Snow White* it certainly was not ready for *Fantasia*. If after his first exposure to the centaur sequence Disney could—as he did—comment, "Gee! This'll make Beethoven," one is entitled to wonder if his organization would ever be ready for a work of this kind. A fawning respect for cultural tradition is not what was required. What was needed here—as in all adaptations of high culture to mass culture—was respect for the integrity of the forms he was seeking to make over. Without that respect, it is impossible to turn out a work with any integrity of its own—and that is why *Fantasia* is such a disturbing jumble.

The experience of Igor Stravinsky is apposite in this respect. He reports getting a take-it-or-leave-it offer from Disney's New York office of five thousand dollars for the use of "The Rite of Spring" in the picture. If he refused the money, he was informed, the music would be used anyway, since it was copyrighted in Russia, which did not protect it here because the U.S. had never signed the Berne copyright convention. Stravinsky decided he was faced with a *fait accompli* and gave permission for the use of the music. "I saw the film with George Balanchine in a Hollywood studio at Christmas time, 1939," he wrote later. "I remember someone offering me a score and, when I said I had my own, the someone saying, 'But it is all changed.' It was indeed. The instrumentation had been improved by such stunts as having the horns play their glissandi an octave higher in the 'Danse de la terre.' The order of the pieces had been shuffled, too, and the most difficult of them eliminated—though this did not save the musical performance, which was execrable. I will say nothing about the visual compliment, as I do not wish to criticize an unresisting imbecility. . . ."

At the time, Stravinsky made no protest; he even submitted to having his picture taken with Disney. The latter apparently mistook resignation for approbation and he recalled Stravinsky's emerging from the screening "visibly moved"—as perhaps he was, one way or the other. He claimed, in any case, that Stravinsky made two visits to the studio and had, on the first one, approved all the changes and cuts that were made, an allegation the composer denied on the ground that he was confined to a tuberculosis sanitorium in Europe prior to and during the recording of his work. The point is that what happened to Stravinsky happened to all the composers represented in *Fantasia*. His misfortune was that he was the only one still alive and able to take offense. He therefore encountered the producer at his entrepreneurial worst. Had Disney actually possessed the soul of an artist—as his more ardent and ignorant supporters liked to claim he did—he could not have treated Stravinsky as he did and he would have approached the entire *Fantasia* project in a different spirit.

Superficially, *Fantasia* seems to put Disney's critic in a contradictory position. If the basic objection to his work in general is that its style and viewpoint too quickly and too narrowly solidified into a minor mold, how can one criticize his only large-scale effort to break out of his self-imposed limitations? The answer, of course, lay in the attitude Disney took to *Fantasia*. Had he been able to see it as a genuine artist sees his failures, as instructive experiences from which one learns something about the limits and possibilities of his art, then he might have gone on from it to create something that fully satisfied the promise that was scattered throughout the picture but never satisfactorily integrated. But, as *Time* later wrote, "though Walt learned a lesson from *Fantasia*, he learned the wrong one. Mistaking for culture what Stokowski and Taylor offered him, he decided that culture was not for him."

Worse, he was denied even his customary technological pleasures. He had planned to shoot *Fantasia* in a wide-screen process, complete with stereophonic sound and to play it on a reserved-seat basis for a long period of time before going into a more customary releasing pattern. Under pressure from his bankers, he was forced to abandon his new projection system entirely, to forego stereophonic sound in all but a few key cities and to shorten the film considerably so it could get quickly into the neighborhood houses, where the bulk of Disney's profits were traditionally made. "The bankers panicked," Disney said later, "*Fantasia* was never made to go out in regular release. I was asked to help cut it. I turned my back. Someone else cut it." Ultimately he was vindicated. Revived several times since the end of World War II, and shown at its full length, the picture has more than returned its $2.3 million investment, earning some $4.8 million in the U.S. and Canada alone.

But it still fared very badly in its initial run. The critics were fairly enthusiastic. Bosley Crowther of the New York *Times*, for example, found that it sometimes wearied the senses, that the color was occasionally too pretty and the dominance of visual imagery over the music could be annoying. Nevertheless he declared that "motion picture history was made last night" by a film that "dumps conventional formulas overboard and reveals the scope of films for imaginative excursion. . . . *Fantasia* . . . is simply terrific." Over at the *Herald Tribune*, Howard Barnes was less ecstatic. Calling the film "a brave and beautiful work," he felt constrained to add an objection that was to become a commonplace in discussion of *Fantasia*. "Being one who does not believe that one artistic medium needs translation into terms of another, it seems . . . Disney is attempting the impossible. There are times when his breaking down of music into animated art strikes me as definitely pretentious . . . the images on the screen are not apt to match with your reactions to the score." Still, he viewed it as a "fascinating new experiment . . . a courageous and distinguished production." The trouble was that the public was uninterested in experiment from Disney. They simply stayed away from *Fantasia*, more frightened by its reputation, perhaps, than they would have been had they tasted its comparatively mild pleasures and perils firsthand.

From Disney's point of view, they could not have picked a worse time to shy away from one of his products. When he undertook *Fantasia* he had not had any reason to suppose that it would be as economically important as it was turning out to be. Indeed, in 1938 he had every reason to believe that he could afford a money-losing experiment. But with *Pinocchio* unable to perform as expected abroad and with the potentially profitable but difficult *Bambi* put aside in order to devote full effort to *Fantasia*, he was suddenly in perilous straits. Far more than should have depended on *Fantasia*.

And the situation worsened even as the film approached completion. Early in 1940 the banks shut down the Disney credit line entirely, and in order to raise operating capital he was forced in April, 1940, to offer stock to the public for the first time, thus diluting ownership—a step he feared would mean dilution of his control over the enterprise. Although the latter never came to pass, it was a psychological blow that was perhaps harder for him to take than it would have been for other men. Indeed, shortly after the stock went on the market Disney was passing through Detroit and was told that Henry Ford liked his work (it was almost inevitable that he would). A meeting between the inspiration of *Fordismus* and one of his great, though unconscious, disciples was arranged. When Disney told the old man that he had just put his stock on sale, Ford commented sourly, "If you sell any part of an enterprise, you should sell it all." He, too, had recently been forced into the clutches of the money market, which he naturally identified with his fantasy of the worldwide Jewish conspiracy and which he had long fought to avoid. He more or less exactly summarized Disney's worst fears about the matter in his single-sentence comment.

Disney tried, at least, to keep some of the stock close to home by announcing that about 20 percent of the shares that had been set aside for himself and his brother would be distributed to employees. The stocks were 6 percent cumulative convertible preferred shares with a par value of twenty-five dollars. This and future distributions were to be in lieu of the holiday bonuses previously awarded employees, and the basis of the distribution was to be what it had been in the case of the bonuses—the length of the individual's service with the company and the value of those services as determined by the management. It is possible, as some former Disney employees have charged, that the gifts of stock were intended to assuage a worsening labor situation in the studio. It is also possible that no such thing was intended. The public record is silent on the point.

What is clear, in retrospect, is that the arbitrary distribution of stock deepened one of the grievances that most rankled the employees. If Disney seems to have been high-handed and insensitive with people like Stravinsky, he was doubly so with those who worked for him. There was no consistent policy in the studio regarding wages, and the ability to obtain raises was based, so far as the employees could see, purely on one's current standing with the boss. (Some say that they were encouraged to think of Roy Disney as the heavy in this situation, with Walt Disney, in effect, hiding behind his brother's skirts when requests for wages or favors were turned down.) The stock distribution was, to the rebellious employees, one more example of this system—or nonsystem—in operation. They saw only the stick in the offer, not the very real carrot. In the quarter century since, the shares have split several times, and the new shares have, as previously noted, soared to a level four times that of the original offering. In the years immediately after 1940, however, the trend was mostly down, and in the aftermath of the bitter strike that afflicted the studio in 1941, the stock slumped to a price as low as three dollars, driven there at least partially because employees dumped them in a gesture of displeasure with Disney. Like many such gestures, it backfired. Disney's faith in the future of his company was unwavering, and he picked up stock bargains by the hundreds during the dark years. Still, the gesture was understandable, for if the strike was not a violent one in its overt manifestations it was a wrenching experience psychologically for all concerned.

26

Come all of you good workers,
Good news to you I'll tell,
Of how the good old union
Has come in here to dwell.
Which side are you on?
Which side are you on?
—Labor song, popular in the 1930s

IF HIS FATHER'S response to the new studio had been disappointing, the reaction of his employees when they moved in in 1940 was an unpleasant shock for Disney. However luxurious their new accommodations, they did not assuage a sense of disappointment that psychological conditions were not similarly upgraded. Indeed, they were worsened by the move, for now the production units were isolated from one another, and at the entrance to each group of offices sat a young lady whose duty was to ask the destination and mission of anyone desiring to leave his work area to visit another. Assistant animators and in-betweeners were isolated on the first floor of the animation building, which became a hotbed of discontent, verging on rebellion. Thus the easy give-and-take camaraderie of the Hyperion Avenue studio was threatened, and there appeared to be no compensation in the offing for its loss. The sense of foreboding that had gripped the studio ever since the completion of *Snow White* was mightily increased. Many of the employees had given Disney large quantities of free overtime when he was driving to complete that film, and instead of getting the bonuses they had been vaguely promised if it succeeded, they were now faced with a string of layoffs. Rumors that time clocks would be installed—despite Disney's previously well-publicized personal dislike of the instruments—did nothing to allay fears. The salary structure remained crazy-quilt, and the only general wage increase Disney granted in these years was self-serving: he brought a number of workers up over the forty-dollar-a-week level, at which point, under the Wagner Labor Relations Act, they ceased being entitled to time-and-a-half for overtime. Worst of all, Disney himself responded gracelessly to the pressures of his increasingly difficult economic situation. The conversational coinage of the story conference and the sweat box had always been rough, but now it seemed to grow unbearably brutal. One animator, working on *Fantasia*, decided to take piano lessons—at his own expense—in an attempt to gain a better understanding of musical structure and thus improve his work on this difficult project. When Disney found out about it, he snarled, "What are you, some kind of fag?" In the increasingly tense atmosphere of the studio, such remarks rubbed feelings very raw.

To the employees the answer seemed to be some sort of union. If Disney was so intent upon rationalizing his productive capacity, and if that process was going to destroy the informal style that had previously sufficed in the area of industrial relations, then it was necessary for them to do a little rationalizing of their own. Disney, in a move to avoid the possible intrusion of labor racketeer Willie Bioff—whose distressing presence, it should be added, in fairness to Disney, was actively encouraged by other producers—had already formed a company union. Like so many of its peculiar breed, the union was devoted mainly to smoothing things over in a manner favorable to management. Now, however, some of the more restive employees began talking to representatives to the new Screen Cartoonists Guild, which was affiliated with the AFL painters and paperhangers union. Under the goad of this competition, the company union took a somewhat more militant stance toward management, and battle lines were drawn for what Disney later preferred to think of as a jurisdictional strike.

Throughout the thirties, of course, Hollywood had been moving toward the unionization of all sorts of crafts not covered by the traditional theatrical unions, as well as toward the strengthening of many weaker labor unions. Indeed, the transformation of one of these, the Screen Actors Guild, from a tiny and ineffectual debating society into a tough and powerful union was not without its moments of high and hilarious drama. Early in the depression the actors had taken voluntary paycuts, believing the producers when they said the cuts were necessary to avoid layoffs. Shortly thereafter, layoffs were instituted anyway, and the actors, led by such worthies as the Marx Brothers, Frank Morgan and Eddie Cantor, fired off a two-thousand-word telegram to President Roosevelt, detailing their grievances and bringing considerable public pressure to bear on the producers. The actors managed to get Guild shops at all the studios by the simple expedient of having a delegation, headed by Robert Montgomery, who was the Guild's president, and Franchot Tone, call on Louis B. Mayer and present him with an ultimatum: grant the Guild shop immediately or face a strike the next day. Mayer had plenty of arguments to muster against his polo-coat proletariat: he was in the middle of a bridge game, he had two hundred guests for lunch, a number of his fellow producers were at the racetrack and therefore unreachable for polling on their intent in the matter. Worst of all, it was Sunday and therefore there was no secretarial help to draw up an agreement. The unionists were made of stern stuff, however—and as many have testified, it took stern stuff to stand up to Mayer—so they eventually came away with a hand-written document of surrender from the biggest of the big bosses. The scene was not exactly reminiscent of Walter Reuther's battle with Ford's goons at the overpass at River Rouge, but each man fights a different battle and in a style suited to his condition.

Enough has been written about the peculiarities of the class struggle in Hollywood to indicate that no one but its participants or a Yahoo congressman investigating it well after the fact could take it seriously. There is, for example, the well-known story of the status-conscious thousand-dollar-a-week screen actor afraid to enter a party given to raise funds for loyalist Spain because it had been made known that only those in the fifteen-hundred-dollar-a-week-and-above bracket were welcome. There was the leftist screenwriter who slipped a few lines of La Pasionara's into a coach's pep talk in a football movie. There were the words of the more hard-bitten comrades who let it be understood that the writers and directors and stars of Hollywood were valuable for nothing but their guilt-ridden cash— "fat cows to be milked," as one of them put it. There was Daniel Bell's contemptuous summary of the mass-media hacks of the 1930s as slightly feeble intellectual drifters "for whom causes brought excitement, purpose, and, equally important, answers." There was Murray Kempton's cruelly truthful observation that "the slogans, the sweeping formulae, the superficial clangor of Communist culture had a certain fashion in Hollywood precisely because they were two-dimensional appeals to a two-dimensional community." There was, finally, as an ironic footnote, the "Labor Hall of Fame in Hollywood" which *The Nation* published without its tongue getting anywhere near its cheek. On it, in addition to Montgomery, who was later to be Eisenhower's television Svengali, were such luminaries as Joan Crawford, later of the Pepsi-Cola board of directors, Adolphe Menjou, whose latter-day politics would match the elegance of his haberdashery, and George Murphy, the dancer who turned out to have two right feet when he was elected to the U.S. Senate.

However loony the forms of its expression, there is little doubt that the attitude toward politics as it was expressed in the upper strata of the movie colony was as genuine as it was ingenuous and that it exerted its influence on others in the industry who were less well situated economically. The farther down the economic ladder one ventured, the more real the grievances became for film workers. The fact is that many were underpaid and insecure in their jobs, just as their cousins in less glamorous industries were. It is also true that Hollywood, like that other great one-industry, company-dominated town, Detroit, was particularly vicious in its anti-union tactics. In such places, where industry leaders have ready access to one another, a solid management front is easy to create, and the blacklist becomes an appealingly easy weapon to use. Its threat seemed very real to the anonymous craftsmen within the industry whose talents were interchangeable and for whose special individual abilities there was no public demand of the sort that protected the stars and directors who took a stand on the great issues. One man who was active in the formation of the Screen Cartoonists Guild recalls parking his car blocks from the places where meetings were scheduled and then cutting through back streets and over fences and across backyards to make sure that he was not followed and would run into no one who might ask him any embarrassing question about his destination. Maybe it was melodramatic and unnecessary, but such was the mood of the place and the time that it seemed a perfectly sensible precaution to him then—and it still seems so to him now as he looks back upon it.

He, along with others who were members of the Guild, freely admits that, in general, conditions at Disney's studio were better than they were in the animation departments of the other studios, where salaries at the lower levels were sometimes as little as eighteen dollars or twenty dollars a week for people who were by no means beginners in the art and where, of course, working conditions were generally not as good and the work itself was not nearly as interesting or as prestigious. But Disney was to animation what Metro-Goldwyn-Mayer was to movies in general. It was, indeed, despite its comparatively modest size, the General Motors of its field. If it could be organized, the rest of the industry would almost automatically follow its lead. Thus, to some degree, Disney was the victim of his own success.

There were indeed issues at stake that were overripe for resolution, and so organization of a Screen Cartoonist Guild unit at Walt Disney Productions proceeded, to the excessive displeasure of Walter E. Disney. He viewed the organization drive by "outsiders" as a personal insult—an attitude quite blatantly provincial. Throughout the 1930s, as unions of all kinds gathered strength, they penetrated areas where trade unionists had rarely been seen, let alone allowed to share power over wages, hours, working conditions, even hiring and firing, all previously considered management's exclusive prerogative. The larger concerns fought them, yielded grudgingly and slowly, and finally depersonalized their relationship with their employees until today, as it is often pointed out, it is difficult to tell the difference between the executives who represent the great corporations from the union executives who sit across from them at the bargaining table. The issues between them can easily be expressed in the statistics contained in the cost of living index, the company's current balance sheet and the productivity tables. From these encounters the human element has largely been excluded. But the transition to this state of affairs has been much more difficult for the smaller concerns, many of which, like the Disney Studio, were managed by their owners and sometimes by their very founders—men who took a certain pride in knowing most of their employees by their first names, and perhaps even their families and sometimes their personal problems as well. For them, a future of depersonalized employee relations was unimaginable, just as the invasion of outsiders to bargain over matters the owners had always decided unilaterally was unthinkable. Despite his public position, even though he could be hurt by bad labor relations in a way that a backwoods widget manufacturer could not be, even though he could easily have had access to far more sophisticated advice about labor relations than many employers, Disney could not help himself. He reacted to the threat of the Screen Cartoonists Guild like a stern father faced by the rebellion of youth. And in the process he became as a child himself.

Years later, one of the leaders of the strike and a man who had suffered much during and, in particular, after it said, "I don't think Disney was a bad person. I think if he had been left to his own devices he would eventually have recognized the Guild. But he was poorly advised, and he was naïve as far as politics and all related subjects were concerned." The studio's lawyer, Gunther Lessing, is often singled out as the man who influenced Disney to fight the union with such ferocity.

It may also be said, in Disney's defense, that he was under far heavier pressure than many of his employees knew. The layoffs of the period were not lightly undertaken. With *Pinocchio* doing slow business and *Fantasia* an outright box-office flop, with no cash on hand with which to begin large-scale future projects, and with no new credit available, Disney had little choice but to cut back and keep a relatively small staff working on *Bambi* and *The Reluctant Dragon*, the latter a hastily conceived promotional effort, in which two preexisting cartoons were stitched into a live-action script that had actor-writer Robert Benchley taking a wondering tour of the new studio. It was obviously no more than an attempt to get feature-films receipts flowing into the studio as quickly as possible—though of course it did allow Disney the pleasure of showing his latest toy to the world. Disney was, quite simply, hanging on, hoping conditions would improve.

255

His pride, though, did not allow him to admit this to his employees. Instead of talking about his troubles and asking his team (or family, or whatever it was) to help out, he listened to advisers like Lessing who convinced him that the weak young union could not hold out long if he took an intransigent line. At meetings with the men he alternately wheedled ("I only make $200 a week myself") and threatened (if they persisted, he would shut down the whole studio and they would all be out of work together). He even hustled around to the ink and paint department, a large one full of politically and economically unconscious ladies, who were only too willing to take tea with Mr. Disney and listen to his dire predictions of what would happen if the Cartoonists Guild should gain a foothold in his newly created paradise.

What he did not understand was that many of his employees were locked into his psychological condition with him. If he was the father figure, they were, in fact, his spiritual sons. "The strike was a psychological revolt against the father figure," one of them recalls. "We were disappointed in him, in the promise of the big happy studio where everyone would be taken care of that was simply not working out in reality." Instead, all the men saw were layoffs, an assembly-line style of production that promised less rather than more creative freedom, and an industrial design that seemed to have as its goal the ending instead of the encouragement of the old sense of community which had pervaded Hyperion Avenue. Another animator recalls that "Daddy wouldn't talk to us. We had the feeling that if he'd really listen to us, the dream of the paradise for artists would have come true." By the same token, he says that if Disney had taken his men into his confidence about the true state of the company's affairs, they would have been willing to compromise many of their differences with him in the hope of better times to come.

The causes for the strike as they went into the public record are ambiguous. Disney claimed that he was willing to put the whole matter to a vote: in a studio-wide election supervised by the National Labor Relations Board, his employees could choose as their bargaining representative either the Cartoonist Guild or the company-encouraged union, and he would abide by the results. It was his contention that this offer was rejected by Herb Sorrell, who was the organizer for the Guild's parent, the painters and paperhangers union. Sorrell, according to Disney, told him that he had once lost such an election by one vote and that he would not submit to the process again. Disney claimed that Sorrell threatened to "make a dustbowl" of the studio by organizing a secondary boycott against its products. It is possible that he did so, since such techniques were not uncommon and, indeed, were not illegal until the passage of the Taft-Hartley Act. It is also quite possible that the Cartoonists Guild would have lost the bargaining election. The Guild later claimed that about 55 percent of the studio's employees went out on strike, but the figure might have been inflated. In any case, before the vote the relative strength of the unions was very difficult to determine since a large number of employees had signed up with both in order to hedge their bets.

It should, of course, be pointed out that however effective the studio union was made to appear, it was still not a true labor union that could genuinely protect its members in the event of serious difficulty or obtain for them, through collective bargaining, significant improvements in the terms of their employment. Such company-created unions are perfectly legal and they were a device much favored by managements in the 1930s as a way of delaying recognition of genuine unions, even keeping them out entirely. They made the employer look as if he were offering his employees a choice of representation and allowed him to appear before the public as a democratically spirited citizen, awaiting the will of his electorate before deciding an important matter. Legality is not, however, morality, and the number of such company unions that ever attained the stature of a true countervailing force to management is small, if it exists at all.

Nor should it be believed, given the covert pressures that an employer can bring to bear on his workers, that a vote in favor of such a union, even when it is duly certified by a government agency, is a truly fair vote.

Nevertheless, the situation at the Disney Studio might still have been resolved had another series of layoffs not begun. Among those put temporarily out of work there were a suspiciously large number of men who had joined the Cartoonists Guild. This heated the atmosphere considerably, and it reached the boiling point when Disney fired Arthur Babbitt, a gifted animator whom Disney had marked down as a troublemaker. Babbitt had been an executive of the company union and instrumental in getting it to take a tougher line with management when the Cartoonists Guild had come on the scene. He had also had a rather serious personal confrontation with Disney in an attempt to get the producer to adjust the wage differential between himself and his assistant. At the time Babbitt had been making in the neighborhood of three hundred dollars a week, while his assistant had been getting no more than fifty dollars a week. Finally, disgusted with Disney's intractability, he had switched allegiance to the Cartoonists Guild. Shortly thereafter Disney fired him and was foolish enough to state, among the reasons for Babbitt's dismissal, his union activities. Dismissal for this cause was specifically prohibited under the Wagner Labor Relations Act, and the union forthwith took a strike vote, which was overwhelmingly carried, and agreed to go out on May 29, 1941. Contributing to the urgency of the occasion was a rumor—it was true, he later admitted—that Disney was going to lock them out.

At first, he surprised them by keeping the studio open. "Since it wasn't an authorized strike, my artists walked right through the picket line. All of the other union men who worked for me walked through those lines, too—even the musicians. But I was afraid some of my people would get slugged," he later said, justifying his subsequent lockout in humanitarian terms. The union people remember it somewhat differently. They claim that a government mediator worked out a compromise by which the union was granted a substantial share of its demands but only on the proviso that it absorb the strikebreakers of the company union into their own Guild. They refused and only then did Disney shut down, hoping the pressure generated by the general unemployment would force a settlement. In two weeks, he was forced to reopen under legal duress, and the strike dragged on until August—nine weeks in all.

In the beginning, there were friendly relations between the members of both unions. After all, they had shared much, and in good spirits, in the past few years. There was more binding them together than separating them. The good relations did not survive the lockout however. The Guild's refusal to let the other workers join it peacefully and their immediate loss of work in the aftermath of that rejection surely soured the picnic spirit, as someone in management well knew it would. When the studio was reopened, the threat of violence on the picket lines was omnipresent. One day, according to Disney's reminiscences, some members of the company union poured a ring of gasoline around a Guildsman who was manning a bull horn and one of their number stood with a lighted cigarette poised over the gasoline, daring the speaker to repeat something he had just said. On another occasion Disney himself attempted to drive his open Packard through the line. Before the shutdown it had been his habit to wave gaily to the demonstrators, showing that he was unbothered by the situation. This time, however, Babbitt called out to him, "Walt Disney, you ought to be ashamed of yourself." Disney, it is reported, made as if to attack Babbitt physically, and the gate police had to intervene quickly. Some recall that Disney later called out to Babbitt, then struck the boxer's fist-up stance, though it is unclear if he was serious, joking or somewhere in between when he made this gesture. In his attempt to keep up an optimistic front, Disney also had an aerial photo taken of the studio and published in the Los Angeles newspapers. For it, he had all the studio's vehicles moved from their garages to the streets to create an air of unabated activity.

Naturally, the strike against the beloved magic maker of the movies, as the public saw him, was well publicized, and the union was at pains to see to it that would-be patrons of Disney films were aware of the situation in his studio. Rather obviously, no lasting damage was done to the Disney image, and it also seems clear that the "worldwide boycott of his films," which some Disney supporters claimed was Communist-inspired, was more of a fantasy than fact. The Cartoonists Guild did manage to get picket lines up around some of the larger theaters in the major American cities where *The Reluctant Dragon* was playing. (At one of them a group of sympathizers crawled into a dragon costume bearing the legend "The Reluctant Disney" and snake-danced down the street.) The picture was not designed to be a very big hit in the first place; in those labor-conscious days a few patrons undoubtedly passed it by, but their number was not significant. If the strike had a serious effect on any group outside the studio it was on the liberal intellectuals who found this demonstration of their idol's political views at surprising variance with the folkish merits of his films. The contradiction, difficult to digest, signaled the beginning of the end of good relations between the producer and his most articulate appreciators, many of whom had seen in the group creativity of the Disney studio the logical extension of their belief in collective political action of all kinds.

In the end, the mediators settled the strike. "The negotiators gave Herb Sorrell . . . everything he wanted," as Disney saw it. And it was true that the Screen Cartoonists Guild became the bargaining agent for the whole studio, though the union was eventually to undergo a debilitating internal struggle to purge itself of a leftist, possibly Communist-dominated, faction. After that it lost the right to bargain for Disney's employees to a union affiliated with the International Alliance of Theatrical Stage Employees, all of which seemed to justify Disney's belief that he had been, as a friend phrased it "one of the first targets of the Communists in Hollywood."

Talking about the strike many years later, Disney was pleased to point out that Sorrell was later "hauled" before the California Un-American Activities Committee (a notoriously unreliable entity in a notoriously unreliable field) and asked "some pointed questions." He added: "At the time I took photographs of those picket lines and studied those photos. I'd never seen half of those faces. They'd never been near my studio. When I showed them to the FBI and to the investigators for the California Un-American Activities Committee, I was told, 'The fellows in those photos have been in every strike Sorrell has called.'" Many employees who were involved in the strike, however, still resent the implication that they were used or taken over by outside agitators. "There weren't more than three or four leftists in the whole studio," one of them said recently. Their grievances were real and their immediate problem was to ameliorate them. They turned to the only source of help available, and as one of them put it, "Disney should have been the adult in the situation. If he had been, all of us could have had what we wanted."

When he was not choosing to regard the strike as part of the great Communist design for subverting the nation (what better place to start?) Disney was inclined to see it as an ill wind that blew him some good, after all. "It was probably the best thing that ever happened to me," he told his daughter, "for it eventually cleaned house at our studio a lot more thoroughly than I could have done. I didn't have to fire anybody to get rid of the chip-on-the-shoulder boys and the world-owes-me-a-living lads. An elimination process took place I couldn't have forced if I'd wanted to. Our organization sifted down to the steady, dependable people. The others have gone."

This, of course, is utter nonsense. A large percentage of his most independent and creative artists left him after the strike. Not all of them did so in overt protest. Some were drafted away from him after the U.S. entrance into the war; some merely drifted away in search of a better atmosphere in which to exercise their talents. Among them were the group who formed the nucleus of UPA, the little studio that wrested leadership in the animation field from Disney at the end of the war; among them, too, were such gifted cartoonists as Walt Kelly, Sam Cobean and Virgil Partch. It is impossible to say if any or all of them might have stayed on with Disney if the climate of the studio had been more salubrious, but it is clear that the strike irreparably widened a breach that had begun to open long before the picket lines went up. As *Time* said in summary: "Whatever the rights of the affair . . . Walt handled it badly and lost the decision gracelessly." Ironically, the one man Disney was forced to retain was the man whose dismissal precipitated the strike— Arthur Babbitt. He went into the armed forces while his case, alleging unfair labor practices against the studio, went through the courts. At the end of the war, Disney was forced to pay him back salary and to give him back his old job. Babbitt was set to work on a film that, he later said, the studio never intended to produce. But he insisted on carpeting for his office and that it be tacked down so it could not be removed in the dark of some night. He stayed on for two years, then left for another studio— the last of the "chip-on-the-shoulder boys" and, in the opinion of his peers, a great animator, too.

PART NINE

The Long Pause

27 *Walt Disney, at his best an inspired comic inventor and teller of fairy stories, lost his stride during the war and has since regained it only at moments.*
—James Agee, *Comedy's Greatest Era*, 1949

DISNEY WAS AWAY from the studio when the strike was finally settled. The office of Nelson Rockefeller, then coordinator of Latin American Affairs in the State Department, approached Disney with a plan for him to make a goodwill tour of South America. His pictures were extraordinarily popular there, and in these early days of the Good Neighbor Policy and these last days of American neutrality, it was thought that he would be an excellent cultural ambassador. Disney at first demurred. He did not want to make a simple grand tour. Typically, he wanted to have some real work in hand to justify the trip. Besides, the strike was in progress. State suggested, in turn, that he might find something down there about which he could make a movie—maybe several movies. The Department would underwrite each film he made up to fifty thousand dollars and would, no matter what happened, underwrite the traveling expenses of Disney and his party (his wife and seventeen people from the studio eventually accompanied him) in the amount of

263

seventy thousand dollars. They also agreed, as Disney remembered it, to put pressure on federal mediators to get the strike settled. All in all, it was a propitious offer—getting Disney away from the strike scene, which had more than once caused him to weep (whether in rage, frustration or genuine sadness is unclear), and giving him a new project to occupy his mind, a sense of new horizons to conquer.

The therapy worked. And while he was away, Disney did work. He told the American ambassadors to the countries he visited not to entertain him, since he was there on a mission that included no time for play. He set up a little, temporary animation shop in Brazil and invited local artists to drop in and see how the work was done. He even entertained the crowds of fans who followed him about by literally standing on his head, his rather engaging substitute for learning the customary few gracious phrases of greeting and thanks that most dignitaries polish up for visits to lands where they don't know the language. After six weeks in Argentina, Brazil, Chile and Peru, El Groupo, as the Disney party took to calling itself, sailed for home on a leisurely voyage through the Pacific that allowed plenty of time for story conferences on the four short films they decided to make, one for each of the countries they visited.

Arriving back home, Disney found the strike settled. There appears to be little doubt that his removal from the scene and his subsequent distraction into new directions cooled his passions considerably. And, removing the object of their love and hate cooled those of the strikers too. There were some personal awkwardnesses to overcome, but by and large the studio was functioning well by the time Disney returned.

Some say, in fact, that it functioned better than it ever had while the boss was absent. They point to the speed and ease and inexpensiveness with which a little film called *Dumbo* was finished in this period. There were no delays while Disney would wrestle over each story point, each gag, each piece of design that flowed from the artists' desks. It may be observed that, whatever the reasons, *Dumbo* was the least pretentious as well as the least costly of Disney's animated features. Only sixty-four minutes long, it told the comic tale of a baby circus elephant rejected by his peers because of his enormous ears, then raised to stardom after he learns to fly by using the outsize appendages as wings. Based on a contemporary children's story (published by the Roll-A-Book publishers), *Dumbo* carried far less cultural weight than the major Disney films that had preceded it, and the artists seemed grateful to be relieved of the burden. Their work recaptured some of the freshness, exuberance and innocence of the short cartoons as well as their pure and simple fun. The gossipy lady elephants who attend the baby's birth and then reject him for being different are good, satirical versions of a standard comic device, the tongue-wagging middle-aged, middle-class feminine moralist. The hugeness of the beasts contrasted gorgeously with the smallness of their souls. The scene in which these giant creatures attempt to construct a pyramid of pachyderms for a new act, all the while bitching to one another about their weight and clumsiness, is a delightfully funny one, a high point in the Disney *oeuvre*.

There was one distasteful moment in the film. The crows who teach Dumbo to fly are too obviously Negro caricatures. It held its terrors, too, among them a storm sequence, a drunken revel by the clowns (some strikers have claimed that the strikebreakers, while working on it in their absence, caricatured strike leaders in the scene) and the enforced separation of Dumbo from his mother. Enraged by the snickers of some little boys observing Dumbo in the menagerie tent, she spanks one of them soundly with her trunk, setting off a panic in the crowd, then turns on the keepers, who try to quiet her. In effect falsely charged with having reverted to the wild state, she is isolated and locked up, away from Dumbo, in a prison-like boxcar. The scene in which she runs amuck is not, in itself, terrifying, but her fate is, particularly to small children who tend to have a strong anxiety about separation from their mothers. A matinee audience, dominated by small children, tends to stir very uneasily over this sequence. It is the most overt statement of a theme that is implicit in almost all the Disney features—the absence of the mother. Very often she is either dead before the film begins or dies while it is in progress (as in *Bambi*) or is represented by a substitute—the cruel stepmother. It is, of course, true that the absence of one or both parents is one of the long-lived conventions of children's literature and that much of the material Disney naturally worked with deliberately dispenses with adults (the freedom implied by their absence is as delicious in a child's fantasy as the terror of their absence in real life is authentic). In this context, it seems fruitless to criticize Disney for adhering to convention. What is worth remarking, in passing, is the similar absence of his own mother in any of his reminiscences. His father, however difficult his relationship with him, was a force in Disney's life. His mother seems rarely to have been present.

Disney resisted blandishments to lengthen *Dumbo* and thus make it easier to sell as a feature. It went out to the world short and modest in its demands, and Bosley Crowther was pleased to note that "this time Mr. Disney and his genii have kept themselves within comfortable, familiar bounds." The result, he thought, was "the most genial, the most endearing, the most completely precious cartoon film ever to emerge from the magical brushes of Walt Disney's wonder-working artists." Otis Ferguson in *The New Republic* was equally enthusiastic. "Every time that you think the Disney studio can't do any more because they have done everything, they turn around and do it again, the new and never dreamed of, the thing lovely and touching and gay. . . ."

Another magazine praised *Dumbo* "for its freedom from the puppeteering of *Snow White*, the savage satire of *Pinocchio*, the artiness of *Fantasia*, and the woolgathering of *The Reluctant Dragon*," and added that "seldom has Disney articulated his characters so aptly." By Christmas Dumbo was, as *Time* put it, "all over the place. His name was up in lights on some 200 cinemansions, he was getting a big play in big city department stores. Toyland was his without a struggle. He was selling giant green peas and bottles of ink, gasoline and women's collars. As a children's book, he was sensational—50,000 copies at $1 each." "Dumbo," the magazine noted, "could only have happened here. Among all the grim and forbidding visages of A.D. 1941, his guileless, homely face is the face of a true man of good will. He may not become a U.S. folk hero, but he is certainly the mammal-of-the-year."

In short, the film was a hit. It was, perhaps, impossible for Disney to score as heavily as he had with *Snow White*, now that the novelty had worn off; and surely the fact that the U.S. entered World War II on December 7, 1941, some six weeks after *Dumbo*'s release, hurt it as well as all the other movies then in release. But in relation to its $700,000 cost—one-half to one-third of the normal cost of a Disney animated feature—it did very well, but not well enough to rescue Disney from his accumulated debts.

Neither did *Bambi*. It finally went into release in the summer of 1942, five years after it was begun. It grossed only $1.23 million, more than a million dollars less than its cost. To date, the film has grossed $28.4 million domestically, but in 1942 its future potential at the box office was no help to Disney. Neither was an extremely perceptive review of the film that appeared in *The New York Times*. Signed only with the initials "T.S." it stated that *Bambi* "left at least one grown-up more than a little disappointed. . . . Mr. Disney has again revealed a discouraging tendency to trespass beyond the bounds of cartoon fantasy into the tight naturalism of magazine illustration. His painted forest is hardly to be distinguished from the real forest . . . in 'Jungle Book' [the film, starring Sabu, was in release at the same time]. His central characters are as naturalistically drawn as possible. The free and whimsical cartoon caricatures have made way for a closer resemblance to life, which the camera can

267

show better. Mr. Disney seems intent on moving from art to artiness . . . in trying to achieve a real-life naturalism [he] is faced with the necessity of meeting those standards, and if he does, why have cartoons at all? One cannot combine naturalism with cartoon fantasy. . . . [It] throws into relief the failure of pen and brush to catch the fluent movement of real photography . . . . In his search for perfection Mr. Disney has come perilously close to tossing away his whole world of cartoon fantasy."

The failure of the film cannot be laid to the indifference of critics. Most reviews were far kinder than the one in the *Times*, and in any case, critical evaluation was—and is—largely irrelevant to the performance of Disney films at the box office. The trouble appears to have been that a woodland fancy was simply too distant from the immediate concerns of his audience. They wanted escape, all right, but they wanted a different variety from what Disney seemed to be offering. They wanted realistic-*seeming* war dramas or, in a lighter vein, the straight comedies and musical comedies that were such staples during the war years. For Disney, *Bambi* represented the end of an era. He made only one animated feature—*Victory Through Air Power*—between *Bambi* and the release of *Cinderella* in 1950, and that, of course, was a special project, far out of his usual line. By the time he got around to *Cinderella* he had transformed his operation from one devoted completely to animation to one in which animation was becoming only a sideline.

The war years, seemingly unkind to Disney, were actually of tremendous significance to him. Surely he lost his original stride, as Agee said, but he was provided with the time to experiment with a new one—live action—and to contemplate entirely new directions, thanks to a flow of government contracts for training and propaganda films of all sorts. These did not return large sums of money to Disney, but they did keep cash flowing through his studio, allowing him to keep it open and to keep his name before the public at a moment when it was entirely possible that, without this subsidy, he might have been forced to shut up shop entirely.

Since he was a man who prized his freedom of enterprise, it cannot be said that Disney suffered gladly his enforced relationships with the various bureaucracies of the government. Indeed, the harassments and the financial embarrassments to which they subjected him undoubtedly deepened his inbred distrust of big government. His first contact with Washington after the outbreak of war was, in fact, downright unpleasant. On December 8, 1941, Disney received a call from his studio manager telling him that the army had commandeered a portion of his studio as headquarters for a seven-hundred-man antiaircraft unit assigned to man the guns protecting the defense plants of the Los Angeles area. Without so much as a by-your-leave they moved in, took over Disney's only sound stage to use as a repair shop (like all such stages it could be sealed against light leaking in or out, which meant the shop could run at full capacity during blackouts) and stayed for seven months. The army is, if nothing else, almost as compulsively neat and tidy about its domiciles as Disney was, but the intrusion of so many strangers on his lot, not to mention the business of identification passes and armed sentries, must have galled him.

The hard-outline precision possible in animation makes it an ideal educational medium, and the government was quick to realize its potential. On the day the soldiers moved out of his studio, Disney received a call from the Navy, offering him a contract to produce a series of films dealing with aircraft spotting, a serious matter for fighting men and an innocent way to occupy small stay-at-home boys whose traditional game of identifying the new model cars had been, perforce, suspended for the duration. Disney managed to make a 10 percent profit on the eighty-thousand-dollar contract, and he was able to deliver the first film within the ninety-day deadline it specified. By the time the series was complete, Disney had contracts for all sorts of educational films—directed at pilots, navigators, nurses, technicians of every kind. He did four hours' worth of films on the care and use of the Norden bombsight alone, a series on the topography of the islands to be hit by the marines in the South Pacific campaign, another on the causes and prevention of airplane crashes. Within a single year he turned out four hundred thousand feet of government films. While he was adapting a tome called *The Rules of the Nautical Road as Applied to Ship Traffic in Harbours and Confined Areas*, he even permitted its technical advisor to sleep in one of the two offices in his suite, where, he later recalled, the officer ran a tight ship—even to the extent of doing his laundry on the premises.

Perhaps Disney's most famous contribution to the war effort was a short subject commissioned by the Treasury Department, which under the new tax law of 1942 and the upsurge of wartime prosperity had suddenly placed fifteen million new taxpayers on its rolls. It was anxious to persuade them into voluntary cooperation rather than alienate them by a heavy emphasis on its powers to force compliance. Disney rushed to Washington for conferences, rushed home to begin planning the film, which starred Donald Duck, then rushed back east with story boards just prior to beginning its animation. These he was obligated to show the Secretary of the Treasury, Henry J. Morgenthau, Jr., acting out the dialogue as he showed the drawings.

Morgenthau was a banker, which did nothing to endear him to Disney, and he was unimpressed. He had hoped Disney would create a new character, a sort of Mr. Average Taxpayer, who would take a somewhat more sober tack than the duck did in dramatizing the virtues of paying taxes in full and on time. Disney argued for making the painful business seem to be as much fun as possible. The Secretary countered by claiming that he did not like Donald Duck. There are at least two versions of Disney's reply, but in effect he noted with some heat that his giving the Treasury the duck was the equivalent of M-G-M's giving the Treasury Clark Gable; he was the studio's top star.

Disney added that because the duck would be going out into the theaters in a film that was to be supplied free, receipts from his other Duck shorts, which had to be rented at full fees, would be hurt. Morgenthau finally yielded, and *The New Spirit* went into production.

At the same time Disney started another Duck film which, though it had no formal connection with the government, was certainly intended as propaganda. Originally titled *In Nutzy Land*, later known as *Der Fuhrer's Face* (after the title song), it won an Academy Award; it contained the studio's major contribution to the country's arsenal of propaganda: the sound of flatulence. "We heil [Bronx cheer], heil [Bronx cheer], right in der Fuhrer's face."

Except for *Der Fuhrer's Face*, Disney's predictions of the effect of *The New Spirit* on the receipts of other Duck cartoons proved more or less correct and, more galling, the Treasury Department did not pay its full bill for the cost of the film. Disney had obtained from it a letter of intent calling for return of his expenses, which came to $47,000, and in addition, Treasury had to pay Technicolor for prints, which raised its total costs to $80,000. Somehow, however, it could not find the money within the funds Congress already had appropriated for its operations and so had to go back to The Hill with a deficiency appropriation bill to cover a number of projects like Disney's. There it encountered a storm of objections to the $80,000 total. Disney was even accused of being a war profiteer. Ultimately—and largely because the Treasury wanted to use his studio's talents for other projects—a payment of $43,000 was squeezed out to him and he glumly pocketed his loss.

It was by no means the only government contract Disney lost money in fulfilling. He remained, as always, a perfectionist within his special definition of the term, and perfection cannot be achieved on anything but an unlimited cost-plus basis. He kept pushing beyond the cost estimates on which contract terms were based. He was not, after all, an anonymous maker of parts for war matériel—he was Walt Disney, and his government films were to be seen by the people who had been the audience for his commercial films and would be his audience again. He had to keep up standards. The outraged congressmen might have been interested in the figures on one of his most popular wartime activities—the creation of comic insignia for military units. He did some four hundred of them at a cost of twenty-five dollars apiece and received not a dime for the service.

About the only government-supported project that worked out profitably was *Saludos Amigos*. The four short films Disney had planned to make after his South American trip were rejected by the State Department, which claimed that, separately, each film could be released only in the country with which it specifically dealt. He was asked to stitch the four into one impressionistic film covering his entire adventure south of the border. He did so by using the sixteen-millimeter film of his party on tour that he had taken for souvenir and promotional purposes. This linked together the animated sequences dealing with the legends, natives and national characteristics of the places he visited. The result was part travelogue, part fantasy, and it was perhaps, as Howard Barnes put it in a notice that was typical of the general response, "singularly beautiful and diverting as well as a striking bit of propaganda for Pan-American unity." James Agee, just beginning his famous tour of duty as *The Nation*'s film critic, disagreed. Save for "a few infallible bits of slap-stick and one or two kitschy ingenuities with color," he found it depressing. Addressing himself to the intent of the propaganda, he stated that "self-interested, belated ingratiation embarrasses me, and Disney's famous cuteness, however richly it may mirror national infantilism, is hard on my stomach." The merit of the argument between this point of view and the more commonly expressed one of Barnes is now impossible to determine for oneself, since prints are unavailable and a childhood memory is notoriously unreliable, but for Disney the film turned out to be one of the few profitable ventures of the period. It returned almost $1.3 million in grosses.

Taken together with his previous experience with *The Reluctant Dragon*, it also suggested a new possibility to him—the combination of live action with animated cartoons. Perhaps he could retain the profit potential of feature production without losing the magical qualities of the increasingly expensive animation process. It did not bother him that this, in effect, took him back to his technical starting point, the Alice in Cartoonland series. That had been long ago, techniques had improved, and there was a whole new generation from whom the combination film might be a technical novelty of the sort that had proved so successful for him in the past. There was another appeal equally potent in Disney's mind—the chance to make a smooth transition for himself and for his audience to live-action feature production, which he sensed would have to come sometime if he was to maintain a steady flow of films and not have to place all his bets on the relatively small number of animated features he could turn out. "I had to grow with them," he said later. "I couldn't make a live-action feature until I had experience."

After *Saludos Amigos* was launched, he began planning another South American feature, this one to combine live and animated characters in the same sequences. Meantime, he fell prey to the influence of Major Alexander de Seversky, the great proponent of and author of *Victory Through Air Power*, by which De Seversky meant strategic as opposed to tactical (or support) bombing. Disney decided that the major's message was so important that he should finance the film himself, however heavy the further strains it would impose upon his budget and the flack he would encounter from the battleship-oriented navy brass with whom he was also doing business at the time. The curious appeal of strategic air power to people of rightist political leanings has never been analyzed, though it is perhaps not amiss to note that any philosophy that views the human aggregate as a mob incapable of choosing its own destiny and therefore in need of totalitarian leadership can easily be stretched to accommodate a certain indifference to massive, unselective destruction of that inconvenient mob. Strategic bombing has, possibly, a special appeal to people of a midwestern background, for whom it may be the reverse coin of the region's traditional isolationism. Peremptory aggressiveness at long range and in a manner that spares one the sight of the suffering it causes is a way of "getting the boys back home" quickly, perhaps even enabling them never to leave. That, of course, is the highest value, and after all, those who suffer from the strategy are only foreigners, not "human beings"—to again borrow the Cheyenne phrase. In the case of Disney there was, in addition, an appealingly novel technological vision in De Seversky's ideas. Air power was, to him, efficient power—and, as we have seen, Disney loved efficiency. War fought by planes would, De Seversky argued, be economically inexpensive compared to the creation and supply of huge land armies. Best of all, it would save more lives than it would cost—a dangerous speculation that is, of course, still accepted as revealed truth by many people, despite the well-known air force studies of the strategic bombing of Germany that tend to prove it is the most wildly inefficient form of mass destruction yet devised.

At any rate, Disney pushed the film out in a hurry, even setting aside his distrust of limited animation under the impress of urgency. There was a certain massive power in the studio's rendering of sky-blackening flights of planes moving across the screen and in its loving treatment of machinery of all sorts in action. But Agee began his review by hoping "Major de Seversky and Walt Disney know what they are talking about, for I suspect an awful lot of people who see *Victory Through Air Power* are going to think they do. . . . I had the feeling I was sold something under pretty high pressure, which I don't enjoy, and I am staggered by the ease with which such self-confidence, on matters of such importance, can be blared all over the nation, without cross-questioning." Beyond this general *caveat*, the critic had some specific reservations, having to do with the film's occasional attempts to be poetic, cute and funny about such a serious subject, in particular "the gay dreams of holocaust at the end."

Agee added: "I noticed, uneasily, that there were no suffering and dying enemy civilians under all those proud promises of bombs; no civilians at all, in fact. Elsewhere, the death-reducing virtues of De Seversky's scheme—if he is right—are mentioned; but that does not solve the problem. It was necessary here either (1) to show bombed civilians in such a manner as to enhance the argument, (2) to omit them entirely, or (3) to show them honestly, which might have complicated an otherwise unhappy sales talk. I am glad method (1) was not used, and of method (3) I realize that animated cartoons so weak—at least as Disney uses them—in the whole human world would be particularly inadequate to human terror, suffering, and death. Even so, I cannot contentedly accept the antiseptic white lies of method (2). The sexless sexiness of Disney's creations have always seemed to me queasy, perhaps in an all-American sense; in strict descent from it is this victory-in-a-vacuum which is so morally simple a matter . . . of machine-cat-machine."

Of the film one could at least say that Disney was, by his own lights, trying to be true to himself and to a belief he held at least strongly enough to back at considerable cost to himself. *The Three Caballeros*, the last of his wartime features, released in January, 1945, did not have even this rudimentary integrity. Agee found, after seeing it, that "a streak of cruelty which I have for years noticed in Walt Disney's productions is now certifiable." Otis L. Guernsey in the New York *Herald Tribune* found it no better than "a variety show with the accent on trick photography and oddities of line and color." There was a certain novelty value in the combination of live actors and cartoon characters in the same scenes, he said, but he found it "difficult to see what Disney has gained in the way of entertainment from this development," particularly since "there is artistic self-consciousness about almost every individual sequence in the film." Even Disney's best critical friend, Bosley Crowther, was somewhat put off by the film. It "dazzles and numbs the senses," he said, "without making any tangible sense."

It is fair to say that the film reflected Disney's own mood. Nothing made sense to him. He had gone, in a decade, from threatened poverty to success of the sort he had dreamed of for many years, then back to a poverty that seemed to promise very little. He now had a large studio, a large payroll and a basic product that was increasingly expensive to create and increasingly dubious as a marketing proposition. He also had stockholders to worry about. Said Walt Disney, "We had to start all over again." Said Roy Disney, "When you go public, it changes your life. Where you were free to do things, you are bound by a lot of conventions—bound to other owners."

"We're through with caviar," Walt Disney declared. "From now on it mashed potatoes and gravy." "His first four post-war features . . . looked like mashed potatoes all right," one magazine commented, "but they didn't bring in much gravy." All were, in one way or another, "musicals," folksy and folkish in a sham sort of way and all contained, as Agee said of the first of them, *Make Mine Music*, "enough genuine charm and imagination and humor . . . to make up perhaps one good average Disney short." The second, *Song of the South*, contained some pleasant animated sequences devoted to Joel Chandler Harris' tales of life in the briar patch, plus a finale in which the darkies gather 'round the big plantation to sing one of Massah's children back to health—a scene sickening both in its patronizing racial sentiment and its sentimentality. Of it, Crowther declared, "More and more, Walt Disney's craftsmen have been loading their feature films with so-called 'live-action' in place of their animated whimsies of the past and by just those proportions has the magic . . . decreased . . . approximately two to one [is] the ratio of its mediocrity to its charm. If he doesn't beware, a huge Tarbaby will snarl his talents worse than poor B'rer Rabbit's limbs." Crowther was undoubtedly unaware of just how snarled Disney was at this moment, but his proportion of mediocrity to charm held for the next two releases, *Fun and Fancy Free* and *Melody Time*, also anthologies of short sequences in which live action and animation were mixed. It was now clear that the tradition that had produced the folk song and story somehow could not be used to vitalize popular art, as many admirers of Disney—and other purveyors of mass art—had hoped might happen. As early as 1944, Louise Bogan, the poet, declared that the folk tradition "has become thoroughly bourgeoisified. At present there is no way for the artist to get at it, for it has been dragged into a region where nothing living or nutritious for its purpose exists." Disney had certainly been one of the bourgeoizifiers and since he had never really known what he was doing culturally, he, particularly, could not find his roots. Between him and his past he had erected a screen on which were projected only his own old movies, the moods and styles of which he mindlessly sought to recapture at cut rates in the bastard cinematic form of the half-animated, half-live-action film.

It must be stated that in the Forties and throughout the Fifties most of the entrepreneurs of mass culture who had any pretensions to the higher things were caught in the same Sargasso Sea, however more or less successful they were than Disney in navigating it. The war years and the postwar years were, preeminently, the years of what came to be called the "middle-brow," and the middle-brow style, which attempted to blend the best elements of folk culture and high culture, nearly always succeeded in meshing their worst elements. These were the years of *Oklahoma!* and *Carousel*, of *Death of a Salesman* and *The Best Years of our Lives*, of "A Ballad for Americans" and the Katherine Dunham dancers, an era in which the rough edges of common speech and experience were poeticized to the advantage of neither our prose nor our poetry. It was an era when our popular writers and performing artists—and many others—groped, with varying degrees of honesty, for what was authentic in the vernacular tradition, but they could not seem to resist smoothing it down, slicking it over, making it just a little more palatable to polite company, never realizing that such ministrations were bound to spoil it and spoil as well the higher forms into which they willy-nilly sought to cram this intractable and essentially formless stuff.

It seemed, one supposes, a patriotic thing to do. And one can say this much for the whole dreary process; it awakened for a few in the audience an awareness that there were any number of precious things in our common heritage. If a few people were led back from the various adaptations of folk material to the original sources, then the business may possibly have been worthwhile. It is also beginning to seem clear that the forced marriage between high culture and low was not quite the calamity that some observers thought it was (and still think it is). The bastardized products continue to roll off the various assembly lines, and many of them continue to achieve an eminence and a prosperity out of all proportion to their merits. But it is becoming increasingly clear that mass popular culture is not necessarily a hybrid form, that it can create forms of its own in its own good time. This is particularly true of movies. The western, the detective story, comedies of certain sorts, even an occasional musical on the order of *Singin' in the Rain*, can transcend whatever backgrounds originally nourished them and become preeminently screen forms, which cannot be rendered as well in any other medium. Similarly, the film has attained, over the comparatively short span of its history, a critical language uniquely its own, one that can be compared only invidiously, in the techniques and effects it describes, to the other arts. A film by an Antonioni, a Truffaut, a Resnais, even a Stanley Kubrick, owes nothing to anything but the history of the cinema. A great deal of the excitement that has stirred our culture in recent years—the McLuhan fad, for example, or the rediscovery of the

great silent comedians, even such cultish affairs as the deifica-
tion of Bogart or the campy delight taken in old Busby Berkeley
musicals—testify to this. Whether a mechanized culture can
produce works of the lasting value of either the folk tradition it
obliterated and now, in fact, succeeds, or of the traditional
culture with which it uneasily coexists and may possibly oblit-
erate in its turn, is a serious and open question.

What can certainly be said is that for a brief moment, in the
early Thirties, Disney had the opportunity to contribute to the
emerging tradition of mechanized, popular culture and that he
did indeed make sizable contributions to it when he was working with the characters he and his employees created. It may
even be argued, though with a little less certainty, that his feature-length animated adaptations of folk and children's stories
were so strongly cinematic in their values that they, too, transcended any sensible comparison with their sources and reached
toward a state of pure cinema, at least in their highest moments.

The fact remains, though, that Disney was an untutored man
with, as we have seen, a fatal attraction toward the intellectual
community in this period. Their regard clearly meant something to him, and prior to announcing for "meat and potatoes,"
he had tried to satisfy them as well as the simpler demands of
his audience and his own primitive taste. He was led astray by
them—or more properly by people who posed as intellectuals—
in several ways. To begin with, those who might have sprung
the mote from his eye—a James Agee, for example—were not
available to him. Instead, he mistook someone like Deems Taylor for the genuine intellectual article. Second, he fell prey to the
depressing tendency of the time to downgrade the film as film
and to think that its elevation as an art form depended on its
being ever more tightly tied, through adaptation, to the literary
forms—a tendency by no means dead in Hollywood, especially
since it has become clear that a hit play or a best-selling novel
has a "pre-sold" audience. This blend of cultural insecurity and
practical economics has been one of the great hindrances to the

development of a more flexible American motion picture style, and Disney, surely, was one of its earliest victims.

At least as important as these practical, immediate factors in the growing estrangement between Disney and the intellectual community was the general intellectual climate of the century. As the English critic Martin Green has summarized it, this climate has grown out of a dialectic between two sensibilities. On the one hand, there is "a spirit of broad general knowledge, national and international planning, optimism about (or at least cheerful businesslike engagement with) the powers of contemporary science and technology, and a philistinism about the more esoteric manifestations of art and religion." Ranged against this possibly majoritarian view of the world is another that "insists on narrow intense knowledge (insights), on the need for personal freedom within the best-planned society, on the dangers of modern science and technology, on the irreducibility of artistic and religious modes." Disney, clearly, held with the first group, the most significant members of the artistic and intellectual community with the second. Looking back, it seems a miracle that they could find common ground for as long as they did. It seems likely that their brief embrace was greatly facilitated by the general interest in science and technology that "scientific" Marxism generated among the intellectuals as well as the rather self-conscious egalitarianism of this group as it attempted to relate itself to the masses, among whom the idea of progress was most widely construed in materialist terms. In any event, this deviation from the true faith was a brief one for the intellectuals, and by the beginning of the postwar era they had returned to their customary suspicion of the values treasured by people like Disney. Disneyland would become a pure expression of his basic faith; their disgusted response to it, a pure expression of theirs. Disney, with his sure grasp of the popular mind, had almost no sense of how the cultivated mind worked. He was dismayed and disgusted by its inability to share his vision and could only bluster at its criticism.

From Disney's point of view, it is fair to say he was unceremoniously dropped by the intellectuals at the very moment he could have most used their help. From being an object of veneration he passed with blinding speed to being an object of scorn, and if, in his later years, he was particularly unpleasant about intellectuals, he may truly be said to have had his good reasons. It was they, not he, who declared his early work to be art; they who informed him that his more self-conscious attempts, as well as the work of his last years, were symbols of all they came to be alienated from in American life. Undoubtedly Disney misunderstood them, but it is equally true that they misunderstood Disney—and never more so than when they most adored his work. This is not an unfamiliar phenomenon; their praise spoiled the great silent comedian Harry Langdon and turned the head of Charles Chaplin, not to mention dozens of other less well-known film makers who had the misfortune to turn out a praiseworthy work or two and the weakness of character to believe their notices. Disney was in a great tradition of mutual misunderstanding.

28 *A winner never quits, a quitter never wins.*
　　　—Old American saying

IN THE FORTIES, it seemed that everyone had deserted Disney—the intellectual audience and the mass audience. So he paused to grope. It was a long pause. Visiting the Disney Studio in this period, Bosley Crowther gathered the impression that Disney was preparing to withdraw from active participation in his business, make his brother caretaker of such assets as it had and "live . . . happily ever after on television residuals."

Added Crowther: "He seemed totally disinterested in movies and wholly, almost weirdly concerned with the building of a miniature railroad engine and a string of cars in the workshops of the studio. All of his zest for invention, for creating fantasies, seemed to be going into this plaything. I came away feeling sad."

There was no need for sorrow, as Crowther himself later came to realize. What he witnessed was not a man preparing to give up but rather a man gathering strength and ideas for another great leap forward. "The late forties," says a veteran of that period in the studio, as he attempted recently to summarize Disney's mood, "was the time when Walt Disney discovered Walt Disney." One of the ideas he was toying with was, for the first time, hiring established movie stars to appear in his films. "We kept trying to tell him that he himself was a bigger star, a bigger name, than any he could hire, but he had trouble believing us." Fortunately for him, Disney had an equal amount of trouble bringing himself to pay star salaries and granting star billing since, until then, his was the only name that had ever appeared above the title of any of his films. In the end, he did begin to hire well-known names to do the voices for his cartoon features, but that was as far as he went, at least at first. (After live-action features worked out so well for him, he began to use slightly bigger names in them but under a policy that was cruelly, if accurately, summarized by the industry in the phrase, "Disney gets them on the way up or on the way down.")

At the end of his long period of brooding, Disney emerged with a plan that called both for a return to first principles and for diversification. For a start, he determined not to abandon animation but to concentrate the studio's efforts in that area of production on one fully animated feature every three or four

282

years. There would be no more mixture of live action and animation in vaguely focused anthology presentations, though animation might be used occasionally to enhance an essentially live-action feature (as it was in *Mary Poppins*) or there might be live-action sequences in essentially animated films. But films of both sorts would concentrate on telling a single story: there would be no more anthologies. As for the live-action films, they would appear on a much more frequent basis than the animated films, their cost would be carefully controlled, at least in the beginning, they would not be contemporaneous in setting or subject matter, thereby protecting them from dating and thus insuring their rerelease value. As far as short subjects were concerned, work on new cartoons would be phased out, the old ones rereleased (as of 1966 the average Disney color cartoon had been reissued sixteen times, and they have, all told, grossed seventy million dollars. If animated features were risky, short animated films were impossible to make profitably, at least in the full animation technique to which Disney was committed. Replacing them on the current production schedule would be the short nature films on which he had long planned to experiment.

Then there was television. "When the industry was cussing television and trying to ignore it," Roy Disney said later, "Walt moved in and worked with it and made it work for him." This took time, and precise plans crystalized slowly over more than a half-decade. The same may be said of an idea that was, at the time of Disney's retreat into himself, no more than another long-held dream, Disneyland. But its development, now moving again to the forefront of Disney's mind, ultimately hinged, according to Disney himself, on the development of an effective television strategy.

Sometime in the late forties, Disney had it out with his brother on the subject of their future in general and of animated features in particular. Roy Disney was frankly fearful of committing the studio to the risky expense of a full-scale animation project of the kind they had done before the war and was happy to hold on to the modest profits that were beginning to accrue again. Walt Disney, however, was adamant on the need for progress. As he remembered his speech to Roy, it went this way: "If we try to coast, we'll go backward. Let's get back into business or sell out." He demanded financing for what amounted to a five-year production plan, including three new animated features and the schedule of nature shorts. Somehow, Roy managed to find the money, principally by denying the stockholders anything but minimal profits and by plowing the slightly expanding grosses of the late forties back into the business. This was undoubtedly one of the periods Roy Disney had in mind when he said to a reporter: "I just try to keep up with him—and make it pay. I'm afraid if I'd been running this place we would have stopped several times *en route* because of the problems. Walt has the stick-to-itiveness."

In any event, Walt, as usual, got his way. And three animated films were soon on the boards—*Cinderella*, *Alice in Wonderland* and *Peter Pan*. He also had another opportunity to prove his stick-to-itiveness. The exact date on which Walt Disney wandered into the little camera store that a man named Al Milotte was running in Alaska is unrecorded, but Milotte and his wife, Elma, who was a schoolteacher, were enthusiastic semiprofessional movie photographers and apparently Disney had heard of their work. "How would you like to make some pictures for me up here?" Disney inquired, as Milotte recalled later. "I said, 'What kind of pictures?' He said vaguely, 'I don't know—just pictures. Movies. You know—mining, fishing, building roads, the development of Alaska. I guess it will be a documentary or something—you know.' "

Milotte did not, but he went to work anyway. He was soon flooding the studio with film and receiving enigmatic wires in return: "Too many mines. Too many roads. More animals. More Eskimos." More footage was ground out. Then Milotte decided to try an idea he had long harbored, which was to visit

the Pribilof Islands, to which fur seals by the thousands mysteriously migrate year after year to fight, mate, bear their children
and, from which, when their convention ends, they disappear as
mysteriously as they arrived. He explained the notion and received one more wire from Disney. "Shoot fur seals," it read.

He did—for a year—during part of which he lived in igloos
with the local Eskimos. Periodically he would ship his footage
off to Hollywood, and periodically he would get a wire in return
saying, "More seals." In the end he shot miles of film, and in the
end Disney emerged with a thirty-minute short he called *Seal
Island*.

There are two versions of what happened next, though there
is basic agreement that no exhibitor would touch the film. The
official version was told by Disney: "My brother Roy phoned
me, heartbroken, from New York," he said, " 'Nobody wants to
buy it. They all say, "Who wants to look at seals playing house
on a bare rock?" ' I said, 'Come back out here, Roy; the sun is
shining.' " There is also agreement on what happened next. In
1948 the Disneys got a Pasadena theater owner to play the film,
where it attracted good audiences and where, more important, it
qualified for a 1949 Academy Award nomination (a film must
play a week in the Los Angeles area within the year preceding
the awarding of the Oscars). When the film won the prize as
the best two-reel short subject of 1948 the exhibitors naturally
clamored for bookings—and for more nature films. The difference between the official version and the one told by old studio
hands is, in the latters' memories, that Roy Disney agreed with
the exhibitors in seeing no future in the project. They say that
the morning after Disney won the Academy Award, he trotted
around to his brother's office, opened the door and flung the Oscar at the wall above his head.

Whichever story of the beginnings of the nature series is
true, it is obvious that Roy Disney quickly began to appreciate
the virtues of the series. By ordinary Hollywood standards nature photographers are low-paid craftsmen, and so the short

films they turned out were ridiculously inexpensive to shoot, even though the studio had to pay field expenses for as much as a year in order to get enough usable footage to make one short film. And if Disney did not pay the photographers very much, he did provide them with the opportunities and the equipment to pursue their difficult craft in a manner to which they were not ordinarily accustomed. The Milottes, for example, were given a specially reinforced truck in which they could live and from which they could safely photograph wild African animals at ranges closer than anyone ever had before. At one point the Milottes stayed in the truck for sixty days without ever leaving it. An elephant charged it on one occasion and might easily have overturned it had he not been distracted by something at the last moment. On the other hand, at least one pride of lions grew so used to the strange vehicle that they even allowed it to accompany them on nocturnal safaris, on which they were photographed by the light of searchlights—something no one had ever accomplished before. Other photographers were provided with telephoto lenses that could photograph subjects over a mile away or with high-powered arrangements of lenses, automatic cameras and strobe lights that could stop action up to 1/100,-000 second and fill the screen with a shot of a beetle so small that fifty of its breed could hitch a ride on a honey bee's leg.

The tales of the patience and the courage of the cameramen are endless. The Milottes, for example, waited six weeks beside an alligator's egg in order to photograph it hatching. Another time they watched herons for forty days in order to photograph one of them catching a fish within camera range. On yet a third occasion Milotte spent five months stalking a mountain lion to get the footage he wanted. (Even he had his limits, however. Working on *Beaver Valley* he waited weeks to get a close-up of a beaver gnawing down a tree, but as he put it, "Contrary to popular notion, beavers are not always busy. Most of the time they just horse around. I was cold and wet and tired after weeks of watching these clowns and ready to give up when I looked out and saw what I'd been waiting for. I grabbed my camera—and it wasn't loaded." He was so angry he dashed out of his blind, grabbed the sapling, jabbed it back in the ground and ordered the astonished beaver "to do that again, you punk." By the time he had his camera loaded, the beaver was obligingly sawing away at the tree.) Another Disney photographer resorted to the old Indian trick of dressing in a buffalo skin in order to crawl among the herd to get close-ups. He reported some curious sniffings from wiser heads among the animals, but no aggressive behavior. One of his colleagues reported a herd of the great beasts thundered when they moved, just as the cliché had it, "But what I had never heard of was the sibilant, silken swish which accompanies the stampeding buffalo. It was even more terrifying than the thunder."

Disney's policy with the nature photographers was, essentially, the same one he had employed with his animators. He gave them the best equipment, all the time they needed (the ratio between footage shot and footage used in the finished film was about 30 to one), and as he had with animation, he insisted on tight storytelling and, of course, anthropomorphism. Said a writer who worked on the series, "Any time we saw an animal doing something with style or personality—say, a bear scratching its back—we were quick to capitalize on it. Or otters sliding down a riverbank—humorous details to build personality. This anthropomorphism is resented by some people—they say we're putting people into animal suits. But we've always tried to stay within the framework of the real scene. Bears do scratch their backs and otters are playful."

The success of the series of animal shorts led Disney, in the early Fifties, on to another series dealing with "People and Places" and, at last, to the generally well-regarded documentary features, *The Living Desert* and *The Vanishing Prairie*, of 1953 and 1954 respectively. Working on them, Disney had two criteria he always insisted upon. He wanted facts and more facts in the narration. How much does the baby elephant weigh? How old is he? How many animals in the herd?—these were questions of the kind he constantly asked as he screened finished sequences. On the framework of this factuality, however, he insisted upon erecting dramatic structures sometimes whimsical, sometimes terrifying. A sampling of his comments on *The Living Desert* was preserved by the studio and set down by Robert De Roos in his 1963 profile of the man:

"In sequence where tortoises are courting . . .: They look like knights in armor, old knights in battle. Give the audience a music cue, a tongue-in-cheek fanfare. The winner will claim his lady fair . . .

"*Pepsis* wasp and tarantula sequence: Our heavy is the tarantula. Odd that the wasp is decreed by nature to conquer the tarantula. When her time comes to lay eggs she must go out and find a tarantula. Not strength but skill helps her beat Mr. Tarantula.

". . . Our other heavy is the snake. . . . With wasp and tarantula it's a ballet—or more like a couple of wrestlers. The hawk should follow. Tarantula gets his and then Mr. Snake gets his. . . . Should be ballet music. Hawk uses force and violence. One could follow the other and have a different musical theme as contrast."

And so on. *Time* saw what Disney was getting at—"the sense that the camera can take an onlooker into the interior of a vital event, indeed into the pulse of the life-process itself," then added: "Thus far, Disney seems afraid to trust the strength of his material. He primps it with cute comment and dabs at it with flashy cosmetical touches of music. But no matter how hard he tries he can't make mother nature look like what he thinks the public wants: a Hollywood glamor girl. 'Disney has a perverse way,' sighs one observer, 'of finding glorious pearls and then using them for marbles.'"

Reviewing *The Living Desert*, Bosley Crowther was even more severe. He decried the "playful disposition to edit and arrange . . . so that it appears the wild life . . . is behaving in human and civilized ways . . . all very humorous and beguiling. But it isn't true to life." He also detected another current, running alongside the cuteness, which he described as a "repetition of incidents of violence and death" that tended eventually "to stun the keyed-up senses" in such a way that his accumulated impression of the film was one of "a sort of zoological morbidity."

The nature films indeed present one of the most difficult problems of critical evaluation in the entire Disney history. On the one hand, no one can doubt that Disney's photographers did bring back the pearls and that they were displayed in a variety and profusion previously unknown on the theater screens of the world. The short documentary has not had a very distinguished history in the commercial cinema: the Pete Smith specialties and the Fitzpatrick travelogues were the rule rather than the exception. In their sheer technical virtuosity, in their ability to put on the screen rarities and oddities from the natural world, the Disney films were so far above their competition as to deserve a category all to themselves. In them he satisfied the simplest, most basic demand of film making: he gave us a chance to see things we might not otherwise have ever seen. And yet he falsified this material in precisely the ways his critics suggested. The charge that he excessively emphasized the violence of nature does not stand up nearly so well as the charge that he prettified it. Nature is, after all, violent, and all one can say about his handling of its life and death struggles is that it is quite wrong to make a tarantula or a snake into a "heavy," even by implication. There is no moral hierarchy among the species, and the business of "cuing" response through music, narration or film editing that leads to this sort of ranking by the spectator is reprehensible. Just as bad is the business of reducing to a joke a mating ritual, or a young bird's attempts to master flight or a young animal's first experience of the hunt. None of these matters, to put it simply, is funny to the participants, and they would not seem funny to us if we were to observe them unedited, with our own eyes, in the field. They become jokes only when the creatures are anthropomorphized for us by the film maker. The business of individuating animals not only falsifies our understanding of them; in the last analysis, it cheapens experience, substituting patronization for the sense of awe that the truly sensitive observer feels in the presence of nature's enigmas. "Nature," said Henry Beston, one of the most sensitive of contemporary writers on the subject, "is a part of our humanity, and without some awareness and experience of that divine mystery man ceases to be man." In the work of the great naturalists, the more knowledge one gains, the greater one's sense of the ultimate mystery contained in the subject. But in Disney's work a little knowledge becomes a truly dangerous thing, shutting down the subject instead of opening it up, either compartmentalizing it by emphasizing its believe-it-or-not freakishness or cheapening it by making it seem that animals are comically trying to imitate, in their little ways, man's mastery of his environment. The tone of a Disney nature film is nearly always patronizing. It is nearly always summoning us to see how very nicely the humble creatures do, considering that they lack our sophistication and know-how. This attitude does no credit to our humanity as Beston uses the word, and certainly it explains away the necessary, endlessly entrancing mystery far too easily. In any case, anthropomorphism is, literally, child's play. Seeking explanations for phenomena that he cannot comprehend, a child, with his innocent egoism, always invests the objects of his play and observation with the only qualities he knows—those of his own personality. As he grows older he learns that there are other forces in the world beyond his own personality, and education rightly conceived is the process by which he learns to value those forces, mastering and turning to his own uses those that he can, respecting those he cannot. Disney never could seem to learn this simple distinction, and confronted with things that were inexplicable to him, he either turned away in disgust or willfully falsified them by reshaping them in terms that he understood and approved. Much the same thing, had happened in his animated films in which all material was re shaped to suit the limited artistic style he insisted upon, how ever incongruous that style might be to the subject matter.

It can be argued that without this kind of reductionism the series would not have been successful and we would therefore have been denied access to many of the treasures it did contain. This, of course, is always the argument of the popularizer; it is a difficult one to deny, however much one disapproves of it What one can say is that we will never really know its truth until someone actually does an unpatronizing set of nature films for popular consumption (just as until someone tries we will never know if there is an audience for a genuinely serious set of television dramas). At this point all we know for certain is that Disney preempted the field in such a way that it will probably be a long time before anyone tries again and that, if they do try, they will undoubtedly be tempted to imitate his proven formula.

It seems worthwhile to observe that Disney's pattern with the nature series exactly followed the one he had previously estab-lished in the cartoon field. Beginning with the relatively modest and unpretentious first efforts, he moved quickly to a greater sophistication of technique and to a more crowded production schedule that had the effect of dominating the market through saturation. Then came the move to features and then, finally, to bastardization. What followed the documentaries were semi-documentaries, the first of which was *Perri*, the biography of a squirrel, in which the animal "stars" had names and virtually human relations with one another. After that came such works as *The Legend of Lobo*, in which a seemingly wild animal was trained by handlers to do tricks that he could not and would not normally do in his wild state but that, of course, the audience did not know were unnatural to him. However entertaining some of these stories were, at the most childlike level, they were fundamentally dishonest—the more so since they partook of the

291

True-Life Adventure series' reputation for not containing any faked footage.

In any case, the nature films did help Disney to reestablish himself and his studio. They brought him back into the public's consciousness, and they made money. The first two features, for example, cost $300,000 and $400,000 respectively and grossed in their first domestic releases upwards of five million and four million apiece. Along with the five other nature features, they stayed in release for years, continuing to play here and there, now and then, almost constantly and without need of any elaborate or expensive promotion and advertising budget as a True-Life Adventure Film Festival.

The year after he won his Academy Award for *Seal Island* Disney at last made his first complete live-action fictional feature, following Roy's suggestion that he use funds that had been blocked in England. The film was *Treasure Island*, and it featured Robert Newton's superbly funny, scenery-chewing performance as Long John Silver. For it, Disney received the best set of notices he had got since *Pinocchio*, and best of all, he brought it in at a cost low enough to insure a high margin of profit. He returned to England each summer for the next three years to supervise personally the production of sequels—*Robin Hood, The Sword and the Rose*, and *Rob Roy*. As *Time* said "They were all amazingly good in the same way. Each struck exactly the right note of wonder and make-believe. The mood of them all was lithesome, modest. Nobody was trying to make a great picture. The settings in the English countryside were lovely, wide swards and sleepy old castles and glens full of light. Best of all, Disney was careful to choose his principals— Richard Todd, Glynis Johns, Joan Rice, Bobby Driscoll—not for their box-office rating or sexual decibel, but rather as friends are chosen, for their good, human faces and pleasant ways. As a result each of the pictures was just what a children's classic is supposed to be, a breath of healthy air blown in from the meadows of far away and long ago." Indeed, it is possible to say that because of the modesty of their aims these pictures came closer to fulfilling their intentions than any films Disney ever did. Feeling no impetus toward afflatus and, indeed, having every reason to control himself, Disney kept the pictures tight in structure, simple in development. There was nothing to tempt him, as there was in animation, toward the frightening or the fantastic. On the other hand, realism here was simple to achieve—being merely a matter of costuming and set decoration —instead of an essentially insoluble problem in illusion making. In all, there was an air of ease and comfort about the films that was most refreshing.

Undoubtedly the atmosphere of their creation was considerably lightened because the last three were made in a period when it was clear that the trend in the studio was, at long last, an upward one. By the summer of 1950, when *Treasure Island* was released and *Rob Roy* was in production, the potential of the nature series was clearly established and, even more comforting, the studio was in the midst of what Roy Disney later called "our *Cinderella* year." By that he meant that the release of that film had untapped the greatest flow of coin into the box offices since *Snow White*. Indeed, the $4.247 million gross it achieved in its first run actually topped the mark set by Disney's first feature in its first domestic run by some $55,000. Over the years its domestic gross has reached $25.5 million. There were a good many critical reservations about the film, talk of its "full-blown and flowery animation," and its "glamorous style of illustration" toward which "the more esthetic may take some degree of offense." Again, there was dubiousness about the ability of the animators to capture human movement in a satisfying way. But *Cinderella* is, of course, the most famous of all fairy tales, and perhaps more important, the generation that had been enraptured by *Snow White* thirteen years before was now anxious to provide their children with a similar thrill of discovery. In that year, after two years in the red, the company showed a profit and the Disney sales chart began its long upward trend, a trend that, just five years later, was to take it past the ten-million-dollar mark in revenues for the first time and that was to provide it with the funds it needed to lay the groundwork for the truly fantastic growth that was to come in the ten years that followed. The long era of trouble was finally over. For the first time in a decade Walt Disney could definitely say that the worst was past and begin to claim the rewards of survival.

PART TEN

Disney's Land

---

29 *The moment of* survival *is the moment of power.*
　　　　　　　　—Elias Canetti, *Crowds and Power*

IT IS A MEASURE of how well things were going for Walt
Disney in the early 1950s that the failure of his next animated feature, *Alice in Wonderland*, was not reflected in a downward dip on the gross sales chart of his organization or in a
failure to register profits in that year. He personally did not like
the film very much, retrospectively grumbling that it "was
filled with weird characters." The more he worked with it, the
more he came to think of Alice herself as a prim and prissy little
person, lacking in humor and entirely too passive in her role in
the story. Audiences, he felt, could not identify with her and he
could not blame them much: neither could he. He made a little
money on *Alice*'s successor, *Peter Pan*, which grossed $24.4 mil-
lion over the years, though it was no better received critically
than *Alice*. In both instances, he later said, he felt trapped by
the literary reputations of the works, unable to Disnify them as
freely as he could a fairy tale. In both cases the styles in which
the tales had been originally told were at least as memorable as
the characters or the stories. These three elements were inter-

twined so tightly that the studio could not successfully substitute its standard style for those of the originals without altering their basic patterns beyond recognition. Somehow, the Disney artists were never able to set the gentle wistfulness of *Peter Pan* on film, somehow they were never able to convey the free, fantastical parody of conventional logic which is the reason for *Alice's* existence and which makes it a work of art. Occasionally the sheer speed of the action in the encounters with Captain Hook, for instance, or the broad humor of the Disney version of the Mad Hatter's tea party, could sweep away—temporarily—one's objections, but the fact is that by this time the studio style was so inflexibly realistic, so harsh and so obviously the product of a factory system, that it was incapable of catching more than the broad outlines of these classics. It simply could not come to grips with their essences, and without those essences, there was very little point either to making the movies or to seeing them.

Indeed, there was something arrogant about the way the studio took over these works. Grist for a mighty mill, they were, in the ineffable Hollywood term, "properties" to do with as the proprietor of the machine would. You could throw jarring popular songs into the brew, you could gag them up, you could sentimentalize them. You had, in short, no obligation to the originals or to the cultural tradition they represented. In fact, when it came to billing, J. M. Barrie's *Peter Pan* somehow became Walt Disney's *Peter Pan*, and Lewis Carroll's *Alice* became Walt Disney's *Alice*. It could be argued that this was a true reflection of what happened to the works in the process of getting to the screen, but the egotism that insists on making another man's work your own through wanton tampering and by advertising claim is not an attractive form of egotism, however it is rationalized. And this kind of annexation was to be a constant in the later life of Disney. The only defense one can enter for him is that of invincible ignorance: he really didn't see what he was doing, didn't know how some people could be offended by it, and certainly could not see that what was basically at fault was his insistence that there was only one true style for the animated film—his style. If he had earlier learned that there are many ways to draw and that some are better suited to a given story than others, then he would not have been forced to jigger every project into a shape his style could handle. Similarly, if he had had some taste for comedy other than the folksy or emotions other than the bathetic, he would not have had to twist stories so radically to fit them for his Procrustean bed. Had he been a more flexible entrepreneur of animation in these later years, he might have fared better with animated features at the box office in the 1950s.

As it was, returns slipped until they reached the nadir in *Sleeping Beauty*, not only the most serious financial failure the organization suffered in this period but the most disastrous artistic failure as well. Paul V. Beckley, writing in the *Herald Tribune*, said that it was "Disney imitating Disney" and "it is not necessarily the best qualities of Disney's earlier work that is [sic] being imitated. . . . The stress is . . . on soft cuteness. Goodness gets itself defined as a form of bumbling innocence with a perky tail-twitching, simpering quality, just as badness becomes unvaryingly sinister, black, slinky, sinuous and grotesque." In short, even returning to the fairy tale form, with which he had been most successful and in which there were few detailed or subtle conventions of literary form to bother him, could not rescue him. Facility had surely grown, but it was a narrow facility and somehow choked off the capacity to feel and the ability—never the strongest element in the studio's work—to express these feelings graphically. In all the later animated products there were plenty of effects but there was little emotional impact.

Of them all, the best was the last that Disney personally had a hand in supervising—his version of *The Jungle Books* of Rudyard Kipling. It is not, of course, for purists. It is based on a selection of stories contained in the original collection, and those stories are compressed and radically rearranged to form a tight, cohesive story. Again, there is no attempt to find a filmic equivalent for the marvelous tone imparted to the original work by Kipling's splendid style (in which mystery and simplicity were blended in a way that never condescended to the child). The jungle animals of India are, in the film, turned into familiar American types making familiar American jokes and singing songs intended for the top of the pop charts. But in terms of the Disney tradition, rather than any literary tradition, the film is remarkably successful. Just 64 minutes long, full of inventive sight gags and innocent high spirits, *The Jungle Books* has a gaiety and a lack of pretentiousness absent from Disney animated features since *Dumbo*. There is almost no striving for effects of high visual art and no attempts—so often disastrous in Disney studio work—to realize any mythic or literary overtones. There aren't even any scary effects of the type parents objected to in the past. It is as if Disney and his craftsmen had finally abandoned the last of their aspirations to make a statement in cultural traditions fundamentally alien to them. In effect, and despite the fact that their starting point was, indeed, a classic of children's literature, they returned here to first principles—they were, like the earliest Disney artists, making just another cartoon.

The performance of the animated features in the Fifties and Sixties seemed to confirm what Disney had suspected before launching them—that their potential return might not be worth the risk. Just five animated features emerged from the Disney studio in the years 1953–68, while in the same period over fifty live-action features were released. To be sure, all five of the animated features appear on *Variety*'s list of the all-time box office champs, but sixteen live-action films from the same period also appear there. What is more significant is that only two of the animated films of this time grossed more than $8 million (*Cinderella*, with $9.25 million in three releases and *Lady and the Tramp*—a 1955 release that was replayed once—with

$8.3 million), he could not forsee that ultimately they would gross as well as his earlier animated features. He could see, though that no less than eight of the live-action films, all made at far less cost made better than $8 million in domestic release. *Sleeping Beauty*, moreover, which grossed $5.3 million in 1960, actually lost so much money that it sent the company into the red that year—the only time that unfashionable color had appeared in the Disney ledgers since 1949. Meantime, live-action features that most people scarcely remember —films like *The Shaggy Dog, The Parent Trap, The Absent-Minded Professor, 20,000 Leagues Under the Sea, Old Yeller, Swiss Family Robinson, Lt. Robin Crusoe, USN, Son of Flubber* and *The Ugly Dachshund*—all performed spectacularly and at virtually no risk. And, of course, *Mary Poppins*, which most people will remember for a long time to come, became Disney's greatest hit, ultimately grossing $45 in the U.S. and Canada alone.

There is really very little point in discussing these movies critically. They are about equally divided between animal stories, children's classics and originals that put a heavy stress on personal inventiveness and family togetherness. *The Shaggy Dog*, released in 1959, has some significance in the Disney canon because it was his first venture into live-action comedy. "I got to thinking," he said later, "'When it comes to comedy, we're the ones'; so we did *The Shaggy Dog*." In its first release it earned nine times its cost of slightly more than one million dollars, a figure that should be compared to the money-losing *Sleeping Beauty* to see why Disney's production schedule changed so radically in the 1950s. The genesis of films like *The Absent-Minded Professor* and its sequel, *Son of Flubber*, was equally simple as Disney told it. "We've always made things fly and defy gravity," he told an interviewer. "Now we've just gone on to flying flivvers, floating football players and bouncing basketball players." Indeed, both of these films, which starred Fred MacMurray, who is to the Disney studio of the 1950s and 1960s what Mickey Mouse was to it in the 1930s (and to whom Disney was devoted as to no other actor since The Mouse), were a little bit better than his average. The hero, an impractically practical, small-town, backyard inventor, is an authentic American folk figure, the sort of person Disney himself admired and may occasionally have thought he was himself, and so there is a little more feeling in the pictures—despite their descent into aimless farcicality and folksiness—than was customary in most of the studio's live-action films. Moreover, the plot line of *The Absent-Minded Professor* is very reminiscent of Disney's own wartime experience with Secretary Morgenthau—the hasty night flight to Washington by the simple, rustic inventor who comes bearing a creation of priceless worth to his government; the rebuff of man and work by uncomprehending bureaucrats, and finally the triumphant moment when the true value of his creation is incontrovertibly proved and the former skeptics tumble over themselves to do him honor. In fact, the entire film might be seen as a symbolic vision of Disney's career as seen from his vantage point.

There were a handful of other reasonably good films in the live-action group. *20,000 Leagues Under the Sea*, for instance, was a superior and, for Disney, quite lavishly produced adventure, with James Mason superbly cast as the mad inventor, Captain Nemo, whose submarine, complete with pipe organ and marvelously plush fittings that combined twentieth-century functionalism with nineteenth-century elegance, was a triumph of the set designer and decorator's art. Nemo was also something of a projection, albeit a dark one, of Disney himself—a lonely man in love with vanished graces and future technology. *Davy Crockett*, a paste-up of three television shows that started one of the great juvenile fads of the decade, turned out to be a pleasantly exuberant adventure story when it reached the theater screens—simple, solid kid stuff—while *The Great Locomotive Chase* was a bit better than that. Based on a true Civil War adventure, in which northern raiders sneaked behind southern lines and made off with a train, it looked to *Herald Tribune* critic William K. Zinsser as if "everybody obviously had a fine time making the movie"—which, given Disney's love of trains, was probably true. Certainly the film had a zestful inventiveness about it, as well as a drive and a tension that one did not often find in the Disney product of this time—possibly because there was no time to pause over the antics of children and animals as so many of his other pictures did.

All these films, however, stood a little bit outside the Disney mainstream. Or Main Street. The place where he operated most comfortably in the late Fifties and early Sixties was the American small town and in the country surrounding it. Admittedly, this locale, as he pictured it, was not unattractive. It was always verdant, and the old houses, with their wide verandas and big front lawns, sitting complacently on their quiet streets, formed an irresistible setting for the little romances and comedies and animal stories that unfolded there at a leisurely pace and in a warm, chuckly tone. Even when he used this setting for a crime story, as he did in *That Darn Cat*, it completely lacked menace. Or modern reality. The small town is today as subject to blight as any other American place: the neon signs blink on and off; cars circle endlessly in search of parking places, causing small-scale traffic jams in the process; the big old houses are converted into apartment buildings or offices for group medical practices; the blare of television and the gunning of hot-rod motors drown out the chirp of crickets in the evening hours. Only in Disney films has it remained unchanged.

Equally unchanged are the characters that inhabit it: the flinty old shopkeeper; the rich, eccentric widow up on the hill; the comically bumbling, heavily impractical, often inventive or idealistic leading man, with his sensible, patient wife or girl friend and their pleasantly bubbling teen-age relatives—children or siblings—who may be prankish but are never delinquent. In these films time has stopped and character has become conventionalized. We feel, when we enter this world, a certain nostalgia and, more than that, a certain comfort. Nothing will surprise us or hurt us or unduly stir our emotions. Comedy is a canoe overturned in the midst of a romantic outing, tragedy is a lost dog, love is a song cue, and villainy is an overpunctilious banker (Disney never tired of satirizing the type).

Disney no more invented the basic situations of the standard situation comedy then he invented animation, but he did take it over as if to the manner born, which, in fact, he was. Indeed, most Americans are. The tradition begins, so far as film is concerned, in the 1930s, with comedies of the sort that Disney's friend, Frank Capra, frequently made—well tooled and tending to demonstrate that you did not have to be as zany or grotesque as the silent comedians and their direct heirs, the Marx Brothers and W. C. Fields, were in order to be funny. Perfectly nice, seemingly normal people could get laughs when they were placed in odd situations. These films were particularly well suited to carrying a light, liberal message about the virtues of the common man—very useful in the depression decade as well as during the war years, when demonstrations of the qualities we were fighting to preserve were in order. Capra's work had the virtue of freshness, while Preston Sturges, the greatest of the directors to work in this tradition, added a note of mordant blackness to it. Other sources also contributed to the conventions of the form—the Andy Hardy pictures, for example, and such excellent radio programs as *Fibber McGee and Molly* and *The Great Gildersleeve*. Within two decades there were few stereotypes left to be cast, few new situations left to be explored. One laughed—if one still attended at all—not in surprise but in recognition of old friends up to their old, familiar tricks once again. Of recent years, shows in this tradition—*The Donna Reed Show, Leave It to Beaver, Dennis the Menace, Hazel*—have even worn out their welcome on television (though the reruns roll on in the daytime and Disney's favorite actor, Fred MacMurray, perseveres in prime time with *My Three Sons*).

The form is kept alive with gimmicks: families that used to be found on Main Street are now *Lost in Space* or they have a son with the second identity of *Mr. Terrific*. This is essentially what Disney did with the form in the movies. The small-town locale, was preserved and many of the subsidiary characters stayed on in it unchanged, but there was usually a gimmick of some sort to enliven things: The Absent-Minded Professor invents Flubber in his garage and thereafter his Model T, fueled on the stuff, can fly; Hollywood's idea of a typical teen-age boy is turned into a shaggy dog; a Siamese cat becomes a detective, and so on. Juvenile fantasies all—but superficially original, "imaginative," as the untutored audience understands that term. And all taking place against a reassuringly ordinary, completely familiar, indeed well-loved, background.

The search for novelty was not always successful. Disney sent his typical American family to Paris in *Bon Voyage* for a change of scene and there Pop (MacMurray) got innocently involved with a prostitute. Somehow Disney let the sequence pass, but it was a disaster: his audience couldn't understand what was happening, he later claimed, and neither could he. The shock to their expectation was simply too rude, and Disney vowed never again to deal in such racy material. Without an enlivening gimmick, however, the films were unbearable instead of merely innocuous—for example, the film released a few days before Disney died, *Follow Me, Boys!* The story of a man who spends some forty years as a small-town scoutmaster though denied children of his own, it is an orgy of sentiment unparalleled in our day and age. The plot is no more than a loose string of incidents; the characters, so bland as to be fully explained the minute they appear on screen. There is nothing to do therefore but exploit the almost saintly goodness of the scoutmaster, his wife and the generations of splendidly upstanding lads he helps to mold. It is incredible, it is smug, it is without any relationship even to historical, let alone contemporary, reality. It is, in short, a travesty—unintentional camp and the *reductio ad absurdum* of Disney's strange need to ingratiate himself with the least common denominator of his audience.

In addition to the intrinsic weaknesses of its plot and characterizations, *Follow Me, Boys!* suffered from the effort to puff it up into a major release for the Christmas trade. What might have been barely bearable at the length of an hour and a half was simply excruciating when spread over better than two hours. The film was not quite an imitation *Mary Poppins*, but it was no longer in the mold of, say, *The Absent-Minded Professor.* To see just how the Disney product could suffer from a fully conscious attempt to have, as Roy Disney put it in 1967, "at least one *Mary Poppins* every year," viewers had to wait until the following Yuletide season when *The Happiest Millionaire* came crashing down on them. Based on Kyle Crichton's modest little play of the same title (which was, in turn, based on the reminiscences of her father that Cordelia Biddle Duke wrote with Crichton's aid), the film was a nightmarish blending of dull songs, flat family comedy, fraudulent period charm and insipid juvenile romance, directed by Norman Tokar—a contract craftsman at the studio—as if twenty years of movie history had never happened. Fred MacMurray, a modestly gifted *farceur* in the romantic comedies of his youth, an amiable enough presence in more middle-class surroundings, was unable to capture the Wodehousian nuttiness of an aristocratic eccentric and received precious little help in the attempt from either direction or script (both of which had been supervised by Disney). Greer Garson played his wife as a curious blend of Mrs. Miniver and Mrs. Roosevelt, Tommy Steele's unquestioned talent and energy were not effectively governed in the role of a supposedly comic butler, and Lesley Ann Warren and John Davidson were simply blah as the young lovers. In short, this would have been a badly blighted product in the best of circumstances, but as a major production—again well over two hours long—which for a time the studio had hoped to release as a hard-ticket, road-show attraction, it became positively offensive. Here, perhaps more than in *Mary Poppins*, Disney and his people could be faulted for rounding off the principal character's eccentricities until he was reduced to a veritable nubbin of ordinariness. Worse, they put before their public something quite rare for the studio—a truly dishonorable failure. In the past, at least, it could be said that films like *Fantasia*, or the nature films failed, despite at least partially honorable intentions because the studio lacked the intellectual sophistication to realize all the values inherent in their subjects, or that the little live-action comedies failed for lack of ambition. But *The Happiest Millionaire* failed for the most cynical of all reasons—it was a wretched attempt to imitate merely the success—not the spirit—of a previous hit.

Again, as in the case of MacMurray's scoutmaster in *Follow Me, Boys!*, the film's chief interest lay in its projection of an aspect of Disney's own character in one of the leading parts. This time we had a vision of Disney as a young man (John Davidson's juvenile lead) monomaniacally pursuing a new technological breakthrough. This youth's dream was not the advancement of the animated cartoon but the automobile, and like the historical Disney, he was capable of ignoring even the girl he loved when a new idea—an innovative braking or carbureting system, for example—suddenly occurred to him. He even had a ghastly song, an unbeat hymn to "Detroit" ("you can hear it humming, you can feel it coming") with which to express his faith in the future. The last shot in the film is, indeed, of the young couple putt-putting along in their car with the towers of the Motor City rising out of the smog in front of them and a great orange sun rising above the pollution. To Americans of 1967, for whom the car has turned from a wonder to a maiming, environment-destroying nightmare, the scene is incredible in its naïveté. Disney films are usually out of touch with reality, but rarely as directly opposed to it as this one was.

Most of his live-action films were not so actively offensive as

The Happiest Millionaire. But they were all, one way or another, too cozy, too bland, too comfortable and comforting. In
the best American culture, high and low, there is always expressed, one way or another, a wild surmise. Sometimes it is, as
music critic Wilfrid Mellers expressed it, "a Whitmanesque
energy and comprehensiveness, an ubiquitous love of humanity
and of every facet of the visible and tactile world." Sometimes
the American creator "is a solitary, alone with Nature, seeking
a transcendental order within the flux of reality." Whichever
direction he turns, however, the effect is the same. One senses in
his work the vast spaces of the land, the loneliness of the individual lost in that space and trying to find meaning in his isolation. True, Disney was only an entrepreneur of low-level popular art, of family entertainment. But one catches a glimpse, an
overtone, a perhaps unthinking hint of the American nightmare
in all kinds of popular work no more elevated in its intentions
than his—in the good westerns and in the adventures of the private eyes, in the surreal comedies of the Marx Brothers, in the
vicious parodies of bourgeois longing that W. C. Fields used to
make, surely in the wanderings of Chaplin's Tramp. How was
it that Disney could never bring himself—or at least allow his
employees—to allude to these great and constant themes of our
culture? Why did he, by preference, huddle in the small town
or on the farm, forted up, as it were, against wild beasts and the
skulking Indians of the American imagination? The answer is,
perhaps, clear by this time, and it is, of course, the clue to his
popularity as purveyor of entertainment as well as to his ultimate failure as an artist after the decent promise of his beginning. He was heir, of course, to the puritan spirit, defined by
William Carlos Williams as "a tough littleness to carry them
through the cold" and as a religious zeal—"not a thrust upward
to the sun, but a stroke in: the confinement of the tomb." In
short, he feared man as he feared nature. As he sought to deny
the mystery of the latter by anthropomorphism—that is, by relating it to the more familiar patterns of human behavior—so

306

he denied the former its infinite possibilities by reducing it to comic clichés, by imitating old imitations of life. And so he passed, at last, beyond—or beneath—criticism. In the last decade and a half, the success or failure of a Disney film in no way depended upon reviews or upon the magic or lack of magic of some combination of stars and story. It existed quite outside such factors, which normally weigh so heavily upon the heads of the film industry's chieftains in these parlous times. Pressed by a reporter to explain what he thought might be the deciding factor in the success or failure of a Disney film, E. Cardon Walker, the tough-minded Disney vice president in charge of merchandising, confessed that he had no easy answers. Titles, he thought, had something to do with it. *Savage Sam*, for example, was a perfectly good dog story of the sort the studio had previously done well with, but it flopped. Walker thought people probably were misled into thinking that it might not be quite the warm, whimsical, wonderful sort of a tale about a pooch they liked to see. Even the Disney name could not pull it over the top. Similarly in the case of *The Moonspinners*, the adventure-mystery sounded unlike Disney's customary concoction, so they skipped it. *Pollyanna*, on the other hand, has come to have an unfortunately oversticky connotation, and all the studio's efforts to convince people that its version of the old story was nothing like the image they had in their heads were to no avail. (Nor indeed did the studio convince the star, Hayley Mills, who played the title role and recalled recently, in a New York *Times* interview, that she enjoyed working for Disney in five films "but those films were very restricting. Conveyor belt jobs. So goody-good, you know? And there was this image I created which was hideous. I wasn't supposed to be seen drinking or buying cigarettes or smoking in public. The reasoning was that the audiences for the Disney films were very young and if they saw me smoking eight cigars a day, why shouldn't they?") Walker was inclined to be philosophical about all this. So long as he has a strong film for Christmas release, usually one that has what are known in the trade as "production values" —*i.e.*, an expensive look—and a film that has a solid appeal to teen-agers for the summer drive-in trade, he can be reasonably sure of having a good year. Around these, the studio tends to tuck one animated "classic," a double feature bill composed of rereleases of the more successful live-action films and one or two minor new pictures done on very tight budgets (often one of these will be a two-or three-part television project that looks strong enough for theatrical release and can always be placed on the TV show a couple of years later anyway).

That this pattern worked was because it was a pattern, a predictable schedule of films on a fairly narrow range of subjects and directed at two well-defined movie markets—the so-called family market, which was essentially a youthful one, still taking its young children to the movies rather than sending them off alone, and the teen-age market, which in these still simple days demanded only broad farce and sanitized romance conducted by actors who looked to be the audiences' contemporaries. Such audiences responsive to gimmicks and twists but not to excessively novel or challenging material—a posture that suited Disney very well. If one of his films flopped in these markets, the loss was not high and the chances were that later in the year something else, for its own somewhat mysterious reasons, would recoup the modest failure that preceded it. From 1950 through 1966 no less than twenty-one Disney features grossed more than four million dollars apiece, an average of slightly more than one a year. It is as risk-free a way of making movies as anyone ever devised.

It did not, however, spring into being overnight. And it is doubtful if it would have become as successful as it did without several developments that occured in 1953 and 1954. The first of these was the formation of Disney's own distribution subsidiary, Buena Vista, named after the street on which the studio's main gate is situated. The company was largely Roy Disney's invention, and he was motivated largely by the reluctance of

Disney's long-time distributor, RKO Radio, to get behind *The Living Desert* with suitable enthusiasm and by a similar lack of interest among RKO's competitors. In addition, the Disneys had long chafed under the heavy percentage of their grosses that went to their distributors, even though the 30 percent RKO was getting was close to rock bottom in the industry. By setting up the lean Buena Vista operation, the Disney people cut distribution costs to 15 percent of the gross and, equally important, gained direct control over the handling of their films in the market. They could keep them off double-feature bills where they might be paired with products deemed unsuitable to them, and they could begin to package entire programs, consisting, for example, of a feature, a cartoon and a nature film, which usually ran a bit longer than the usual short subject and was thus often hard to wedge into other people's programs. Compared to the other events of this decade in Disney's land, the founding of Buena Vista was not very glamorous or exciting, but it did represent, for the Disneys, the final step in gaining complete control of their own destiny, complete freedom from interference by outsiders in the creation and exploitation of their products. It also, of course, symbolized their rise out of the ranks of the independents to a status in every way coequal with the major Hollywood production companies, which had all along had their own distribution arms. Most important of all, the studio was able to time the release of its films so that they could most effectively be coordinated with ancillary activities (*i.e.,* the big film is released in a period where television-viewing is at a peak, and more people are likely to see the commercials for it). The studio also gained a very valuable form of self-protection: no one will ever be able to sell out a Disney product at a discount or in a hurry merely to improve the distributor's cash flow, a problem that constantly besets the independent producer.

The other important development of 1954 was Disney's entrance into television. Unlike other Hollywood producers, he was quite unafraid of the new medium. It was, after all, only a new technology applied to the business of entertaining large numbers of people, and new technologies in this area had never been anything but good for Walter E. Disney. His immediate goal, however, was not the exploitation of his movies but of what one observer called "the world's biggest toy for the world's biggest boy"—an amusement park of a quality and a dimension no one but Disney and a few associates could quite envision.

Through the years he had continued to nourish his notion of a new kind of amusement park, and in the outwardly fallow postwar period he had begun to doodle plans for what his brother continued to regard as another of "Walt's screwy ideas," principally because Disney, as usual, had trouble in logically articulating just what he had in mind and just how his amusement park would be different from all other amusement parks. All he knew for certain was that he did not want to imitate the existing pattern because, as he had discovered when visiting them with his daughters years before, they were "dirty, phoney places, run by tough-looking people." There was, he said, "a need for something new, but I didn't know what it was."

By 1952 he had some ideas roughed out, and using ten thousand dollars of the studio's money—all, apparently, that Roy would give him—and such money as he could raise on his life insurance, he had commissioned "plans, drawings, designs and models" for a park, which he had already christened "Disneyland." In that same year he set up WED Enterprises and recruited for it a small staff of designers, quickly, and with a typically Disneyian disregard for the niceties of language, dubbed "Imagineers." In that same year the Stanford Research Institute was commissioned to conduct an analysis of the sites available for the park within the Los Angeles area. Among the criteria it drew up to aid in its search, the most important was that the parcel be large enough to accommodate a preexisting ground plan, without crowding. In addition, the Stanford peo-

ple were told to study the desirability of the neighborhood, the
price of the land, available utilities, topography, population
trends, freeway patterns, summer and winter temperatures, tax
rates, zoning and even the number of days per year when the
smog might settle so heavily on the magic kingdom as to rob it
of its magic (Pacific Ocean Park, the largest traditional amusement park in Los Angeles, lost far too many profitable nights
of operation, in Disney's opinion, because of smog).

The Stanford study ruled out anything within the more
densely populated areas of Los Angeles because land costs were
too high, and the San Fernando Valley was climatically all
wrong—too hot too much of the time. That left Orange County,
south of the main population center, as the best bet. Land was
still cheap there—it was given over mostly to orange groves—
and the maps of the highway planners showed the half-
completed, half-projected Santa Ana Freeway reaching out in
the right direction and, indeed, passing along one of the borders
of a site the researchers thought met most of their other criteria
reasonably well. This was a parcel of small orange groves near
Anaheim, a town that owed its limited fame to nightly mention
in the frost warnings broadcast on local radio stations for the
benefit of fruit ranchers and to Jack Benny's radio writers who
made a national gag out of an imaginary train route running
through the heart of the citrus country—"Anaheim, Azuza and
Cucamonga." The advantage of the site was that it would
be no more than twenty-seven minutes from the Los Angeles
city hall once the freeway was finished. Its disadvantage was
that the 160 acres it encompassed—it has since grown to 185
acres—were owned by no less than twenty families.

Negotiations with them consumed the better part of two
years, and as is customary in assembling properties piecemeal,
the asking prices of the owners immediately shot far above
what was customary in the neighborhood. One family even
struck a unique deal with Disney, demanding as a condition of
sale that he preserve in perpetuity two stately palm trees of sen-

timental value to them. (Luckily they stood in an area designated for "Adventureland" and so form part of the tropical
background for the well-known African riverboat ride there.)
Another family home, a pleasant example of the Spanish mission style of architecture, was also preserved and became an administration building for the park.

But even as the land was being purchased, Disney was running into other problems. Early in 1954, the year he had scheduled for the beginning of building, Disney sent four staffers off
on an around-the-country tour looking for ideas at established
amusement parks and the firms that manufactured equipment
for them. The only idea on which they found general agreement
was that Disney was crazy. You could not, their informants said,
have an amusement park without a roller coaster, a ferris wheel
or barkers to shill the customers into the various attractions.
Nor were they optimistic about Disney's plan to bar outdoor hot-
dog stands and the sale of beer just because he did not like the
smells these enterprises created on hot, still days. The equipment
manufacturers were especially frustrating, according to one of
the surveyors. They wanted to sell Disney their standard-model
thrill machines and were totally indifferent to the Disney
group's pleas for something new. "I can remember only two or
three of the long-time amusement operators who offered any encouragement at all," one member of the team said later.

That was two or three more visionaries than Disney encountered in the financial community. Everywhere he went in search
of the minimum of seventeen million dollars he needed to start
building, he was reminded that, if anything, the outdoor amusement business was a cultural anachronism that had already declined into senility. Even with his plans laid before them, the
bankers could not see what Disney was talking about. Once
again, in his phrase, "they stepped on my neck."

Some funds were acquired by a unique leasing arrangement
with concessionaires. Thirty-two firms, among them some of

the great names in American industry as well as some of the best-known purveyors of food and drink, expressed a willingness to open exhibitions and restaurants in the projected park, which proved again Henry Ford's dictum that managers have a sense of imagination infinitely more venturesome than that of the financial community. Disney insisted that they sign five-year leases and pay the first and fifth year's rent in advance. This provided him with some ready cash, and he found it easy to get more by turning the unpaid but safely anchored portions of the leases over to the banks as collateral for loans.

This was still not enough, however, to begin construction. And that was where television entered the picture. "I saw that if I was ever going to have my park," he told an interviewer, "here, at last, was a way to tell millions of people about it—with TV." It was perhaps a little more complicated than that. What happened was that he concluded a deal with American Broadcasting Company–Paramount Theatres, Inc., under which he agreed to produce, for seven years, a weekly one-hour television program, to be called "Disneyland," on which he would be free to promote liberally not only his amusement park but his films. Distinctly the third network at the time, ABC desperately needed Disney's name and skills, so much so that as an obvious *quid pro quo*, but under an entirely separate arrangement, ABC–Paramount agreed to purchase 34.48 percent of the shares of Disneyland, Inc., a new firm chartered in 1951. Western Printing and Lithographing, the large firm that has published almost all the books and comic books bearing the Disney imprimatur since the 1930s, took 13.8 percent of the new firm's stock, and by this time, Walt Disney Productions was rich enough to be able to afford 34.48 percent. Disney himself retained the twenty-four hundred shares that had been, in 1953, the entire outstanding stock of Disneyland. They had a par value of $100, represented 16.55 percent of the new firm and were considered recompense for his independent work in developing the concept of Disneyland. Another one hundred shares went to WED Enterprises in return for that holding company's license to use the Disney name on the park and the TV show.

Even in this tight spot, the Disneys drove a hard bargain. Both ABC–Paramount and Western Printing had to give Productions an option to buy their shares at par value plus 5 percent per year, an option that Productions began to exercise in 1957, just two years after the park's official opening. The last of ABC-Paramount's holdings were bought out in 1961, but at the considerably higher price of $1,500 a share. Meantime, the ABC network quickly discovered it had two hits on its hands— its newly acquired stock and its new television program. When *Disneyland* went on the network in the fall of 1954 it quickly climbed into the top ten of the Nielsen ratings, and it and its successor on NBC, *Walt Disney's Wonderful World of Color*, have stayed there, or close to it, ever since. In his first year on television, Disney won an Emmy for *Underseas Adventure*, hardly more than a promotion piece for his most expensive live-action film to date, *20,000 Leagues Under the Sea*, describing the process and problems of underwater photography. By no means coincidentally the film became the biggest grosser to that time of any of Disney's live-action features, totalling $6.815 million in its first release and adding another $5 million or so in various releases. As for Disneyland itself, after a full season of promotion on television, followed by an opening day that was also televised nationally, it was almost instantly a success. Very quickly it came to be regarded as one of the wonders of the modern world, both by visitors to the premises who have included, according to its publicity department, which is maniacal about such matters, eleven kings and queens, twenty-four heads of democratic states and twenty-seven royal princes and princesses—and by the once-indifferent financial community, which is impressed by properties capable of grossing $195 million in ten years, as Disneyland did, and capable of financing, out of its own pocket, $33 million worth of expansion in the same period. Grosses and similarly financed expansions in the same amounts are projected for the five-year period ending in 1971.

There are many humorous stories about the opening of Disneyland—of how the toy Dumbos that were supposed to fly on wires over the park weighed a skein-snapping 700 pounds apiece when they were delivered; of how the rivers snaking endlessly through the park ran dry immediately after the water was pumped in and had to be rebottomed with clay to hold the moisture, of how extra crews had to be worked around the clock in the two weeks before the park opened in order to have it ready on time and of how concrete was still being poured as the TV cameras were set up to record the great day (there were thirty thousand invited guests in the park for it). So it must have been a fretful Disney who stepped before the microphones and cameras on July 18, 1955, to promise that "Disneyland will never be completed, as long as there is imagination left in the world."

Still, one can also imagine a deeper satisfaction underlying whatever minor impatiences and disappointments he felt that day. For he was opening something that was far more than a whimsy or a modest sideline to his other endeavors. Disneyland, to him, was a living monument to himself and his ideas of what constituted the good, true and beautiful in this world. It was a projection, on a gigantic scale, of his personality. If not quite an extension of man, in the McLuhanesque sense of the term, then surely an extension of a man in the way that the pleasant grounds of Versailles were an extension of The Sun King. It has none of the discreet impersonality of, say, Rockefeller Center. It was, and is, a statement containing, in general and in particular, conscious expressions of everything that was important to Disney, unconscious expressions of everything that had shaped his personality and of a good many things that had, for good and ill, shaped all of us who are Americans born of this century.

Very simply, it was, in every sense, the capstone of a career and a life. Very possibly it is the capstone of a time and a place as well.

30 *Shows there will certainly be in great variety in the modern civilization ahead, very wonderful blendings of thought, music and vision; but except by way of archeological revival, I can see no footlights, proscenium, prompter's box, playwright and painted players there.*

—H. G. Wells,
A Modern Utopia (1905)

THE CAPSTONE OF A career very quickly became the cornerstone of an empire. In the first six months of its existence in 1955 slightly more than a million people visited Disneyland, and Productions that year grossed over $25 million for the first time in its history. In 1956, Disneyland's first full year, about three million people passed through its gates, producing gross revenues of $10 million, its share of which helped send Walt Disney Productions past the $30-million mark in revenues for the first time. Thereafter, word of mouth and the publicity machinery spread the news of the glories of Disneyland, and with the park itself generating the capital to finance a vast expansion (abetted by tax laws that encourage fast depreciation), the curves on all the charts shot upward. By 1959, Productions was trending close to the $60-million line in its grosses. Every time the park expanded its capacity, revenues increased more than proportionately to the added capital.

The figures were undoubtedly gratifying to Walt Disney— an omen of a stability he had not known before. But they were, it seems, of no consequence in comparison to the joys he found in the park itself. Stephen Birmingham, the novelist, captured

Disney's feeling for the park very well when he wrote: "Walt feels about Disneyland the way a young mother feels about her first baby. He coddles it, pampers it, fusses at it, bathes it, dresses it, undresses it, peers at it from all directions—and boasts of its latest accomplishments to anyone who will listen. He has even fitted himself out with a tiny apartment in Disneyland—above the fire house on Main Street—decorated it like a jewel box in red plush and velours and furnished it with gaslight era antiques, just so, if necessary, he can be near Disneyland when it sleeps. He keeps his own twin-engine plane and pilot and as often as twice a week, flies over Disneyland to check on his darling from the sky."

The stories of his compulsion to keep the place perfectly groomed at all times are legion. All day long sanitation men dressed in costumes appropriate to their assigned area prowl the streets, picking up litter, so that the slovenly habits of the "guests" (who are never called customers) will not be apparent. Every night every street and walkway is thoroughly hosed down, and crews armed with putty knives get down on hands and knees to scrape up carelessly discarded chewing gum. Even the targets in the shooting galleries, dulled by the previous days shots, are repainted. Every year some 800,000 plants are replaced because Disney refused to put up signs asking his "guests" not to trample them. All of this cuts profit margins as a cost accountant normally calculates them, but it adds immeasurably to the appeal of the place and undoubtedly contributes to the astonishing fact that 50 percent of the people who enter the gates are returnees.

The staff is as well scrubbed as the grounds and is so polite and well mannered that by the end of the day a visitor's face muscles begin to ache from the effort of returning so many smiles and murmuring so many return pleasantries. There are twenty-three hundred permanent employees; in the peak summer months that figure doubles. Known in the ineffable jargon of the park as "people specialists," all must spend a few hard days in the "University of Disneyland," which trains them in the modern American arts forms—pioneered by the airlines—of the frozen smile and the canned answer delivered with enough spontaneity to make it seem unprogramed. Many of the summer people are college and high school students who bring a certain natural exuberance to their work and whose salary expectations are no higher than were those of the apprentices in Disney's old studio art school. All are carefully schooled in good grooming and good manners, as defined by local custom, and crammed full of the facts, figures and folksy anecdotes that comprise the lore of the place. Some measure of the prevailing institutional tone may be gathered from the university's textbook: "We love to entertain kings and queens, but the vital thing to remember is this: *Every* guest receives the VIP treatment." Or, "Disneyland is a first name place. The only 'Mr.' here is 'Mr. Toad.' . . . It's not just important to be friendly and courteous to the public, it's essential." Or, in a more philosophical vein: "Show business is fun and fulfilling and rewarding. But it is also an exacting endeavor which requires the toughest form of personal discipline. . . .

"At Disneyland we get tired, but never bored, and even if it is a rough day, we appear happy. You've got to have an honest smile. It's got to come from within. And to accomplish this you've got to develop a sense of humor and a genuine interest in people. If nothing else helps, remember that you get paid for smiling. . . ."

The Disneyland staff has a place for every human variety except the traditional carnie type, but as its director of personnel told one visiting reporter, every effort is made to scrub the rough edges off: "No bright nail polish, no bouffants. No heavy perfume or jewelry, no unshined shoes, no low spirits. No corny raffishness, yet the ability to call the boss by his first name without flinching. That's a natural look that doesn't grow quite as naturally as everybody thinks."

As a result, the people specialists tend to present a rather standardized appearance. The girls are generally blonde, blue-eyed and self-effacing, all looking as if they stepped out of an ad for California sportswear and are heading for suburban motherhood. The boys, who pilot vehicles and help you on and off rides, are outdoorsy, All-American types, the kind of vacuously pleasant lad your mother was always telling you to imitate. If the interviews and the personality tests every Disneyland job applicant must undergo suggest that a young man is more than usually outgoing—more of a ham—then he may get a job as a talker on one of the rides like the jungle boat ride, on which the spiel both enhances the drama of the occasion and also, because it has an ironic twist to it, constantly reminds the overimpressionable that all is artifice. Even the cops at Disneyland are a new breed—generally moonlighting schoolteachers, with physical-education instructors predominant among them.

Disney himself once commented: "The first year I leased out the parking concession, brought in the usual security guards—things like that. But I soon realized my mistake. I couldn't have outside help and still get over my idea of hospitality. So now we recruit and train every one of our employees. I tell the security officers, for instance, that they are never to consider themselves cops. They are there to help people. . . . It's like a fine restaurant. Once you get the policy going, it grows."

The aim of the staff is to keep everyone in a spending mood without ever once overtly suggesting that Disneyland is, in the last analysis, hardly a charitable enterprise. The trick is not to harass the visitor into spending but rather to relax him to the point where the inner guardians of his frugality are lulled into semiconsciousness. It works; the average expenditure is $6.50 per person per visit—making a day at Disneyland no cheap excursion for the family. The spending process is eased by books of discount tickets that are effectively and extensively promoted. Priced at various levels, they are available at the entrance and all contain tickets of different denominations that may be evenly exchanged for rides and attractions.

Thus, money, well known to be dirty stuff, is not exchanged every time the family decides to embark on a new adventure, and Dad is freed of the gathering annoyance of repeatedly groping for his hip pocket. Indeed, quite the opposite effect is created. The family feels an impetus to get the full benefit of its bargain by using all its tickets before leaving the park. Officials say that the only consistent complaint they receive comes from those whose spirits are willing but whose flesh weakens before they have used up their tickets. But that, of course, is their problem, not Disneyland's. Since they have received an automatic 20 percent discount by buying the ticket books and they probably have fewer than 20 percent of their tickets left at the end of the day, the chances are that they have come out a little bit ahead anyway. The unspent tickets, in any case, remain redeemable throughout the year.

The park's general design further enhances one's willingness to contribute to its economic well-being. There are plenty of shade trees and much greenery and, psychologically most important of all, lots of cool water in ponds and lakes and rivers, all of which are man-made. As many rides as possible move on water. As a result, one is never oppressed by heat or by the dust and dirt that inhabit most amusement parks. Equally pleasant, all the walks are wide enough to accommodate large numbers of people without any sense of crowding or jostling, and there is little anxiety, therefore, that a small child may be lost or hurt in the press of people. The walks curve gently under the trees and beside the waters, encouraging a strolling pace, a sense of ease and well-being that is utterly unique in the context of an amusement park. The layout also encourages a sense of discovery that, in turn, encourages impulse buying. One is forever turning a corner and coming upon some previously unnoticed marvel that cries out for investigation.

Once committed to such an investigation, the visitor will generally find that the line at the entrance is far shorter than he is used to finding in public places of amusement. The "imagineers" have broken the usual single line into many short lines by an arrangement of railings that neatly, and without a word needing to be spoken, divide the patrons into small groups who are then kept moving through a sort of maze until they reach their destination. If it is a ride, they will discover that the vehicle, instead of having one or two entrances, has one for each row of seats and that the rows exactly match the number of exits from the waiting maze. An attendant gives you a firm hand into the vehicle, which helps prevent accidents, seems a polite thing to do and, incidentally, speeds the loading process. The whole thing is a marvel of technology applied to mass psychology. People simply feel better if the line they are in is short and if it is constantly moving. And the management increases turnover, and therefore its take, with no need to resort to the depressing advice to step lively, keep moving, don't push, that so often marks the effort to find recreation in these overcrowded times. Disneyland is, on this basic level, one of the most intelligently conceived pieces of architecture in America and one well worth the study of anybody faced with the problem of creating structures to serve large numbers of people comfortably but with no loss of efficient revenue production.

But one has come to expect nothing less than the best from Disney in technical matters. Once one yields to the rage for order, however, it is difficult to know when its demands become excessive. This is particularly true in the case of Disneyland, where the audience is an integral part of the show and where the proprietor has gone to extraordinary efforts to provide for its convenience. The next logical step is some sort of visa control at the frontier of the magic kingdom, assuring that visitors meet the same standards of dress and decorum that the staff maintains. A small start in this direction has been made on those nights when the park remains open after its normal closing hour to serve as the host for all-night high school graduation parties. A way of turning idle hours to profit, and also providing youngsters with a safe, supervised environment for their revels, the idea has added to the park's revenues and has been gratefully received by parents. But boys are required to wear coats and ties on these occasions, and anyone eccentrically dressed is turned away at the gates. The young men are also frisked to see that they are carrying neither weapons nor liquor into the Magic Kingdom. One understands the problem facing the management and yet one resents its solution. Why should the innocent majority be subjected to this humiliation, however briefly and discreetly it is carried out, because of excessive corporate fear of a group it acknowledges is a distinct minority in the teen-age subculture? Why, indeed, is the American middle class so preoccupied by the current fantasy of mass juvenile delinquency? Why does it insist upon alienating the law-abiding majority of youngsters by treating them as if they, too, are potential revolutionaries? Why this excessive concern with such superficialities as hair style and dress when the creation of genuine understanding and sympathy between the generations receives so little concern? These questions are beyond the scope of this book, but the unthinking embrace of the conventional wisdom about youth by Disneyland is further evidence of its almost maniacal desire to keep the eccentricities of individualized expression—right down to the details of dress and verbal expression by its staff—away from its door. If it could, one suspects that it would try to force all its visitors into patterns its founder regarded as suitably nice.

The contents of the park present a more difficult critical problem than its style of hospitality. Though Disneyland is a circle divided into five sections, each representing an area of thought or fantasy of great significance to Walt Disney, there is only one entrance to the park, and its location compels the visitor to pass through a turn-of-the-century American Main Street, an idealized vision of Disney's boyhood environment. Beyond it is a central plaza with main avenues leading to four lands. Tomorrowland is, of course, a monument to Disney's delight in technological progress. Fantasyland allows one to step into the environments made familiar by Disney's animated films and which, like them, remind one that, imaginatively, Disney has remained always something of a child. Adventureland is, of all the areas, the one least closely tied to Disney's own psychology and creations, but it nevertheless represents an ageless and universal dream of exploration. Finally, Frontierland, with its western street and its river, not only captures the standard American nostalgia for free land and a free life but also contains two attractions that are obviously derived from Disney's boyhood fantasies in Missouri. One is a river steamer, setting of *Steamboat Willie*, and, of course, a dream of grandeur shared by many midwestern boys and often movingly expressed by the writer after whom Disney named his large scale model, Mark Twain. The *Twain*, along with some Indian war canoes, some "Mike Fink" keelboats, some "Tom Sawyer" rafts and the strangely out-of-place model of the *circa* 1790 sailing ship *Columbia*, circumnavigates (on tracks hidden under the water) "Tom Sawyer's Island." This, according to employees, is the only attraction in the park that Disney designed completely by himself. Everything on it—a play fort, log cabins, a string of pack mules you can ride—is free. "I put in all the things I wanted to do as a kid—and couldn't," Disney once told a reporter. "Including getting into something free."

Throughout the park there is, as Aubrey Menen, the Anglo-Indian novelist, observed, a masterful matching of scales and proportions to the psychological content of the fantasy environments. For example, Main Street. "It's not apparent at a casual glance that this street is only a scale model," Disney said once. "We had every brick and shingle and gas lamp made five-eighths true size. This costs more, but made the street a toy, and the imagination can play more freely with a toy. Besides, people like to think their world is somehow more grown up than Papa's was." Menen noted that there was even more subtlety to Main Street's design than this simple reduction in scale. A trick was played with proportions as well. "Everything on the first floor was as it should be. . . . But above them were windows that were too small; above them, again, were gables that were smaller still. The reduction in size as one's eye traveled upward was so beautifully done that it was almost imperceptible."

Moving on through Frontierland and Fantasyland, he observed a similar sensitivity to the uses of proportion in different ways. The bears and the Indians on Tom Sawyer's Island are, he noted, full size; the fairytale creations in Fantasyland are on the tiniest possible scale. He comments: "There are two kinds of legends: with one sort we can get inside them; with the other we are all spectators. I suppose there can be no American male who's not, at some time in his life, found himself alone in the countryside and explored Tom Sawyer's Island or fought Indians or crept on his belly up to a paleface fort. But nobody, I think, at any age plays water rat and toad, or goes in a mole's house, or plays Prince Charming or Cinderella (unless driven to it by sentimental elders). These stories are too complete to have room for the outsider. We should know what to say to Pinocchio if we met him, or the Three Ugly Sisters. But we do not imagine ourselves as being these people. A lesser man than Disney would not realize this. But here Tom Sawyer's island is big enough for children to play on; and Pinocchio's village is so small there is not even room in its street to put one's foot. Once again Disney shows himself a master of the use of proportion."

It must not be assumed that the pleasures of Disneyland are only a matter of such artifices. The bobsled run around and through the scale model Matterhorn (which is big enough to contain in its bowels a basketball court for the use of off-duty employees) is as thrilling a ride as anyone is ever likely to take in an amusement park, a vast improvement on the rollercoaster, which is strictly an up-and-down thing. The Matterhorn ride produces not only dips and rises but hairpin turns as well. The submarine ride in Tomorrowland gives a generally satisfactory impression of a trip underwater: a giant squid and an octopus fight it out just beyond the portholes, the ship scrapes along beneath a polar ice cap and past a sunken pirate treasure ship and even rocks alarmingly when a squall comes up suddenly. Throughout there is realistic-sounding dialogue between captain and crew on the intercom; the oily smell of machinery at close quarters pervades the air, and there is a deliciously claustrophobic quality to the experience—not enough to frighten you, but just enough to jerk you out of the attempt to remain a detached, nonparticipant observer of the fantasy.

It is possible, in a single day in Disneyland, for the visitor not just to see but to enter into time and experience past and time and experience future, to recapitulate not only his own memories and fantasies but those of the race as well. He can visit, impressionistically, every continent of the globe, its mountain heights and ocean depths as well as almost any historical epoch, including the prehistoric. In the revised Tomorrowland he can now even gain an impression of interstellar travel on a ride that simulates such astronautical experiences as weightlessness. It is therefore sometimes possible to feel that Disneyland is best summarized as a model of the "global village" Marshall McLuhan is always talking about, a place where one can literally touch, smell and see, in an instant, and almost as easily as turning around, some representation of the thoughts and experiences that have made us what we are, some representation of the thoughts, traditions and styles that made our fellow villagers.

If that is too fanciful a description of the place or too weighty a one, it is certainly possible to see in Disneyland one of the best mixed-media shows ever devised, an unconscious—and, of course, simplified—vision based on the tradition that, in the higher arts, has been known as modernism. In this tradition, as Arnold Hauser puts it, "The dream becomes the paradigm of the whole world-picture, in which reality and unreality, logic and fantasy, the banality and sublimation of existence, form an indissoluble and inexplicable unity. . . . Art is seized by a real mania for totality. It seems possible to bring everything into relationship with everything else, everything seems to include within itself the law of the whole. . . . The accent is now on the simultaneity of the contents of consciousness, the immanence of the past in the present, the constant flowing together of the different periods of time . . . the impossibility of differentiating and defining the media in which the mind moves."

Disney surely wanted Disneyland to be perceived as a work of art. He was very careful to protect it from intrusions by the outside world. He threw an earthen bank fifteen feet high around it so that nothing of its surroundings could be seen from inside; he spent fourteen thousand dollars removing telephone poles from sight and burying the lines. He even found himself temporarily dismayed by the free-enterprise system as a result of his insistence on keeping the world out. He had originally planned to buy power for the park from the local, municipally owned power plant, but then he was approached by a utilities magnate who observed that Disney ought to be true to his economic principles, avoid the taint of socialism and buy his electricity from a privately owned firm. Disney agreed and the company began bringing in lines from a considerable distance. To his horror Disney discovered that they were mounted on huge, ugly poles that seemed to be marching, marching, marching down upon his art object. "I stood in the middle of *my* park and all I could see were high tension lines," he recalled. "It was awful." The company refused to bury the lines unless Disney paid the cost, which he naturally did.

To Disney, Disneyland was the greatest of all his statements, and he could never let it rest until the end of his life, because it was his. His. His. In the first decade of the park's existence he spent thirty-six million dollars over and above the initial investment, and there were plans to spend another forty-five million dollars over the next five years. By 1966 the number of major attractions at Disneyland had risen from twenty-two to forty-eight, and the end was not yet in sight. "For twenty years I've wanted something of my own," Disney said in 1963. "I worry about my pictures, but if anything goes wrong in the park, I just tear it down and put it right."

It is perhaps a measure of his unfamiliarity with the ways of the artist that Disney took this attitude toward his great work. The drive of the artist must always be toward completion, toward an end to the agony of creation. It is impossible, psychologically, to sustain creativity endlessly. But Disney, obsessed with the perfection of this peculiar vision of his, could not help but go on and on trying to improve it. Indeed, in his mind, the great virtue of Disneyland might have been that it could be ever improved. Unlike his movies, which perforce he had to let go out into the world regretfully, Disneyland was pinned down in time and place and could not avoid his endless ministrations, any more than he could avoid being reminded of its existence. It could not avoid him, he could not avoid it. The dream was thus a nightmare. Indeed, it was worse than a nightmare, the effects of which the light of day can sometimes dispel and the passage of time can surely blunt. So he huddled over the thing, changing it, expanding it, improving it, if not absolutely, then in his own eyes. As a result, what was genuinely good about it was somewhat advanced, but what was bad about it grew more and more obvious.

From the start there had been occasional critiques of the place from wandering literary minds who, as usual, saw the place darkly, through the glasses of their attitudes, and therefore quite lost their balance. For example, Julian Halévy's 1958 piece in *The Nation* correctly summarized the elements in the park that must make any objective observer a little queasy, then noted:

"As in the Disney movies, the whole world, the universe, and all man's striving for dominion over self and nature, have been reduced to a sickening blend of cheap formulas packaged to sell. Romance, Adventure, Fantasy, Science are ballyhooed and marketed: life is bright colored, clean, cute, titivating, safe, mediocre, inoffensive to the lowest common denominator, and somehow poignantly inhuman. The mythology glorified in TV and Hollywood's B films has been given too solid flesh. By some Gresham's law of bad art driving out good, the whole of Southern California and the nation indivisibly is affected. The invitation and challenge of real living is abandoned. It doesn't sell tickets. It's dangerous and offensive. Give 'em mumbo-jumbo. One feels our whole mass culture heading up the dark river to the source—that heart of darkness where Mr. Disney traffics in pastel-trinketed evil for gold and ivory.

"But the overwhelming feeling that one carries away is sadness for the empty lives which accept such tawdry substitutes. On the river boat, I heard a woman exclaim glowingly to her husband, 'What imagination they have!' He nodded, and the pathetic gladness that illuminated his face as a papier-maché crocodile sank beneath the muddy surface of the ditch was a grim indictment of the way of life for which this feeble sham represented escape and adventure."

There is truth in these observations, but it is not a carefully observed truth. It is important to note that the alligator is, technologically far advanced over papier-maché—it is a very, very good imitation of the real thing, and one can respond, legitimately, to the quality of its craftsmanship. Nor is the ditch muddy. It is sparkling clean—and that represents progress over Coney Island. Nor can one blame Disney for the impoverishment of our national mental life. It is indeed sad that we generally do not have the time, money and courage to venture up the real Amazon in search of authentic experience, but what age has? Finally, it is an absolute error to impute cynicism to Disney in the creation and operation of his empire. He always wanted to make money with anything he did. But he always gave fair value, so far as he could see, in return. Indeed, to many the best part of his success was precisely that he did make money without seeming to compromise his vision. It proved, to these impressionable souls that there was something magical in the American air causing art and commerce to blend most wonderfully.

In a later piece in *The Nation* John Bright caught perfectly the secret of Disney's success with mass man. "To call him a genius, as his sycophants do, is not only absurd; it is unenlightening. I think the man's unique success can be understood only by reference to his personal non-uniqueness. Of all the activists of public diversion, Uncle Walt was the one most precisely in the American midstream—in taste and morality, attitudes and opinions, prides and prejudices. The revealing clue is his familiar (and utterly sincere) statement that he never made a picture he didn't want his family to see. His competitors made pictures they thought, or guessed, the public wanted to see. Disney operated through maximal *identification* with John Doe; the others seek to discover what John Doe is like in order to cater to him.

"The celebrated Disney inventiveness is the x-factor in the success story. The key to this might be found in his immaturity, or not realized maturity—not used here in the pejorative sense. Walt, growing from infant to child to youngster, to adult, to uncle, and granduncle, never abandoned the delights and preoccupations of each stage of development, as most of us have done, at least in part. That was his 'genius.' Disneyland could have been created only a by a man-child who never tired of toys or shed the belief that animals and insects have human attributes."

Time once put this curious case of psychological arrest, and its results, in a good metaphor: "He has no mind or time for the niggling refinements of taste. There is too much to be seen and done, too many wonderful things in the world. . . . Away he rushes with his intellectual pockets full of toads and baby bunnies and thousand leggers, and plunges eagerly into every new thicket of ideas he comes across. Often enough he emerges in radiant triumph, bearing the aesthetic equivalent of a rusty beer can or an old suspender. They are treasures to Walt and somehow his wonder and delight in the things he discovers make them treasures to millions who know how dearly come by are such things as wonder and delight."

The point is that he had the courage to proclaim the childlike quality of his imagination for all the world to see, and that, frankly, was more than his audience ever did. They hide their happiness over the opportunity he provided for controlled regression behind middle-class styles and attitudes and thus avoid damaging admissions about the true nature of the Disneyland experience. The lady that Halévy overheard praising the "imagination" of the jungle ride's alligator was using a favorite method of disguising the truth of what she felt. There is nothing at all imaginative about the ride itself: we have all seen it in a hundred movies and television shows, fictional and documentary. The ride gives us nothing deeper intellectually or emotionally than we get from either the big or the tiny screen. To go beyond that Disney and his "imagineers" would have had to create, in three dimensions, an objective correlative to the mood of terror Conrad invokes, through means that are in the last analysis inexplicable and therefore inimitable, in *Heart of Darkness*. Not only is art of that quality unreproduceable technologically, it is also quite unwanted by most people.

When the lady on the boat ride spoke of imagination she meant something quite different from what a literary critic does. She meant that the technical imitation, the piece of machinery, if you will, is "imaginatively" put together—which it is (all home craftsmen and backyard tinkerers purely love Disneyland for precisely this reason). She does not mean by the word an intensification, through art, of experience, either inner or worldly, to an exciting and therefore agitating level. What the average, middle-class American wants and has always wanted of art and of the objects he mistakes for art, is the fake alligator that thrills but never threatens, that may be appreciated for the cleverness with which it approximates the real thing but that carries no psychological or poetic overtones. "By golly, that Disney—he sure could have fooled me," chuckles Dad as the boatman-guide blasts the steel and plastic thing with a blank cartridge and they all head for Schweitzer Falls just around the bend, where a few passengers will be lightly touched by spray that is at least created out of real water. It's all over in a minute, just the way the scary part in the movie is, and you carry away not some dark phantom that may rise up someday to haunt you but an appreciation of the special-effects man's skill.

At that simple level Disneyland is as harmless as any other amusement available in the country (whether we have entirely too many amusements available is another question entirely). What is frustrating about it is that it is not better than it is, that just when something has about captured you, caused you willingly to suspend disbelief, the "imagineers" rudely nudge you awake and whisper, "Just kidding, folks." On the submarine ride, for example, all is going reasonably realistically, when the ship suddenly glides past a mermaid and the trip is spoiled by the intrusion of this obviously fictional creature. (One cannot help but notice also that though she is bare-breasted, she lacks nipples.) Another example: there is a train ride—an adaptation of the New York World's Fair "Magic Skyway"—that takes you past extremely artful dioramas, showing various geological ages. When the age of the dinosaurs is reached, there are excellent moving models of the great creatures, far more artfully done than any you are likely to encounter in a museum. But then you notice the baby tyrannosaurs represented as just breaking out of their eggs, and, wonder of wonders, they are cuddly and adorable. The dear round bottom of one is wiggling comically as he shakes off his shell. The message, apparently, is that cuteness existed as an ideal in nature long before man appeared.

The word "dream" is often associated with Disneyland, particularly by its promotion and publicity people. It is, as they have it, "Walt's Dream" and a place that awakens the desire to dream in the visitor. But the quality of the dreams it represents is most peculiar—no sex and no violence, no release of inhibitions, no relief from real stresses and tensions through their symbolic statement, and therefore no therapeutic effect. It is all pure escapism, offering momentary thrills, laughs and nostalgic pleasures for the impressionable; guaranteed safety for that broad spectrum of humanity whose mental health is predicated on denying that there is any such thing as mental ill health or, indeed, a mental life of any significance beneath the conscious level; guaranteed interest for the technologically inclined; guaranteed delight for those who like to prove their superiority to the mass of men by making fun of their sports.

These levels of enjoyment remain available at Disneyland, but since 1963 there has been something else there, Walt Disney's last great advance in animation, a last mighty thrust toward the ability to simulate physical reality technologically. And this grotesquery, which he called Audio-Animatronics, threatens to change the essentially harmless character of the place.

The story of Audio-Animatronics begins at least as early as 1945 when one of his New York-based executives visited the studio and found Disney "all wrapped up in something mysterious that he was doing with Buddy Ebsen, who in those days was a fine dancer. He had Buddy doing a dance number and there were men in the room punching holes in what seemed to be a mechanical roll. Then I saw Walt playing with little Buddy Ebsen dolls which were attached by electric wires to a huge console-type machine. The men would feed the piano rolls into the console like a continuous IBM card and the little Buddy Ebsen dolls would repeat the dance steps I had seen Ebsen himself do. It didn't work because those were the days before transistors and the equipment was too cumbersome, so Walt put the whole thing aside. But now it is nineteen years later and I go to the World's Fair and I see Walt's Lincoln and fighting dinosaurs and dancing children. It's been refined by electronics but it's the Buddy Ebsen idea all over again. He carried it around with him all of those years until, finally, he was able to make it work."

In fact, Disney did more than carry the idea around with

him: he obsessively fought for its development. By 1955 WED engineers began making 1½-inch model figures that had cam and lever joints. In a short while an electronic-hydraulic-pneumatic approach to the problem of movement was added to the cam-lever devices, and the simple movements required, for example, of that alligator in the jungle ride were made possible. Then came the big breakthrough—the activation, through sound impulses, of pneumatic and hydraulic valves within a performing figure. These impulses, recorded on a magnetic tape, can activate anything in a figure from a simple, gross arm movement to a facial tic and can coordinate them in patterns of amazingly lifelike subtlety. By recording these impulses—as many as 438 of them at any given second—on a thirty-two-channel magnetic tape, they can be perfectly synchronized with music, dialogue, sound and light effects to present an entire show at the touch of a button, a show that, barring a breakdown in the machinery, will never suffer from a human error: no light or sound cue can ever be missed, no actor will ever blow his lines.

There is a certain irony in the development of Audio-Animatronics in that the sophisticated tape technique that forms the basis for its highest development was first perfected as a means of controlling the launching of space rockets. But there is no record that Disney, the Goldwater Republican, hesitated to use such government-sponsored research for his own ends any more than he hesitated to use tax advantages to his own ends. If anything, he probably saw it as some sort of compensation for the spiritual discomfort big government caused him. Certainly, the irony of putting space-age technology to use in an amusement park is not much stressed by Disney's people. Indeed, to them there is a certain implicit fitness about this meeting and mingling of the great forces of the age, Disneyism and electronic scientism, in the former orange groves of Anaheim.

The new Audio-Animatronics system was introduced discreetly in a Disneyland attraction known as "The Enchanted Tiki Room," where audiences group themselves around a bunch of handsomely feathered birds that perch in a small jungle setting and sing and tell jokes. It's a fantasy and the question of how it works is really quite insignificant, since the effect is not an especially arresting one. The same may be said of the use of Audio-Animatronics to give life to three of the exhibits WED ran up for the New York World's Fair of 1964–65. Pepsi-Cola's It's a Small World had a nightmarishly insistent theme song, sung by dolls in all the tongues of the world, but the dolls which seem to sing it are, after all, only toys. The Magic Skyway in which Ford cars moved riders past glimpses of the world in various epochs was morally neutral as was the General Electric Progressland. One was free to enjoy, or not enjoy, according to taste, their applications of Disney's latest technological marvel. Even his use of the technique to create mechanical men to serve as guides to exhibitions was somehow acceptable. They were only a little more disconcerting than a human guide who has been brainwashed and programed by a corporate training program and whose manner and response to stimuli is therefore almost as routinized as that of a robot. But as George Orwell once observed, "so long as a machine *is there*, one is under an obligation to use it." And so Disney's "imagineers" could not resist taking the logical next step.

Which was Great Moments with Mr. Lincoln, the Illinois Pavilion's prime attraction, in which the outer aspects of a man, a man who once lived and walked the earth and who had a unique personality—a soul if you wish—was mechanically reproduced with astonishing fidelity to his known physical dimensions and mannerisms. To Disney, as to most of his audience, Lincoln was the greatest of our folk heroes, the common man raised to the highest level of achievement, both worldly and spiritual. To place his wonderful new machinery in the worshipful service of this mythic figure was, for Disney, an unparalleled opportunity. His love for the Lincoln Legend knew few

334

bounds—he could quote long passages from his speeches—and as for Audio-Animatronics, he thought it was an art comparable in stature to Lincoln's stature as a human being. Says a WED press release: "Walt has often described 'Audio-Animatronics' as the grand combination of all the arts. This technique includes the three-dimensional realism of fine sculpture, the vitality of a great painting, the drama and personal rapport of the theater, and the artistic versatility and consistency of the motion picture." Obviously so great an achievement must not be wastefully confined to the representation of tropical birds and prehistoric monsters.

So Disney labored over Mr. Lincoln as he had not labored to bring forth his Mouse. "Imagination had to be tempered with authenticity," he told a reporter. "Drama must intertwine with serenity. Fantasy would be entirely abandoned, since its presence would defeat our purpose. Reserve was demanded but would have to take the form of subdued excitement. And dignity would have to be the constantly sounded keynote." Thus was the sixteenth president, martyr, hero and summary of the virtues of democratic man turned into a living doll, "capable of forty-eight separate body actions as well as seventeen head motions and facial expressions" (at least in 1965: there was a new, improved model a year later). Disney, caught in the grip of his technical mania and protected by his awesome innocence about aesthetic and philosophical matters, had brought forth a monster of wretched taste which, for all the phony reverence and pomposity surrounding its presentation, leaves one in a state of troubled tension. Are we really supposed to revere this ridiculous contraption, this weird agglomeration of wires and plastic, transferring to it, in the process, whatever genuine emotions we may have toward Lincoln in particular, toward mankind in general? If so, we are being asked to abjure the Biblical injunction against graven images and, quite literally, we are worshipping a machine that is no less a machine for having the aspect of a man. Perhaps, then, it is a form of art? But art is not imitation; its strength lies precisely in the art object's inability to speak or to move and the transcending compensations the artist makes for his inability. Take, for crude example, the Lincoln Memorial in Washington. The huge size of the figure, so often described as brooding, overpoweringly reminds the visitor of his own puniness of size and that Lincoln's spirit was alleged to have been similarly outsized in comparison to the ordinary. The silence of that massive figure encourages one's own reflections on the enigma of greatness. Surely, the atmosphere cues a quasi-religious response, but it does not force a choice between the varieties of religious experience on the spectator as the Disneyland show attempts to do with its effects of lighting and music. Disney's Lincoln, for all its mechanical sophistication, reminds us that Lincoln was only a man; Daniel Chester French's Lincoln suggests that he might have been something more. But perhaps modern man is made uncomfortable by such suggestions. It is easier for him to live with the smaller "uneasiness" Ortega described some years ago when writing about the dummies at Madame Tussaud's. "The origin of this uneasiness lies," he wrote, "in the provoking ambiguity with which wax figures defeat any attempt at adopting a clear and consistent attitude toward them. Treat them as living beings, and they will sniggeringly reveal their waxen secret. Take them for dolls, and they seem to breathe in irritated protest. They will not be reduced to mere objects. Looking at them we suddenly feel a misgiving: should it not be they who are looking at us? Till in the end we are sick and tired of those hired corpses. . . ." He adds: "The mob has always been delighted by that gruesome waxen hoax."

He should only have lived to see Abraham Lincoln rise up from his chair to mouth his carefully selected platitudes, with gestures—and is that a tear glistening in his eye as a mighty chorus of "The Battle Hymn of the Republic" swells behind him? Ortega would have been interested to hear of the curious fact the "imagineers" discovered, quite by accident, about the

336

plastic Mr. Lincoln's skin is made out of; after a while it excretes oils just as the human skin does and so "takes makeup wonderfully," as the press agent eagerly informs you. He would also like to know of the interesting protection afforded Mr. Lincoln's electronic guts down in the basement. Should a fire break out, huge metal doors clang irrevocably shut in thirty seconds and the temperature drops instantly to thirty degrees below zero. A man tending the machinery could get killed, but the dummy would live—or exist, or whatever it does.

What can be said of Mr. Lincoln and of all the other human simulacra that are following him into Disneyland (not far away a group of pirates loots a town and rapes its women, who enjoy the process because, it is explained, they're all old maids) is that he stirs the observer to thought in a way that nothing else in the Magic Kingdom does. Here *is* the dehumanization of art in its final extremity, paradoxically achieved by an ignorant man who was actually, and in good conscience, seeking its humanization and who had, indeed, arrived at this dreadful solution, after a lifetime search for a perfect means of reproducing the reality of human life. At this point the Magic Kingdom becomes a dark land, the innocent dream becomes a nightmare, and the amusement park itself becomes a demonstration not of the wondrous possibilities of technological progress, as its founder hoped, but of its possibilities for horror.

PART ELEVEN

A Final Balance

31
Sometimes I feel there is a conspiracy to seduce and ruin our youth. I'm sick of sick pictures, offensive sex and bad taste. . . . The movies these days seem more interested in turning your stomach than in warming your heart. And over and over again: What's going to happen now that Walt Disney is dead?

> —Lenore Hershey, quoting responses to a poll of readers in "What Women Think of the Movies," *McCall's*, May, 1967

HE WAS MR. CLEAN NOW. All reference to him as a serious film artist had long since ceased and in a period when traditional middle-class values— indeed, the traditional middle class itself—were being questioned as they had not been since the 1920s, Walt Disney became, in the minds of his public, something more than a purveyor of entertainment. He became a kind of rallying point for the subliterates of our society, the chosen leader for the desultory—and ambiguous—rear-guard action they were trying to fight against a rapidly changing cultural climate. At the very outset of this book it was noted that the period of Disney's greatest economic advance coincided with the years of the greatest economic trials among his competitors.

His solution to their common problem was based primarily not, as some simple souls believed, on the moral uprightness of his films but rather on his ability to convert his operation from a narrow base to a broad one, drawing its revenues not merely from films but from all the other areas he penetrated—right down to a glorified bowling alley, known as the Celebrity Sports Center, in Denver. Just before he died, still greater diversification was in the works: the Mineral King project in northern California was to be not merely a place of entertainment but a place of recreation. The best solution offered by his competitors to the problem of the vanishing audiences in the crisis years was a greater daring in subject matter and a greater frankness in the way they treated love, sex, social problems, language and, indeed, the human form in its unclothed and semiclothed states. The age of what the industry proudly called "the adult film" thus finally arrived

And it did succeed in drawing audiences. But it also confused them. For the age of the adult film was also an age of sensationalism, and not only in the movies, and so it became difficult for ordinary and untutored people to tell the difference between honest realism and dishonest titillation. (A curious instance: in the *McCall's* poll, cited above, the good ladies managed somehow to choose *Who's Afraid of Virginia Woolf?* as both their least favored film and their fourth most favored one. As usual, the mass audience is a hypocritical one, buying tickets to entertainments they later pretend to despise.)

In any case, one of the first casualties of the new Hollywood was the old program feature aimed at the family market—modest situation comedies, chaste adventure films, romances, animal pictures and so on. Products in this vein could be seen any night on television. By default, whatever movie-theater market was left for this sort of thing was left to Disney. And that meant that the confused middle class expected him not merely to cater to their values but to articulate them as well.

It might have been hard for him at first. He had a broad

streak of vulgarity in him: one screen writer quit the studio for a brief time because Disney kept trying to slip bathroom jokes into her scripts, and he insisted on calling a one-time associate and latter-day competitor, "I. P. Freely"—an eight-year-old's crude pun on the man's name. But Disney soon learned the value of the public figure that had been foisted on him. He could hide behind it. He gave freely of his opinions on contemporary screen morality to interviewers, a topic that gave him a ready-made theme for public appearances. His exalted standing thus diverted attention from matters that were of infinitely more interest to him—like converting prosperity into certifiable tycoonship. It has been observed that practical joking, to which Disney had once been addicted, is a last, despairing effort by the fundamentally shy person to establish communication. But the creation of a public personality, behind which the insecure individual can hide and that ultimately may subsume the poor, stunted, incomplete private self, is an even better solution for some people. And so it was for Walt Disney.

This book began with a set of figures, for somehow figures more accurately summarize Disney's achievements than any other data about him can. Along the way, the bench marks of his career have been set forth in statistics about box-office grosses, stock prices, cost figures and so on, for those were the measures he and his associates applied when they spoke most frankly of his progress. As the end of his life approached, the figures piled up higher and higher, and they almost completely obscured the private man. We are left, in the end, not with an individual to try to summarize but with a celebrity, which is quite a different thing.

Even his closest associates found Disney more enigmatic, more difficult to describe or to communicate with in the 1960s. No one spoke of him any more as a father figure. Instead, he was known almost universally to those who worked for him—but always behind his back—as "Uncle Walt." There is a considerable difference between a father and an uncle; one of the prime characteristics of the latter is that he tends to be a rather distant and emotionally neutral figure. As late as the mid-Fifties, associates reported that Disney sat on the lawn near his office after lunch, chatting amiably with employees; he was not observed in public in so relaxed a posture afterward, though strenuous efforts were made to maintain the old down-to-earth image that had served him so well for so long. "Before I met him," Aubrey Menen reports, "every effort was made by his aides to impress me that Walt Disney was, in fact, avuncular. He was open and affable, they said, and easy to talk to. Instead, I met a tall, somber man who appeared to be under the lash of some private demon . . . I remember him smiling only once and he is not at ease." The slender, smooth-faced young man of the 1930s with the thin, retail clerk's mustache and the slick hair that formed a widow's peak had become a portly, some-what rumpled figure in later life. His face was pouchy, the mus-tache, like the waistline, had thickened; the smile, when it was offered, flashed too briefly and often seemed to involve only a baring of the teeth; the eyes remained almost hidden beneath the heavy-hooded lids. Listening, Disney had a way of seeming to stare through visitors as if fascinated, as one associate ob-served, by the sight of "something very small and very ugly at the back of your skull."

The mood of the studio, though outwardly serene, had a cer-tain tension about it. One visiting journalist recalled being warned by a secretary to be sure to accept the glass of tomato juice Disney habitually offered his luncheon guests in his office before taking them over to the commissary, where he, of course, had a corner table commanding a view of the whole dining room where stars and executives eat (there is a cafeteria for the others). The man inquired what would happen if he turned down the proffered refreshment. "I suggest you just take the tomato juice," the girl said with cool finality. The same visitor detected a distinct change in the atmosphere when Disney en-tered the dining room a little while later. "Everyone's voice shot up about two octaves," he recalled. "All the women began sounding like Minnie Mouse, all the guys started sounding like Porky Pig." He may have exaggerated, but only slightly. For when Uncle Walt came around, anxiety came with him, and like army privates who feel the eyes of the sergeant upon them, his people tended to get very, very busy, very, very preoccupied, hoping they would not be singled out for his attention.

Among the countless stories of his forays into the lower depths of his organization, there is one about the time he set up on a sound stage for one last inspection prior to shipment, all the exhibits WED did for the New York World's Fair. He watched silently as Mr. Lincoln did his turn, then said mildly that he thought it would be wonderful if the effigy could be programed to take one step forward before beginning his spiel. This necessitated, of course, an enormous—and highly complex —effort at electronic revision against a deadline. But Walt's whim was his employees' command, and the task, not strictly necessary to the success of the enterprise, was undertaken. On another occasion Disney dropped in at the Tahitian Terrace in Disneyland, the chief feature of which is an enormous tree placed in the center of the room, where it shelters both spectators and South-Seas-style dancers. Disney took a seat in the top tier and found that the tree's branches partially obscured his view of the floor show. The tree was forthwith ordered pruned, but its proportions were destroyed. So new branches were installed to restore its symmetry. Another time, a visiting magazine editor, allowed to see Lincoln before his public debut, made an innocent inquiry about exactly how precisely the figure's lip movements, when he spoke, matched those of a human being saying the same phrases. Disney immediately ordered an underling to hire a lip reader as a consultant—to assure accuracy in this matter and, perhaps, to reassure himself of his power. No detail was to small for Disney to catch. He ordered Annette Funicello the one-time star of the Mickey Mouse Club who had matured into adolescent roles, never to wear low-cut dresses in order to preserve the modesty of her image—and his. He was forever telling Disneyland guides who erred on the side of formality to be sure to call him Walt. Once at the park he took the jungle-boat ride and to his dismay found that the trip only lasted five minutes. "It should take seven," he told the man in charge. "Do you know how much those hippos and elephants cost? I hardly saw them." Needless to say, the ride was slowed down.

This incessant tinkering with the machinery of his empire was, naturally, costly. But Disney compensated for it by extraordinary stinginess in other matters. In particular, he was notorious for the bargains he and his organization drove with outsiders—actors and, above all, the authors of material on which screenplays and television shows might be based. As strangers, they could claim none of the prerogatives of citizenship in the Magic Kingdom, where, though salaries were low by Hollywood standards, there was security that was unavailable at competing studios (Disney had a sentimental regard for those who had demonstrated their loyalty despite temptations from outside and could scarcely bear to retire them, let alone fire them.) These old retainers, of course, became virtual unemployables elsewhere in Hollywood and so grew each year more dependent on him. The free-lancers, naturally, had no opportunity to build up a symbiotic relationship with Uncle Walt, and to them the studio was often cavalier. Actors got nothing like the top Hollywood salaries, and a participation deal on a Disney film was unheard of. The stars of a Disney movie are utterly interchangeable with others of their types. It was his name, and now, in his absence, it is the studio's name that draws the people. As for the original authors of the stories filmed by Walt Disney Productions, they were often caught in a curious bind. Few other studios were interested in acquiring stories about children or animals, and so there was no one to bid up the prices for them as there often is when a hot best seller goes on the Hollywood market. Running the only game in town,

Disney could often obtain worldwide film rights to a children's book for as little as one thousand dollars, and his alert story department often found short stories and magazine articles that everyone else had passed but that could serve as the basis of a Disney picture and for only a few hundred dollars. Almost never did the original authors get a cut of the merchandising bonanza the studio habitually generated when the story finally reached the screen. (Irony of ironies, some booksellers reported Disney versions of *Mary Poppins* outselling the original Travers novels by 5 to one.) It was Disney's frugal habit to acquire low-cost literary properties in job lots, then put his staff of writers to work Disnifying them. If this could not be done, the loss was small, and no grudge was held against the local talent who bought the works in the first place or failed to translate them into acceptable scripts. There were always plenty of stories where they came from—literature's bargain basement.

On the Disney lot, the only strong personality was Uncle Walt's. He rarely hired a forceful director, a man who would place a strongly individual stamp on a film. He liked men who could work efficiently within the house style—which is why the Disney films tended to look alike in an age when independent production had erased the stylistic trademarks by which one used to be able to tell the difference between a Warner film, for instance, and one from Fox.

Disney's tinkerings were not, of course, confined to Disneyland or his movies. He occasionally toyed with people, too. He used to tell the story of how he had got exactly the revisions he wanted on a script from a writer by dangling before him the opportunity of directing the finished product, a breakthrough the man was particularly anxious to make. Even in Disney's telling, the story had a faintly sadistic ring, however joking Uncle Walt's tone. He also made little attempts to improve, by his lights, relationships between his workers. One executive reported that on his first day at the Disney studio he hollered out his window to ask a gardener to stop mowing the lawn, since it was interfering with his concentration. An hour later, the executive got a call from one of Disney's secretaries, telling him to report to the boss's office immediately. The man rushed to answer the summons and was told, "You spoke harshly to that man. He's been with me for twenty years. I don't want it to happen again." The executive managed a contrite, if startled, "Yes, sir." To which Disney responded, "And there's another thing I want you to remember. There's only one s.o.b. at this studio—and that's me."

So it must have seemed to his employees at times. They always tried to place in the best light his sudden, generally unpremeditated intrusions into their routines. If not seen as the harmless foibles of a genius, they were regarded as evidence of that genius's infinite capacity to take pains with its work. But one must wonder. The incidents described here are all clustered at the end of Disney's life, when his power and his fame were at their height, when, in fact, there were left no real threats to his economic survival. One imagines that all of them were manifestations of a new form of anxiety, the anxiety of absolute command. Elias Canetti points out that "the satisfaction which follows a successful command is deceptive and covers a great deal else. There is always some sensation of a recoil behind it, for a command marks not only its victim, but also its giver. An accumulation of such recoils engenders a special kind of anxiety. . . ." The ultimate sanction of command is, of course, the threat of death, if not physically, then—as in Disney's case—in terms of one's professional life. All commands leave their victims with the memory of this threat, the commander knows this and sees, growing up around him, an ever-increasing number of people who he knows nurture this memory in common; there is always the possibility of revolt should they unite against him. He has no choice but to do "everything he can to make such a reversal impossible."

If this seems a somewhat melodramatic description to apply to an entrepreneur of entertainment, it should be remembered that Disney did everything possible to build a closed world, an empire masquerading as a Magic Kingdom. The construction of such a place represents, of course, a recognition that in the larger world a man's power is necessarily limited, that even absolute dictators cannot control everything or have everything their hearts desire. At the opening of one of Disneyland's newest attractions, a New Orleans square, the mayor of the real New Orleans thought to compliment the creator of the imitation by saying, "It looks just like home." To which the imitator replied, with more than a trace of self-satisfaction, "Well, I'd say it's a lot cleaner." That was, to him, the insuperable advantage his little world had over the great world. He could order it precisely as he wanted to. Compared to the mayor of a modern city, who must feel his way through the traps set for him by what one municipal executive calls "the power brokers," ruling Disney's land—or at least its outer aspects—was easy.

Not that citizenship in the Magic Kingdom was an easy matter. Where could you go to discharge those accumulated memories of threat? They naturally tended to pile up around the master of the domain. That is, perhaps, the reason why Disney always resisted giving the appearance of being a mogul in the more familiar Hollywood style. It was as if he was aware of—and afraid of—just the situation in which he found himself at the end. The informality of his working habits in the beginning, his attempts at creating a family atmosphere, his dream of the utopian studio—were these manifestations of his desire to avoid the anxiety of command while enjoying its privileges? The insistence on carrying over at least some of the forms of the old days—the first naming, the lack of conspicuous consumption in dress, food and style of living, the reams of careful publicity about the creativity of his "team"—were these attempts to mask the "endless, torturing awareness of danger" implicit in the assumption of absolute power such as Disney held in his last years?

Perhaps. Certainly he seemed to fear that suddenly his organization—always in his eyes an extension of himself—might have grown too large for him to control personally as he always had in the old days. This undoubtedly accounts for his insistence on involving himself in the most minute details of script revision, in his attendance at the daily rushes of footage from films in progress, in his fretful fussings over each detail of the construction of new attractions for Disneyland. Everything had to funnel through his office, but even so, things got away from him, things were done without his approval, and worst of all, he was convinced that there were things going on in his world, and in the great world beyond it, about which he knew nothing but that conceivably would harm him and his works.

Talking for public consumption, Disney liked to cite "the four Cs" as the secret of his success—"curiosity, confidence, courage and constancy." But to these surely a fifth must be added—the compulsion to control. The word was constantly on his lips. "We must have control," he would say in discussing a project. Or he would reply to a suggestion, "Oh, no, we'd never do that—we couldn't control it." The writer heard him use the word no less then twenty-five times in the course of one half-hour briefing session with some journalists. Perhaps his morbid fear of death was a manifestation of this need for control—for what is death but the ultimate loss of control over one's destiny? Surely, his stubborn intransigence toward outside financial interests was still another manifestation of this compulsion. At the end, with his organization prospering as never before, he maintained a policy of extremely low dividends—40 cents per share and a 3 percent stock increment—even though the company's earnings per share were the highest in the industry and even though far less profitable entertainment concerns were paying out anywhere from 80 cents to $1.20 per share on stocks selling at lower prices. Apparently, he still resented giving up control of "his" money to stockholders, and also apparently, he feared being caught short as he had been in the past. Said one of his executives, "Everything Walt does today is conditioned by his past problems. When he makes one of his tough deals, he negotiates like he's afraid someone might take another Oswald the Rabbit away from him."

There should be no mistake on this matter. Disney in his last years was capable not only of the small, nervous sadisms that marked his tours of his plant but of larger cruelties as well. The sensitive writer of the book on which Disney based one of his most popular recent films recently said in correspondence that any public discussion of the adaptation was impossible, "for I fear and feel distaste for the streak of brutality that is in him and the ferocity with which he wields the power the contract gives him."

Perhaps he would not have used that power; it is hard to see *how* he might have used it. Anyway, Disney had an image to protect, and the harassment of essentially undefended individuals would not have suited it very well. But it is true that in the last years there was no one to say him nay (which makes his anxieties all the more ironic). His films had passed beyond serious comment by critics, and our social critics had little that was relevant to say about his amusement park or his merchandise or his television program. The only serious criticism of Disney came from that small band of specialists who concern themselves with literature for children. They are not, themselves, always as venturesome as they might be in their critical standards, but they could see just what a perversion the Disney literary product was. The lady who emerged as their leader was Frances Clarke Sayers, gray-haired, semiretired, one-time director of children's services for the New York Public Library, latterly a lecturer in library science at U.C.L.A. The instrument of her emergence was the incredible Dr. Max Rafferty, the right-wing Superintendent of Public Instruction in California, who also writes a weekly newspaper column in Los Angeles. One day in 1965 Rafferty was moved to do a piece extolling Disney as "the greatest educator of this century—greater than John Dewey or James Conant or all the rest of us put together." To

Rafferty Mickey Mouse and his gang were merely the spring-
board "from which he [Disney] launched into something
unprecedented on this or any other continent—compensatory education for a whole generation of America's children. The classics written by the towering geniuses out of the past who had
loved children enough to write immortal stories for them to
begin to live and breathe again in the midst of a cynical, sin-
seeking society which had allowed them to pass almost completely into the limbo of the forgotten." There was more in this
dubious vein, including praise for the Disney live-action films
as "lone sanctuaries of decency and health in the jungle of sex
and sadism created by the Hollywood producers of pornography," an observation that "beatniks and degenerates" find Disney "square," and a wish that *Mary Poppins* makes $100 mil-
lion for Disney, so that others would see the profitability of
cleanliness, and a peroration as delicious as one could ever hope
to find in columnar journalism: "Many, many years from now
—decades, I hope—when this magical Pied Piper of our time
wanders from this imperfect world which he has done so much
to brighten and adorn, millions of laughing shouting little
ghosts will follow in his train—the children that you and I once
were, so long ago, when first a gentle magician showed us wonderland."

It was all too much for Mrs. Sayers, who fired off a letter to
the editor that briskly summarized the case against Disney as
follows:

> Mr. Disney has his own special genius. It has little to do
> with education, or with the cultivation of sensitivity, taste
> or perception in the minds of children.
>
> He has, to be sure, distributed some splendid films on sci-
> ence and nature, but he has also been a shameless nature
> faker in his fictionalized animal stories.
>
> I call him to account for his debasement of the traditional
> literature of childhood, in films and in the books he pub-
> lishes.

He shows scant respect for the integrity of the original creations of authors, manipulating and vulgarizing everything to his own ends.

His treatment of folklore is without regard for its anthropological, spiritual or psychological truths. Every story is sacrificed to the "gimmick" (Dr. Rafferty's word) of animation.

The acerbity of Mary Poppins, unpredictable, full of wonder and mystery, becomes, with Mr. Disney's treatment, one great, marshmallow covered cream-puff. He made a young tough of Peter Pan, and transformed Pinocchio into a slap-stick, sadistic revel!

Not content with the films, he fixes these mutilated versions in books which are cut to a fraction of their original forms, illustrates them with garish pictures, in which every prince looks like a badly drawn portrait of Cary Grant, every princess a sex symbol.

The mystical Fairy with the Blue Hair of the Pinocchio turns out to be Marilyn Monroe, blonde hair and all.

As for the cliche-ridden texts, they are laughable. "Meanwhile, back at the castle . . ."

Dr. Rafferty finds all this "lone sanctuaries of decency and health." I find genuine feeling ignored, the imagination of children bludgeoned with mediocrity, and much of it overcast by vulgarity. Look at that wretched sprite with the wand and the over-sized buttocks which announces every Disney program on TV. She is a vulgar little thing, who has been too long at the sugar bowls.

Mrs. Sayers extended her remarks in an interview picked up by several publications, and it drew an interesting response. About half her correspondents, a sort of informal anti-Disney underground of parents, educators and librarians praised her for taking a stand on a matter that had long troubled them but that, in the general atmosphere of idolatry surrounding Disney, they had been unable to articulate; the other half reacted with an emotional intensity that was startling. The lines of these correspondents tended to be remarkably similar. "God bless Walt Disney," one of them cried, "in this day of misguided educators . . . and the peddlers of smut and obscenity to whom his artists are repugnant." "In my estimation," chimed in another, "he [Disney] stands for 'Decency—U.S.A.,'" while a third just said, "NUTS."

Disney himself never entered the fray, nor did his studio take any position in the Rafferty-Sayers controversy. In truth, Mrs. Sayers said nothing, in her letter or in the interview that followed, that had not been said before, though few had said it with such delightful acerbity. What is interesting is the intensity of the response she garnered and the fact that even Rafferty had little to say about the artistic quality of Disney's work. He, as well as Disney's other defenders, rested their case entirely on the crudest of moral grounds, and these, in turn, rested on the distinctly unprovable assumption that the nation has entered upon a period of prolonged moral decline. This is not the place to rehearse the arguments for honestly portraying, in works of art, the potential difficulties and pleasures of the sexual encounter or for the unblinking portrayal of the broad streak of violence that exists in American life. It is worth suggesting, however, that in an era of radical change such as we are now experiencing, the most dangerous immorality is to ignore these matters. Nor does it seem necessary to labor the point that Disney's defenders are hopelessly blind to the content of his work, which contained a fairly high quotient of violence—as has been pointed out by his critics from the beginning of his career—and that his work also has a strong, though displaced, sexual element. So far as his audience was concerned, out of sight was out of mind.

None of the foregoing should be construed as an indictment of Disney for not venturing out of the realm of escapist film making in his later years. Few American producers do. What is interesting is that by the end of his career, such argument as

Disney's work excited was carried on strictly in terms that he and his organization had chosen. Vulnerable on aesthetic grounds, they emphasized the wholesomeness of the Disney product and found a ready response to this emphasis among the untutored. One of the most depressing aspects of American life is that nine-tenths of what should be an aesthetic discussion ends up as a totally inapposite moral argument. There can be no serious discussion of the quality of popular culture until there is a general recognition that the significant moral issues it raises are far more subtle than questions of décolletage.

Disney, in short, was fundamentally uninterested in the issues Mrs. Sayers raised. Of all the many states of his realm, literature—or more properly, subliterature—had the smallest claim on his attention. Though he was an avid, even scholarly, reader of screen treatments and scripts, he rarely read the original works on which they were based. As for the reductions of children's classics that went out unobserved by Disney to dimestore and drugstore book racks, he seemed to care nothing about them so long as someone in the organization saw to it that they were clean and decent: these works were merely part of the "total marketing" concept of which the organization was very proud when it was talking to its stockholders or its fellow tradesmen. The books helped create the three billion "impressions" in all media for which the company estimated it achieved through its marketing techniques when it was releasing, for example, its double feature of *The Ugly Dachsund* and *Winnie the Pooh*. Indeed, the nineteen Disney books based on these subjects were rather less important than the tie-in that was engineered with the Kal-Kan dog food company for the same films. Peter and Dorothy Bart, who wrote on Disney and literature, suggested it was possible that "a flowering of Disney books might take place were Walt Disney himself to emerge from his award-lined office and take a more direct and active interest." But, they added, it seemed unlikely that he would, since "the big money is in film and the big gimmickry is in Disneyland, and money and gimmickry have always exerted a spell over Walt Disney."

If there is any consolation for the bookish in all this, it may lie in the Disney organization's not taking a more serious interest in literature. As it is, there is some reason to believe that most of the reading matter it causes to be produced is irrelevant to children. If it is not the positive force it might be, it is probably not quite the disaster many of his critics thought it was. New York schoolmaster Donald Barr summed up this view when he wrote that in a typical Disney book "what has vanished is motive and temperament, all the pulse of life under the skin of events, all the wild hope of someday understanding. Disney's world is not a child's world at all, for a child is a human reading for his future; it is an oldster's world, for an oldster is a human relaxing into his past." Indeed, Barr's words are applicable to almost all Disney's works in the 1950s and 60s.

Disney agreed as to intent, if not as to valuation. "I don't make films exclusively for children," he said. "I make them to suit myself, hoping they will also suit the audience. . . . I've proved, at least to myself and our stockholders, that we can make money, lots of money, by turning out wholesome entertainment. My belief is that there are more people in America who want to smile than those who want to be artistically depressed." He admitted that the works of Tennessee Williams, for example, might be great art, though they were not for him. Sometimes he could even be funny about his prudery, as when pressed for an exact description of the relationship between Mary Poppins and Bart, her street-artist companion, he replied, "They're just good friends." Often, however, he was extremely emotional on the subject. "I don't like depressing pictures," he growled to another reporter. "I don't like pestholes. I don't like pictures that are dirty. I don't ever go out and pay money for studies in abnormality. I don't have depressed moods and I don't want to have any. I'm happy, just very, very happy."

Does one detect in the last statement a note of hysteria creeping into the tone? Perhaps. But a note of hysteria tends to creep into this subject, whoever is discussing it. However enlightened some people seem to be on the subject of sexuality, however clear it is that titillating as well as honestly stated examinations of the subject often do extremely well at the box office, it is also clear that while the screen as well as our literature has grown more frank in its treatment of the once forbidden topic, this hysteria has grown. Does it represent the view of an alarmed majority seeing its numbers beginning to shrink? Or does it represent a minority growing increasingly paranoid as it sees its "islands of decency" going under? It is impossible to say. But the tone of the discussion is increasingly shrill.

Disney never addressed himself to this subject, for example, until the 1960s, when it appeared to offer some publicity mileage and when, of course, it suited his new role of elder statesman. One does not deny the sincerity of his beliefs on the matter: they fit too well with the other aspects of his personality. One notes them, finally, as yet another example of his uncanny and unstudied identification with his audience. Indeed, had he not at the same time been so avidly concerned to keep on expanding his empire, one might be tempted to believe that this sudden readiness to take every available opportunity to speak out about morality in art was evidence that he was himself "relaxing into his past."

The public man spent a good deal of time in public-relations activities, from squiring VIP guests around his domain (they always got a signed photo of themselves and Disney, with the wrinkles in Disney's face carefully smoothed by the retoucher's air brush) to accepting spurious awards (a cub scout pack, for instance, made him a replica of the Great Seal of the United States out of forty-five pounds of beans, rice, mustard and sunflower seeds, millet, peas, barley and popcorn kernels). He even received some worthwhile awards, among them the President's Medal of Freedom, the citation for which read "Artist and impresario, in the course of entertaining an age, he has created an American folklore." (When the President pinned the medal on him in 1964, Disney wore a small Goldwater button under his lapel, where Mr. Johnson could not miss it, and he had to be talked out of wearing one big enough to show in news photos.) At the time of his death, a small, informal but worldwide group was promoting—with the covert assistance of his publicity department—his nomination for the Nobel Peace Prize.

But for all these diversions, he continued to work at his customary pace—arriving early and staying late at the studio. In 1964 he ventured away from small-risk film production for the first time in many years, turning out *Mary Poppins* and receiving for it awards and rewards beyond the dreams of avarice. We have seen that Mrs. Sayers, among others, objected to Disney's treatment of the beloved Nanny, but it must be said that, for all the studio's tampering with her character, all its attempts to explain her magic in rationally comprehensible terms, the picture had a charm of its own. The fresh-scrubbed, youthful good looks of Julie Andrews in the title role might have been offensive to *Poppins* purists, but they were welcome to others. If she was not quite the deep-dyed, very English eccentric that P. L. Travers originally portrayed, she was hardly the cream puff that Mrs. Sayers described, either. She had, in the film, a spunky independence to which any child could respond, and she remained acceptably nonconformist in her ways and views and therefore a useful model for children to contemplate. It should be remembered, moreover, that good as the Travers novels are, they are a long way from being literary achievements on the level of *Alice in Wonderland* or the Grimms' fairy tales in the version those strange savants originally set down. They have, at times, a queasy, old-maidish patronization in their writing, and at the most basic structural level, the original books present a serious problem for the adapter in that they are essentially collections of loosely related short stories. They have little cumulative narrative strength (though they have a cumulative emotional attractiveness, which is a different matter). Disney's adapters did quite an artful job in maintaining the major incidents of Miss Travers' first book yet weaving them into a sensible, reasonably suspenseful story. The original books, to be sure, are generally painted in pastel hues, while the Disney film has a bright, sharp-edged definition. But some of what is lost from the original—that sense of wonder and mystery that Mrs. Sayers spoke of—is more than compensated for by the energy and high spirits of the film. In short, it is a likable piece of entertainment, and in its musical numbers, especially the picnic that takes place inside Dick Van Dyke's sidewalk painting and the chimney-sweep dance, it is rather more than merely likable. Indeed, these sequences have a cinematic excitement entirely missing from most film musicals of recent years and far in advance—as the whole film is—of something like *The Sound of Music*, to which it is superior musically, directorially, thespically and even intellectually. It is difficult to explain how, after years of routine production, the Disney organization accomplished this breakthrough. Obviously the willingness to gamble on relatively heavy production expenses has something to do with it; clearly the intrinsic quality of the basic literary material does, too. But it has a sense of style, often lacking in the live-action films, a lack of corniness and sentiment all too often excessively present in the other Disney pictures that sets it apart. The film may have been too sweet for some devotees of the Travers stories, but compared to other Disney work it was a tartly flavorsome venture, with some of the good nostalgic quality he achieved in his early live-action adventures made in England.

It should not be imagined that Disney had no setbacks in this period. As noted, the studio never recovered its early lead in animation, some of its live-action films performed feebly at the box office as well as critically, and the expensive effort to develop a projection system that could entirely surround the viewer—introduced at the Brussels World's Fair in 1959 and also visible at Disneyland and at Expo 67—was a failure. Technologically it worked fine, but there was no way of telling a story or even making a coherent documentary production utilizing a 360-degree screen. Again possessed by his desire to encompass all reality with a quasi-art form, Disney exceeded sensible limits. One can see views—living, superpostcards in the special theater required for this device—one can even be briefly entranced by the sheer marvel of the thing. But it is only a novelty, rather like 3-D movies, not a genuine breakthrough, though one can imagine its being used more imaginatively than Disney's people have, perhaps by some purveyor of mixed-media environments or by an exponent of synthetic psychedelic experience.

All of these setbacks were, however, minor compared to realms of glory opened up by the profitability of *Mary Poppins* and of Disneyland. With the latter able to finance almost unlimited expansion out of its own funds, Disney rushed ahead with even more grandiose real estate ventures. He resented that he had not been able to protect Disneyland from the fast-buck oprators who clustered around it, and he would have liked to have a piece of the action obtained by motel and restaurant owners whose business boomed after the park opened. In ten years a quarter of the nation's population had made its way to Disneyland, and the number of available hotel rooms in Anaheim rose from 100 to 4,300; some 250 new businesses located in the area; new sports and convention facilities arose, and the tourist business of this one county is claimed by the locals to exceed that of every other state in the union as well as that of all California's other counties combined. Of the riches that poured in, $273 million went to Disneyland, but another $555 million was spent outside its gates—just outside them.

Disney vowed to rectify that error. And so the organization quietly began buying up land near Orlando, Florida, 27,500 acres of it—twice the size of Manhattan island. He called his new project Disneyworld, and he explained that he needed all that land "to create an environment around the project which would be in keeping with its character and purpose." In short, he would have complete control of it. At its center would be an amusement park, which he originally planned to place under a great dome so that the weather inside could also be controlled. Nearby would be EPCOT (experimental prototype community of tomorrow), housing Disneyworld's employees and any industry that cared to locate north, south, west or east of Eden. The telegram inviting editors to the press conference at which the project was formally announced called it the most significant event in the state's history since its discovery by Ponce de León. Since the organization estimated the capital expenses involved in developing a tract twice the size of Manhattan at perhaps $100 million—who could say the press agent was wrong?

Meantime, Disney pressed ahead with his Mineral King project in northern California—a year-round outdoor resort in a mountain valley, where the "imagineers" were toying with the idea of placing, among other amenities, a restaurant on the top of each of four surrounding peaks. The project's launching was delayed and ultimately killed by a coalition of conservationists after Disney died. And after several millions had been spent on acquiring and planning the site. But that was only one of Disney's latter-day projects.

Before he died he had given land from the studio ranch to serve as the site of a new university, to be devoted to the arts and familiarly known as Cal Arts (he had rejected the idea of putting his name on the place on the ground that students and potential faculty members might be put off by its nonscholarly associations). According to those who worked with him on this project, he was planning for it as he had planned for his commercial ventures—shrewdly purposefully, eagerly. It was, on a grand scale, his old dream of an artist's utopia reconstituted; it was the old studio art classes grown up. A good share of Disney's estate went to this, his last monument, which he saw as a place where all the arts might mingle and stimulate one another, as he had once hoped they might in his movies and as he fondly believed they did in Audio-Animatronics.

In all his last projects one sees clearly "the real mania for totality" so characteristic of our age. One senses also that in these projects Disney was reaching toward a kind of satisfaction he had never found in his motion pictures enterprises. Not long before he died, he told one of his writers that he was no longer interested in "little pictures," that in the future he wanted to concentrate on potential blockbusters of the *Mary Poppins* variety, and given the scale of his operations in late years, this was reasonable. He could afford to risk more now, and there was no need to hedge his bets as he had in the past with movies. Somewhat more saddening was his remark, upon seeing some rushes for *The Jungle Book* which displeased him: "I don't know, fellows, I guess I'm getting too old for animation." In sum, he was leaving behind him—emotionally at least—two careers that would have been enough for most men and beginning to reach out for yet a third.

32 *I count my blessings.*
         —Walt Disney, on numerous occasions

IN EMBARKING on that third career, it seems that once again Walt Disney had intuitively sensed the shift of his audience's tastes. Indeed, it is pointless to undertake a book-length investigation of Walt Disney if the author does not believe that, in his link with his audience, Disney was something more than a mere teller of childish stories, a mere maker of childish tricks. To see him as merely the proprietor of a mouse factory is to miss entirely his significance as a primary force in the expression and formation of the American mass consciousness. One must, at last, take him seriously, because whatever the literary content of his works, however immature his conscious vision of his own motives and achievements was, there was undeniably some almost mystic bond between himself and the moods and styles and attitudes of this people. He could not help but reflect and summarize these things in his almost every action. We have lately been taught to understand that the medium is the message, and it seems indubitable that there was a message in each of the media that Disney conquered, a message that transcended whatever he thought—or we thought—he was saying. When we seemed to demand an optimistic myth he gave us the unconquerable Mouse. When we seemed to demand the sense of continuity implicit in reminders of our past, he gave us fairy tales in a form we could easily accept. When we demanded neutral, objective factuality, he gave us the nature films. When, in a time of deep inner stress, we demanded another kind of unifying vision, he gave us a simplified and rosy-hued version of the small town and rural America that may have formed our institutions and our heritage but no longer forms us as individuals. When we demanded, at last, not formal statements but simply environments in which to lose ourselves, he gave us those in the amusement park built and the amusement parks still to come.

His statements were often vulgar. They were often tasteless, and they often exalted the merely technological over the sensitively humane. They were often crassly commercial, sickeningly sentimental, crudely comic. They were easy ones to criticize, and, over-all, they had little appeal to anyone of even rudimentary cultivation. But the flaws in the Disney version of the American vision were hardly unique to him. They are flaws that have crept into it over decades, and they are the flaws almost universally shared by the masses of the nation's citizens.

Balanced against them must be the virtues he shared with his countrymen as well—his individualism, his pragmatism, his will to survive, his appreciation of the possibilities inherent in technological progress, despite the bad odor it often gives off today. Above all, there was the ability to build and build and build—never stopping, never looking back, never finishing—the institution that bears his name. There is creativity of a kind in this, and it is a creativity that is not necessarily of a lesser order than artistic creativity. "The American mechanizes as freely as an old Greek sculpted, as the Venetian painted," a nineteenth-century English observer of our life once said, and it is sheer attitudinizing not to see in our mechanized institutions one of the highest expressions of this peculiar genius of ours. These industrial institutions may ultimately be our undoing, but it is being blind to history not to see that they were our making as well. It is culturally blind not to see that Disney was a forceful and, in his special way, imaginative worker in this, our only great tradition. The only fitting honor to be paid him is to associate him firmly with it and not with some artistic tradition that was fundamentally alien to him and invoked standards inappropriate to the evaluation of of his accomplishments. The industrial and entrepreneurial tradition that both moved and sheltered him was neither more nor less flawed than he was.

And so there is a certain appropriateness about his last works. Film is, after all, a transitory thing, as all works of art are in comparison to that thing that every American knows to abide—the land. Or as we prefer to think of it, real estate. At heart, Disneyland is the very stuff of the American Dream—improved land, land with buildings and machinery on it, land that is increasing in value day by day, thanks to its shrewd development. With its success Disney realized the dream that had been denied his father and that had driven Disney as it drives so many of us. The day before he died, he lay in his hospital bed, staring at the ceiling, envisioning the squares of acoustical tile as a grid map of Disneyworld in Florida, saying things like, "This is where we'll put the monorail. And we'll run the highway right here." Truly, he had found The Magic Kingdom.

Typically, this was a secret he kept to himself. To be sure, he told anyone who bothered to inquire that he was not a producer of children's entertainment, that in fact he had never made a film or a television show or an exhibit at Disneyland that did not have, as its primary criterion of success, its ability to please him. And he often admitted that his great pleasure was the *business* he had built, not the products it created. But he—and most especially his organization—did nothing to discourage the misunderstanding of his work and his motives. And so much did we want to believe that he was a kind of Pied Piper whose principal delight was speaking, for altruistic and sentimental reasons, the allegedly universal language of childhood, so much did we *need* an essentially false picture of him, that the public clung to this myth almost as tightly as an eager Wall Street hugged to its gray flannel bosom the delightful reports on the recent economic performance of Walt Disney Productions. The commercial statistics were powerful enough to send the price of the stock to the highest levels in history *after* Disney died, even though the only foreboding its analysts expressed about it in the 1960s was based precisely on the fear of the founder's death and the inability of his executives to carry on in the great tradition. The myth of the Disney personality became similarly inflated after his passing. It was well expressed by the correspondent of *Paris Match* who, in an obituary cover story (Mickey was displayed in full color, crying), reached an unprecedented—and quite possibly fictive—mawkishness, with his description of the scene at the hospital on the day of Disney's death:

"At St. Joseph's Christmas party, the delighted children were applauding the trained dog act and singing 'Oh, Tannenbaum.'

"The hospital chaplain stepped out of the room where the successful life of Walt Disney had just come to an end.

" 'Lying there, he looks like Gulliver with his dream menagerie gathered around him,' he whispered."

One likes to think that, however pleased Disney might have

been by the sentiment, there was a secret recess of his mind where he kept a rather precise accounting of who he was and what he was. Ray Bradbury, the writer, once conceived the notion of having Disney run for mayor of Los Angeles on the not completely unreasonable ground that he was the only man with enough technological imagination to rationalize the sprawling mess the megapolis had become. He journeyed out to the studio to put the idea to Disney, who was flattered but declined the opportunity. "Ray," he asked, "why should I run for mayor when I'm already king?"

And so he even managed to speak his own best epitaph, the one he always feared someone would speak for him. Like all his other works it seemed fanciful only at first glance. At heart, it was completely realistic.

Disney Without Walt

33 *Our financial condition continues to remain sound. Our company's overall business performance during the quarter is also an encouraging sign that we are making progress towards key goals. . . .*

I am confident that we have the vitality and the commitments in place to make this a very good year and to build new momentum for the future.

—Ron Miller
President and Chief Executive Officer
Walt Disney Productions
Quarterly Report
Nine Months Ended June 30, 1984

WHISTLING IN THE DARK. Or anyway, whistling in the gloaming. For on Friday, September 7, 1984, a little more than two months after he had signed this optimistic interpretation of his recent stewardship of the Magic Kingdom, his board of directors forced Ronald W. Miller to resign. It was a historic moment, partly because no previous leader of the concern had been removed abruptly, under publicly visible pressure; partly because the replacement of Miller, who was Walt Disney's son-in-law, and fifty-one years old at the time, brought to a climax a family feud that had been simmering for something like a decade, leaving his rival, Roy E. Disney (Roy O.'s son) as the most significant family voice in the councils of the corporation. Moreover, Miller's ouster brought an end to the line of CEOs who had been old Disney hands, men whose careers, like his, had been spent largely within the company and whose business philosophies had been shaped by close daily contact with its founder. The men who were appointed to fill the posts of Miller and Raymond Watson (the company's chairman) two weeks later as president and chairman respectively—Frank Wells, formerly of Warner Communications, and Michael Eisner, late of Paramount—were both "outsiders," entertainment executives whose ways of doing business had been shaped by the exigencies of life in a show business that had changed, in large ways and small, almost totally in the eighteen years since Walt Disney's death.

As they took over, they (and everybody else) were aware that the troubles that had come upon Ron Miller were not entirely of his making. There had been, since the early Seventies, a failure by the entire Disney leadership to recognize the extent and the permanence of the revolutions taking place in two of the firm's principal lines of business: the movies and what we now refer to as "home entertainment," meaning video in its various new forms. It was the failure of the company to act imaginatively and decisively in these areas that slowed its once much-admired pattern of rapidly growing earnings; cost it its reputation as an innovative and aggressively managed company; dulled the luster of its stock and rendered thinkable what once would have been unthinkable, the threat of a takeover by unfriendly forces.

It was, indeed, the response of Miller's management to such a threat—a response deemed inept by Wall Street analysts, whose frequent pronouncements in the press on corporate doings have taken on oracular force in recent years—that

appears to have been the immediate cause of his dismissal.
Whether or not that was entirely fair to Miller (he was in
part the scapegoat for decisions that were surely shared with
the board of directors that ultimately forced him out), this much
is certain: in the past decade Walt Disney Productions has lost
much of its former glamour in the investment community and,
perhaps to a lesser degree, in the general population as well.
What had once been esteemed as a company ever on the alert
for technological breakthroughs and uncannily in touch with its
audience's tastes—with the aggressive marketing and promotional skills to exploit both these advantages to the fullest—had
become a company widely regarded as ingrown, out of touch
with current cultural reality, even something of a joke. Its
largest successes were as a merchant of nostalgia for its own
past creations; its most humiliating failures came when it
attempted to reclaim the position it had held in Walt
Disney's day, as a creator of new forms and styles of popular
entertainment.

In short, by 1984 Walt Disney Productions had lost its
uniqueness, that powerful singularity of identity that, if we
look back, is what had previously preserved its independence
during hard times, and had been the source of its irresistible
power in good times. That company, with its famous (and
beloved) founder guiding it (and guarding it), was impossible
to imagine as a branch of a huge conglomerate. Even more improbable was the thought of someone taking it over with the
intention of breaking it up—selling off its more profitable arms
to the highest bidders and closing down the less profitable ones.
But that was the pass that Disney reached in 1984. To the cold,
calculating eye of a financial manipulator, it had become at last
a company like many others, one that had seen better days and
still had left over some worthwhile assets that might be turned
to account. Even if it could not be taken over, it gave off an
air of stolidity and perhaps a certain stupidity, leading avari-
cious men to think that in its innocence and otherworldliness,

the company might be terrorized into paying them what people had taken to calling "greenmail," money paid to investors for surrendering large, threatening blocks of stock they have acquired. Which, as it turned out, is precisely what it did in the spring of 1984, to the surprise and consternation of many innocent souls who could not believe that Mickey Mouse's proprietors could behave in a manner so . . . Mickey Mouse.

But Walt Disney Productions did not arrive at this startling point quickly or by a simple path. Neither did it arrive there in an entirely unambiguous condition. Some of its troubles were products of strength, not weakness; some of those strengths could be traced to a management that, if it was not very venturesome, had at least a good understanding of where some of the company's enduring values lay and the wit to defend them. No, the story of Walt Disney without Walt is not without its complexities. And it is not to be understood without reference to the play of other forces in the society where it lived and made, even in its weaker years, a rather handsome living.

34 *Disney's problems are those of a child prodigy now grown to maturity but still following the script written by its parents.*
—Irwin Ross, *Fortune*, October 4, 1982

IN A NUTSHELL, YES. There can be no disagreement with this evaluation by a respected veteran of business journalism who, sixteen years after Walt Disney's death, found that "the master's presence still pervades the Disney studio. . . . His picture is everywhere—in the entrances to buildings, in the hallways, in executives' offices. His name and obiter dicta are invoked on all occasions; he is always referred to as Walt. . . ."

In the years after the founder's death this was a phenomenon everyone who passed through the Disney orbit noticed. Whenever a problem presented itself at the story conference or the marketing meeting, someone was bound to ask, "What would Walt have done?" And, sure enough, there was always someone present who could cite an analogous situation and recall Walt's response to it.

At first this was natural enough. The power of his personality and of his accomplishments assured that. So did the presence of Roy O. Disney. In the company chairmanship Roy Disney oversaw the execution of the long-term plans that he had worked out with his brother before he died. Indeed, there was so much old business to be completed in the late Sixties and early Seventies—notably opening Florida's Disneyworld in 1971, and then compensating for the loss of Roy, who died in the same year—that Donn Tatum, a financial man, and E. Cardon Walker, a sales and marketing executive, who became chairman and president respectively, could easily be forgiven for making no radical alterations in the company's course. All it seemed they had to do was keep the ship trimmed up and sailing placidly from one profitable port of call to the next. Internally there was no stress, and from the outside there was nothing but admiration from both the show business community and the investment community. Indeed, once Disneyworld opened, the company attained unprecedented levels of revenues and earnings. In the first eight years after Walt Disney's death, for example, revenues increased by 230 percent, profits by 285 percent. From 1973 until 1980, when they began to level off, earnings (partly fueled by inflation, to be sure) rose another 183 percent. The meaning of these figures can perhaps be best understood by reference to another, simpler set of figures: The annual gross revenues of Walt Disney Productions in recent years have flattened out in the neighborhood of $1 billion, which is around ten times what they were in Walt Disney's last year; profits have been running around $120 million a year, which is roughly equal to the company's total annual revenues when Walt died.

This achievement and, perhaps more glamorous, the potential of the company were clearly visible to the most casual observer. In the early Seventies, when the outside limits of the theme-park business could only be guessed—and many guessed that they were infinitely expandable—Disney became a stock market favorite as a company that had achieved stability, diversification and growth potential in one neat package. The parks, along with the company's steadily growing activities as a merchandiser of some of the best-known and best-beloved characters in popular culture, promised steady profits no matter what troubles might afflict its activities in the ever-risky movie business. But even in that realm Disney seemed to have significant advantages over its competitors. Rereleases of its classics provided a steadily profitable basis for operations, while its new productions also seemed to carry a lower risk factor than pictures coming out of other studios. Frugally produced and aimed at what seemed then to be the most reliable of markets— the middle-class, middle-American family, a Disney film, it seemed, could not fail. And in those days the hope of another blockbuster like *Mary Poppins* remained lively.

In addition, the company was both land rich and cash rich. The bitter lessons of Anaheim having been well and truly learned, Walt Disney had insisted on acquiring far more acreage than was needed for the initial development of Disneyworld. Before he died, the company had accumulated, often through subterfuge (and, it was charged, through the occasional use of stronger tactics), most of the 28,000 acres around Orlando that it owns today. Bought for an average price of $200 an acre, the land has increased in value to the point where some experts (perhaps overoptimistically) place its average valuation at $5,000 an acre. At the same time, profits, mainly from the theme parks, piled up at a near unspendable rate. Perhaps if Disney had been able to proceed with its Mineral King development, the money would have flowed out much more quickly than it did. But finally, the environmentalists—aided by the protectionist consensus that developed throughout the society in the late Sixties and early Seventies—won the fight against that resort. So much of this cash was profitably parked until the late Seventies, when it helped build the $700 million EPCOT center, which (at least for a time) increased traffic at Disneyworld after its opening in the fall of 1982. In the meantime, Disney's land and cash holdings added to the attractiveness of its shares among sophisticated investors, precisely because they seemed to increase the attractiveness of the company as a takeover prospect. There was a possibility, however remote, that someone could make a determined bid for these treasures (to which, as we shall see, its film library had to be added). If that remotest of possibilities came to pass, the ensuing proxy fight could be counted on to inflate the value of Disney stock.

But with large blocks of stock closely held by the Disney family and by trusted executives, with the company performing extremely well in its normal operations (and with, eventually, a corporate bylaw on the books requiring 80 percent of the stockholders to approve a wholesale change in the board of directors), management proceeded throughout this period on an anxiety-free basis. Walt Disney's ten-year plan for the company he had left behind was proving to be as successful without him at the helm as it had been when he was present, fussing and growling and keeping everyone up to the mark.

Except for one thing. Movies. Somehow, as the Seventies wore on, they ceased doing as well as they once had. It was not a sudden thing. It was just that on the routine releases, the gimmicky comedies and the unelectrifying adventures, there was a steady, though unalarmingly slow, drop in box-office receipts. Some of them made a little money, some of them lost a little money, but few of them broke through to the status of a major hit. And the studio's best efforts in the live-action field, pictures on which more than usual amounts of time, money and talent were spent (*e.g.*, 1971's *Bedknobs and Broomsticks*,

1974's *The Island at the Top of the World*, 1977's *Pete's Dragon*, 1979's *The Black Hole*, 1982's *TRON*), either failed, or at most returned very modest profits—and then only after the last receipts from the least of the foreign territories were toted up.

Surprisingly—considering their loss of prestige around the Disney company in the 50s and early 60s, when larger profits at less risk were being run up by the live-action features—it was the animated films that, overall, did best for the studio. *Jungle Book*, so loose and gaggy, succeeded at the box office, and in 1970 so did *The Aristocats*, in which some critics detected at least a partial return to the old Disney richness of design and execution. *Robin Hood* in 1973 and *The Rescuers* in 1977 got very mixed critical receptions, but *The Fox and the Hound*—a dark tale of an innocent friendship that develops between the title figures and is betrayed as their natures assert themselves in maturity—received an extraordinarily warm critical welcome in 1981. Though directed by an old Disney hand, Wolfgang "Woolie" Reitherman (who had supervised all the animated films since Walt's death), it was largely the work of a new generation of animators. A movie that "confronts the Dostoyevskian terrors of the heart"—in critic Richard Corliss' apt phrase—it represented, as he also said, "a return to primal Disney, to the glory days of the early features when the forces of evil and nature conspired to wrench strong new emotions out of toddlers and brooding concern from their parents."

The studio caught this drift and rode it intelligently. With Walt gone it was possible, at last, to single out for acclaim the elves in the animation department and to identify their individual contributions to sequences in the classic films everyone knew and loved so well—and to lift the curtain on the techniques by which they had accomplished their magic. The beginnings of this process can perhaps be dated from 1973, when Christopher Finch's *The Art of Walt Disney* was lavishly published by

Harry Abrams, a firm mainly associated with high-priced works on artists working out of a grander tradition than movie animation. The book was a huge success, and in its wake came others that were equally careful to identify the craftsmen responsible for notable Disney work. Two Disney animators, Frank Thomas and Ollie Johnston, wrote *Disney Animation: The Illusion of Life* in 1981, and their publishers brought out an equally elaborate sequel, *Treasures of Disney Animation Art*, a year later. Abrams followed a year after that with John Culhane's *Walt Disney's Fantasia*, a carefully detailed discussion of that film's creation.

As a result of these works and other publicity efforts, a knowing cult of animation buffs developed, people who could identify the specialties of the studio's "star" draftsmen and discuss their work with affectionate regard. Attention focused largely on the "Nine Old Men" (so-called by Disney himself), artists who had been with him through all his ups and downs since the 1930s. By the 1970s four of them—Ward Kimball, Les Clark, Marc Davis and Eric Larson—had either retired or moved on to other assignments in the organization. And it was generally understood that 1977's *The Rescuers* would be the finale for the remaining five, with Milt Kahl, Frank Thomas and Ollie Johnston scheduled to retire when they completed their work on it. John Lounsbery, who shared directorial credit on the film, died before it was completed, leaving his co-director Woolie Reitherman, to finish one last assignment as co-producer of *The Fox and the Hound*.

By this time critical enthusiasm for its work was, if anything, rarer around the Disney studio than profits on them, but, as always, it was the bottom line that was most significant. All of the animated features made between 1967 and 1980 had been profitable to the studio, returning worldwide grosses in the neighborhood of $25 to $30 million on their initial releases, but *The Fox and the Hound* is said to have done around twice as well: in the $50-million range. Its success—coming as the rest of Disney's motion picture business reached an all-time nadir—confirmed (if any confirmation was needed) that animation was far from an indulgence, an idle obeisance to Disney tradition. The steady success of these films, coupled with the acceptance of the newer animated features in the rerelease market (*Jungle Book* brought in around $15 million when it was most recently recycled in the summer of 1984), powerfully suggested that animation, far from being a whimsical fringe benefit of history, might just be central to the revitalization of the company's operations.

If that proves to be the case, it may incidentally prove that Disney management in the 70s and early 80s was not quite as lackluster as its critics have charged. In this field, at least, it proved itself to be competitively shrewd and tenacious. To begin with, Disney obviously observed that among the younger audience there was developing a reverence for animation as an art form that began in the film schools, where it was closely studied, and spread out from there to the rest of the youthful audience. To a degree, those youngsters were indulging an increasingly common phenomenon of the time—accelerated nostalgia. They began mourning for what they saw as the lost innocence of childhood and its pastimes almost before it had passed. Equally important is the fact that the typical animated film—fantastic in concept, but dependent on mysterious techniques and technologies for execution—fitted well with youthful tastes in movies; in design and the stories they told, the characters they featured and, above all, the values they upheld, the Disney films were analogous to the megahits of the day—the *Star Wars* saga, for example, or *E.T.* Finally, the intense stylization of the animated films, together with Disney's historical prominence in the field, made attendance at them socially acceptable—intellectually respectable—which was not the case with the live-action pictures, so resolutely and so unresonantly kid stuff.

Coincident with the release of that film a grand retrospective of Disney animation was organized at New York's Whitney Museum. Entrancingly mounted, it was, like the handsome books that preceded it, a revelation. For however pleasing the elemental anarchy of Ub Iwerks' rubbery stick figures from the early days remains, however delightful his minimalist backgrounds are to the postmodernist sensibility, both exhibition and books conclusively demonstrated that the drive for three-dimensional plasticity and ever greater subtlety of character animation was not misplaced. The short films got funnier and funnier, and in virtually every long film there were sequences in which the orchestration of comic effects became dizzying in its complexity, delicious in its appeal. Similarly, the range of the studio—displayed when a bit from one film was placed in close proximity to something of quite a different character from another film—became vividly impressive. Finally, to study the exuberant roughs of a slapstick comedy sequence or a series of ever-more-complicated character analyses or just a handsome uninhabited background was to gain a new appreciation of what the Disney artists accomplished. Abstracted from the often banal story in which it was buried, freed of the often vulgar musical score that distracted from its draftsmanship, the best work of the animators could be seen, at last, for what it was—an exercise in pure cinema, untrammeled by the demands of narrative or (for that matter) middle-class morality. It was, at best, a simple delight, arrived at through the most painstaking sophistication of technique.

There was, it must be said, a pleasing justice about this belated public recognition of the contributions of dozens of individuals to what will most likely be the Disney studio's abiding contribution to the history of cinema and, perhaps, to the history of art as well. But it must not be supposed that the generally increasing respect for its work in this field was the exclusive product of Disney's efforts to increase public understanding of its historical accomplishments, tasteful and intelligent as those efforts generally were. The circle of people who both know and care about the art of animation remains, after all, comparatively small. The most significant accomplishment of Disney animation—an accomplishment the studio could not manage in the live-action field—has been to hold on to, perhaps better say reconvene, the family audience of yore, an audience that has fragmented as the American family itself has fragmented over the last two decades. To accomplish this, while at the same time exerting an appeal for the adolescent audience (nowadays crucial to the success of any movie), was no small matter.

Of course, the Disney organization was greatly aided in this realm by the quality—or rather the lack of it—among its competitors. There was, for example, much animation on television from the 1960s onward, especially on the Saturday-morning children's shows. But these action-adventure fantasies were very poorly and cheaply done. Their limited animation techniques contrasted vividly with the richness of the Disney tradition. They may perhaps be said to have created a hunger for animation that they themselves could not satisfy. Nor could the feature-length works made by competitors for theatrical release. Some of these proved that the spare, sketchily backgrounded manner so interestingly pioneered by UPA in the immediate postwar years did not work at length on a large screen; they did not adequately feed the hungry eye. As for Ralph Bakshi and his wretched X-rated animated features, vulgarly drawn, muddily colored and as vulgar in concept as they were in execution, they were momentary sensations but no long-term threat. Others working more traditionally learned to their sorrow that in this realm, at least, the Disney name was sovereign, that the public simply would not accept even reasonable, facsimile animation from other sources. Hanna-Barbera, a power in TV animation, offered *Charlotte's Web* in 1972, and though it was closer to the Disney manner than to the studio's television work and was based on one of the most beloved of modern children's classics, it was unsuccessful. In 1977 no less a conglomerate than ITT backed a production of *Raggedy Ann and Andy*, directed by the gifted Richard Williams, whose animated sequences for *The Charge of the Light Brigade* had brought him deserved acclaim, and it, too, disappointed critically and at the box office. The following year an English unit headed by Martin Rosen filmed a version of the Richard Adams best seller *Watership Down* and came as close as any outsider had at that point to imitating the high Disney style of the late 30s and early 40s; it was not a major success, however. Nor was another Rosen adaptation of another Adams novel, *The Plague Dogs*.

Perhaps the most serious challenge to Disney came from inside, when a Savonarola arose among its own animators, charging that parsimonious practices at the studio were betraying the best in the Disney tradition, preventing him and his colleagues from doing the kind of careful detailing that had been the greatest glory of the glory days. This purist, Don Bluth, left the studio in 1979, taking with him sixteen of the best younger members of the animation department. His departure and his outspoken criticism of the company that had nurtured him shocked and hurt. And finding replacements for the departed artists delayed, by six months, *The Fox and the Hound* and delayed, by perhaps two years, work on what was, when it was released in 1985, Disney's most expensive animated feature, *The Black Cauldron*, a 70 mm. production estimated to cost $25 million.

Bluth and friends achieved backing for an excellent animated feature, *The Secret of NIMH*, which offered animation that was certainly comparable to Disney's best and a story that was perhaps preferable to most of the more recent animated releases, in that it restored an element of terror to the mix. It was well received critically when it was released in the summer of 1982 but did not score well at the box office. In short, the record of

377

recent years indicates that in this field at least, the Disney brand name retains a potency against which it is almost feckless to compete.

This competitive edge consists of something more than consumer confidence; it has an economic element as well. Very simply, the large and expensive technical support system that animation on the feature-length scale requires is permanently in place at Disney. The studio does not have to staff up or freshly equip itself in order to mount an animated feature. Its artists and artisans simply move on from one to the next. Indeed, although the defection of Bluth and his group obviously shook the studio (Ron Miller publicly called Bluth a "son of a bitch," a most un-Disneyish phrase), it had the resources to expensively recruit and train replacements for them and to carry both *The Fox and the Hound* and *The Black Cauldron* until the newcomers were able to complete them. In fact, by the autumn of 1984 Joe Hale, the film's producer, was telling a reporter that although two years previously he would have rehired the entire Bluth crowd, he now would not take any of them back. "We've got a new group and we've laid them over an anvil and beaten the Disney style into them," he said, none too felicitously. The major problem with the new people is that they are slower than the old hands were, turning out perhaps two and a half feet of animation per week against the five feet an experienced artist could create in that time.

Even so, the lesson is clear. In this one peculiar realm of film production, the old studio system, with craftsmen ever on duty, ever functioning at their specialties, still makes sense. Animation is not and cannot be a sometime thing, for it is finally a product of teamwork that is created over a long period of years. And that, of course, requires a patient, long-term commitment by studio management, something that is increasingly hard to come by in modern Hollywood, where the short-term deal is the norm and management turnover is constant.

But if the Disney organization's patient philosophy preserved and protected a unique area of strength—the soul of the company as it might be argued—it was everywhere else disastrously conservative in its decisions. Or perhaps, one might better say, in its lack of decisions for change or at least experiment. Indeed, looking back over the last decade, one sees in the realm of motion pictures so many misperceptions and miscalculations—about the movie audience's changing nature and the movie industry's changing business style—that one is hardpressed to determine which of management's failures—of observation, of imagination, of nerve—was the most significant. All one can say with any certainty is that no studio in the entire history of the movies has for so long been so misguided. If one considers that what Walt Disney left behind him was—whatever the shortcomings of the films his company turned out—a brilliant economic construct, this is no small achievement in the annals of ineptitude.

It is arguably possible that if the problem of the film division had been kind enough to present itself dramatically in the form of a clear-cut crisis, it might have elicited a more energetic response from management. But there was no crash; there were even isolated successes. It was merely that a tide was running against Disney, gently pushing the film division toward the shallows. However, such was the thrusting power of the engine that Walt Disney had created and set in motion that his organization was able, if not to make headway against the tide, then to yield to it very, very slowly and come to ground with the ship essentially undamaged and—with a little luck—refloatable.

It was really astonishing. For consider, when Walt Disney passed from the scene, the war in Vietnam was at its height, as was the generational disaffection and the counter-cultural ferment it had stirred. Consider, too, that within a year of his death, that almost perfect representation of the Sixties spirit, *Bonnie and Clyde*, was released; that in the years immediately following, such culturally portentous works as *The Graduate*, *2001: A Space Odyssey*, *Easy Rider*, *M*A*S*H*, *The God-*

father, and *American Graffiti*—plus a dozen other titles symbolizing the shift in American preoccupations—also were released. But even setting aside the larger cultural ramifications—what they may or may not say to the social historian examining them for clues to our mood as a nation in a transitional moment—these movies assuredly told everyone in the movie business that something new was afoot. For example, the range of subject matter included on the brief list above obviously indicates that the old steadfast reliances on a relatively limited number of immediately identifiable genres was no longer a guarantee of success. Rather, the opposite was true; novelty, or the air of novelty, was essential. Equally important, an air of individuality—a sense that a movie reflected not corporate values but the visions of the director, the *auteur*—had to be projected. Above all, these movies (sometimes by accident, sometimes by design) made their first strong impression on what amounted not so much to a new audience but to an audience newly conscious of itself as a force in the world, an audience that might best be characterized for our purposes as autonomously youthful.

Now all movie audiences have tended to be young since the beginnings of the medium. But this new audience was quite different from those of the previous generations. The company's strength in the film world had always been, as we have seen, in the power of its name to guarantee safe, sane, chucklesome entertainment to young families—Mom, Dad and a couple of small children—looking for some innocent merriment, something the parents wouldn't be embarrassed to share with the kids, on a night or an afternoon out together. To this vital core, in its best days, the studio could count on adding significant numbers of slightly older children allowed to attend a movie on their own recognizance. The trouble was that as of the early Seventies, the number of American families that had fragmented reached what was for Disney a critical mass. The divorce rate had begun its climb to the present level, where close to one out of every two marriages ends in separation. The

380

number of women who stayed in or reentered the work force after the kids were born rose astonishingly. In the meantime television was pressed into service as an electronic baby-sitter, and taken in seven- or eight-hour daily doses, it offered children a surfeit of bland—not to say mind-numbing—entertainment. As a result of these and other factors the family ceased, by and large, to go to the movies as a group, and at earlier and earlier ages children started making their own decisions about entertainment in general, the movies in particular. This foreshortening of childhood, the rush toward an illusory maturity, the creation of an adolescent subculture that has almost no reference points with adult culture—these are all developments to mourn, and so they have been—by sociologists, psychologists and moralists—at length.

What need concern us here is merely that these developments steadily eroded the reliable old market for Disney movies. Indeed, the studio had only one solid success with this audience. That was the 1969 rerelease of *Fantasia*, in which adolescents discerned, not entirely erroneously, psychedelia: an unintended but effective head trip. The smell of pot wafted out of certain theaters where it established itself. The picture, for a time, was lifted out of the cyclical rerelease pattern governing the availability of Disney animated classics and was generally made available to any theater that wanted to book it.

But setting that anomaly aside, and always bearing in mind that the other animated features could generally be counted on to reassemble the old Disney core crowd, one has to think that the slowdown in business attributable to the studio's new live-action features should have been clearly perceptible to management by the mid-Seventies. If this may have seemed to it only a temporary aberration, a calm that would eventually be steamed through, that is understandable; corporations give up habits no more easily than individuals. The fact that many of these films still did well in foreign territories, where both families and the traditional concept of childhood were still in fashion, must have been a consolation—and, perhaps, a rationale for not abandoning the basic conceits that had served the studio so well for so long. Equally important in evaluating the studio's performance in this period was the astounding success of *The Love Bug*, about a little Volkswagen with a comical mind of its own. Taking advantage of the nation's sudden infatuation with basic transportation that had a lovably cute air about it, the picture was a huge, surprise success. It was 1969's top grossing film, ultimately returning over $23 million to the studio in domestic rentals, more than any nonanimated feature other than *Mary Poppins*. It encouraged Disney to make three increasingly listless sequels, concluding with 1980's *Herbie Goes Bananas*. It too was a lucky anomaly, but it seemed to argue the viability of the basic Disney formula.

So, for a variety of defensible, or at least rationalizable, reasons, the company clung to motion picture concepts and styles of execution long past the point it should have. And despite warning signals. As early as 1973, for example—or perhaps, one should say, as late as 1973—a man named David Marlow, who served a few months as an East Coast story editor for the studio and wrote up his experiences for *New York* magazine, observed that Disney was routinely making only two kinds of nonanimated films. One was an outdoorsy sort of thing, in which children or other simple, generally rural, souls got involved with animals, either wild or domesticated (or in transition from the former to the latter state). These pictures tended to celebrate traditional American self-sufficiency and, as Marlow observed, they almost never showed their central male figures (whether man or boy) involved in an intense relationship with a member of the opposite sex. Very often there was no relationship with any female figure—not even mother or sister. There was, instead, a good deal of huntin' and fishin' and aw-shucksin'. These pictures generally did less well at the box office than the other genre in which the studio specialized: the gimmick comedy. These included tales of scientific accidents or wacky inventions in which, say, a pet duck suddenly starts to lay golden eggs, or a wayward electrical current passes through an otherwise ordinary citizen who temporarily turns into an eccentric genius, comically inconveniencing himself and dismaying his friends before he is returned to normalcy. Or perhaps, one should say, ordinariness, which was the point of the exercise, since what Disney was offering in these films were little time capsules of 1950s bourgeois life, when it was thought that the worst thing that could happen to an individual was to be singled out, separated from the crowd.

There was what might be termed a compassionate reason for continuing to make what amounted to sequels to *The Absent-Minded Professor*. Looking back on this period from the vantage point of 1984, an anonymous industry observer of Hollywood folkways called Disney "a cottage industry," and he was not far from the mark. Within the cozy and comfortable confines of the studio, Disney maintained at its workbenches far from the beaten path a group of aging craftsmen—writers, directors, producers—who were either tired of the hurly-burly of independent production, which was now the norm in the movie business, or had never been exposed to it. Secure in the long-term contracts that Disney alone of the studios continued to offer (and in some cases quietly possessed of stock options that would make them wealthy), they tinkered away at their lasts, leisurely writing and rewriting, planning and replanning their increasingly quaint products, for a market steadily shrinking to the size of a special-interest group. If there was anything human or eccentric in their original concepts, it was squeezed out in endless preproduction fussing or by the academic manner in which the typical Disney film was shot and edited. Eventually, of course, even the more adventurous writers and directors came to understand implicitly that challenging the studio's conventionalities of content and style was a waste of energy, that the essence of their task was always to seek the formulaic solution. On its part, the studio felt an obligation to these old hands.

They were agreeable men, and some of them had given the best years of their careers to Disney. Their Disney credits were without market value elsewhere, which meant they were virtually unemployable elsewhere in the industry.

Not that Disney, in the Seventies, was actively unhappy with these pliable employees; they could handle what the studio turned over to them to make. For true to its tradition of frugality, Disney still operated at the lowest levels of the story market, avoiding the bidding wars for hot literary properties that preoccupied its upper levels. Stop to think about it: something like *Jaws*, had it been shot a trifle less bloodily, might have been a perfectly acceptable Disney product—its shark no scarier than *Pinocchio*'s whale. But there was no question of Walt Disney Productions entering the bidding for a best-selling title. If it bought a book as the basis for a film, it bought something that no other studio was interested in—some naturalist's reminiscences of life in the wild, say, or some chipper children's story for which, absent of meaningful competition, it could pay bottom dollar.

The same frugal, not to say miserly, policy held in its relationships with the Hollywood creative community, and this was perhaps more devastating to its fortunes in the Seventies. By this time the best people would work only under an independent contract and sign one picture deal (at most two or three) at a studio which was expected, in turn, to entirely finance whatever dream possessed them. Financing included, of course, handsome salaries for themselves, the support of their staffs, and at the back end, "points," that is, a percentage of the profits. Never mind that most films—perhaps 80 percent of them—never turned a profit; never mind that most independents claimed they rarely saw their fair share of the money that a successful film took in. Points were—and are—a matter of prestige, almost of honor. And Disney's adamant refusal to award them to independents effectively cut itself off from people of talent and from the younger crowd specifically—writers and directors who were to prove as the decade developed that whatever their other abilities, they had a strong sense of what their contemporaries and the generation following them—the only reliable audience—wanted to see at the movies. Disney would talk to most of them during this period and would discover that the majority of them, raised on Disney films and enormously respectful of the studio's work in animation, could not be brought to terms with the studio.

But serious as these problems were, they were perhaps not as crucial in determining the company's fate as was a more amorphous one that revolved around its relationship with the public at large. This had its beginnings in 1968 when the Motion Picture Producers Association, responding to pressure from both within and without the industry, finally promulgated a ratings code. This ostensibly replaced its longtime practice of prior censorship of films before they obtained its "approved" seal, the only moral endorsement it offered. Now all films produced by a member company (and such other concerns that wished to pay a fee for the service) would carry a rating, expressed as an alphabetical grade, broadly suggesting to consumers how "adult" a picture was in terms of language, nudity, sexual attitudes, degree of unconventional or antisocial attitudes and all the other vague what-have-yous that constitute the fundamentally misstated question of what constitutes "morality" in films. The G rating suggested that there was nothing in a picture that could harm the innocent child; M (for mature, but later changed to GP and then to PG) suggested that parental guidance was needed to explain to younger children what was going on up there on the screen—why the nice lady was glimpsed half-naked for a second or two or why the nice man was suddenly using language that Daddy used only when he hit his thumb with a hammer—and suggested a ludicrous vision of theaters filled with parents and children in earnest, whispered confabulation about the subtext of the images flashing before them; R required that an adult accompany anyone under sixteen (the age was later raised to seventeen) who wanted to see a film so classified; X barred everyone under the specified age from seeing the film, even if an adult was willing to purchase his or her ticket.

As with most things emanating from the motion picture business, the motives behind the code were mixed. There was a general sense that something less monolithic than the old approve-disapprove system was needed so that serious film makers could be more flexible in their approach to contemporary reality. There was also a sense that even under the liberalized codes promulgated in 1956, 1961 and 1966, excesses of hypocrisy prevailed in their administration. Behind the scenes, producers and censors were constantly involved in an absurd trading process—cutting the swear words in reel two, let's say, to retain the flash of bare breasts in reel five—so that approval could be obtained. Serious film artists were sickened by this childishness and aware of what their colleagues abroad could show and tell in their pictures without a by-your-leave from anyone. A significant segment of the public, the more articulate and sophisticated members of the movie audience, agreed with them although their opinion on the matter was not of huge economic significance.

What was significant was that the mass of the public, from the early Sixties onward, began gathering the impression that something funny was going on at the movies. It was only an impression on most people's part, the typical American adult having lost the movie-going habit by this time, venturing forth to see a film no more than four or five times a year. But yet there was talk in the press, on television, over the back fence, of licentious doings on the screen, things that one would not like the children to see. The children, especially the teenaged children, were, of course, delighted to slyly partake of these pleasures—which were in the vast majority of instances mild and very occasional.

Be that as it may, protectionist sentiment ran strong and

continues to do so. For the question of what should or should not be shown in the movies (and on television) became part of a much larger debate about morality, both public and private, in the United States. Religious fundamentalists and political conservatives saw the new freedom of the screen as yet another threat to traditional values, perhaps not as serious as that posed by legalized abortion or militant feminism or the homosexual-rights movement but of a piece with them somehow. And somehow linked to the rising divorce and crime rates, not to mention the general casting off of sexual restraint in every realm of society. On the other hand, the liberal community, far more concerned with what appeared to be a rising tide of violence in the country, found the increasingly bloody visualizations of conflict in the movies (not to mention the "victimization" of women in many screen stories) equally offensive and a cause for alarm. That mainstream movies, emanating from studios that were, as the result of conglomeratization, often the subsidiaries of major American corporations, were carelessly lumped with the growing—and increasingly visible—industry devoted to hard-core porn (which was mostly mob controlled) was a bitter irony to the legitimate industry.

But what is truly important to bear in mind is that the 1968 ratings code fully satisfied no one. Behind the scenes, censors and film makers continued to trade unkind (and often unwise) cuts, generally to avoid an R rating for films aimed at the younger adolescent market, occasionally to avoid an X (though, in practice studios rarely developed projects that would take them close to the dangerous country). Out front, meanwhile, the people who were passionate on this subject continued to complain about the ratings. The traditionalists still found young people falling into bad company at R-rated (and sometimes even at PG) films; the liberal minded, conversely, still found quite a different sort of film unsuitable for young eyes and ears. In this grumbling atmosphere the ratings system might be said to have worked only in this limited sense: it prevented, for the most part, the recrudescence of state and municipal censorship boards, which had plagued the movies almost from their beginnings, and the new system promoted good public relations for the motion picture industry, demonstrating a concern for an allegedly serious criticism of endlessly suspect Hollywood. And for the majority of citizens—those whose interest in the subject was not very passionate—the ratings did provide quick, crude guidelines they could consult in their desultory and feckless attempts to regulate their children's movie-going habits. It was just that for those who were most deeply involved in this problem—the film makers strongly needing to say something on the one hand, the professional moralizers on the other— vexatious questions continually arose. Stirred and reheated all the time, the pot kept simmering.

And, of all things, the hot water spilled onto Disney's operations. For the law of unintended consequences is (this side of Murphy's law) the most mischievous of all the predictably unpredictable rules that govern our daily lives in the late Twentieth Century. The problem lay in Disney's reputation and in its constituency—small townish, middle class, unsophisticated—the self-same Middle America that commanded so much attention from the politicians in the Seventies, the self-same Middle America where a paranoid sense that its oldest and best values were under contemptuous assault now ruled. These people demanded in letters to the studio and at stockholder meetings that Disney hold the line against the smut peddlers; and thanks to the new ratings system, they knew precisely where that line was drawn—between the G and PG rating.

There was a cruel and unusual specificity in this prohibition. Until the code was passed it was obvious that whatever its failures of art and imagination, however closely it eyed the balance sheet, Walt Disney Productions operated from an unexceptionable moral position. Who knows? It is possible that had a ratings code been in operation in former days, *Snow*

White or *Pinocchio*, for example, might have been given a PG
rating for the childish terrors they evoked in the more impressionable members of the audience. Maybe the bare-breasted
"centaurettes" (as they were known around the studio) in
Fantasia might have been given a PG, too. But with no seemingly concrete standard to repair to, moral critics were forced
to take the studio's word as to its intentions and to believe the
evidence of their own common sense as to the overall effect of
its work on its intended audience. And that inescapably led to
the most benign interpretations of its work. If the studio could
be faulted, it was on aesthetic grounds, and as this writer knows
as well as anyone, criticism at that level carried no weight
whatsoever with the mass audience or its moral tribunes. Walt
and his helpers could roam at will within the preserve his
corporation had staked out for itself.

Now, however, there suddenly existed what looked to be a
well-defined border around its territory, and if Disney attempted
to cross it, that action would be judged weighty with symbolism, an annunciation that one of American capitalism's few
truly beloved institutions had gone over to the enemy. And
there were broad hints that this defection would not go un-
punished, that Middle America, the prime audience for the
most prosperous areas of Disney's operations, Disneyworld and
Disneyland, might boycott the amusement parks if the studio
surrendered to commercial pressure and moral laxity.

Yes, it seems hard to believe—all those decades of good will
being wiped out by one or two or a handful of PG movies or,
heaven forfend, an R rating. Equally difficult to imagine is
someone of Jerry Falwell's ilk mobilizing a boycott that would
reach beyond his lunatic fundamentalist fringe and attract to
its cause sufficient numbers of adherents to actually harm
Disney. But Disney's managers did indeed believe. "Sure I'm a
hypocrite," Ron Miller admitted to journalist Sally Ogle Davis
as late as 1980. "I let my children see everything—Rs, PGs—
the lot. But I have a responsibility to this company. One racy

picture could do incredible damage to a name built up over fifty-five years." What he did not say was that his lack of gumption in this matter was already doing the company "incredible damage." And not merely or primarily in the realm of aesthetics or in its relations with the Hollywood creative community. No, it was with its old prime audience, the kids. What they had determined, after a few years' experience with the ratings system, was that it was distinctly uncool to be caught at a G-rated movie unless glumly accompanying a parent or guardian who insisted on such niceties. They understood that anything so entirely inoffensive was bound to be a bore if you were more than about four years old. Even if, by some miracle, the thing was actually not bad, the kids simply could not be lured to something so clearly marked as kids' stuff. Indeed, it came to pass that shrewd producers of films threatened with a G rating would insert a few swear words or some modest sexual frisson so that they could display a PG stamp in their ads. Poor Miller. He just couldn't bring himself to such expediencies. "They'll have to blindfold and gag me before I'll let them do anything more than a soft PG." Even a hardbitten public relations consultant, Charlie Powell, cynical enough in his other comments on Disney and its problems at this juncture, agreed with Miller on this point. "Miller's protecting *all* of Disney's assets," he asserted.

Miller's fears in this case cannot be attributed solely to Disney inbreeding. Hollywood, in general, throughout its history has paid a disproportionate amount of attention to its most strident moral critics. Whether we are discussing its feeble response to earlier pressures for censorship or its lack of backbone when the political witch hunters came around in the Forties and Fifties, Hollywood had never trusted the basic passivity of its customers—despite its own astonishing record in making them swallow mile upon mile of nonsensical celluloid without complaint. Perhaps because of its isolation from the main currents of American thought, perhaps because it is a business largely founded by immigrants—strangers in a strange land who ever remained outsiders, suspicious, not to say paranoid, about the natives—cowardice has characterized its relationship, as an industry, with the general public. Hollywood can be further characterized by its mistaken belief that the loudest critical voices must have a genuine constituency behind them, when, in fact, it is the movies themselves that have the most reliable constituency—a vast audience that, even after television, remains in *Time*'s ineffable neologism, "cinemaddicts," in need of their fix and impatient with anyone who dares to interfere with it.

Perhaps none of this—Disney's devotion to its traditional style of doing business, its fear of its basic audience's resistance to change—would have made any difference if after the mid-Seventies the activities of other producers had not offered such a vivid contrast to its dithering impotence. Indeed, without this contrast, the studio could have made a better case for its conservatism. After all, with its animated films and its rereleases performing well at the box office, with even its new live-action pictures continuing to find modestly profitable markets abroad, the film division continued to contribute something to the company's overall profits. The trouble was that it was only holding its own; it was not contributing to growth. And growth, relentless, year-to-year, quarter-to-quarter growth, was the name of the game in modern American enterprise. If overall its steadily rising net continued to make Disney stock attractive to Wall Street, its analysts could not help but reflect, after a time, how much more spectacular that figure might be if the company's films were pulling their weight along with the amusement parks. This became painfully true after 1977. The spectacular success of George Lucas' *Star Wars* proved that it was entirely possible to make at relatively modest cost ($12 million) and to market under a PG rating a movie of unimpeachable moral quality that supplied safe and sane entertainment for preadolescents and, at the same time, exercised a powerful hold on their older siblings and, for that matter, on their parents as well. The picture racked up a domestic gross in excess of $200 million, doubled that figure (at least) around the world and became a merchandizing bonanza for its producer. Looking at it and reflecting on the fact that Lucas, an admirer of Disney, had discussed the project with the studio when he was looking for financing, one could not imagine why the Disney people had passed on it. Nor could one imagine why they had not themselves been exploring similar possibilities in-house.

Nor was *Star Wars* a one-shot phenomenon. Within a year Lucas' friend and soon-to-be collaborator, Steven Spielberg, enjoyed a similar success with *Close Encounters of the Third Kind*. After that came two *Star Wars* sequels from Lucas, his collaboration with Spielberg on *Raiders of the Lost Ark* and Spielberg's phenomenal *E.T.* Meanwhile, away from the Lucas-Spielberg axis, all kinds of innocent merriments, ranging from the *Superman* series to *The Black Stallion* and *Chariots of Fire*, came forth, coined fortunes and proved that it was possible to work within the parameters of the PG rating without, as a rule, risking opprobrium from any self-appointed guardian of the nation's morals. All of these pictures could have been made at Disney, and some of them should have been made there. If it was not unreasonable to wonder why none had been, it was also perhaps not unreasonable to wonder whether Miller's concern over the effect of a PG release on his company's other activities was not diversionary, an excuse to cover up a large failure that was more of the imagination than it was of nerve. No one expected him to make dirty pictures or even "racy" ones. What they did expect was an exploration, alive to its near-endless narrative possibilities, of the general-audience film— the film that (like almost every American film prior to 1960) spoke on several levels to several different audiences no matter what their demographic status. In a voice that, though amplified by technology, retained an authentic human tone and offered symbolic representations of recognizable human beings in situations that were, whatever their settings, not entirely improbable. In short, all that was required of the studio was that it cease to think of its troubles as problems in marketing and reconceive of them as opportunities for artfulness. Artfulness, mind, not art.

An effort of this sort was made. The Disney mold was to be broken. But how difficult that would prove to be for Ron Miller and the rest of the studio! Hollywood people would refer to the studio as "the Mickey Mausoleum" or "the land that time forgot," or they would say such things as "some of us are in the film business and some work for Disney." Agents would send over material to the men in "bow ties and brogans" only after all the other studios had passed on it, and they would send writer and director clients over to discuss their projects only when they were desperate. Even Spielberg, greatly respecting the Disney tradition, which had helped to shape his sensibility when he was a child, would not in the end entrust *E.T.* to the studio, though it apparently offered him a deal comparable to those its competitors proposed. He may well have sensed what others sensed—that Disney's sales and promotion departments were also entrapped in the patterns of the past and were not truly competitive in the modern marketing world.

By this time a certain amount of nervous attention within the film and financial communities was beginning to focus on Disney's film division, and it could no longer quietly bury its mistakes. People kept looking for signs that it had regained its stride—and publicly noting its stumbles. For example, consider *The Black Hole*. Released at the Christmas season in 1979, it was an obvious attempt to gain a share of the market *Star Wars* had commandeered two years earlier. Opening in competition with *Star Trek*, an even more expensive (and inept) space fantasy, was bad luck; there wasn't room for both of them in the same marketplace. But despite some handsome special-effects work and production design that recalled the wit and style of *20,000 Leagues Under the Sea*, the picture was rather too obviously an imitation of the George Lucas film, including obvious knockoffs of *Star Wars*'s comic relief, R2D2 and C3PO. The picture opened strongly, and on the basis of its first week's business, a domestic gross of perhaps $50 million was projected. But its characterizations were unoriginal, its story poorly developed, its conclusion visually spectacular but emotionally unresonant. Word of mouth—the most important element in promoting any picture, especially in the youth market—was poor, and business dropped quickly and alarmingly thereafter.

The following spring there was more bad, not to say embarrassing, news. Ever since 1973, when *The Exorcist* had opened and successfully exploited the market for the horrifically supernatural, there had been a steadily profitable action to be found in this genre. Stephen King could not turn out the basic material fast enough to satisfy the demand, but there were plenty of novelists willing to take up the slack—and plenty of movie makers eager to buy their work. Disney ought to have known it could not deliver the goods as others could, since success in this realm demanded heavy psychosexual undertones and, at the right moments, buckets of gore splashed vividly about. Still, the studio would try, with *The Watcher in the Woods*, starring Bette Davis. The picture seemed unable to make up its mind as to whether it wanted to be cute or scary; it ended up being neither. Worse, when it opened in New York, the *Times*'s movie critic, Vincent Canby, wrote: "I challenge even the most indulgent fan to give a coherent translation of the end." He was not the only viewer to be mystified in entirely the wrong way by the film. Ron Miller ordered the picture withdrawn from release for reediting and some reshooting. He now found himself in the same awkward spot that a few months earlier had been occupied by the United Artists executives who had been forced to withdraw *Heaven's Gate* after its initial run, when critics, led by Canby, had hooted that expensive ego trip by Director Michael Cimino off the screen. The cost of *Watcher* was not comparable, but the humiliation was, and the revised version fared poorly when it finally reached the theaters. The studio eventually took a $6.5 million write-off on it. Miller might praise his own decision as "courageous," but everyone knew that what Walt would have done in this situation was not to get into it in the first place. Maybe he fussed the life out of some films, but there is much to be said for getting a script right before you shoot it. Or abandoning it if it cannot be worked into proper shape. The writing stage is the only stage in film production where revision is cheap, and shelving a failure is no outsider's business. In any event, this unfortunate affair reinforced the general feeling that, as far as film production was concerned, Disney was running out of effective control.

Two more late Miller starters—*Condorman* (a spoof in which the mild-mannered cartoonist who drew the birdman of the title becomes involved with real-life spies) and *Night Crossing* (a dour little drama about a family trying to escape East Germany in a homemade balloon)—failed almost as badly financially but with less public ignominy attached. Equally disappointing, but perhaps less costly to the studio, were its first two ventures into co-production with Paramount Pictures. One of them was *Popeye*, an ugly musical fable attempting to exploit what was thought to be a nostalgic interest in bygone cartoon characters. Charmlessly produced by Robert Evans and eccentrically directed by Robert Altman, who could not mesh his highly stylized manner with the antithetical stylizations of the old cartoon, it was witless and confused in intent as well as execution. Rather better was *Dragonslayer*, by the writer-director-producer team of Matthew Robbins and Hal Barwood. Full of marvelous special effects and featuring a vastly comic portrayal of an aging sorcerer by Ralph Richardson, it even had a discreet flash of nudity. But it was marred by a too blandly juvenile love story at odds with its basically dark tonalities. And, like the majority of Miller's choices, it was an attempt to follow a trend—not lead it—coming at the end of a short-lived cycle of sword-and-sorcery films in 1981. It also fared poorly at the box office.

Around this time the press fell into the habit of identifying Miller as a former football player (he had put in a year as a tight end with the Los Angeles Rams after starring at USC) as well as Walt Disney's son-in-law. The implication was that he was a dumb jock who had married his way to power, if not success. The innuendo was doubtless cruel and unfair; and Miller kept pointing out that his family connections had merely helped him to get his first job at the studio, that thereafter he had risen through the hierarchy strictly on merit. There was something in what he said. The evidence is that he was a sound, line producer of movies, conscientious about costs and craftsmanship within the established Disney tradition. The father of a numerous family, he was a man who enjoyed his skiing holidays, took pleasure in a small vineyard the family had acquired as a hobby and was generally conservative in his politics and his view of the world. These are not inconsiderable virtues, but they are not high on the list of qualifications for success as a motion picture executive. An imaginative sense of how the volatile movie market may be jumping a year or a year and a half ahead, is (this is how long it takes for a movie to move from conception to theater screen). And, despite what critics of movie people insist, the successful have a certain conviction—a sense of what one's own and one's company's best range is—what is worth going after with all the energy and resources at one's command. During Miller's regime people kept saying that there was no movie or idea that Disney would kill for, which in a town full of killers was a fatal defect. Indeed, in the final analysis, what was wrong with the majority of Miller's major efforts was that they were imitations of films that others, rightly or wrongly, had killed to make—trend followers instead of trend setters. Beyond that, it is said that Miller first surrounded himself with people who would not argue with him and then, when he did start to open up to new ideas, found very few people in the organization with the nerve to challenge him.

It has been said that Miller's promotion to presidency of the company in 1980 was an example of the Peter Principle—which holds that in complex hierarchies people are promoted to the level of their incompetency—in action. After all, during the years he ran the film division, its contribution to Disney's income had declined from about 50 percent of the total to something like 10 percent. On the other hand, it has also been argued that the world had changed so much in that time that Miller's had been an impossible task, one that anyone might have been hard-pressed to master. In any case, the rest of the company had prospered so mightily that the film division's contribution to overall grosses, when rendered as a percentage, was bound to be smaller now. Finally, however, one suspects there was a sense among top management and board members that Miller's essentially conservative talents might be better utilized away from the fast-moving, highly competitive film world as the steward of a company that had virtually no competition in what were now its principal lines of business. In 1980, the year of Miller's ascendency, Walt Disney Productions had enjoyed its largest income and best profits ever. The amusement parks were flourishing in their accustomed manner, and creative planning for its largest new venture—that permanent world's fair adjacent to Disneyworld, EPCOT—was long since finished. It remained only to be completed and opened with a flourish, and those were jobs well within Miller's range, especially given the company's established strengths technologically and promotionally. There were also plans afoot to launch a new Disney Channel, a subscription service for cable TV; that, too, looked to be a relatively risk-free undertaking. The old Disney audience, the family, if it was to be found at all, was likely to be found gathered around the television set, looking for an oasis of true and moral entertainment in the wasteland.

The recycling of old Disney products and the creation of new works pretty much like them were also still well within the company's capabilities.

Both of these enterprises were, indeed, well begun under Miller. EPCOT, as everyone who is not a permanent resident on an alien planet knows, opened in the autumn of 1982 and attracted admissions in its first year somewhat in advance of projections. It cost $900 million to build, and it is a far cry from the Utopian dream (Experimental Prototype Community of Tomorrow) for which Walt Disney coined the acronym by which it is called. It is, in fact, a collective of corporate pavilions, designed by WED Enterprises, in which projections of a technologically bright future are offered through state-of-the-art exhibition techniques. But congeries of great names, General Electric and General Motors, American Express and AT&T, are contracted to pay rents of $300 million apiece for a decade to present their messages; and ordinary citizens are obliged to pay $15 apiece to view their commercialized visions of hardware to come, in the process probably adding a day, at least, to their stay in the Orlando area, with much of their additional hotel and restaurant charges flowing toward Disney operations. Some critics have deemed EPCOT "a mediocre product." It may not be quite the huge winner that Disneyland and Disneyworld have been, but it is hard to see how it could be a big loser.

As for the Disney Channel, the number of its subscribers has exceeded projections, and it entered upon profitability on schedule in the second quarter of fiscal 1985. Again, we are not talking the kind of growth potential that glamorizes a stock. It may be that in Disney's case we are never again going to be talking that kind of growth. But it is only under the peculiar terms by which Wall Street nowadays understands corporate performance that all other forms of growth and profit are to be sneezed at. In short, aside from the movies, the first three years of Ron Miller's term as president and CEO were respectable, by the reasonable standards one might apply to an enterprise that should have been evaluated not as what it recently had been—a sprouting adolescent—but as what it had rather suddenly become—a mature corporate citizen.

Indeed, even in the troubled movie realm, some halting progress was made. Even before he took over his presidency Miller had begun to share his power in this field with an able, quiet-voiced young man named Tom Wilhite, who in his early twenties had taken over Disney's doddering public relations department and infused it with new life. When he was twenty-seven, Miller gave him formal control over "creative affairs" in both film and television, while retaining for himself control of animated films—largely an administrative task, given the long lead time required to produce these pictures.

Wilhite's task, from the time Miller had placed him at his right hand, was to open lines of communication to film makers of his own generation, and in aid of that effort he was heard, at times, publicly if politely disagreeing with his mentor. "We don't live in a G-rated world," he told one reporter. "It's time to start taking some risks." He allowed that if in order to portray the realities of, say, contemporary adolescent life, a few bad words or a little less than innocent sex was requisite, he would willingly accept a PG rating. And one suspected that dreams of the forbidden R also danced in his head. Be that as it may, he wanted to establish among his contemporaries the notion that nothing was sacred at Disney anymore, that he would consider any reasonable proposal for a film and award points for the right projects. The strategy worked. Some quite good notions began flowing toward the studio, and some quite good people, too. Eventually, in order of their appearance on the screen, the slate of pictures for which Wilhite would take the credit—and the blame—included: *TRON, Tex, Trenchcoat, Something Wicked This Way Comes, Never Cry Wolf* and *Splash.* Among them were three critical successes, two commercial successes (including the biggest hit the studio had enjoyed since *The Love Bug*) and one major technological experiment. Only one was a horrendous commercial failure, and only one could be termed a Disney picture pretty much like any other. Or, to put the matter simply, Disney became, during Wilhite's short tenure, a respectable—if still small-scale—studio, with a balanced schedule of releases and something like the normal ratio of hits to flops. If one considers the record of its immediate past, this was a sterling achievement. The trouble was that the most troubled pictures reached the marketplace first, and Wilhite was not around to enjoy the successes that came later.

It was the biggest and most daring of his films, *TRON*, that rolled out first in the summer of 1982. The concept, by writer-director Stephen Lisberger, thirty-one at the time of the film's release, was both witty and original: a computer nerd, a leading designer of video games, is zapped via electronic hocus-pocus into the innards of one of his inventions, where he must play the game not for recreation but for his life. The world down the electronic rabbit hole is realized almost entirely through computer graphics, and so are all the special-effects perils he and his friends must confront. The film, if nothing else, is a bold, often breathtaking review of the state of that new art—its virtues and its limitations. And it placed Disney for the first time in years on the cutting edge of film technology. Its defect—and it was a substantial one—was a lack of warmth in its characterizations. One simply did not identify strongly enough with any of its leading figures. Partly this was because work on the graphics and effects so consumed the picture's creators that they lacked the time and energy to spare for its people. Partly it was because, as both Lisberger and Wilhite were to admit, they feared Disnification of the story, feared *TRON* would be criticized, as so many Disney films had been, for cuteness.

As a result, the picture aroused curiosity but not affection.

But the film also, perversely, suffered from the studio's pride in it. Overhyped as a technological breakthrough and as its comeback picture, Disney let it be known that it expected *TRON* to be a huge box-office success as well. But then things started to go wrong. The special effects, as so often happens, were slow to complete, and the picture was unable to get to the theaters in time for either the Memorial Day or Fourth of July weekend, crucial summertime playdates, when the blockbusters establish themselves on the now widely publicized charts of the top grossing films and, therefore, as "must see" items. Worse, in its pride, Disney invited some stock analysts to see the film at prerelease screenings, offering it as prime evidence of the company's new spirit. Several of them were unimpressed by the film's commercial prospects, and their "reviews" (which took the form of advising customers through their newsletters to dump their Disney shares) were widely reported in the general press, where Disney's troubles in the motion picture field had become an ongoing story. No promotion or advertising could rekindle enthusiasm for the picture after that damper had been clamped down over it. *TRON* lost money although it is said that loss was not serious, given the results from abroad, the licensing fees from the video game created from the movie and the royalties from cassette sales.

The fate of the studio's next release was, if anything, even sadder, for *Tex* was an unambiguously good movie about a rebellious teenager coming to terms with life and himself in a small Texas town. Based on one of S. E. Hinton's popular novels for adolescents and starring Matt Dillon, a hot young actor among the younger teenage audience, this modestly budgeted picture was exactly the sort of thing Disney should have been doing for years. And director Tim Hunter, a screen-writer making his directorial debut, did a fine job with it. But the marketing department insisted on opening it in the summer of 1982 without benefit of national reviews at several hundred theaters in the Southwest. To this uninformed audience, it looked like just another Disney kids' picture (despite its PG rating), and it died. When it was belatedly brought into the New York Film Festival—and gathered what was surely the best set of reviews a Disney picture had enjoyed in well over a decade—it was too late. Again, no money was lost, but none was gained either.

Trenchcoat, an unfunny spy comedy, a little racier than the usual Disney venture into this genre, died a quiet death, unmourned and unattended. Not so *Something Wicked This Way Comes*. Based on Ray Bradbury's well-liked "classic" (but not, alas, cultish) novel in what might be termed the vein of the domesticated occult, this was a troubled production. The director, Jack Clayton, was one of those competent, impersonal English movie makers whose best work (*Room at the Top*, *The Pumpkin Eater*) was accomplished by realizing a strong script with a certain efficiency. His previous effort in a vein akin to Bradbury's, *The Innocents*, was more notable for a certain understated elegance rather than for the evocation of terror. In short, he was the kind of reliable second-rater that was comforting to a cautious studio management. In any event, he delivered an unsatisfactory cut of the film, and well over a year was spent trying to fix it, while costs mounted inexorably, unconscionably. The picture reached the theaters in 1983, but on little cat feet. It was written off by the studio at $21 million, virtually its entire cost. By that time Wilhite was on his way out, replaced in the top spot by Richard Berger, up to then a low-profile production executive at Twentieth Century-Fox.

Wilhite left behind two films. One was *Never Cry Wolf*, an adaptation of Farley Mowat's witty recollections of his adventures as a young government naturalist studying lupine life deep in the Canadian wilds. Directed by Carroll Ballard, who had scored a great commercial and critical success with *The Black Stallion*, this was the first Disney picture in which the director was awarded points, but it, too, was much delayed. Its principal photography had been completed, under incredibly trying conditions, in 1980, but it was the fall of 1983 before Ballard finished cutting and recutting his film. It was, however, worth the wait. The landscapes are awesome; the Mowat character (played by Charles Martin Smith) is human and funny and ultimately inspiring in the understanding he comes to of the delicate interrelationships between the creatures and the country he briefly, but often harrowingly, experiences. Opening slowly, the picture caught on with the younger, ecologically aware audiences. As 1983 turned into 1984 the movie was still playing quietly—and beginning to turn a modest profit, of all things.

And then came *Splash*. For any studio other than Disney this would have been a routine production, what has come to be known as a "high concept" comedy, a phrase that means exactly the opposite, since it is always used to describe pictures that can be summarized in a single sentence. *Splash* was, of course, the one about the mermaid who falls in love with a mortal and comes to New York to visit him. It was all frothy and innocent, but it stirred consternation in Ron Miller's heart. Mermaids traditionally go topless. She obviously wasn't in New York to catch a couple of shows and see the museums. Still, Wilhite argued forcefully for the project, and Miller acceded glumly to his insistence. But Miller decided, at last, to put into effect a long-discussed plan—the creation of a new releasing entity, Touchstone Films, to release this and other non-G films—in time to handle the picture.

The rest, as they say, is history. Or anyway, an occasion of significance in Disney's corporate history. *Splash*, opening in midwinter, when it had the adolescent market pretty much to itself, was a mighty hit, eventually grossing around $75 million at the U.S. box office alone. It is impossible for an outsider to say whether, combined with those of the more modestly successful *Never Cry Wolf*, its profits wiped out the write-downs Disney had taken on failed films the previous two years, which totaled $48.7 million. But it came close enough so that one could plausibly argue that as 1984 began and the company began laying elaborate, chucklesome plans to celebrate Donald Duck's fiftieth anniversary, a turnaround might have been very close at hand. In late February there was every reason for Ron Miller to feel confident about the year to come—the years to come. But of course, one reckons without the lurking presence of the Big Bad Wolf—or in this case, a whole pack of them— at one's peril. By late March they were baying at his heels.

35 *Management conducted an utter scorched earth policy. They were a client out of control who did not do what their advisers suggested, and who did things that in the final analysis led to their doom.*

—David Kay
Wall Street Merger Specialist
Manhattan, inc.
November 1984

Some days you win and some days you lose....

—Irwin Jacobs
Speculator
Ibid.

ONE HAS TO SYMPATHIZE. The situation in which management found themselves was, for Disney, unprecedented. And the big bad wolves they were up against were not nice, clean, funny wolves. They were not Disney wolves. They were not the sort of people who buy into businesses to improve their management or their long-term profitability; they are the sort of people who buy businesses to sell them—preferably quickly but always at a profit, which is sometimes realized by selling off the assets piecemeal.

These creatures are always circling around America's

corporations, though their number has increased in the last couple of decades. Their fundamental character, which is entirely amoral when it comes to money, was immortally anatomized seventy years ago by Theodore Dreiser in those great and now largely unread novels *The Financier* and *The Titan*. Normally, however, the respectable businessman only hears these intellectual heirs of Dreiser's Frank Cowperwood baying from the distant hills. Even while the Disney film division floundered, even as attendance at the amusement parks flattened, their shadowy forms did not become visible. No, it was not until other troubles, troubles of a much more emotionally painful sort, began to surface that the circling pack drew close enough to the fire to be recognized and to instill an almost unreasoning fear in Ron Miller, his board of directors and his top-management aides.

This trouble was family trouble, and its roots went back over a decade, perhaps longer. Ron Miller was not the only promising young man with close family ties to the business who went to work at Disney when he was in his twenties. The other, a couple of years older, was Roy Edward Disney, Roy O. Disney's boy. Both young men worked in film, with Roy concentrating on documentary work but eager, of course, to move into features. He did not move ahead as quickly as Ron Miller did, perhaps because his father outlived Walt and so bold a show of nepotism would have been anathema to him, perhaps because the Walt Disney side of the family held more stock than the Roy Disney side, and thus more power. (When the brothers first issued stock, Walt held 60 percent of it, Roy 40 percent, and that discrepancy persisted through many years of splits, bonuses and sales.) At least as important to Miller was his alliance with E. Cardon Walker. Walker, a straitlaced Mormon who had come to the studio as a messenger boy in 1938, was the self-appointed guardian of Walt's legacy. An abrupt, bristling man, instinctively conservative and without creative flair, he was the guarantor, to Walt Disney's widow and daughters, that his legacy would be preserved—at least in the realms of faith, morals and tight business practice. In return for their support of his rise in the company, it is clear that Walker brought Ron Miller along with him. And, of course, while the going was good, no one paid the slightest heed to the fact that the most significant aspect of Walt Disney's genius, his instinct for the creative, commercial idea, had no heirs, no guarantor anywhere in the upper reaches of the company's councils.

It is not at all clear that Roy E. Disney possessed any of that instinct. Possibly he did not. But *Fortune* reported that Walker treated him as "the idiot nephew," and that he assuredly was not. It is also said that he and Ron Miller clashed repeatedly, especially after Ron took over management of the motion picture division, and Roy was obliged to submit his ideas for projects to him. Miller apparently liked none of them. Nevertheless, and as if to make this awkwardness yet more uncomfortable, Roy had a seat on the board of directors, and he became increasingly restive with the direction (or lack of direction) the company was finding first under Walker, then under Miller. The two branches of the Disney family had not been particularly close after the death of the patriarchs. One of Ron Miller's sons told a reporter in 1984 that it had been "years" since he had seen Roy. Miller, of course, saw him at board meetings—both at the company and at the California Institute of the Arts, to which Roy O. Disney, like Walt, had left a very substantial legacy and on whose board of trustees both Ron Miller and Roy E. Disney continued to sit. No one has ever testified to open expressions of bitterness between the two on these occasions. But a very persistent anecdote has Roy appearing suddenly at the studio one day and finding his reserved-parking space occupied by an interloper. It seems he so seldom used it that the security people had fallen into the habit of parking visitors' cars in it when all other spaces were filled. Roy, it is said, raised such a ruckus that Ron was aroused from his presidential lair. He appeared in the parking lot, ripped Roy's nameplate from its moorings, tore it up and stomped back to his office, steaming. It's possible. In the status-conscious movie world a man's parking slot is his castle, and its location and the amount of respect it is accorded is a significant telltale of corporate prestige. Nor does the incident ring false to Miller's temperament. He was known to be grouchy on small points; it is only on the large ones that he earned one journalist's description of him as "skittish." It does however ring false to Roy's personality, which is said to be shy and inarticulate in public.

He may not have inherited his uncle's creative skills, but he assuredly inherited his father's gift for finance. He proved this when after leaving Disney's employ—but not its board—he placed much of his wealth and some portion of his side of the family's wealth in a company he founded. Called Shamrock Holdings, its assets were most conservatively stated at $23.4 million in a 1983 filing. But since they came to include seven radio stations, mostly in major markets, two television stations, a 10,000-acre cattle ranch, substantial real estate investments in Southern California and elsewhere and, at one time or another, sizable stakes in various corporations (including, naturally, Disney), one is more inclined to accept *Forbes* magazine's estimate of Roy E. Disney's net worth, which it placed at $220 million. Shamrock, moreover, was a company busily churning its portfolios and on occasion not above slipping into wolf's clothing itself, as it did, for example, when it launched a raid on Fabergé, the cosmetics manufacturer, selling out its stake to that firm's eventual acquirer for a neat $7-million profit.

In short, Roy E. Disney, abetted by Stanley Gold, his tough-talking lawyer and financial adviser, was an agile financial operator and, as it turned out, a prescient one. When he suddenly resigned his seat on the Disney board of directors on March 12, 1984 after seventeen years of service, explaining his action with a vague "personal reasons," he initiated the high-stakes game that would preoccupy the company, the avidly interested press and a public that would extend well beyond the financial community for six months to come—the biggest, bloodiest struggle for control of a major American corporation since the notorious battle for Bendix two years earlier.

Roy Disney was aware, of course, that the company had been a tempting takeover target for at least two years—ever since attendance at the amusement parks had flattened and ever since it became clear that the expensive effort to turn around the film division was not going to produce quick results. Given his talents and interests, it is reasonable to suppose that his sense of the danger the company was in might have been livelier than that of management, which was naturally preoccupied with the company's day-to-day concerns. It is also reasonable to suppose that he might have also had a more acute realization of what the financial ramifications of a sad and troubling family problem might be. It was at this juncture, after thirty years of marriage, that Ron Miller and Diane Disney Miller quietly separated for reasons never publicly stated. Roy Disney must have felt that this was the sort of distress that would send the financial sharks into a feeding frenzy. Mrs. Miller and her mother, Lillian, Walt's widow, were believed to control, between them, at least 10 percent of Disney's stock—perhaps more if one counted in the holdings of Diane's sister, Sharon, who has always kept the lowest profile of all the low-profile Disney clan. Suppose, Roy Disney may have supposed, a takeover specialist imagined that the Disney women were sufficiently alienated from Miller to withdraw their backing for his management. That would constitute an unprecedented break in the ranks of the family and the family's corporate retainers which had for so long made Disney unassailable from the outside. Neither woman ever publicly indicated, in the months ahead, any intention of deserting Miller in his hour of professional need, but it is characteristic of takeover types to believe that everyone has his or her price and that Mrs. Miller's and Mrs. Disney's might just be within reasonable range.

Be that as it may, it would appear that Roy E. Disney, sensing a crisis impending, was opening up options for himself by resigning. If he were not a member of the board, he would not be obliged even to pretend a show of loyalty to a management in which he had no confidence. He had, contrary to rumors that floated at the time, no interest in mounting a takeover campaign of his own. But he was free now publicly to throw the weight of his influence in two directions—against outsiders who threatened to change the fundamental character of the company, a matter about which he had very strong feelings, and against defensive strategies by management that might also alter its nature. He could also, of course, be alert to support those forces, should they emerge, which promised the not-necessarily-antithetical goals of retaining Disney's traditional shape while ridding it of his old rival, Ron Miller. Indeed, of all the board members it was Roy E. Disney who kept insisting that the failure of the motion picture division affected more than the balance sheet. It was he who kept arguing that the failure of creative vitality in that realm was actually harming the performance of other divisions as well—robbing them of new characters and concepts that might have been boons to marketing and to the theme parks, for example, and generally lowering the company's visibility and loosening its grip on the national imagination. It appears, according to Myron Magnet, whose *Fortune* article on the Disney crisis is definitive, that even before he resigned from the Disney board Roy was in touch with Miller's eventual successors, Frank Wells and Michael Eisner, ascertaining their interest in coming aboard should Miller's management be forced out. Later, it seems, both men were kept constantly abreast of developments in the fight for control of the organization and may possibly have contributed to Roy E. Disney's strategic planning.

In this game, Roy Disney commanded considerable resources. Through the spring of 1984 he increased his own and Shamrock's stake in the company to 4.7 percent of the outstanding stock. And it is reasonable to suppose that he had a powerful influence over the large block of shares his ninety-four-year-old mother retained. It is possible that he had, perhaps, 10 percent of the company's shares under his command. In a tight contest for control, that was enough to be the key to the future. And, of course, if he could enlist his aunt and his cousin in his cause, representing himself as the true defender of the faith as opposed to outsiders or the man from whom they were now alienated, he would be in effective control of, by far, the largest amount of the company's stock. As it would happen, Ron and Diane Miller would reconcile before the drama finally played itself out. But as it would also happen, Roy Disney would achieve both the goals he sought—preservation of the company in its traditional form and Miller's removal.

As of late March, of course, that end was not in sight. The only thing that was clear was that a couple of weeks after Roy Disney's resignation, the long-feared challenge to management began to take concrete form. Just after Roy's resignation, rumors about Reliance Financial Services Corp.—a division of Reliance Group Holdings Inc., corporate entities controlled by Saul Steinberg of New York—began to circulate. Steinberg is the kind of financial manipulator who gives corporate executives the jimjams. He toils not, neither does he spin. He merely buys and sells stock, but not as you and I do. He buys in huge quantities the stocks of mature companies—companies whose greatest period of growth appears to be over and which have settled down to modest, steady profitability—with one of two intentions in mind. The first is to gain control, spin off the best assets to other companies, and then cash in his chips, leaving, at best, a shell where a once functioning organization existed. Failing this—and usually he does fail—he lets it be known that his substantial position in the victim concern is for sale. At a premium. Which the shaken management has often shown itself glad to pay to get rid of him. Both *The New York Times* and Chemical Bank had felt the chill of his shadow falling upon them, and he was in the process of collecting close to $50 million—a 28 percent premium—for selling back to Quaker State, the oil company, the 8.9 percent of his stock that he briefly held when he began moving against Disney. The scariest thing about him is that he cloaks himself in mystery. He gives no interviews to the press, and he refuses to meet with the managements of the firms he is pressing. Money is, to him, the purest of abstractions, and he appears to care nothing for the social values a corporation may represent, or the jobs it provides, the lives it affects.

At first, Disney executives paid him scant heed; for on the face of it, with a hit movie working for them for the first time in years, with the heavy start-up costs of EPCOT and the Disney Channel behind them, they were entitled to bullish sentiments. Especially with regard to the latter enterprise. In 1984 it would continue to lose money, though at a slower rate. The important thing was that it was the industry leader in growth, adding a million subscribers (one-third of all Americans who signed up for cable service that year) and gaining a 5.2 share of the cable market. In short, the channel had found its niche, that niche was going to be profitable and some people were saying that, in time, it might become a launching pad for a yet more profitable assault on the sizable adult, prime-time market that was still interested in wholesome, family-style fare.

Beyond these pleasing prospects management was strengthened by their corporate bylaw requiring 80 percent of the stockholders—an almost impossible figure to obtain—to approve a change in management. Indeed, the company's chairman, Ray Watson (who had been president of the Irvine Corp., a Southern California real estate concern), Miller and other top executives were passing through New York on their way to France—where they would discuss the creation of a

European Disneyland along the lines of the licensed facility in Japan that had just opened successfully—when they met socially and confidently with representatives of their investment banker, Morgan Stanley, and of the law firm they had engaged to advise them in the realm of corporate acquisitions, Skadden, Arps, Slate, Meagher and Flom.

What they heard sent a fearful frisson through them. Even distant rumors that Steinberg was on their trail could have an adverse effect on their line of credit with the Bank of America. Never mind that they had been doing business with it since the days of *Snow White*. Banks, in their nervous conservatism, have a way of shortening credit lines when they think a company may be under serious assault. It is, of course, at that very moment that a company needs plenty of credit, principally to dilute its stock, forcing the raider to buy more and more of it in order to achieve control. On their return from France the Disney people arranged to triple their line of credit with Bank of America—and fifteen other institutions—to $1.3 billion. And not a minute too soon. Just two days after the Disney board approved this plan, on March 29, Steinberg announced that he had acquired 6.3 percent of Disney. On April 3 he had 7.3; on April 10 he had 8.3. Two weeks later Reliance Financial Services filed notice with the Securities Exchange Commission that it intended to acquire up to 25 percent of Disney's stock. And on May 1 it was almost halfway to that goal, holding 12.1 percent of the shares.

That board meeting took place on May 17. Though Steinberg would five days later receive a go-ahead from the Federal Trade Commission and the Justice Department's antitrust division to proceed with his plans to acquire 25 percent of Disney, the Arvida deal appears to have forced him to reconsider his strategy. On May 29 he announced that he would launch a proxy fight to remove Disney's present board of directors. To this end, he began putting together the financing for a tender offer for 49 percent of Disney's shares at $67.50 per share, attractively above the market price. Among his allies in this manoevre were Kirk Kirkorian, another mysterious financier who controls M-G-M, the Fisher brothers, who control a New York management and construction firm and—just along for the ride (for the moment)—Irwin Jacobs, the Minneapolis operator a.k.a. "Irv the Liquidator," most famous for his greenmailing assaults on Pabst Brewing and Kaiser Steel. A rather informal and apparently amiable man, known for answering his own phone, he got involved with Disney initially because he judged its stock a good market play, undervalued for the moment and thus ripe for a modest short-term rise on which he might make a few hundred thousand dollars. No big deal. It was too bad for Disney that his interest in the company could not be contained at that casual level. But he was recruited for a $35-million share in his consortium by Steinberg, who intended to grab the company's developed real estate for himself, while promising the Fishers development rights on the unimproved land.

Disney, in the meantime, was not idle. It had decided on a classic riposte to Steinberg, namely the acquisition of another company, to be paid for by a new issue of Disney stock—dilution, in short. A team from Morgan Stanley, constituting itself as Project "Fantasy," started looking into likely candidates. The one that finally looked the likeliest was the Arvida Corp., a Florida-based developer of planned communities and resorts throughout the sunbelt. The concern was 70 percent controlled by the Bass brothers of Fort Worth, themselves grand acquisitors and also fascinating figures to the financial community. The sons of Perry Bass—nephew and partner of Sid Richardson, the legendary oil wildcatter—they have diversified his petroleum holdings in dozens of directions, from computers to underwear to fried-chicken franchising. But if they are quick, smart and tough in their dealings and are not above walking away from a short-term position in a company with some buyout money in their own pockets, they are not regarded as wreckers or as greenmailers. They have the reputation of not meddling unduly with companies in which they have a stake, so long as management is doing its job competently. It might be said that beyond profits, they seek influence in companies, not control. Youthful, pleasant in manner and WASPy, they were the sort of white knights Disney was looking for. And Arvida was a plausible, easily defensible acquisition, its interests and competence fitting well with Disney's real estate operations. Indeed, company spokesmen were at pains to insist that buying Arvida had nothing to do with its fight against Steinberg. It was, they insisted, part of a plan to begin realizing profits on their long-standing real estate holdings. For $200 million in new Disney shares, the price—for a growing company that had turned a profit of $10 million in 1983—was right. And it did dilute Steinberg's Disney holdings by one-twelfth.

The trouble was that it diluted everyone else's holdings as well, and Roy Disney and Stanley Gold, Shamrock's CEO, publicly objected on the grounds that the deal saddled Disney with still more undeveloped land and a large new debt. Said Gold: "They needed those acres like they needed another asshole." They may also have feared that Arvida represented not an end to but the beginning of an aggressive acquisitions policy at Disney. Indeed, one of the strongest implications of a good, richly detailed journalistic study of the case, by John Taylor in *Manhattan, inc.* magazine, was that Disney's strategy was to respond to each significant purchase of its stock by Steinberg with an acquisition that would instantly dilute it. Obviously the company felt that in a war of attrition its resources were greater than Steinberg's. But, just in case, at the same meeting that approved the Arvida acquisition, the Disney board granted "golden parachutes" to fourteen top executives—the rip cords of which could be yanked whenever an outsider acquired 25 percent of the company's stock. Miller had been given a similar deal a month earlier.

414

As for Kirkorian, if the deal went through, he would get the cable channel and, most interesting to him, the film library. Through M-G-M he already controlled the biggest such library in the United States, including not only that company's "vaulties" (as *Variety* calls them) but those of United Artists, with which his concern had recently merged to form the M-G-M/UA Entertainment Co. Among UA's assets were all the films Warner Brothers had produced prior to 1948. All in all, this entity controlled 4,459 titles at this time. But they had all been heavily exposed in the various video markets for many years. The Disney library was tiny by comparison— only 169 features, plus its short subjects and some old TV shows—but the theatrical films, as we have seen, were entirely unexposed on television. Thus, although it was only one-twenty-sixth the size of Metro's library, it had been independently valued at $500 million—two-thirds the worth of the latter's holdings. Since it was estimated that 49 percent of Disney could be picked up, via Steinberg's tender offer for less than $900 million, with Kirkorian in for less than half of that, it can be seen that a genuine bargain was in the offing for him.

Disney, however, hung tough. Or did it hang panicky? It is a matter of some dispute. What is known is that it prepared a tender offer of its own for the 51 percent of the shares that would be remaining after Steinberg's forces had finished their acquisitions. Disney's idea was to offer an astounding $80 a share, financing the purchase with borrowings. Once it had them in hand, it let it be known, Disney would retire this stock. This would leave Steinberg and friends with 100 percent of the company, all right. But it would be a company burdened with a two-billion-dollar debt—in other words, an inoperable shell. Steinberg would have to sell off its components just to relieve it of its debts. Which would leave him and his associates with a certain loss of almost everything they had put into their tender offer. Scorched earth indeed!

Would Disney have actually carried through this plan? Or

was it merely a bluff? It is impossible for an outsider to say. But one has to admire the high hard one it threw at the famous hardballer, Steinberg. In the aftermath of this exchange there were, predictably, cries of outrage—mostly from people allied with Steinberg. Why, they piously declaimed, the Disney people were threatening to wreck the company. But wasn't that precisely what Steinberg was trying to do? In fact, in the course of doing a humorous piece on plans for Donald Duck's birthday celebration, in addition to a TV special, there was to be a twelve-city tour of a little show featuring Clarence "Ducky" Nash, who had been Donald's voice since 1934 (and who died in February 1985)—and, finally, the implantation of the world's most famous webbed feet in concrete at the Chinese theater in Hollywood—Stephen J. Sansweet of *The Wall Street Journal* found an anonymous observer who opined that the way to settle this struggle was to "lock Steinberg and the duck in a room and see who emerges." In effect, that's what management did, stepping out of their Mickey Mouse costume to emulate the studio's more erratic and temperamental star.

As so often happens, the threat of anarchy turned out to be the best answer to anarchy, for Steinberg folded his hand. For a price. Which turned out to be $325.5 million for his shares, or $70.83 each, plus what worked out to be $6.67 apiece for "expenses." "Greenmail" the world cried, although Steinberg would claim that he had been quite simply blackmailed by the company's threat to self-destruct rather than let him take it over. Later, when Miller was fired, observers would theorize that what cost him his job was embarrassment over this incident. It is probably true that Disney underestimated the extent of the furor its payoff to Steinberg would cause.

The trouble was that its action came at the end of a string of such incidents elsewhere in American enterprise: these had, since 1982, cost a lot of people a lot of money and irritation. Worse, Disney was not some anonymous widget maker, but rather a concern that had always seemed to investors as well as to the general public as clean and upstanding in its dealings as it was in its products. People did not like to see it getting down into the muck with the likes of Steinberg. And they couldn't help wondering, what next, who next, if a stop was not put to the greenmail ploy. But that view underestimates management's satisfaction at beating back this challenge and its own sense of righteousness. It also ignores the fact that the majority of those hurt—when Disney stock fell back to normal levels after the takeover bid ended—were people who had lately bought the stock because they expected to be able to sell it back shortly at a handsome profit to whoever turned out to be the highest bidder. The objective observer can muster but small sympathy for people whose greed was turned back on them.

In determining the immediate future of the company, the outrage of the investment community and the general public was of small importance. The significant outrage was Roy E. Disney's. And it was considerable. Before the Steinberg buyout was consummated, Gold (accompanied by Frank Wells) had gone to Miller and Watson and urged them not to pay Steinberg's price. Instead, they proposed arranging a leveraged buyout of the company by (it would seem) a group they would put together. They promised that current management would retain a significant voice if that eventuality came to pass. But their idea was rejected, and Gold and Disney at last decided on an open fight—a suit to rescind the Steinberg deal, an injunction to block another proposed acquisition (see below) and a proxy fight to throw out the present board. But there was a leak—not planted by Roy's camp, it is said—and the corporation's board sued for peace. They offered Roy, Gold and Roy's brother-in-law, Peter Dailey, vice-chairman of the Interpublic Group of Companies, seats on the board. Together with Arvida's representative, Charles Cobb, they would represent a dangerous—ultimately fatal—minority for Miller and Watson to contend with.

And going in, these potential dissidents had a ready-made

issue on which to stand in the aforementioned acquisition. During the great hunt for properties to buy for stock dilution purposes, Gibson Greeting Cards had caught the Disney managers' eyes. And they fell in love with it. Maybe it was just auction fever running hot in their blood, but maybe it was, indeed, as Ron Miller later called it, "a damn fine piece of business." Gibson was the third largest concern in its field, and it was a field that "synergized" (a word that would keep appearing in connection with this deal) extraordinarily well with Disney. The essence of the greeting card/gift wrap/party supplies business is characters. Gibson was already licensed to produce jolly paper products emblazoned with the likenesses of Garfield the cat, Big Bird, and something called Kirby Koala—worthy figures all, but not yet immortals on the order of Mickey, Donald, Pluto *et al.* Disney, meanwhile, had long licensed its gang to Hallmark. But if it owned its own company, might it not enhance profitability? For example, with Gibson it could support its current Donald Duck promotion through a campaign in the nation's card shops and dime stores. And what about the timely release of merchandise to support the release of such pictures as *The Black Cauldron*, which would be introducing many new creatures who would benefit from familiarization through retailing and which, in turn, might generate who knows what new merchandising lines? This was, in short, just the kind of deal Walt might have gone for. And given the fit of Disney's interests and Gibson's, it was every bit as defensible as the Arvida deal had been. Perhaps it was even more so.

The trouble was the cost and the timing. The price finally arrived at was announced in the press at $337.5 million, to be met with more newly issued Disney stock. But since the price of that stock varied from day to day, and there were both upper and lower limits on the amount of Disney stock that could be traded for the Gibson shares, that price was only approximate, and there would come a time when critics would estimate the acquisition as costing around $100 million more. That was one basis for criticism of the deal: Gibson, in the eyes of its opponents both on and off the Disney board, was overpriced. The other problem was timing. The Gibson proposal was placed before the board on the same day the board was asked to finalize the Arvida deal, and while Steinberg was still in the field, the dimensions of his threatened tender offer still unrevealed. For the moment, the idea of thinning out its stock still further and placing the new issue in friendly hands carried the day against the doubters.

But in the next few weeks, as we have seen, Steinberg was beaten back, and suddenly the Gibson deal did not look so attractive. For one thing, the Disney treasury was depleted and had used up around $500 million of its new credit line acquiring Arvida and buying out Steinberg. Worse, in the wake of the Steinberg deal the value of Disney stock had dropped sharply, meaning more shares of it would have to be exchanged for Gibson shares in order to fulfill the deal. If it went through as originally planned, Gibson, whose principal owners were former Secretary of the Treasury William Simon and Raymond Chambers, through their Wesray Corp. (which had picked up the company for a paltry million in cash on a highly leveraged deal, out of which they had already made $70 million by taking the company private and then public again), would become Disney's largest stockholder. The passage of time did nothing to make their deal any more attractive to Roy Disney and his faction, and, it has been said, having these two Eastern sharpshooters represented on the board did not sit well with him either. Meantime, he had come to know and like the Bass brothers, and they him, especially since Miller and Watson had treated them dismissively. Thus Roy had a new ally in Charles Cobb, Arvida's man on the board, who liked the Disney-Gibson fit but thought the price of the acquisition out of line.

Their objections might have amounted to no more than a grumble fit without a new intervention by Irwin Jacobs, who had been away on vacation while his sometime ally Steinberg was being escorted to the door. Ever alert for a bargain, Jacobs decided Disney's stock was once again undervalued in the wake of Steinberg's departure and decided to buy in. This was probably true. But the explanation was too bland. What Jacobs cleverly saw was that if he could gather a new team of raiders and make a plausible-looking bid for control of the company, the market in Disney stock would rally again, and he hoped that the company would again pay greenmail. Soon enough he and his group had 6 percent of Disney's stock in hand. Better still, the Gibson deal gave them an issue on which to challenge management, especially after Raymond Watson admitted to Jacobs that under terms of the agreement with Gibson, Disney could back out by simply paying a $7.5 million kill fee. That, said Watson, he was not about to do. Disney's board had now approved the transaction, and anyway, among men of honor, a deal was a deal.

A new show of outrage was now staged, with Jacobs writing a letter to all members of the Disney board claiming that the price it intended to pay for Gibson, $41.83 per share, grossly overvalued stock of which the book value, he said, was $6.61, the per-share earnings $2.16, in 1983. His estimate of the valuation Disney was placing on its shares for the purposes of this transaction appears to have been slightly off, but it was close enough to support his contention that the transaction should be put to a stockholder vote, and he requested that the board call such a meeting and notify him that it had done so by July 30. When the board did not respond, Jacobs and friends launched a stockholder's suit against the board, charging "breach of fiduciary duty and waste of Disney's corporate assets."

This was the 400th blow. The company could not stand more adverse publicity, further diminution of stockholder and stock market confidence, more public questions about the quality of its leadership. On August 17 it simply abrogated the deal and paid Gibson its $7.5 million kill fee. Jacobs professed himself pleased and withdrew his suit—but not yet his stockholdings, which now amounted to 7.7 percent of the company. It would be another month before the Bass brothers bought him out—and provided him with a profit said to be $28.5 million. His takings were enough to make one wonder how seriously his criticism of the Gibson deal was meant. He, too, had handsomely capitalized his nuisance value.

Now Disney had a new principal owner—the Bass family, with 28.83 percent of Disney. It was the first time in the company's history that an outsider had owned more stock in the company than the Disney family did, taken together. Not, of course, that it made much sense to take them together any longer—if, indeed, it ever had since the passing of the patriarchs. Looking back from the vantage point that September finally provided, it is clear that an informal arrangement between Roy Disney and the Bass brothers took place. To be sure, the brothers got something they had never had before: major influence in, perhaps *de facto* control of, a famous frontline American company—no chicken franchiser, but rather one with a once and perhaps future power to shape substantially American popular culture and, therefore, mass sensibility. To be sure, the company received a boon in return—the removal of the thorny Jacobs from its side. But of all the players in this game, it was Roy Disney who got the most of what he wanted. The company, though it had sacrificed some financial strength, was still recognizably the company, still functioning in its traditional lines of business. It had not distorted itself permanently to survive, and his holdings in it and those of his family were not seriously diluted. But much more to the point, he was finally, finally, a powerful voice in its councils, the vice-chairman of the newly constituted board, head of animation at the studio and the only board member who could speak authoritatively, out of lifelong experience, for the Disney tradition, with weight added to that voice by his own successes in finance.

And with no rival to challenge him. For even before the Basses rid the company of Irwin Jacobs, they had joined Roy Disney in ridding it of Ron Miller.

It was not difficult to do, actually. It could be said, for instance, that especially in the Gibson deal, Miller and Watson had moved, with a quickness that looked like panic, to a commitment at a price that, in all objectivity, was too high. It could be argued as well that the so-called scorched earth defense against Steinberg appeared, to knowledgeable citizens of the financial world, perhaps excessive, amateurish. And the same argument might be applied to the settlement that was finally reached with Steinberg. In short, it could be said that Miller and Watson were simply not experienced enough, therefore not steady enough, to pilot the company through the deep and tricky waters they suddenly found themselves forced to navigate. It could also be argued that even if there were good reasons for some of the courses they took, they were too difficult to explain —and would not alter the damaging perception of them that had grown in the financial world. So it was cleaner and better-looking simply to replace them as quickly as possible. Finally, it has to be admitted that in the summer of 1984, *Splash* began to look more like an accident than an augury. Richard Berger's first production, *Country*, which had had its troubles during production, was available to see, and it was dreary—unlikely to be a winner critically or commercially. Worse, *Return to Oz*, a sequel of sorts to the *Wizard of Oz*, had got out of control and was well over budget. In short, in the area of Miller's expertise, where the excuse of inexperience did not apply, things looked no better than they ever had.

And now, coincidentally, a shake-up appeared to be in the offing at the top of what had for several years been the most profitable studio—Paramount. Two valuable pieces of manpower, Barry Diller and Michael Eisner, were looking for new situations. The former had virtually committed himself to Twentieth Century-Fox, and was in any case probably not

Disney's sort of chap. But his creative partner, Eisner, obviously was, as Roy E. Disney had long ago sensed. Indeed, even before he approached Eisner the latter had happily accepted an appointment to the Cal Arts board. As a kid, he had loved the studio's pictures, and Disney still had great glamour for him. All he needed was a formal invitation to come aboard. And an opening, of course. The weary board, anxious to do something right, something bold at last, needed only the slightest urging to provide it. When the Basses' man, Charles Cobb, sided with Roy Disney and his allies on the board, the deed was quickly done. On September 7 they unanimously accepted Ron Miller's resignation.

There remained some hesitation among other board members about Eisner and Wells. A certain sentiment developed among the more conservative members of the board—the term is strictly relative—for Dennis Stanfill, sometime head of Fox, whose background prior to that raffishness was in a more conventional business, General Mills. But some of the best and brightest of the young movie people, people who had, in fact, been Disney's most powerful and successful competitors— George Lucas, Steven Spielberg, Jim Henson (creator of the Muppets)—campaigned for Eisner, who was of their generation and spirit. On September 22, 1984, he replaced Watson as chairman and chief executive officer of Walt Disney Productions, and Wells became president and chief operating officer. At last, the long agony was over. All that remained was to transform the company.

36 *God knows, this is a fickle business. . . . I believe we will have some great successes and have some failures. Without the failures, you can't have the successes.*
—Michael Eisner
Chairman and Chief Executive Officer
Walt Disney Productions
September 24, 1984

THERE IS AN ARGUMENT, often advanced by the unconventionally wise, that takeover attempts are good for a business. They hold that often it is only under stress that valuable, but arteriosclerotic, companies can be shaken from old habits—traditional ways of doing business. That argument must finally be applied to Walt Disney Productions. As an economic entity it had been, in its day—that is to say, in its founder's day—a work of genius. But, as the years since proved, it required the active presence of a genius at its head to continue functioning at its best. Men beholden to that genius, either through family ties or through residual emotional commitment to the revered shade, could not, try as they might, function freely, adaptably, as conditions—social, financial, moral—changed. Ron Miller certainly tried as hard as he could, as hard as he dared. He simply did not have that force of personality, that weight of success, that his father-in-law could have mustered, when pressed, to change the course of his company. In the end, one feels nothing but sympathy for him.

Indeed, one must wonder whether, had they lived, Walt and Roy O. Disney could have mastered the changing movie market —or the changing money market. The solid values they represented are not the values of the Seventies and the Eighties. One thinks, of course, that Walt Disney would have seen his way clear to making pictures like those of Lucas and his friends—

424

although one doubts if he could have permitted such strong competing personalities on his lot. But about the financial marketplace, one is less certain. The Disneys, economically speaking, believed in the old, middle-western virtues of building solidly for the long-term future. They would have been uncomfortable in a climate where the largest value was instant success, continued quarter-by-quarter improvements in profits, no backsliding allowed. They believed, too, in keeping their own counsel, in not permitting the opinions of, let's say, stock analysts, to sway them from their course. It is possible that they might not have fared much better than Miller and his people in today's world.

Be that as it may, it is clear, finally, that only an outsider— someone decently respectful of the company's traditions but un-beholden intellectually or emotionally to each and every detail of policies past—could make the changes that are required at Disney if it is to survive into a new century. Before his first month in office had passed, Michael Eisner was saying things that no one at Disney had ever dared say before. For example, he was looking at video in a new way, speculating that care-fully controlled exposure in these markets could perhaps en-hance the value of some of the studio's classics, teasing the public into a new awareness of their virtues—while contributing mightily to cash flow. He was proposing, as well, that there might be valid ways for the studio to use the less costly and long-eschewed techniques of limited animation to penetrate the Saturday-morning television cartoon market, which Disney had abandoned to competitors less caring about what was said and shown. If that required some slight diminution of Disney's artistic standards, it could also represent some distinct social good. At any rate, within six months two such shows were sold.

What about prime-time television? CBS, the third and last network to program a Disney hour, had canceled the program because of low ratings in 1983, and with that valuable show-case (and the cash flow it generated) denied it, the company was represented in television's most visible realm only on an occasional basis. But there was no reason to regard that situation as immutable. Surely Disney was capable of producing comedies at least as commercially effective as *Happy Days* or *Laverne and Shirley* had been for Paramount when Eisner was there. Nor was the possibility of excellence within TV's standards—*Taxi, M*A*S*H*, even *Hill Street Blues*—theoretically beyond the studio's capabilities. Such a program, *Golden Girls*, was sold early in Eisner's reign, and premiered to strong ratings and reviews. And in the feature area, if the company were willing to treat with independents, the range of possibilities was endless. What reasonable person could dispute Disney's right to produce something as fine as Paramount's *Terms of Endearment* had proved to be during Eisner's tenure there? Would any but the most primitive religious fanatic focus entirely on the film's language instead of upon the values it upheld? It seemed unlikely. And if, as Eisner proposed, the company could eventually produce ten to fifteen such films annually, instead of three or four, was there not potential for growth that would compensate for the inevitable slackening of its take at the amusement parks? That said nothing about what might be done with all that unglamorous real estate it owned.

The first Eisner signed up, Paul Mazursky's adaptation of the Jean Renoir classic *Boudu Saved from Drowning*—potentially an R-rated picture as well as the sort of intelligent, but black, comedy that Disney had never before attempted—seemed an admirably clear symbolic statement of the new regime's taste and intent. Frankly, one cannot imagine anyone who had previously managed Disney even knowing about—let alone respecting—the project's source.

What was being signaled in various ways during the first weeks of the new regime was very simple, really. It was that Walt Disney the man and Walt Disney the symbol were finally being permitted to pass into history, where, one imagines, both will find a respected (and, one hopes, controversial) place.

What will continue as a living legacy will be the work with which it all began: animation, and that is appropriate and heartening. What will be lost over the course of time will be the singularity—perhaps better call it eccentricity—that characterized the institution when it maintained itself primarily as the lengthened shadow of one man. It is something that is inevitably lost as time passes and corporations take on lives of their own. That does not mean that a corporate character, a character tinctured by, but not ruled by, memories of the founding father, need necessarily be lost. Maintaining that character while maintaining economic viability in a radically altered world is the continuing challenge before this company. But Disney has the potential now (which it did not have while operating under familial and historical constraint) to be something it has not been since its earliest days: a force not merely for idle nostalgia or empty traditionalism but for common decency; for reasonableness and a lack of sensation; for, if you will, the morality and the aspirations of that once confident middle class that nurtured Walt Disney but which has, in recent decades, found itself spluttering and confused and, in its own mind, besieged. It would be no small thing if this company could use its accrued goodwill to speak to, and for, this multitude, in some intellectually valid, artistically arresting, commercially viable way. Now, briefly, that opportunity opens. It will be interesting—and important—to see what Disney makes of it.

Acknowledgments

IT IS A PLEASURE to thank the people—many of them strangers—who helped me gather the material on which this book is based. It is a frustration not to be able to acknowledge publicly the assistance of an almost equal number of people, connected in one way or another with the Disney studio, who were generous with their reminiscences and opinions but whom I can best serve by preserving their anonymity. I have no firm reason to suppose that the organization would take reprisals against them, but late in 1966, well after I had begun my research and shortly before Mr. Disney died, I was given to understand that the studio did not approve of this study, and I have no wish to compromise those who spoke to me, many of them before this opposition was formally stated.

I must, however, extend my gratitude to Walt Disney Productions for generously allowing me to join a group of journalists who were given a week-long tour of the studio, WED and Disneyland and who were extended every courtesy, allowed to ask any question and to obtain detailed answers to all of them. I am sorry only that the mood of that trip did not extend a little bit farther, and I remain somewhat mystified by the sudden closing of the studio's doors, particularly after I had already been made privy to so much of what goes on behind them. I feel obligated, moreover, to report to my readers that no pictures of Disney's creations are in this book because the studio denied my request for permission to include such copyrighted material. But I do want to thank all the very kind people who made my trip so pleasant and so deeply informative.

ACKNOWLEDGMENTS

As to those people whose cooperation I can acknowledge, I really do not have words to express my gratitude adequately. All are busy people, but many found time to spend several hours being interviewed. Still others responded with alacrity to queries, dug into their files or memories to supply me with missing facts and, in the process, often volunteered valuable information that I did not know existed. Among them were a group of animators—Arthur Babbitt, Stephen Bosustow, James Culhane, William Hertz, John Hubley, Lou Keller—who conducted a short course in their art with patience, wit and insight. The following people provided me with anecdotes, theories, introductions, references and other passports to The Magic Kingdom: Ray Bradbury, Robert Crichton, Josette Frank, Arthur Knight, Joe Morgenstern, Carol Morton, Harold Rand, Maurice Rapf, Frances Clarke Sayers, Willard G. Triest and William K. Zinsser. Willard Van Dyke kindly arranged for me to screen the Disney short subjects in the collection of the Museum of Modern Art. Mary De Marzo undertook the jobs of researcher and typist and handled both with intelligence and dispatch far above the usual call of those duties.

My editor, Richard Kluger, undertook the lengthy task of encouraging me to attempt this book, fretted with me over its problems and then, having got me into it, had the grace to get me out again, relatively unscathed, by the application of a firm yet sensitive editorial pencil.

I am grateful for the care, concern and interest of two editors, Ruth Kozody of Simon and Schuster and Colin Webb of Pavilion Books.

R.S.

Bibliographical Note

THE FOLLOWING is by no means a complete listing of the books and articles consulted in the preparation of this book. It does, however, name all the works from which direct quotations were taken—with one exception. I have not given citations here for the reviews of individual Disney films. My method regarding reviews was to sample the relatively small number of American newspapers and periodicals that have a tradition of reliable movie reviewing and to select for quotation in the text those reviews that seemed to me most representative of the general critical response to a particular film or the ones that were especially interesting in their own right. All told, a couple of hundred reviews were thus checked, and to list all of them would unnecessarily burden this bibliography. For the record, the publications consulted were: *The Nation*, *The New Republic*, *Newsweek*, *The New York Times*, The New York *Herald Tribune* and *Time*. Also excluded from this listing are a handful of short news items, many of them anonymously written, culled from a file of clippings placed at my disposal by the Disney organization. Though useful to me, I doubt if they would be of interest to the general reader. In any case, my primary purpose here is to acknowledge —with gratitude—my largest debts and to suggest to that handful of readers who may be interested in pursuing this subject still further the directions in which they might most sensibly head. In furtherance of this last objective I have annotated the bibliography with a few critical comments that may prove helpful.

I. Books About Disney or Containing Substantial References to Him

AGEE, JAMES. *Agee on Film.* McDowell, Obolensky, 1958.
BECKER, STEPHEN. *Comic Art in America.* Simon and Schuster,

1959. Contains a short chapter on film animation, with some useful material on the early days of the art.

FEILD, R. D. *The Art of Walt Disney*. Macmillan, 1942. Concentrates heavily on the technical aspects of production at the Disney studio, and is decidedly uncritical in its approach. Nevertheless, it contains some material unobtainable without the cooperation of the Disney Studio, which the author had.

HALAS, JOHN, AND ROGER MANVELL. *Design in Motion*. Visual Communications Books—Hastings House. 1962. An illustrated survey of animation around the world which helps to put Disney's achievements in a wider-than-usual perspective.

JACOBS, LEWIS. *Introduction to the Art of the Movies*. Noonday, 1960. Contains Gilbert Seldes' first review of Disney's short subjects, "Disney and Others," written for *The New Republic* in 1932.

JACOBS, LEWIS. *The Rise of the American Film*. Harcourt, Brace. Now outdated, this survey remains a model of sound historical writing about the period it covers, and its chapter on Disney is an excellent reflection of the esteem in which he was held in the late 1930s.

KNIGHT, ARTHUR. *The Liveliest Art*. Macmillan, 1957. The best of the more recent historical surveys.

LINDSAY, CYNTHIA. *The Natives Are Restless*. Lippincott, 1960. An informal, witty survey of manners and morals in Southern California in the period of Disneyland's rise.

MILLER, DIANE DISNEY (AS TOLD TO PETE MARTIN). *The Story of Walt Disney*. Holt, 1957. An admittedly prejudiced, daughter's eye-view of the man but still the most complete firsthand record of his life and works that we have.

STRAVINSKY, IGOR, AND ROBERT CRAFT. *Expositions and Developments*. Doubleday, 1962. The composer's version of his quarrel with Disney is included.

TAYLOR, DEEMS. *Walt Disney's Fantasia*. Simon and Schuster, 1940. An early example of the nonbook, but it does contain a few useful anecdotes as well as a convenient summary of the film.

TALBOT, DANIEL. *Film: An Anthology*. Simon and Schuster, 1959. Includes Irwin Panofsky's "Style and Medium in the Motion Pictures," originally written in 1934 and later revised. It contains an extremely interesting note on Disney's work.

THOMAS, BOB. *The Art of Animation*. Golden, 1958. Undertaken at the behest of the Disney Studio and therefore somewhat narrow in its viewpoint, the book is nonetheless clearly written, generously illustrated and full of anecdotes about the creative life in Disney's shop.

II. Magazine and Newspaper Articles About Disney and His Works

ALEXANDER, JACK. "The Amazing Story of Walt Disney." *The Saturday Evening Post*, Oct. 31, Nov. 7, 1953. Approving, anecdotal, superficial.

ALPERT, HOLLIS. "The Wonderful World of Walt Disney." *Woman's Day*, Oct., 1962.

BARR, DONALD. "The Winnowing of Pooh." *Book Week*, Fall Children's Issue, Oct. 31, 1965. A short, sharp, sensible critique of Disney's reduction of the Milne classic to fit mass tastes.

BART, PETER AND DOROTHY. "As Told and Sold by Disney." *The New York Times Book Review*, Children's Book Section, May 9, 1965.

BIRMINGHAM, STEPHEN. "The Greatest One Man Show on Earth." *McCall's*, July, 1964. A solid profile in the slick magazine vein.

BREWER, ROY. "Walt Disney, RIP." *National Review.* Jan. 10, 1967. A eulogy in the form of a letter from a determinedly antileftist union leader.

BRIGHT, JOHN. "Disney's Fantasy Empire." *The Nation.* Mar. 6, 1967. An intelligent attempt at a balanced survey of the man's career and his place in our cultural history.

Business Week. "Disney's Live-Action Profile." July 24, 1965.

CHURCHILL, DOUGLAS. "Disney's Philosophy." *The New York Times Magazine*, Mar. 6, 1938. An interview with the master at the height of his acclaim over *Snow White*.

———. "Now Mickey Mouse Enters Art's Temple." *The New York Times Magazine*, June 3, 1934. More about the how of animation than the why of Disney's success with it.

CORLISS, RICHARD. "The New Generation Comes of Age." *Time*, July 20, 1981.

COMPTON, NEIL. "TV While the Sun Shines." *Commentary*, October, 1966. A discussion of animated films for children on television, a very useful appraisal of the current state of the art.

CROWTHER, BOSLEY. "The Dream Merchant." *The New York Times*, Dec. 16, 1966. A good, short appraisal of Disney's career, appearing in the same edition as that publication's extensive and useful obituary.

DAVIDSON, BILL. "The Fantastic Walt Disney." *The Saturday Evening Post*, Nov. 7, 1964.

DAVIS, SALLY OGLE. "Wishing Upon a Falling Star at Disney." *The New York Times Magazine*, Nov. 16, 1980.

DE ROOS, ROBERT. "The Magic Worlds of Walt Disney." *National*

Geographic, August, 1963. An extraordinarily complete trip through The Magic Kingdom, conducted by a coolly admiring guide.

DISNEY, LILLIAN (WITH ISABELLA TAVES). "I Live with a Genius." *McCall's*, February, 1953. Mrs. Disney's only extensive public reminiscence of her life with Walt Disney.

DISNEY, WALT. "The Cartoon's Contribution to Children." *Overland Monthly*, October, 1933. Obviously ghost-written but still an interesting insight into the young Disney's mentality.

————. "The Life Story of Mickey Mouse." *Windsor* (London), January, 1934. Another useful curiosity.

FARLEY, ELLEN. "Disney Heirs' Stock May Be Key." *Los Angeles Times*, April 1, 1984.

FERGUSON, OTIS. "Walt Disney's Grimm Reality." *The New Republic*, Jan. 26, 1938. A sizable essay on *Snow White*, marked by fine taste and perception.

Forbes. "Disney Without Walt . . ." July 1, 1967.

FORSTER, E. M. "Mickey and Minnie." *The Spectator* (London), Jan. 19, 1934. Mr. Forster in his lightest, most charming mood.

Fortune. "The Big Bad Wolf." November, 1934. Anonymous but marvelously complete study of Disney's artistic and financial techniques of the time.

HALÉVY, JULIAN. "Disneyland and Las Vegas." *The Nation*, June 7, 1958. An annoyingly attitudinizing piece.

HARRIS, KATHRYN. "Takeover Talk Adds Pressure as Disney Tries to Snap Back." *Los Angeles Times*, April 1, 1984.

HAYES, THOMAS C. "The Troubled World of Disney." *The New York Times*, Sept. 25, 1984.

————. "Trouble Stalks the Magic Kingdom." *The New York Times*, June 17, 1984.

————. "Disney's Chief is Forced Out." *The New York Times*, Sept. 8, 1984.

HERSHEY, LENORE. "What Women Think of the Movies." *McCall's*, May, 1967.

JOHNSTON, ALVA. "Mickey Mouse." *Woman's Home Companion*, July, 1934. Old-fashioned, slick journalism by a reporter better than his medium.

LOW, DAVID. "Leonardo da Disney." *The New Republic*, Jan. 5, 1942. An expression of enthusiasm by the great British cartoonist.

McDONALD, JOHN. "Now the Bankers Come to Disney." *Fortune*, May, 1966. Perhaps the best single piece about the total Disney operation—certainly the best of recent years.

BIBLIOGRAPHICAL NOTE

McEvoy, J. P. "Of Mouse and Man." *This Week*, July 5, 1942.

———. "McEvoy in Disneyland." *The Reader's Digest*. February, 1955. Concentrates on the nature films.

Magnet, Myron. "No More Mickey Mouse at Disney." *Fortune*, Dec. 10, 1984. The best single source on recent Disney doings.

Main, Jeremy. "The Kempers of Kansas City." *Fortune*, April, 1967. Contains interesting materials on the city as Disney knew it as a young man.

Mano, D. Keith. "A Real Mickey Mouse Operation." *Playboy*, Dec., 1973.

Marlow, David. "Working for Mickey Mouse." *New York*, April 6, 1973.

Menen, Aubrey. "Dazzled in Disneyland." *Holiday*, July, 1963. Perhaps the best single piece about Disneyland.

Miller, Jonathan. "Another Wonderland." *The New York Times Book Review*, Children's Book Section, May 7, 1967. A fine piece on the art of adaptation.

Morgenstern, Joseph. "Walt Disney (1901–1966): Imagineer of Fun." *Newsweek*, Dec. 23, 1966. A sensitive obituary.

———. "What the Kids Should See." *Newsweek*, Sept. 18, 1967.

Nathan, Paul. "Rights and Permissions." *Publishers' Weekly*, Jan. 2, 1967. Anecdotes about Disney's dealings with the literary world.

Newsweek. "Fifty Million Customers." Mar. 14, 1955. A comprehensive report on the studio in midpassage.

Paris Match. "Farewell to Walt Disney." Dec. 24, 1966. A tearjerker.

Rafferty, Max. "The Greatest Pedagogue of All." *Los Angeles Times*, Apr. 18, 1967.

Reddy, John. "The Living Legacy of Walt Disney." *The Reader's Digest*, June, 1967.

Ross, Irwin. "Disney Gambles on Tomorrow." *Fortune*, Oct. 4, 1982.

Rouse, James. "Rouse on Problems and Wifely Help." *Life*, Feb. 24, 1967.

Russel, Herbert. "Of L'Affaire Mickey Mouse." *The New York Times Magazine*, Dec. 26, 1937. A summary of the varmint's worldwide impact when he was at the height of his fame.

Santora, Phil. "Disney: Modern Merlin." New York *Daily News*, Sept. 29, 30, Oct. 1, 2, 1964. An anecdotal newspaper series.

Sayers, Frances Clarke. "Letters to the *Times*." Los Angeles *Times*. April 25, 1965.

BIBLIOGRAPHICAL NOTE

SCHICKEL, RICHARD. "In Computerland with TRON." *Time*, July 5, 1981.

————. "The Great Era of Walt Disney." *Time*, July 20, 1981.

SELDES, GILBERT. "No Art, Mr. Disney?" *Esquire*, September, 1937. The best critical statement about the short cartoons ever written —thorough, sympathetic, sensibly critical.

SHEARER, LLOYD. "What Kind of Motion Pictures Do You Really Want?" *Parade*, Jan. 7, 1962. Contains a long statement from Disney on the need for clean and uplifting films.

VAN DOREN, MARK. "Fairy Tale in Five Acts." *The Nation*, Jan. 22, 1938. Less interesting for what it says than for who is saying it.

Time. "Father Goose." Dec. 17, 1954. A cover story that really covered its subject—critically, sociologically, financially. In all, a superior piece of journalism, ranking with the best treatments of Disney ever written.

————. "Mouse and Man." Dec. 27, 1937. Cover story on the occasion of *Snow White*'s release.

————. "Walt Disney: Images of Innocence." Dec. 23, 1966. An obituary.

TAYLOR, JOHN. "Project Fantasy." *Manhattan, inc.*, Nov., 1984.

WALLACE, KEVIN. "Onward and Upward with the Arts. The Engineering of Ease." *The New Yorker*, Sept. 7, 1963. A piece concentrating on the creature comforts of Disneyland.

WOLFERT, IRA. "Walt Disney's Magic Kingdom." *The Reader's Digest*, April, 1960.

WOLTERS, LARRY. "The Wonderful World of Walt Disney." *Today's Health*, April, 1962.

Variety. "All-Time Boxoffice Champs." Jan. 4, 1967; Jan. 16, 1985.

Other Books Consulted

ARNHEIM, RUDOLF. *Film as Art*. University of California, 1957. Contains several chapters from his earlier *Film* (1933), in which he made the most coherent summary of his theoretical views.

BENDINER, ROBERT. *Just Around the Corner*. Harper & Row, 1967. An excellent short social history of the 1930s, concentrating heavily on the quality of ordinary, day-to-day experience in the depression decade.

CAMPBELL, JOSEPH. *The Masks of God: Primitive Mythology*. Viking, 1959. The first volume of his monumental study of world mythology, the opening chapters contain an excellent summary of this great scholar's theories on the subject.

CERAM, C. W. *Archeology of the Cinema.* Harcourt, Brace & World, 1965. The best study of the inventions that preceded the development of the motion picture camera and projector as we now know them.

CANETTI, ELIAS. *Crowds and Power.* Viking, 1963. Idiosyncratic, insightful study of the will to dominate and the urge to be dominated.

CROWTHER, BOSLEY. *The Lion's Share.* Dutton, 1957. A history of M-G-M, invaluable to any student of Hollywood's greatest age of power.

Disneyland. Arnoldo Mondadori Editore. No date.

ERIKSON, ERIK H. *Young Man Luther.* Norton, 1958. Indispensable to anyone wishing to understand the Protestant mind in any age.

GALBRAITH, JOHN KENNETH. *The Great Crash.* Houghton Mifflin, 1955. Social history at its wittiest and most incisive.

Grimm's Fairy Tales. Pantheon, 1944. The folkloristic edition, with an excellent afterword by Campbell.

HARPER, RALPH. *Nostalgia.* Western Reserve, 1966. A fine little essay on a subject often discussed, rarely studied.

HAUSER, ARNOLD. *The Social History of Art* (Vol. 4. *Naturalism, Impressionism, The Film Age*). Vintage, 1958.

HILLEGAS, MARK R. *The Future as Nightmare: H. G. Wells and the Anti-Utopians.* Oxford, 1967.

HOFFER, ERIC. *The Ordeal of Change.* Harper & Row, 1963.

———. *The Temper of Our Time.* Harper & Row, 1967. Mr. Hoffer's essay on "The Juvenile Temperament" could have been written with Disney and his audience in mind.

HOFFMAN, FREDERICK J. *The Twenties.* Viking, 1955. Literary and social history of the highest order.

HUGHES, JONATHAN. *The Vital Few.* Bantam, 1965. Short studies in the American tradition of entrepreneurship, excellent for placing Disney in the industrial context.

JONES, ERNEST. *Papers on Psychoanalysis.* Beacon, 1961. Contains his classic study of anality.

KAYSER, WOLFGANG. *The Grotesque in Art and Literature.* Indiana University, 1963.

KEMPTON, MURRAY. *Part of Our Time.* Simon & Schuster, 1955. Has a good chapter on the Hollywood leftists of the 1930s and 40s.

LEWIS, C. S. *Of Other Worlds.* Harcourt, Brace & World, 1966. Contains several excellent essays on juvenile taste, on writing for children and on the fairy tale as an art form.

McLUHAN, MARSHALL. *Understanding Media.* McGraw-Hill, 1964.

BIBLIOGRAPHICAL NOTE.

MELLERS, WILFRID. *Music in a New Found Land.* Knopf, 1964. Though the metaphor is modern American music, this is a profound study in national character as well.

MENNINGER, KARL. *A Psychiatrist's World.* Viking, 1959. The Kansas doctor includes in this huge collection of articles several bearing directly on the country and society that shaped Disney's personality.

ORTEGA Y GASSET, JOSÉ. *The Dehumanization of Art and Other Essays.* Princeton University Press, 1948.

ORWELL, GEORGE. *The Road to Wigan Pier.* Harcourt, Brace, 1958. Has some of his most interesting reflections on the mass mind in an industrial age.

SCHLESINGER, ARTHUR M., JR. *The Age of Roosevelt* (Vol. 1. *The Crisis of the Old Order*). Houghton Mifflin, 1957.

VEBLEN, THORSTEIN. *Absentee Ownership.* Beacon, 1967. The chapter on the country town is a great essay in descriptive sociology and invaluable to an understanding of a little-studied force in our culture—and certainly in Disney's.

WHALEN, RICHARD J. *The Founding Father.* New American, 1964. Good on film finance.

WILSON, EDMUND. *The Shores of Light.* Farrar, Straus & Young, 1952. His review of Alva Johnston's *The Great Goldwyn* contains his thoughts on Disney as well as the Hollywood mogul as a type.

Other Material

WALT DISNEY PRODUCTIONS. Proxy Statement, Annual Meeting of Stockholders, Feb. 2, 1965. This meeting was asked to approve the purchase of WED Enterprises by Productions, and the statement contains a lengthy section on the history of both companies as well as the proposed purchase agreement itself. Both are invaluable to the student of the corporation's history.

———. Walt Disney Productions Presentation to New York Society of Security Analysts, Mar. 18, 1966. The transcript of the remarks by Roy Disney, by treasurer Laurence Tryon and vice presidents Donn B. Tatum and E. Cardon Walker is the most detailed statement of the company's recent history, its worth and its prospects available anywhere—a treasure trove for the researcher.

———. *The Disney World*, Vols. 3, 4, 5, 1966–67. The firm's six-times-a-year house organ is a handsomely produced, well-written and—naturally enough—somewhat narrowly focused record of life in The Magic Kingdom. It is, however, full of the most fascinating tidbits.

Index

INDEX

INDEX

INDEX

INDEX

INDEX

INDEX

INDEX

INDEX

447